Advances in Spatial Science

For further volumes:
http://www.springer.com/3302

Antonio Páez · Julie Le Gallo
Ron N. Buliung · Sandy Dall'erba
Editors

Progress in Spatial Analysis

Methods and Applications

Springer

Editors

Professor Antonio Páez
School of Geography
and Earth Sciences
1280 Main Street West
McMaster University
Hamilton, Ontario L8S 4K1
Canada
paezha@mcmaster.ca

Professor Julie Le Gallo
Université de Franche-Comté CRESE
45 D, Avenue de l'Observatoire
25030 Besançon Cedex, France
jlegallo@univ-fcomte.fr

Professor Ron N. Buliung
Department of Geography
University of Toronto at Mississauga
3359 Mississauga Road North
Mississauga, Ontario L5L 1C6
Canada
ron.buliung@utoronto.ca

Professor Sandy Dall'erba
Department of Geography
and Regional Development
University of Arizona
P.O. Box 210076
Tucson, AZ 85721, USA
dallerba@email.arizona.edu

Advances in Spatial Science ISSN 1430-9602
ISBN 978-3-642-26149-7 e-ISBN 978-3-642-03326-1
DOI: 10.1007/978-3-642-03326-1
Springer Heidelberg Dordrecht London New York

Softcover reprint of the hardcover 1st edition 2010

Cover design: SPi Publisher Services

Printed on acid-free paper

Springer is part of Springer Science+Business Media (www.springer.com)

For Patricia, Leonardo, and Luanna (AP)
For Tara, Meera, and Emily (RB)

Foreword

Space is one of the fundamental categories by means of which we perceive and experience the world around us. Behaviour takes place in space, and the geographical context of behaviour is important in shaping that behaviour. While space by itself explains very little, spatial processes and the spatial patterning of behaviour have long been viewed as a key to understanding, explaining, and predicting much of human behaviour.

Whether or not spatial analysis is a separate academic field, the fact remains that, in the past 20 years, spatial analysis has become an important by-product of the interest in and the need to understand georeferenced data. The current interest in the mainstream social sciences to geography in general, and location and spatial interaction in particular is a relatively recent phenomenon. This interest has generated an increasing demand for methods, techniques, and tools that allow an explicit treatment of space in empirical applications. Thus, spatial analysis tends to play an increasingly important role in measurement, hypothesis generation, and validation of theoretical constructs, activities that are crucial in the development of new knowledge. The fact that the 2008 Nobel Prize in economics was awarded to Paul Krugman indicates this increasing attention being given to spatially related phenomena and processes. Given the growing number of academics currently doing research on spatially related subjects, and the large number of questions being asked about spatial processes, the time has come for reflecting on the progress made in spatial analysis.

As an editor of the book series, I highly welcome the present edited volume on *Progress in Spatial Analysis* with a focus on theory and methods, and thematic applications across several academic disciplines. The effort is a worthy intellectual descendent of previous volumes in the series, including Anselin and Florax's *New Direction in Spatial Econometrics* in 1995, Fischer and Getis' *Recent Developments in Spatial Analysis* in 1997, and Anselin, Florax, and Rey's *Advances in Spatial Econometrics* in 2004.

I am pleased to realize the mixture of very well-established leaders in the field of spatial analysis and a new generation of scholars who, one hopes, will continue to carry the torch ignited more than 50 years ago at the dawn of *Quantitative Geography and Regional Science*. In this spirit, it is my hour to formally proffer the welcome to this edited volume, and to the effort poured into bringing major developments and applications into a single source representing recent publications in spatial analysis. I anticipate that this volume will become a valuable reference, as the previous offerings in the series.

Vienna
May, 2009

Manfred M. Fischer

Contents

Part V Regional Applications

List of Figures

Dealing with Spatiotemporal Heterogeneity: The Generalized BME Model
Hwa-Lung Yu, George Christakos, and Patrick Bogaert

Local Estimation of Spatial Autocorrelation Processes
Fernando López, Jesús Mur, and Ana Angulo

"Seeing Is Believing": Exploring Opportunities for the Visualization of Activity–Travel and Land Use Processes in Space–Time
Ron N. Buliung and Catherine Morency

Evolution of the Influence of Geography on the Location of Production in Spain (1930–2005)
Coro Chasco Yrigoyen and Ana M. López García

Comparative Spatial Dynamics of Regional Systems
Sergio J. Rey and Xinyue Ye

List of Tables

Gender Wage Differentials and the Spatial Concentration of High-Technology Industries
Elsie Echeverri-Carroll and Sofia G. Ayala

Fiscal Policy and Interest Rates: The Role of Financial and Economic Integration
Peter Claeys, Rosina Moreno, and Jordi Suriñach

Spatial Models of Health Outcomes and Health Behaviors: The Role of Health Care Accessibility and Availability
Brigitte S. Waldorf and Susan E. Chen

Immigrant Women, Preventive Health and Place in Canadian CMAs
Kelly Woltman and K. Bruce Newbold

Is Growth in the Health Sector Correlated with Later-Life Migration?
Dayton M. Lambert, Michael D. Wilcox, Christopher D. Clark, Brian Murphy, and William M. Park

Evolution of the Influence of Geography on the Location of Production in Spain (1930–2005)
Coro Chasco Yrigoyen and Ana M. López García

Comparative Spatial Dynamics of Regional Systems
Sergio J. Rey and Xinyue Ye

Contributors

Patrick M. Andrade 150 North Martingale Road, Schaumburg, IL 60173, USA, Patrick.Andrade@nielsen.com
and
150 North Martingale Road, Schaumburg, IL 60173, USA

Ana Angulo Department of Economic Analysis, University of Zaragoza, Gran Via 2-4, Zaragoza 50005, Spain, aangulo@unizar.es

Sofia G. Ayala IC^2 Institute, University of Texas at Austin, 2815 San Gabriel, Austin, TX 78705, USA, sofia_ayala@mail.utexas.edu

Patrick Bogaert Department of Environmental Sciences & Land Use Planning, Université Catholique de Louvain, ENGE – Croix du Sud, 2, bte. 16 à 1348, Louvain-la-Neuve, Belgium, patrick.bogaert@uclouvain.be

Ron N. Buliung Department of Geography, University of Toronto at Mississauga, 3359 Mississauga Road North, Mississauga, ON L5L 1C6, Canada, ron.buliung@utoronto.ca

Susan E. Chen Department of Agricultural Economics, Purdue University, 403 W. State Street, West Lafayette, IN 47907-2056, USA, sechen@purdue.edu

Seong-Hoon Cho Department of Agricultural Economics, University of Tennessee, 321 Morgan Hall, Knoxville, TN 37996-4511, USA, scho9@utk.edu

George Christakos Department of Geography, San Diego State University, 5500 Campanile Dr., San Diego, CA 92182-4493, USA, christak@geography.sdsu.edu

Peter Claeys AQR Research Group-IREA, University of Barcelona, Avinguda Diagonal 690, 08034 Barcelona, Spain, peter.claeys@ub.edu

Christopher D. Clark Department of Agricultural Economics, University of Tennessee, 321 Morgan Hall, Knoxville, TN 37996-4511, USA, cdclark@utk.edu

Sandy Dall'erba Department of Geography and Regional Development, University of Arizona, P.O. Box 210076, Tucson, AZ 85721, USA, dallerba@email.arizona.edu

Elsie Echeverri-Carroll IC[2] Institute, University of Texas at Austin, 2815 San Gabriel, Austin, TX 78705, USA, e.carroll@mail.utexas.edu

Steven Farber Centre for Spatial Analysis/School of Geography and Earth Sciences, McMaster University, 1280 Main Street West, Hamilton, ON L8S 3Z9, Canada, farbers@mcmaster.ca

Bernard Fingleton Department of Economics, Strathclyde University, 130 Rottenrow, Glasgow, Scotland G4 0GE, UK, bernard.fingleton@strath.ac.uk

Julie Le Gallo Centre de Recherche sur les Stratégies Economiques, Université de Franche-Comté, 45D, Université de Franche-Comté, 25030 Besançon Cedex, France, jlegallo@univ-fcomte.fr

Ana M. López García Dpto. Economía Aplicada, Facultad de Ciencias Económicas y Empresariales, Universidad Autónoma de Madrid, Carretera de Colmenar Viejo Km. 15.500, Madrid 28049, Spain, ana.lopez@uam.es

Frank Goetzke Department of Urban and Public Affairs, School of Urban and Public Affairs, University of Louisville, 426 W. Bloom Street, Louisville, KY 40208, USA, f0goet01@louisville.edu

Rachel Guillain LEG-UMR 5118, Université de Bourgogne, Pôle d'Economie et de Gestion, BP 21611, 21066 Dijon Cedex, France, guillain@u-burgogne.fr

Seung Gyu Kim Department of Agricultural Economics, University of Tennessee, 321 Morgan Hall, Knoxville, TN 37996-4511, USA, sgkim@utk.edu

Wilfried Koch Laboratoire d'Economie et de Gestion, LEG-UMR 5118, Université de Bourgogne, Pôle d'Economie et de Gestion, BP 21611, 21066 Dijon Cedex, France, wilfried.koch@u-bourgogne.fr

Dayton M. Lambert Department of Agricultural Economics, University of Tennessee, 321 Morgan Hall, Knoxville, TN 37996-4511, USA, dmlambert@utk.edu

James P. Lesage McCoy College of Business Administration, Department of Finance and Economics, Texas State University-San Marcos, McCoy Hall 504, San Marcos, TX 78666, USA, jlesage@spatial-econometrics.com

Fernando López Department of Quantitative Methods and Computing, Technical University of Cartagena, Paseo Alfonso XIII, 50 Cartagena 30203, Spain, fernando.lopez@upct.es

Catherine Morency Department of Civil, Geological and Mining engineering, École Polytechnique de Montréal, P.O. Box 6079, Station Centre-Ville, Montréal, QC H3C 3A7, Canada, cmorency@polymtl.ca

Rosina Moreno AQR Research Group-IREA, University of Barcelona, Avinguda Diagonal 690, 08034 Barcelona, Spain, rmoreno@ub.edu

Darla K. Munroe Department of Geography, The Ohio State University, 154 N. Oval Mall, Columbus, OH 43210, USA, munroe.9@osu.edu

Jesús Mur Department of Economic Analysis, University of Zaragoza, Gran Via 2-4, Zaragoza 50005, Spain, jmur@unizar.es

Brian Murphy Department of Agricultural Economics, University of Tennessee, 321 Morgan Hall, Knoxville, TN 37996-4511, USA, bmurphy7@utk.edu

K. Bruce Newbold School of Geography and Earth Sciences, McMaster University, 1280 Main Street West, Hamilton, ON L8S 3Z9, Canada, woltmak@mcmaster.ca

R. Kelley Pace College of Business Administration, Department of Finance, Louisiana State University, Baton Rouge, LA 70803-6308, USA, kelley@pace.am

E.J. Ourso College of Business Administration, Department of Finance, Louisiana State University, Baton Rouge, LA 70803-6308, USA, kelley@pace.am

Antonio Páez Centre for Spatial Analysis/School of Geography and Earth Sciences, McMaster University, 1280 Main Street West, Hamilton, ON L8S 3Z9, Canada, paezha@mcmaster.ca

William M. Park Department of Agricultural Economics, University of Tennessee, 321 Morgan Hall, Knoxville, TN 37996-4511, USA, wpark@utk.edu

Sergio J. Rey Department of Geography, San Diego State University and School of Geographical Sciences, Arizona State University, P.O. Box 875302, Tempe, AZ 85287, USA, Sergio.Rey@asu.edu

Roland K. Roberts Department of Agricultural Economics, University of Tennessee, 2621 Morgan Circle, Knoxville, TN 37996-4511, USA, rrobert3@utk.edu

Jordi Suriñach AQR Research Group-IREA, University of Barcelona, Avinguda Diagonal 690, 08034 Barcelona, Spain, jsurinach@ub.edu

Erik Volz Department of Epidemiology, School of Public Health, University of Michigan, 109 Observatory Street, Ann Arbor, MI 48109, USA, erik@erikvolz.info

Brigitte S. Waldorf Department of Agricultural Economics, Purdue University, 403 W. State Street, West Lafayette, IN 47907-2056, USA, bwaldorf@purdue.edu

Joan L. Walker Department of Civil and Environmental Engineering, University of California, Berkeley, 760 Davis Hall, Berkeley, CA 94720-1710, USA, walker@ce.berkeley.edu

Michael D. Wilcox Department of Agricultural Economics, University of Tennessee, 321 Morgan Hall, Knoxville, TN 37996-4511, USA, mwilcox2@utk.edu

Ian Sue Wing Department of Geography & Environment, Boston University, 675 Commonwealth Avenue, Boston, MA 02215, USA, isw@bu.edu

Kelly Woltman School of Geography and Earth Sciences, McMaster University, 1280 Main Street West, Hamilton, ON L8S 3Z9, Canada, woltmak@mcmaster.ca

Xinyue Ye Department of Geography, Joint Doctoral Program of Geography, San Diego State University and University of California-Santa Barbara, 5500 Campanile Drive, San Diego, CA 92182-4493, USA, xinyue.ye@gmail.com

Coro Chasco Yrigoyen Dpto. Economía Aplicada, Facultad de Ciencias Económicas y Empresariales, Universidad Autónoma de Madrid, Carretera de Colmenar Viejo Km. 15.500, Madrid 28049, Spain, coro.chasco@uam.es

Hwa-Lung Yu Department of Bioenvironmental Systems Engineering, National Taiwan University, 1, Section 4, Roosevelt Road, Taipei 10617, Taiwan, R.O.C., hlyu@ntu.edu.tw

Edmund J. Zolnik Department of Geography and Geoinformation Science, George Mason University, Fairfax, VA 22030, USA, ezolnik@gmu.edu

Progress in Spatial Analysis: Introduction

Antonio Páez, Julie Le Gallo, Ron N. Buliung, and Sandy Dall'Erba

1 Background

With its roots in geography and regional science spatial analysis has experienced remarkable growth in recent years in terms of theory, methods, and applications. The series of books, that in the past decade have collected research in spatial analysis and econometrics, provide both documented evidence and a powerful platform to further this upwards trend. Among the collections that have done so stand those compiled by Anselin and Florax (*New Directions in Spatial Econometrics*, 1994), Fischer and Getis (*Recent Developments in Spatial Analysis*, 1997), and Anselin, Florax and Rey (*Advances in Spatial Econometrics*, 2004). In the spirit of this series of volumes, the present book aims at promoting the development and use of methods for the analysis of spatial data and processes.

Traditionally, the core audience for the spatial analysis literature has been found in the Quantitative Geography and Regional Science communities, but also increasingly within the allied disciplines of Spatial and Regional Economics, Urban and Regional Planning and Development, Civil Engineering, Real Estate Studies, and Epidemiology, among others. Previous edited volumes, in particular the two spatial econometrics collections cited above, tended to emphasize, in addition to theoretical and methodological developments, economics and regional economics applications. In this book, we have made an attempt to capture a broader cross-section of themes, to include fields where spatial analysis has represented in recent years a boon for applications, which have in turn encouraged further technical developments. Besides the disciplines represented in previous collections of papers, up-and-coming areas that are seen to be making more extensive use of spatial analytical tools include transportation and land use analysis, political and economic geography, and the analysis of population and health issues. In order to provide a faithful picture of the

A. Páez (✉)
Centre for Spatial Analysis/School of Geography and Earth Sciences, McMaster University, 1280 Main Street West, Hamilton, ON L8S 3Z9, Canada,
e-mail: paezha@mcmaster.ca

A. Páez et al. (eds.), *Progress in Spatial Analysis*, Advances in Spatial Science,
DOI 10.1007/978-3-642-03326-1_1,

current state of spatial analysis it is also our wish to present recent theoretical and methodological developments. Together, this collection of theoretical and methodological papers, and thematic applications, will project, we hope, the image of a thriving and dynamic field, with wide-ranging intellectually stimulating challenges, and rich opportunities for applied research that promises to promote and advance data analysis in a variety of fields.

In terms of the contributions collected for this volume, the papers represent a selection of research presented at the 54th North American Meetings of the Regional Science Association International celebrated in Savannah, Georgia, in November of 2007, as well as a small number of invited papers. All contributions were subjected to a strict process of peer review; the outcome is a set of papers that have been organized, in addition to a section on *Theory and Methods*, into four thematic sections: *Transportation and Land Use Analysis*, *Population and Health Issues*, *Political and Economic Geography*, and *Regional Applications*. Some of these areas have traditionally been associated with the use of spatial analytical tools (e.g., regional applications). Others represent nascent opportunities for the development and use of spatial analysis (e.g., transportation and land use, population and health). It is our hope that this edited volume will simultaneously help to consolidate the reputation and value of spatial analysis established by previous titles in the series, and to increase awareness about the utility of spatial analysis in other application domains.

2 Theory and Methods

Five chapters comprise the section on theory and methods. Pace and LeSage, in chapter "Omitted Variable Biases of OLS and Spatial Lag Models", address a question that has received relatively little attention in the spatial econometrics literature, namely, the effect of omitted variables in regression analysis. This research is motivated by the conjecture that omitted variable bias is less severe in spatial models than in ordinary regression approaches. One of the bases for this conjecture is that the additional components in a spatial model are perhaps sufficiently capturing missing relationships to offset the effect of bias. The problem of omitted variables in spatial analysis, on the other hand, is complicated by the fact that spatial variables often display non-negligible amounts of spatial autocorrelation. Most likely, this will be the case for both the included and the omitted variables. In order to sort out what the impacts of this are, the authors develop a very general framework that allows them to derive results for a wide range of situations likely found in applied research. The analytical derivations presented in the chapter are backed by extensive simulation experiments that help to give a feeling for the magnitude of bias under different cases. The results indicate that, contrary to the original conjecture, omitted variable bias is magnified by the presence of spatial dependence. Several implications lead to useful guidelines for applied research.

Chapter "Topology, Dependency Tests and Estimation Bias in Network Autoregressive Models", by Farber, Páez and Volz, also deals with a specification issue in spatial modeling, namely the definition of spatial weights matrices, the instrument used to specify how spatial cross-sectional observations are connected. While this matrix is usually defined based on geographic criteria (e.g., contiguity, distance-based matrices, nearest-neighbors matrices etc.), there has recently been increasing interest in using a network-based connectivity specification. The subject of this chapter is the structure of the weights matrix and the effect of network topology on the estimation of network autocorrelation models and statistical tests of dependence. The authors investigate, both analytically and through extensive Monte-Carlo simulations, the power of the likelihood ratio (LR) tests for network dependence in SAR and SEM models. They first show that for all the model specifications, the level of network dependence is the most significant factor in predicting the power of the LR test, albeit with a non linear effect and differently for SAR and SEM models. Second, the effects of network density and clustering on the power of the LR test are analyzed. Finally, the relationship between bias and the various topological properties of networks are graphically illustrated. In sum, the various results unambiguously show that the topology of the weights matrix used in autocorrelation models has a strong impact on statistical tests and the accuracy of maximum-likelihood estimates.

Fingleton and Le Gallo, in chapter "Endogeneity in a Spatial Context: Properties of Estimators", are concerned with the important issue of identifying appropriate estimators when dealing with endogeneity in a spatial econometric context. While the appropriate treatment and estimation of the endogenous spatial lag has received a good deal of attention, the analysis of effects related to other endogenous variables has been less popular. Based on their previous work, the authors focus on the case where endogeneity is induced by the omission of a (spatially autoregressive) variable. They show the inconsistency of the usual OLS estimators induced by omitting a significant variable that should be in the regression model but which is unmeasured and hence is present in the residual. A simulation experiment is implemented that demonstrates how an augmented spatial Durbin model with a complex error process is a reasonably appropriate estimator. This is estimated using 2SLS (2 Stage-Least-Square) and SHAC (Spatial Heteroskedasticity and Autocorrelation Consistent) estimator for the variance-covariance matrix. This estimator performs better in terms of bias and Root Mean Square Error than the OLS-SHAC estimator. They reach the same conclusion when they modify the properties of the omitted variable used in their Monte-Carlo simulations. The discussion moves on to the case where endogeneity is a consequence of simultaneity and errors in variables. The authors conclude again that the 2SLS-SHAC estimation of the spatial Durbin model is better than an OLS-SHAC estimation of a single equation model where the endogeneity problem remains untreated.

Chapter "Dealing with Spatiotemporal Heterogeneity: The Generalized BME Model" by Yu, Christakos and Bogaert, discusses a stochastic approach for studying physical and social systems and their attributes, when these systems are characterized by heterogeneous space-time variations under conditions of multi-sourced uncertainty. The proposed Generalized Bayesian Maximum Entropy approach emerges from the fusing together of generalized spatiotemporal random field theory and a Bayesian Maximum Entropy mode of thinking. The result is a versatile approach to conduct spatiotemporal analysis and mapping that exhibits a number of attractive features, including the following: the approach makes no restrictive assumptions concerning estimator linearity and probabilistic normality (i.e., non-linear estimators and non-Gaussian distributions are naturally incorporated); it can be used to study natural systems with heterogeneous space-time dependence patterns; it can also account for various kinds of physical knowledge (core and case-specific) concerning the system under study; and it provides a general framework from which mainstream methods can be derived as special cases. The proposed space-time approach is applicable in a variety of knowledge domains (e.g., physical, health, social and cultural). Numerical experiments provide key insights into the computational implementation and comparative performance of the approach.

Along the lines of spatial heterogeneity, a long standing question refers to instability or nonstationarity in spatial models. Although this issue can be traced back to the development of Casetti's expansion method in the early 1970s, it has claimed renewed attention in light of newer methods for exploring local variations in spatial autocorrelation patterns and multivariate relationships (e.g., LISA, Getis-Ord statistics, geographically weighted regression or GWR). The problem of spatial instability is important as it refers to the well-known problem of the complex relations between spatial heterogeneity and spatial autocorrelation. The last chapter in this section by López, Mur, and Angulo, approaches this issue and investigates models where the intensity of spatial autocorrelation depends on the geographical location of each observation. In this respect, the chapter first presents a simple LM test of parameter instability for the spatial autocorrelation coefficient in a spatial lag model. Second, an extensive Monte-Carlo exercise is undertaken to study the distortions affecting the usual cross-sectional diagnostic measures (spatial autocorrelation LM tests, Jarque-Bera, Breusch-Pagan, White and RESET tests), when the assumption of constant spatial autocorrelation does not hold. Third, a local estimation algorithm labeled "zoom estimation", which can be considered an extension of the SALE model (Pace and LeSage 2004) is suggested and its performance with regard to the "zoom size" is evaluated with Monte-Carlo experiments. Finally, a strategy to identify spatial regimes in the spatial autocorrelation coefficient is proposed and compared to four other strategies based respectively on the k-means algorithm, Gaussian mixture models for multipolarity, Getis-Ord statistics, and trimmed mean classification rule. By providing novel information regarding the effect of spatial instability on usual diagnostic measures and specification search strategies, as well as giving suggestions to identify the presence and form of spatial regimes, this chapter represents a valuable step toward increasing our understanding of spatial instability in spatial econometric models.

3 Thematic Applications

3.1 Spatial Analysis of Land Use and Transportation Systems

The impressive visual qualities of transport and land use systems and processes in the real world have arguably not been matched by an equally impressive and constructive exercise in abstract data visualization. The availability of both proprietary and open environments for data analysis and visualization, coupled with the implementation of innovative approaches for data visualization presents an opportunity to advance the state-of-the-art with regards to the visual communication of spatial, temporal, and social qualities of transport and land use processes. Moreover, progress in automatic data collection through onboard GPS, cellular phone traces, or smart cards increases requirements for useful approaches and tools for summarizing and communicating the complexity and relevance of emerging modalities for communication and spatial interaction. Chapter "Seeing Is Believing": Exploring Opportunities for the Visualization of Activity–Travel and Land Use Processes in Space–Time, by Buliung and Morency, has as its objective to introduce recent innovations with regards to both platforms and approaches for the visualization of transportation and land use processes. To draw a parallel with "the arts," visualization can be compared to an anamorphosis interpreter wherein the act of visualization makes use of specialized devices (e.g., computer programs, statistical tools, GIS, interactive spreadsheets), or compels the viewer to occupy a specific perspective (e.g., spatial, temporal or social feature), with a view to reconstituting the "original" for the purpose of developing a clearer understanding of "process." Using examples drawn primarily from Montreal and the Greater Toronto Area, Canada, this chapter demonstrates how visualization techniques and tools can be used, often in a complementary way, to clarify transport- and land use-related spatial, temporal and social processes.

In addition to the exploration of transportation and land use processes, through visualization techniques, there has been considerable recent work on the confirmatory analysis, via multivariate techniques, of transport and land use phenomena. Four papers in this section apply spatial analytical techniques to the investigation of different aspects of land development and travel behavior. The first contribution in this group, by Munroe, is concerned with the expansion and rapid growth of urban areas, a process that can occur unevenly across space and through time. The availability of detailed spatial and temporal data describing land use, combined with the application of spatial and temporal modeling approaches (e.g., spatial logistic regression, hazard models), facilitates, in Chapter "Pattern-Based Evaluation of Peri-Urban Development in Delaware County, Ohio, USA: Roads, Zoning and Spatial Externalities", the detection and description of the global and local spatial properties of urban growth – i.e., dispersion, decentralization, fragmentation. The more abstract conceptualization of urban sprawl, as a somewhat even and regular expansion of urban areas into rural or peri-urban places, can be replaced by a more detailed, empirically informed view of key development processes and outcomes.

Investigation of peri-urban development in Delaware County, Ohio, is based on a discrete time-to-event model for Delaware County, one of the fastest growing counties in the US, located north of the state capital of Columbus, Ohio. Overall, the results suggest that the process of urban expansion/dispersion has simultaneously included an increase in the local clustering of development. A simulation experiment examines the sensitivity of predicted patterns of residential growth to policy and/or market-based drivers of growth processes including: density (intensification), access to roads, and development externalities. Controlling for the timing of development, avoidance of development, maximum density zoning policies, and distance to major roads emerge as factors contributing to the fragmentation of residential development in the county. From a policy perspective, the findings suggest that cooperative land use management at the township level, and open space preservation, are potentially useful approaches to control the growth processes described within the chapter.

Also related to the topic of sprawling development, "Chapter Demand for Open Space and Urban Sprawl: The Case of Knox County, Tennessee" by Cho, Lambert, Roberts and Kim, is concerned with the demand for open space. While there is limited consensus in the literature regarding the conceptualization and measurement of urban sprawl, scholars, practitioners, and governments, consider the study and implementation of growth management to be an important intellectual and practical exercise. The research reported in this chapter makes use of a two-step spatial modeling approach to examine the efficacy of open space conservation as a policy tool for managing urban sprawl. The conceptualization of sprawl chosen by the authors includes processes of expansion or encroachment into rural areas, and the leapfrogging of development. The case study of Knox County presents an interesting situation because the county has experienced rapid growth overall, with some local heterogeneity (spatially and temporally) in the pace of residential development. Analysis is supported by a very detailed spatial database of the region obtained from secondary sources, and the use of GIS techniques and remote sensing data to quantify household access to open space. The spatial modeling task combines hedonic price modeling with geographically weighted regression. Comparative analysis of model results indicates that the GWR (spatial error) model provides an important complement to the global (OLS) alternative. With regard to policy, the results appear to be open to several interpretations; acting freely in the market, affluent households may be willing to pay (i.e., buy into policy) to preserve open space, on the one hand, or demand open space at the edge, potentially giving rise to additional and perhaps undesirable patterns of growth – particularly in the absence of an appropriate regulatory framework.

The next two chapters are concerned with issues in travel behavior. One theme that has interested geographers and planners for some time is the existence of differential patterns of mobility by gender. The tenth chapter is entitled *Multilevel Models of Commute Times for Men and Women*. In his contribution, Zolnik examines, from a spatial perspective, the well-documented issue of the commuting-time gender gap. Research has often presented evidence suggesting that women typically have shorter commutes than men. Sociological and economic explanations have been advanced, with some recent evidence suggesting some convergence in the generalized cost of commuting, particularly at the margins of male/female income distributions, and within certain occupational or ethnic groups. The research presented in this chapter draws independent (male, female) and pooled (male and female) samples from the 2001 US National Household Travel Survey, which are used to estimate multi-level models of self-reported journey to work commute time. The samples included individuals who worked and commuted (within a single Metropolitan Statistical Area) by private vehicle the week prior to the survey. Income and occupation effects appeared to be stronger for women, while access to private vehicles appeared to have a stronger positive influence on commute times for male workers. Interestingly, Zolnik concludes that his findings lend little support to the household responsibility hypothesis. Apart from strong congestion effects differentiable by sex, his findings suggest only marginal commute time savings with changes in development intensity and the mixing of land uses.

The final chapter in this section, by Goetzke, is concerned with two topics of current interest from the spatial analysis and travel behavior perspectives: the role of spatial effects in choice models, and the possibility that information spillovers may lead to interdependent choice processes. The research reported in chapter "Walkability as a Summary Measure in a Spatially Autoregressive Mode Choice Model: An Instrumental Variable Approach" is motivated by the difficulties posed by the non-linear functional form of spatially autoregressive binary choice models (logit or probit models), especially if the analyst does not wish to assume a conditional spatial structure, which has the disadvantage that it imposes a strong restriction on the model. On the other hand, a linear probability model (LPM) can easily be extended to a spatially autoregressive model with few additional difficulties. However, a LPM exhibits by definition always heteroskedasticity, which makes the estimation inefficient. Empirically, the model proposed is demonstrated using the 1997/1998 New York Metropolitan Transportation Council comprehensive regional household travel diary survey, in analysis that aims at determining whether social spill-over effects exists for walking commutes in Manhattan (i.e., a large enough sample size to adequately capture pedestrian behavior). The spatial process is modeled using the instrumental variable 2SLS method. In a third step, the LPM is additionally corrected for heteroskedasticity using a weighted least square approach with the assumption of a binomial distribution in the error term. The estimation method proposed is extended to a probit/logit model where the spatial process is also modeled using the instrumental variable approach (spatially autocorrelated IV probit/logit model). The results of both models are compared with the results of a conditional spatially autoregressive probit/logit model. This application shows that the instrumental variable method for estimating spatially autoregressive probability models is able to overcome the shortcomings of a conditional spatially autoregressive binary choice model, besides being relatively straightforward to implement.

3.2 Economic and Political Geography

The second thematic section is comprised of four chapters dealing with various topics in economic and political geography. In Chapter "Employment Density in Ile-de-France: Evidence from Local Regressions", Guillain and Le Gallo address the issue of suburbanization in the Ile-de-France region in France. With the development of peripheral employment centers, the spatial organization of the region's activities does not necessarily correspond to the traditional monocentric model. The aims of this chapter are first to understand whether the Central Business District (CBD) does still influence the employment distribution in Ile-de-France, and secondly, if so, whether this effect differs by sector. In order to answer these questions, the authors identify first the location of employment centers by measuring the spatial agglomeration of economic activities, with global and local spatial autocorrelation statistics. Second, they conduct an in-depth analysis of the centers by identifying their sectoral specialization and their attractiveness for strategic activities. The authors use various spatial econometric specifications of the density functions and perform local regressions, using geographically weighted regression, where the rate at which density falls with distance from the CBD is estimated for each observation. The local regressions facilitate the detection of changes in density by distance (heterogeneous distribution) and direction (anisotropic distribution) from the CBD. The main results of this study indicate that the CBD still influences total employment in Ile-de-France but that its influence varies by sector, distance, and direction from the CBD. From a political viewpoint, their conclusions provide new insights about the location strategies of households and economic activities in Ile-de-France.

Chapter "The Geographic Dimensions of Electoral Polarization in the 2004 U.S. Presidential Vote", by Sue Wing and Walker, is motivated by the apparent divisiveness of the 2004 US presidential election. This observation gave rise to the exploration of the hypothesis that the U.S. electorate is geographically polarized. Using spatial econometric analyses, these authors investigate the effects of the characteristics of populations and places on voter turnout in favor of George W. Bush. Specifically, the authors identify key factors affecting Bush's odds of success at the national level, and demonstrate how these aggregate effects vary over finer spatial scales. The results provide an intriguing first look at overall spatial patterns in the correlates of voting behavior, and argue for a new way of thinking of polarization as a phenomenon which occurs within individual sub-groups across space, with geography playing a crucial role at both local and regional scales, but in ways which are not easily categorized or explained.

The topic analyzed in chapter "Gender Wage Differentials and the Spatial Concentration of High-Technology Industries", by Echeverri-Carroll and Ayala, deals with gender wage differentials in cities and is relevant for the study of the issues of agglomeration, the productivity of cities and the existence of localization and urbanization economics. From a methodological perspective, the work deals with specific econometric problems linked to the analysis of spatial microeconomic data: heteroskedasticity and endogeneity. Previous studies have found that male workers attain higher wages in cities (high-tech cities in particular) with a large endowment of human capital than in those with a low endowment. New Economic Geography models maintain that the higher wages of males are linked to productivity-enhancing effects from the (formal and informal) exchange of knowledge that characterizes high-tech cities. The authors question whether women enjoy similar productivity-enhancing effects. A large sample is drawn from the 5% PUMS of the 2000 Census of Population, and is used to estimate regressions separately for a sample of male and female workers, accounting for arbitrary clustering, heteroskedasticity in the error terms, and endogeneity. The estimates show that after controlling for individual- and city-level variables that affect wages, male workers that live in a high-tech city and work in a high-tech industry, holding other factors fixed, indeed earn more than comparable female workers. The results support the view that women might benefit less from knowledge networks that are predominant among high-tech industry workers in high-tech cities and from the demand for talent exercised by these industries.

Among the wide range of spatial econometric applications, fiscal and monetary economic applications remain quite scarce. Chapter "Fiscal Policy and Interest Rates: The Role of Financial and Economic Integration", by Claeys, Moreno and Suriñach, fills this gap by analyzing the role of spatial spillovers in the crowding-out effects of fiscal expansion on interest rates. The chapter parts from the common belief that fiscal expansion raises interest rates. However, the crowding-out effects of deficits have been found to be small or non-existent. One explanation is that financial integration offsets interest rate differentials on globalized bond markets. As a result, the authors measure the degree of integration of government bond markets, using spatial modeling techniques, with a view to taking this spillover effect on financial markets into account. Using a panel of 101 countries and annual data on interest rates and fiscal policy covering the period 1990–2005, the main finding is that the crowding out effect on domestic interest rates is significant, but that it is reduced by spillover across borders. The detected spillover effect is important in major crises or in periods of coordinated policy actions. The result is generally robust to various measures of cross-country linkages, and indicates strong spillover effects among EU countries.

3.3 Spatial Analysis of Population and Health Issues

The next three chapters in the volume are concerned with the spatial analysis of various aspects of population and health. In chapter "Spatial Models of Health Outcomes and Health Behaviors: The Role of Health Care Accessibility and Availability", Waldorf and Chen address the question of whether poor spatial accessibility to health care providers leads to poor health outcomes. Their work focuses on the 80 counties of Indiana, a state that, as many other across the US, experiences a spatial mismatch between the location of supply and demand for medical care as well as spatial variations in the quality of medical facilities. This problem presents a challenge for policymakers who need to determine how to equitably allocate medical resources to improve public health in general and help medically underserved rural areas in particular. The study is grounded on the measurement of accessibility (as opposed to availability) of health care providers in order to better capture the distance-cost faced by patients wishing to receive treatment. It is worthwhile to note that while accessibility to health care has been extensively studied, the approach presented in this chapter is innovative for two reasons. First, accessibility to health care is linked through a modeling framework to health outcomes. Second, the models are estimated after the inclusion of spatial dependence effects. In the case of health outcomes, spatial dependence may be a statistical artifact, but it can also be grounded in behavioral processes such as imitation behavior and the spatial diffusion of cultural norms influencing health care utilization. These effects could also be a result of underlying factors such as poor labor market conditions which affect people's access to health insurance and thus ultimately people's health. The models reported in the chapter are estimated for six health outcome variables relating to the health status of infants and the elderly, and four health behavior variables. The results indicate that the impact of health behaviors, health care accessibility, and spatial dependence varies across the various health outcomes investigated. The authors conclude that, from a policy perspective, it is important to recognize that efforts to improve health behaviors in one county could impact health behaviors in neighboring counties, eventually trickling down through an entire state.

While the research of Waldorf and Chen is concerned with the effect of *accessibility* to health care *on* health outcomes, the work of Woltman and Newbold in chapter "Immigrant Women, Preventive Health and Place in Canadian CMAs" is related to the *utilization* of health services, with a particular focus on immigrants in Canada. While the health status of immigrants has been studied extensively, the health service challenges facing immigrants are perhaps less understood. This chapter advances current thinking on the use of health care services by immigrant women in Canadian Census Metropolitan Areas (CMAs), and more specifically, the utilization by immigrant women of cervical cancer screening. Analysis is conducted by examining the multilevel factors associated with Pap (smear) testing in native-born and immigrant women. Cross-sectional multi-level logistic regression analysis is then used to detect individual and neighborhood level correlates of lifetime uptake (i.e., ever had a Pap test), and regular use (i.e., a test within the last three years) of Pap testing. Individual data are drawn from the Canadian Community Health Survey (Cycle. 2.1, 2003) for the population of interest, namely immigrant and native-born women between the ages of 18 and 69, living in the Montréal, Toronto, and the Vancouver Census Metropolitan Areas. Contextual factors are constructed by linking individual level data with census tract profile data from the 2001 Census of Canada. The results indicate the presence of between-neighborhood variation in uptake. Immigrant status and cultural origin appear to be significantly associated with lifetime uptake, although uptake appears to be less common amongst recent immigrant women and women of Chinese, South Asian and other Asian backgrounds. The results also suggest that neighbourhood disadvantage (i.e., a composite index) and immigrant concentration are positively associated with regular Pap testing. Findings concerning the role of culture and immigration status, coupled with the reported neighbourhood effects, lend support to the development of neighbourhood level interventions focused on increasing the awareness of recent immigrant women of the availability of cervical cancer screening services.

The last chapter in this section, by Lambert, Wilcox, Clark, Murphy and Park, combines in the most explicit way the two themes of population and health. The question posed for this chapter is the extent to which new-generation retirement communities are responsible for agglomeration within the health care sector. Demographers estimate that over the next 18 years at least 400,000 retiring baby boomers will migrate beyond their state borders each year, carrying with them an average of $320,000 to spend on a new home. It is no surprise then that migrating retirees can stimulate economic growth and development in their host rural communities. Factors of import to migrating seniors with respect to residential site selection include health care service availability, recreational amenities, affordable housing, low taxes, and proximity to friends and family. The geographical focus of this paper is the Southeastern US, an area that has experienced an extraordinary influx of retiring seniors since 1990. As more retiring seniors choose a particular residential location, demand for health services will presumably increase, creating new employment opportunities. On the other hand, migrating seniors may be attracted to communities with a wider array of health care services. This problem is reminiscent of the "jobs-to-people/people-to-jobs" conundrum. In order to tease out these relationships, the authors draw from recent developments in the spatial econometric literature to develop a regional adjustment econometric model that accounts for endogeneity and heteroskedasticity. The results of the analysis suggest that rural communities, able to support diversified health services are at a comparative advantage with respect to attracting retirees, whereas provision of such services in counties near metropolitan centers appears to be of reduced importance. In addition, there is evidence that retiree in-migration is correlated with overall growth in the health sector.

3.4 Regional Applications

The last set of papers in the book includes applications of spatial analysis to questions focused on regional systems. Chapter "Evolution of the Influence of Geography on the Location of Production in Spain (1930–2005)", by Chasco and López, is concerned with the relative importance of geographic features on the location of production in Spain. Based on a panel of 47 Spanish provinces and the 1930–2005 period, they quantify how much of the spatial pattern of GDP can be attributed to only exogenous *first nature* elements (physical and political geography) and how much can be derived from endogenous *second nature* factors (man-made agglomeration economies). The authors employ an analysis of variance (ANOVA) to infer the unobservable importance of first nature indirectly in a stepwise procedure. In order to disentangle the two net effects empirically, as well as their mixed effect, they control for second nature because every locational endowment will be reinforced and overlaid by second nature advantages. In a dynamic context, they also estimate how much agglomeration can be explained by both gross and net second nature with the aim of isolating the importance of first nature alone. The authors stress the fact their results could be biased if some potential econometric questions (multicollinearity, relevant missing variables, endogeneity and more particularly spatial effects) were not properly taken into account. They conclude that production is not randomly distributed across Spanish regions: 88% of the GDP's spatial variation can be explained by the direct and indirect effects of geography. After controlling for agglomeration economies and interaction effects of the first/second nature, the net influence of natural geography goes from 20% in 1950 to 6–7% nowadays. On the other hand, while second nature agglomeration forces (e.g., transport and communications) were dominant in the 1930s, they were overcome by first nature geography by the end of the period. These results also differ across the two spatial regimes that characterize the country: the coast plus the Madrid metropolitan area, and the hinterland. Overall, the research presented in this chapter represents an innovative way to measure the extent to which regional policies are able to favor agglomeration in areas without clear geographic advantages.

The spatial dynamics of regional systems is the topic of chapter "Comparative Spatial Dynamics of Regional Systems", by Rey and Ye, and in particular, the dynamics of income convergence. These authors note that the focus of research, having shifted from the national to the regional perspective in the early 1990s, continued to be dealt with using the same theoretical and technical frameworks underpinning national growth research. By the end of the 1990s, however, the geographical dimension of convergence issues had already attracted substantial attention. This chapter contributes to the literature on income convergence by considering two of the world's largest and deeply entangled economies, the US and China, at different developmental stages, and by bringing to bear some of the most recent tools in exploratory spatial data. In addition to their use for convergence analysis, the new set of statistical measurements introduced in this chapter open up new opportunities for scientific visualization and the generation of hypothesis in other fields that deal with dynamic space-time processes.

The closing paper, contributed by Koch, examines regional growth and convergence. The literature focusing on issues of growth and convergence from the specific perspective of spatial econometrics techniques is today extensive. The studies in this area focus on the interdependence between nations and regions, highlighting how the economy of one country or region is not independent of the economies of neighbouring countries or regions (and perhaps non-neighbours as well). However, a common feature of these papers is that the spatial econometric specifications are introduced in an ad hoc way, i.e., spatial lag or spatial error models are estimated, and the choice of the specification tends to be based on statistical criteria. Recently, theoretical foundations of spatial dependence have been suggested. Chapter "Growth and Spatial Dependence in Europe" is representative of this trend since it presents an augmented Solow model that includes spatial externalities and spatial interdependence among economies. A spatial econometric reduced form allows the testing of the effects of the rate on saving, the rate of population growth and location on per worker income, and on the conditional convergence process in Europe. Based on a sample of European regions, the econometric model leads to estimates of structural parameters close to predicted values while Marshallian externalities are found to be non-significant.

The articles in this book therefore highlight the importance of spatial effects in various themes and applications. Each of them opens new research areas and we hope that they will foster further advances in spatial statistics and spatial econometrics.

Part I
Theory and Methods

Omitted Variable Biases of OLS and Spatial Lag Models

R. Kelley Pace and James P. LeSage

1 Introduction

Numerous authors have suggested that omitted variables affect spatial regression methods less than ordinary least-squares (OLS; Dubin 1988; Brasington and Hite 2005, Cressie 1993). To explore these conjectures, we derive an expression for OLS omitted variable bias in a univariate model with spatial dependence in the disturbances and explanatory variables. There are a number of motivations for making this set of assumptions regarding the disturbances and explanatory variables. First, in spatial regression models each observation represents a region or point located in space, for example, census tracts, counties or individual houses. Sample data used as explanatory variables in these models typically consists of socioeconomic, census and other characteristics of the regional or point locations associated with each observation. Therefore, spatial dependence in the explanatory variables seems likely, motivating our choice of this assumption. Note, the literature rarely examines the spatial character of the explanatory variables, but this can affect the relative performance of OLS as shown below. Second, application of OLS models to regional data samples frequently leads to spatial dependence in the regression disturbances, providing a justification for this assumption. Finally, there are a host of latent unobservable and frequently unmeasurable influences that are likely to impact spatial regression relationships. For example, factors such as location and other types of amenities, highway accessibility, school quality or neighborhood prestige may exert an influence on the dependent variable in hedonic house price models. It is unlikely that explanatory variables are readily available to capture all of these types of latent influences. This type of reasoning motivates our focus on the impact of omitted explanatory variables. Since the omitted and included explanatory variables are both likely to exhibit spatial dependence based on the same spatial connectivity structure, it seems likely that omitted and included variables will exhibit

R.K. Pace (✉)
E. J. Ourso College of Business Administration, Department of Finance, Louisiana State University, Baton Rouge, LA 70803-6308, USA,
e-mail: kelley@pace.am

A. Páez et al. (eds.), *Progress in Spatial Analysis*, Advances in Spatial Science,
DOI 10.1007/978-3-642-03326-1_2,

non-zero covariance. The expression we derive for OLS bias in these circumstances shows that positive dependence in the disturbances and explanatory variables when omitted variables are correlated with included explanatory variables magnifies the magnitude of conventional least-squares omitted variables bias.

We extend the considerations above to also include models where the dependent variable exhibits spatial dependence, following a spatial autoregressive process. LeSage and Pace (2009) provide a number of different motivations for how spatial dependence in the dependent variable arises in spatial regression relationships. It is well-known that spatial dependence in the dependent variable leads to bias in OLS estimates (Anselin 1988). We show that this type of spatial dependence in the presence of omitted variables exacerbates the usual bias that arises when applying OLS to this type of sample data. In particular, the bias is magnified, with the magnitude of bias depending on the strength of spatial dependence in: the disturbances, the dependent variable, and the explanatory variable included in the model.

Our derivation shows that the combination of an omitted variable, spatial dependence in the disturbances, dependent and explanatory variables leads to an implied model specification that includes spatial lags of both the dependent and explanatory variables. This type of model has been labeled a spatial Durbin model (SDM) in the literature (Anselin 1988). Estimates based on the SDM specification which matches the implied DGP in this set of circumstances shrinks the bias relative to OLS.

In the following section, we consider the implications of omitted variables in the presence of spatial dependence for OLS estimates. Next we demonstrate that the SDM model specification matches a reparameterization of the DGP that results from various assumptions on omitted variables and spatial dependence. We consider an expression for the omitted variables bias that arises when the SDM model is used to produce estimates, and compare this to the bias expression for OLS estimates. We show that the magnitude of omitted variable bias for the SDM model does not exhibit the magnification of OLS and it no longer depends on the magnitude of spatial dependence in the disturbances, dependent, or independent variables. These desirable properties of the SDM model provide a strong motivation for use of this model specification in applied practice.

2 Spatial Dependencies and OLS Bias

We begin with a frequently used spatial econometric model specification shown in (1) and (2). Equation (1) represents a spatial autoregressive process governing the dependent variable and (2) adds the assumption of spatial autoregressive disturbances. This model has been labeled SAC by LeSage (1999) and a spatial autoregressive model with autoregressive disturbances by Kelejian and Prucha (1998). It should be noted that we will work with a model involving simple univariate explanatory and omitted variable vectors for simplicity. There is no reason to believe that the results we derive here would not extend to the more general case involving matrices of explanatory variables in place of the univariate vector.

$$y = x\beta + \alpha Wy + \varepsilon \tag{1}$$
$$\varepsilon = \rho W\varepsilon + \xi \tag{2}$$
$$\xi = x\gamma + u \tag{3}$$
$$x = \phi Wx + v \tag{4}$$

In (1) through (4), the n by 1 vector y represents observations on the dependent variable, x represents an n by 1 vector of observations on a non-constant explanatory variable, ε, ξ, u, and v represent various types of n by 1 disturbance vectors, $\alpha, \beta, \rho, \phi$, and γ represent scalar parameters, and W is an n by n non-negative symmetric spatial weight matrix with zeros on the diagonal. We assume that u is distributed $N\left(0, \sigma_u^2 I_n\right)$, v is distributed $N\left(0, \sigma_u^2 I_n\right)$, and u is independent of v. For simplicity, we exclude the intercept term from the model.

We extend the conventional SAC model specification using (3) that adds the assumption of an omitted variable correlated with the explanatory variable x. The strength of correlation is determined by the parameter γ and the variance of the noise vector u, σ_u^2. The last equation, (4) adds the assumption of a spatial dependence in the explanatory variable x, which is governed by a spatial autoregressive process with dependence parameter ϕ. We focus on non-negative spatial dependence, by assuming $\alpha, \phi, \rho \in [0, 1)$. We note that in the case where $\gamma = 0$, there is no covariance between the included explanatory variable x and the omitted variable ξ. In the case where $\phi = 0$, the explanatory variable does not exhibit spatial dependence, and when $\alpha = 0$, the dependent variable does not exhibit spatial dependence. Similarly $\rho = 0$ eliminates spatial dependence in the disturbances.

The weight matrix has positive elements W_{ij} when observations i and j are neighbors, and we assume each observation has at least one neighbor. The symmetry of W contrasts with the usual lag operator matrix L from time series, since L is strictly triangular containing zeros on the diagonal. Powers of L are also strictly triangular with zeros on the diagonal, so that L^2 specifies a two-period time lag whereas L creates a single period time lag. It is never the case that produces observations that point back to include the present time period. In contrast, W^2 specifies neighbors to the neighbors identified by the matrix W, and since the neighbor of the neighbor to an observation includes the observation itself due to symmetry, W^2 has positive elements on the diagonal. This results in a form of simultaneous dependence among spatial observations that does not occur in time series analysis, making spatial regression models distinct from time series regressions. We use the same spatial weight matrix W to specify the pattern of spatial dependence in the explanatory variable, which seems reasonable since this matrix reflects the spatial configuration of both the dependent and independent variable observations.

We assume that W is symmetric and real, so the n by 1 vector of eigenvalues λ_W is real. For simplicity, we assume the eigenvalues are unique, the principal eigenvalue of W equals 1, and this is the maximum eigenvalue as well. This is not a restrictive assumption since dividing any candidate weight matrix by its principal eigenvalue would yield a weight matrix with a principal eigenvalue of 1.

Since the sum of the eigenvalues ($tr(W)$) equals 0, the minimum eigenvalue is negative, but the minimum eigenvalue is not the principal eigenvalue, and $\min(\lambda_W) > -1$. Therefore, for some real scalar $\theta \in (\min(\lambda_W)^{-1}, 1)$, $I_n - \theta W$ will be symmetric, positive definite, and thus $(I_n - \theta W)^{-1}$ exists. Clearly, $\theta \in [0, 1)$ is sufficient for positive definite $I_n - \theta W$. Finally, since the maximum eigenvalue equals 1, $tr(W^{2j}) \geq 1$ (all eigenvalues of even powered matrices are non-negative and the largest eigenvalue equals 1) and 0 for any integer $j > 0$ (traces of non-negative matrices are non-negative).

We rewrite (1) to solve for y using $F(\alpha) = (I_n - \alpha W)^{-1}$ and this yields (5). We rewrite (2) to solve for ε using $G(\rho) = (I_n - \rho W)^{-1}$ and substitute $x\gamma + u$ for ξ via (3) to yield (6). Similarly, we rewrite (4) to isolate x using $H(\phi) = (I_n - \phi W)^{-1}$ to produce (7). Equation (8) summarizes the definitions.

$$y = F(\alpha)x\beta + F(\alpha)\varepsilon \tag{5}$$

$$\varepsilon = G(\rho)(x\gamma + u) \tag{6}$$

$$x = H(\phi)\nu \tag{7}$$

$$\begin{aligned} F(\alpha) &= (I_n - \alpha W)^{-1} \\ G(\rho) &= (I_n - \rho W)^{-1} \\ H(\phi) &= (I_n - \phi W)^{-1} \end{aligned} \tag{8}$$

Taken together, (5), (6), and (7) lead to the DGP shown in (9).

$$y = F(\alpha)\mathrm{H}(\phi)\nu\beta + F(\alpha)G(\rho)H(\phi)\nu\gamma + F(\alpha)G(\rho)u \tag{9}$$

Given the assumptions made concerning the matrix W, the matrix inverses: $F(\alpha)$, $G(\rho)$, $H(\phi)$ exist. We refer to (9) as a DGP since this expression could be used with vectors ν, u of random deviates to generate a dependent variable vector y from the model and assumptions set forth. Given the structure of the model set forth in (1)–(4), the parameters $\alpha, \rho, \phi, \gamma$ allow us to generate dependent variable vectors that reflect varying combinations of our assumptions. For example, setting $\gamma = 0$ and maintaining positive values for α, ρ, ϕ would produce a vector y reflecting no covariance between the included and omitted variable vectors x and ξ. Similarly, setting $\phi = 0$ while maintaining positive values for the other parameters (α, γ, ρ) would produce a vector y from a model having no spatial dependence in the explanatory variable x.

OLS estimates $\hat{\beta}_o = (x'x)^{-1}x'y$ represent "best linear unbiased" estimates when the DGP matches that of the ordinary regression model: $y = x\beta + \varepsilon$, and the Gauss–Markov assumptions. These require the vector x to be fixed in repeated sampling and the disturbances to have constant variance and zero covariance.

However, suppose that the true DGP is (9) and we apply the least-squares expressions to produce the estimates in (10). That is, we apply least-squares in the circumstances considered here involving spatial dependence in the dependent variable, disturbances and the model contains an omitted variable that is correlated with the spatially dependent included variable.

$$\hat{\beta}_o = \frac{\nu' H(\phi) F(\alpha) H(\phi) \nu}{\nu' H(\varphi)^2 \nu} \beta + \frac{\nu' H(\phi) F(\alpha) G(\rho) H(\phi) \nu}{\nu' H(\phi)^2 \nu} \gamma$$
$$+ \frac{\nu' H(\phi) F(\alpha) G(\rho) u}{\nu' H(\phi)^2 \nu} \tag{10}$$

This expression can be further simplified. To do so, we turn to some additional results. We begin by defining (11),

$$R(A) = \frac{d' A d}{d' d} \tag{11}$$
$$Q(A) = \frac{d' A r}{d' d}$$

where d, r are distributed $N(0, \sigma_d^2 I_n)$, $N(0, \sigma_r^2 I_n)$ with r independent of d, and A is a n by n symmetric real matrix. Using different techniques, both Barry and Pace (1999), and Girard (1989) show that:

$$E(R(A)) = \frac{tr(A)}{n} \tag{12}$$
$$\sigma^2_{R(A)} = \frac{2\sigma^2_{\lambda(A)}}{n}$$
$$E(Q(A)) = 0$$

where *tr* denotes the trace operator and $\sigma^2_{\lambda(A)}$ is the variance of the eigenvalues of matrix A. Obviously, $E(d'Ar) = 0$ due to the independence of r and d, while $d'd > 0$ so that $E(Q(A)) = 0$.

Consider a variation of (11) involving n by n symmetric real matrices A and B.

$$R(A/B) = \frac{d' A d}{d' B d} = \frac{(d'd)^{-1} d' A d}{(d'd)^{-1} d' B d} \tag{13}$$

From (12), the expectation of the numerator of (13) equals $tr(A)/n$, and the expectation of the denominator of (13) equals $tr(B)/n$. Also, an implication of (12) is that as $n \to \infty$, the variance of the numerator and denominator go to 0. Therefore,

$$plim_{n \to \infty} R(A/B) = \frac{tr(A)}{tr(B)} \tag{14}$$

Applying these results to expression (10), results in the third term of (10) vanishing asymptotically via (12). Applying result (14) to the first two terms of (10), and using the cyclical properties of the trace, produces expression (15) and its equivalent abbreviated form in (16).

$$p\lim_{n\to\infty}\hat{\beta}_o = \frac{tr\left[H(\phi)^2F(\alpha)\right]}{tr\left[H(\phi)^2\right]}\beta + \frac{tr\left[H(\phi)^2F(\alpha)G(\rho)\right]}{tr\left[H(\phi)^2\right]}\gamma$$

$$= T_\beta(\phi,\alpha)\beta + T_\gamma(\phi,\alpha,\rho)\gamma \quad (15)$$

$$T_\beta(\phi,\alpha) = \frac{tr\left[H(\phi)^2F(\alpha)\right]}{tr\left[H(\phi)^2\right]}$$

$$= T_\gamma(\phi,\alpha,\rho) = \frac{tr\left[H(\phi)^2F\left(\alpha\right)G\left(\rho\right)\right]}{tr\left[H(\phi)^2\right]} \quad (16)$$

Naturally, as the factors $T_\beta\left(\phi,\alpha\right)$ and $T_\gamma\left(\phi,\alpha,\rho\right)$ rise above 1, the bias of using OLS to produce estimates for a model with a dependent variable y generated using our spatial DGP from (9) can increase. This will be especially true when β and γ have the same signs. We will show that $T_\beta\left(\phi,\alpha\right) > 1$ for $\alpha > 0$ and $T_\gamma\left(\phi,\alpha,\rho\right) > 1$ when spatial dependence in the dependent variable y or disturbances exists ($\alpha > 0$ or $\rho > 0$), and that spatial dependence in the explanatory variable $\phi > 0$ amplifies these factors. Our strategy involves showing that when no spatial dependence in the explanatory variable exists ($\phi > 0$), $T_\beta\left(0,\alpha\right) > 1$ when $\alpha > 0$ and $T_\gamma\left(0,\alpha,\rho\right) > 1$ when $\alpha > 0$ or $\rho > 0$. We then show that $T_\beta\left(\phi,\alpha\right) > T_\beta\left(0,\alpha\right)$ and that $T_\gamma\left(\phi,\alpha,\rho\right) > T_\gamma\left(0,\alpha,\rho\right)$ when $\phi > 0$.

We begin by showing that $T_\beta\left(\phi,\alpha\right) > 1$ when $\alpha > 0$, $\phi > 0$ and $T_\gamma\left(\phi,\alpha,\rho\right) > 1$ when α or ρ are positive and $\phi = 0$ (no spatial dependence in the explanatory variable). To see the first assertion, let θ_m represent some positive scalar parameter. Since $(I_n - \theta_m W)^{-1} = I_n + \theta_m^2 W^2 + \cdots$, and since $tr\left(I_n\right) = n$, $tr\left(W^{2j}\right) \geq 1$, and $tr\left(W^{2j-1}\right) \geq 0$, $tr\left(I_n - \theta_m W\right)^{-1} > n$. To generalize this, let $\theta_1 > 0$ or $\theta_2 > 0$ and consider $P\left(\theta_1,\theta_2\right) = \left(I_n - \theta_1 W\right)^{-1}\left(I_n - \theta_2 W\right)^{-1} = I_n + \pi_1 W + \pi_2 W^2 + \cdots$ where $\pi > 0$. Since products and sums of positive parameters (θ_1, θ_2) are positive, $tr\left[\left(P\left(\theta_1,\theta_2\right)\right]\right. > n$ because $tr\left(I_n\right) = n$, $tr\left(W^{2j}\right) \geq 1$, and $tr\left(W^{2j-1}\right) \geq 0$. When $\phi = 0$, this describes the numerator of both $T_\beta\left(0,\alpha\right)$ and $T_\gamma\left(0,\alpha,\rho\right)$ and the denominator of both terms is n. Consequently, $T_\beta\left(0,\alpha\right) > 1$ and $T_\gamma\left(0,\alpha,\rho\right) > 1$.

We now turn to the effect of positive spatial dependence in the explanatory variable ($\phi > 0$) on $T_\beta\left(\phi,\alpha\right)$ and $T_\gamma\left(\phi,\alpha,\rho\right)$. We show that $T_\beta\left(\phi,\alpha\right) > T_\beta\left(0,\alpha\right)$ and $T_\gamma\left(\phi,\alpha,\rho\right) > T_\gamma\left(0,\alpha,\rho\right)$ for $\phi > 0$. Let Ω and Ψ be monotonic functions of similar symmetric matrices so that both Ω and Ψ are symmetric positive definite and are not proportional to an identity matrix. Given these assumptions, consider the assertion in (17). Multiplying both sides by the positive scalar $tr\left(\Psi\right)/n$ does not change the direction of the inequality and this leads to (18). Since Ψ and Ω are based upon the same eigenvalues (similar matrices) and are monotonic functions of these eigenvalues, the eigenvalues of Ψ and Ω have the same ordering. Moreover, the eigenvalues of $\Psi\Omega$ are the product of these ordered eigenvalues as shown in equation (19).

$$\frac{tr\,(\Omega)}{n} < \frac{tr\,(\Psi\Omega)}{tr\,(\Psi)} \tag{17}$$

$$\frac{tr(\Omega)}{n}\frac{tr(\Psi)}{n} < \frac{tr(\Psi\Omega)}{n} \tag{18}$$

$$\left[n^{-1}\sum_{i=1}^{n}\lambda\,(\Psi)_i][n^{-1}\sum_{i=1}^{n}\lambda\,(\Omega)_i\right] < \left[n^{-1}\sum_{i=1}^{n}\lambda\,(\Psi)_i\,\lambda\,(\Omega)_i\right] \tag{19}$$

In fact, (19) is a restatement of the Chebyshev sum inequality from Gradshteyn and Ryzhik (1980). Expression (19) holds true as a strict inequality, since the eigenvalues are not all the same (because Ω and Ψ are not proportional to the identity matrix). Substitution of $\Psi = H(\phi)^2$ and $\Omega = F\,(\alpha)$ or $\Omega = F\,(\alpha)\,G\,(\rho)$ proves the assertion that $T_\beta\,(\phi,\alpha) > T_\beta\,(0,\alpha)$ and $T_\gamma\,(\phi,\alpha,\,\rho) > T_\gamma\,(0,\alpha,\rho)$ for $\phi > 0$, where the strict inequality arises because the eigenvalues of W and the monotonic functions of the eigenvalues of W are not similar to the identity matrix.

As already indicated, our expression (9) for the DGP allows us to consider special cases that arise from various settings of the control parameters α,ρ,ϕ,γ. We enumerate how some of these special cases impact omitted variables bias in various applied situations using our results applied to the expressions from (15).

1. Spatial dependence in the disturbances and explanatory variable, but no covariance between the explanatory variable and omitted variable. This results from setting the parameters $(p{>}0,\phi{>}0,\alpha = 0,\gamma = 0)$. In this case, $plim_{n\to\infty}\hat{\beta}_o = \beta$, and there is no asymptotic bias.
2. Spatial dependence in the explanatory variable in the presence of an omitted variable that is correlated with the included explanatory variable but no spatial dependence in the dependent variable or disturbances. This results from setting the parameters $(\phi > 0,\gamma > 0,\alpha = 0,\rho = 0)$. In this case, $plim_{n\to\infty}\hat{\beta}_o = \beta + \gamma$, and we have the standard omitted variable bias that would arise in the least-squares model.
3. Spatial dependence in y and the explanatory variable with no correlation between the explanatory and omitted. This results from setting the parameters $\big(\alpha > 0,\ \phi > 0,\ \gamma = 0\big)$. In this case, $plim_{n\to\infty}\hat{\beta}_o = T_\beta\,(\phi > 0,\alpha > 0)\,\beta$, and OLS has asymptotic bias amplified by the parameter α reflecting the strength of spatial dependence in y and by ϕ representing the strength of dependence in the explanatory variable.
4. No spatial dependence in y, spatial dependence in the disturbances and the explanatory variables with an omitted variable that is correlated with the included explanatory variable. This results from setting the parameters $\big(\rho > 0,\ \gamma > 0,\ \phi > 0,\alpha = 0\big)$. In this case, $plim_{n\to\infty}\hat{\beta}_o = \beta + T_\gamma\,(\phi > 0,\alpha = 0,\rho > 0)\,\gamma$, and OLS has omitted variables bias amplified by the spatial dependence in the disturbances and in the explanatory variable reflected by the magnitudes of the scalar parameters ρ and ϕ.

The first result is well-known, and the second is a minor extension of the conventional omitted variables case for least-squares. The third result shows the bias from applying OLS when the true DGP produces spatial dependence in the dependent variable y, and there is spatial dependence in the included explanatory variable. The bias for this case exceeds that shown in Anselin (1988) due to the spatial dependence in the explanatory variable. The fourth case shows that the usual result that spatial dependence in the disturbances does not lead to bias in OLS estimates does not hold in the presence of an omitted variable (that is correlated with the included explanatory variable). We find that spatial dependence in the disturbances (and/or in the explanatory variable) in the presence of omitted variables leads to a magnification of the conventional omitted variables bias.

To obtain some feel for the magnitudes of these biases, we conducted a small Monte Carlo experiment. In the computations, we simulated a square random set of 1,000 locations and used these locations to compute a contiguity-based matrix W. The resulting 1,000 by symmetric spatial weight matrix W was standardized to be stochastic (doubly stochastic). We set $\beta = 0.75$ and $\gamma = 0.25$ for all trials. The setting for y reflects a relatively low level of correlation between the included and omitted variables. Given W and a value for α , ρ, and ϕ we used the DGP to simulate y for 1,000 trials. For each trial we calculated the estimate $\hat{\beta}_o$ and recorded the average of the estimates. We did this for 27 combinations of α, ρ, and ϕ. For each of these 27 cases we also computed the theoretical $E(\hat{\beta}_o)$. Table 1 shows the empirical average of the estimates and the theoretically expected estimates for the 27 cases. The theoretical and empirical results show close agreement, and the table documents that serious bias can occur when omitted variables combine with spatial dependence in the disturbance process. This is especially true if there is spatial dependence in the regressors, a realistic prospect in applied use of spatial regression models that seems to have been overlooked in the literature. For example, OLS estimates yield an empirical average of 3.9984 (expectation of 4.0221) when ρ, α, and Ψ equal 0.8, even though $\beta = 0.75$ and $\gamma = 0.25$. That is, we have a fivefold bias in the OLS estimates.

3 A Comparison with Spatial Lag Models

We consider the contrast between the above results for least-squares estimates and those for estimates from spatial lag models that match the DGP arising from the presence of omitted variables in the face of spatial dependence.

We begin with the DGP (9) which we repeat in (20). In (21) we substitute in x for $H(\phi)v$ as we condition upon x in this analysis. We introduce the identity $G(\rho)G^{-1}(\rho)$ in (22), rearrange terms in (23) using the linearity of $G^{-1}(\rho) = I_n - \rho W$, and arrive at the final expression in (24).

$$y = F(\alpha)H(\phi)v\beta + F(\alpha)G(\rho)H(\phi)v\gamma + F(\alpha)\,G(\rho)u \tag{20}$$

$$y = F(\alpha)x\beta + F(\alpha)G(\rho)x\gamma + F(\alpha)G(\rho)u \tag{21}$$

Table 1 Mean $\hat{\beta}_o$ and $E\left(\hat{\beta}_o\right)$ as function of spatial dependence ($\beta = 0.75, \gamma = 0.25$)

Case	Parameter			Empirical	Theoretical
	ϕ	ρ	α	Mean $\hat{\beta}_o$	$E(\hat{\beta}_o)$
1	0.0000	0.0000	0.0000	1.0020	1.0000
2	0.4000	0.0000	0.0000	0.9999	1.0000
3	0.8000	0.0000	0.0000	0.9993	1.0000
4	0.0000	0.0000	0.4000	1.0382	1.0376
5	0.4000	0.0000	0.4000	1.1366	1.1363
6	0.8000	0.0000	0.4000	1.3401	1.3438
7	0.0000	0.0000	0.8000	1.2531	1.2540
8	0.4000	0.0000	0.8000	1.6176	1.6161
9	0.8000	0.0000	0.8000	2.5559	2.5666
10	0.0000	0.4000	0.0000	1.0076	1.0094
11	0.4000	0.4000	0.0000	1.0361	1.0341
12	0.8000	0.4000	0.0000	1.0862	1.0860
13	0.0000	0.4000	0.4000	1.0604	1.0592
14	0.4000	0.4000	0.4000	1.1918	1.1915
15	0.8000	0.4000	0.4000	1.4722	1.4771
16	0.0000	0.4000	0.8000	1.3121	1.3080
17	0.4000	0.4000	0.8000	1.7348	1.7361
18	0.8000	0.4000	0.8000	2.8559	2.8723
19	0.0000	0.8000	0.0000	1.0638	1.0635
20	0.4000	0.8000	0.0000	1.1570	1.1540
21	0.8000	0.8000	0.0000	1.3897	1.3917
22	0.0000	0.8000	0.4000	1.1450	1.1458
23	0.4000	0.8000	0.4000	1.3759	1.3762
24	0.8000	0.8000	0.4000	1.9511	1.9552
25	0.0000	0.8000	0.8000	1.4952	1.5006
26	0.4000	0.8000	0.8000	2.1363	2.1475
27	0.8000	0.8000	0.8000	3.9984	4.0221

$$y = F(\alpha)G(\rho)G^{-1}(\rho)x\beta + F(\alpha)G(\rho)x\gamma + F(\alpha)G(\rho)u \tag{22}$$

$$y = F(\alpha)G(\rho)x\beta + F(\alpha)G(\rho)Wx\,(-\rho\beta) + F\,(\alpha)\,G(\rho)x\gamma + F(\alpha)G(\rho)u \tag{23}$$

$$y = F(\alpha)G(\rho)x\,[\beta + \gamma] + F(\alpha)G(\rho)Wx\,[-\rho\beta] + F(\alpha)G(\rho)u \tag{24}$$

We can transform the DGP in (24) to arrive at an estimation model in (26) containing spatial lags of the dependent and independent variables, which we label the spatial lag model (SLM).

$$G^{-1}(\rho)F^{-1}(\alpha)y = x\beta + Wx\Psi + \upsilon \tag{25}$$

$$(I_n - \rho W)\,(I_n - \alpha W)\,y = x\beta + Wx\Psi + \upsilon \tag{26}$$

$$y = x\beta + Wx\Psi + (\alpha + \rho)\,Wy - \alpha\rho W^2 y + u \tag{27}$$

For the case where there is no spatial dependence in the disturbances so that $\rho = 0$, we have the SDM in (28).

$$
\begin{aligned}
(I_n - \alpha W)\, y &= x\beta + Wx\Psi + \upsilon \\
y &= x\beta + Wx\Psi + \alpha W y + u
\end{aligned}
\tag{28}
$$

The SLM model result in (27) points to a potential problem that has been discussed in the spatial econometrics literature. This model specification could lead to what is known as a *label switching identification problem*, if we do not impose the theoretically implied restriction on the estimated parameters α and ρ. In part, this potential for identification problems arises from our use the same spatial weight matrix W in the specification for dependence in both y and x as well as the disturbances for purposes of simplicity. Kelejian and Prucha (2007) show that in the absence of omitted variables the model is identified when using the same spatial weight matrix W for the dependent variable and disturbances, provided that the parameter $\beta = 0$. However, the absence of omitted variables in their consideration results in a simpler model that does not include the two expressions containing spatial lags of the dependent variable, $(\alpha + \rho)\, Wy$, and $-\alpha\rho W^2 y$.

We proceed by working with the SLM model, but assume that the restrictions are used to avoid the potential identification problem. Unlike many restrictions, the restrictions on label switching will not affect the value of the likelihood. Assuming consistency of maximum likelihood estimates for the spatial lag model parameters β, Ψ, ρ, and α, these estimates from the SLM model will equal the underlying structural parameters from the DGP in large samples (Kelejian and Prucha 1998; Lee 2004; Mardia and Marshall 1984). In other words, the asymptotic expected values equal the corresponding parameters in the reparameterized DGP (27), so that $E\left(\tilde{\beta}\right) = \beta + \gamma$, $E\left(\tilde{\Psi}\right) = -\rho\beta$, $E\left(\tilde{\rho}\right) = \rho$, and $E\left(\tilde{\alpha}\right) = \alpha$. There is no asymptotic bias in estimates of α and ρ for the SLM model that arise from omitted variables. (This would also be true for the SDM model that would arise in cases where $\rho = 0$.)

However, the asymptotic bias in this model's estimates for β that arise from an omitted variable is $E\left(\tilde{\beta}\right) - \beta = \gamma$. Unlike the results for OLS in (15), the bias for the SLM does not depend on x, eliminating the influence of the parameter ϕ reflecting spatial dependence in the included variable x, nor does it depend on spatial dependence in the disturbances reflected by the parameter ρ. Instead, the SLM has a constant bias that depends only upon the strength of relation between the included and omitted explanatory variables reflected by γ. (The same holds true for the SDM model which arises in the case where $\rho = 0$.)

4 Conclusion

The nature of omitted variables bias arising in OLS estimates versus spatial lag model estimates was explored. We assumed that the DGP reflected a situation where spatial dependence existed in the disturbances, the dependent variable, and the explanatory variables, and we assumed that the omitted variables were correlated with the included explanatory variable. We established that spatial dependence in the explanatory variable exacerbates the usual bias that arises when using OLS to estimate a model relationship generated by a typical spatial econometric model specification that includes dependence in both the disturbances as well as the dependent variable.

Unlike the standard least-squares result for the case of omitted variables, the presence of spatial dependence magnifies conventional omitted variables bias in OLS estimates. We derived expressions for the amplification in bias showing that this depends on the strength of spatial dependence in the disturbances, dependent variable, and explanatory variables. In contrast, we show that using spatial econometric model specifications containing spatial lags of both the dependent and explanatory variables (that we labeled SDM and SLM) produces estimates whose bias matches the conventional omitted variables case. Our results provide a strong econometric motivation for using spatial econometric model specification such as the SDM and SLM in applied situations where the presence of omitted variables are suspected. The theoretical results presented here also confirm conjectures made by number of authors that omitted variables affect spatial regression methods less than OLS (Brasington and Hite 2005; Dubin 1988; Cressie 1993).

To summarize our findings from the standpoint of a practitioner, we make the following observations. If only the disturbances and explanatory variables exhibit spatial dependence and there is no omitted variable that is correlated with the included explanatory variable, OLS and spatial models should both yield similar regression parameter estimates for large data sets (Pace 1997). This theoretical result is interesting in light of empirical studies that continue to uncover examples where the spatial and OLS estimates differ materially in large samples. The differential sensitivity to omitted variable bias set forth here may account for these observed differences between least-squares and spatial regression estimates reported in applied work. For example, Lee and Pace (2005) examined retail sales and found that OLS estimates for the impact of store size on sales had a significant, negative effect while the spatial model produced a positive significant estimate. In addition, they found that spatial estimates reversed the sign of a number of other counterintuitive OLS parameter estimates. Similarly, Brasington and Hite (2005) in a model of demand for environmental quality found that OLS produced positive and insignificant estimates for the price of environmental quality, whereas a spatial lag model resulted in negative and significant estimates.

Finally, the method used here may aid in understanding other spatial model specifications such as the matrix exponential, conditional autoregressions, and moving average autoregressions in the presence of omitted variables and spatial dependence (LeSage and Pace 2007; LeSage and Pace 2009). Related work considers the issue of omitted variables in a spatial context using a combination of GMM and HAC estimation procedure applied to models involving right-hand-side endogenous variables (Fingelton and Le Gallo 2009).

Acknowledgements Kelley Pace would like to acknowledge support from NSF SES-0729259 and from the Lousiana Sea grant program, while James LeSage is grateful for support from NSF SES-0729264 and the Texas Sea grant program. Both authors would like to thank Shuang Zhu and David Brasington for helpful comments on earlier versions of this chapter.

References

Anselin L (1988) Spatial econometrics: methods and models. Kluwer, Dordrecht

Barry R, Pace RK (1999) A Monte Carlo estimator of the log determinant of large sparse matrices. Lin Algebra Appl 289:41–54

Brasington DM, Hite D (2005) Demand for environmental quality: a spatial hedonic analysis. Reg Sci Urban Econ 35:57–82

Cressie N (1993) Statistics for spatial data, Revised edition. Wiley, New York

Dubin R (1988) Estimation of regression coefficients in the presence of spatially autocorrelated error terms. Rev Econ Stat 70:466–474

Fingelton B, Le Gallo J (2009) Endogeneity in a spatial context: properties of estimators. In: Páez A, Le Gallo J, Buliung R, Dall'Erba S (eds) Progress in spatial analysis: methods and applications. Springer, Berlin: 59–73

Girard DA (1989) A fast Monte Carlo cross-validation procedure for large least squares problems with noisy data. Numer Math 56:1–23

Gradshteyn IS, Ryzhik IM (1980) Table of integrals, series, and products. Corrected and enlarged edition, Academic, Orlando

Kelejian HH, Prucha IR (1998) A generalized spatial two-stage least squares procedure for estimating a spatial autoregressive model with autoregressive disturbances. J R Estate Finance Econ 17:99–121

Kelejian HH, Prucha IR (2007) The relative efficiencies of various predictors in spatial econometric models containing spatial lags. Reg Sci Urban Econ 37:363–374

Lee LF (2004) Asymptotic distributions of quasi-maximum likelihood estimators for spatial econometric models. Econometrica 72:1899–1926

Lee ML, Pace RK (2005) Spatial distribution of retail sales. J R Estate Finance Econ 31:53–69

LeSage JP (1999) Spatial econometrics. Unpublished manuscript available at www.spatial-econometrics.com. Sept. 2009

LeSage JP, Pace RK (2007) A matrix exponential spatial specification. J Econom 140:190–214

LeSage JP, Pace RK (2009) Introduction to spatial econometrics. Taylor & Francis, New York, FL

Mardia KV, Marshall RJ (1984) Maximum likelihood estimation of models for residual covariance in spatial regression. Biometrika 71:135–146

Pace RK (1997) Performing large-scale spatial autoregressions. Econ Lett 54:283–291

Topology, Dependency Tests and Estimation Bias in Network Autoregressive Models

Steven Farber, Antonio Páez, and Erik Volz

1 Introduction

Regression analyses based on spatial datasets often display spatial autocorrelation in the substantive part of the model, or residual pattern in the disturbances. A researcher conducting investigations of a spatial dataset must be able to identify whether this is the case, and if so, what model specification is more appropriate for the data and problem at hand. If autocorrelation is embedded in the dependent variable, the following spatial autoregressive (SAR) model with a spatial lag can be used:

$$\begin{aligned} \mathbf{y} &= \rho \mathbf{W}\mathbf{y} + \mathbf{X}\boldsymbol{\beta} + \boldsymbol{\varepsilon}, \\ \boldsymbol{\varepsilon} &\sim N(0, \sigma^2). \end{aligned} \tag{1}$$

On the other hand, when there is residual pattern in the error component of the traditional regression model, the spatial error model (SEM) can be used:

$$\begin{aligned} \mathbf{y} &= \mathbf{X}\boldsymbol{\beta} + \mathbf{u}, \\ \mathbf{u} &= \rho \mathbf{W}\mathbf{u} + \boldsymbol{\varepsilon}, \\ \boldsymbol{\varepsilon} &\sim N(0, \sigma^2). \end{aligned} \tag{2}$$

In the above equations, **W** is the spatial weight matrix representing the structure of the spatial relationships between observations, ρ is the spatial dependence parameter, **u** is a vector of autocorrelated disturbances, and all other terms are the elements commonly found in ordinary linear regression analysis.

The spatial models above are used in situations where the spatial structure of a set of observations can be represented by a generalized weight matrix conforming to certain constraints which ensure some desirable asymptotic properties

S. Farber (✉)
Centre for Spatial Analysis/School of Geography and Earth Sciences, McMaster University, 1280 Main Street West, Hamilton, ON L8S 3Z9, Canada,
e-mail: farbers@mcmaster.ca

A. Páez et al. (eds.), *Progress in Spatial Analysis*, Advances in Spatial Science,
DOI 10.1007/978-3-642-03326-1_3,

(Anselin 1988b). Most applications to date are concerned with the type of areal data common in socio-economic research. Recently, in addition, there has been increased interest in a network-based conceptualization of the weight matrix, with both applied and technical issues receiving some attention in the literature (Dow et al. 1982; Leenders 2002; Anselin 2003; Páez and Scott 2007). In the network autocorrelation model, the weight matrix is a numerical representation of the network structure, the links connecting observations in the network. It is analogous to the adjacency matrix borrowed from graph theory. In a comparison of spatial and network weight matrices, Farber et al. (2009) provided evidence that topological properties of real-world networks differ from those of graphs representing regular and irregular spatial systems. Further, using simulations they found that statistical tests to identify network dependence in a model of substantive autocorrelation were related to the topological characteristics of the networks. However they considered only one type of model (SAR) and one type of network definition (Poisson-generated degree distribution). The main objective of this chapter is to extend the line of research initiated by the work of Farber et al. (2009) by investigating the properties of tests used to identify spatial error autocorrelation in a SEM model, and by using network structures generated with an exponential degree-distribution function.

The work reported in this chapter on the power of tests for spatial autocorrelation effects draws and benefits from recent research by Smith (2009) and Mizruchi et al. (2008) who investigated the impact of weight matrix characteristics on estimation bias of the spatial dependence and regression parameters. Specifically, Smith (2009) shows analytically that the negative bias observed in simulations by Mizruchi et al. (2008) is a function of weight matrix density. In this chapter, the effect of bias on the likelihood ratio test of significant spatial dependence is investigated analytically, and the relationship between bias and network topology is further explored through visualizations and a regression analysis of simulation results.

The chapter is structured as follows. The relevant literature is reviewed in the next section, followed by a description of the experimental design in Sect. 3. The simulation results are presented graphically and discussed in Sect. 4, and further organized and explored within a modelling framework in Sect. 5. This is followed by concluding remarks and a discussion of outstanding issues in Sect. 6.

2 Literature Review

2.1 *Monte Carlo Simulation and the Properties of Tests for Dependence*

Due to the complex functional specifications of tests for spatial dependence, analytical descriptions of test properties are difficult to obtain. In response, a rather voluminous collection of simulation studies has arisen as a tried and tested method for obtaining practical guidelines regarding the properties of tests. Florax and de Graaff (2004) provide a chronological summary of the progression of simulation studies exploring tests for spatial autocorrelation. Early on, simulation was used to describe the properties of statistical tests of autocorrelation as applied to raw cross-sectional data – statistics such as Moran's Coefficient (Cliff and Ord 1973, 1975, 1981; Haining 1977, 1978) – and to model regression residuals (Bartels and Hordijk 1977). In the late 1980s, following Anselin's (1988a) work concerning maximum likelihood estimates and their associated Lagrange Multiplier (LM) tests, a stream of simulation studies have explored the properties, in many cases the small sample properties, of dependence tests (Anselin and Rey 1991; Anselin 1995; Cordy and Griffith 1993). These studies consistently conclude that the power of tests increases with the magnitude of the spatial parameter or parameters, and with sample size.

Simulation has long been used to explore the impact of weight matrix specification on spatial models. Stetzer (1982) used simulation methods to determine that weighting functions impact spatial parameter estimates, and Anselin later identified simulation methods as a means for exploring weight matrix specification in general (Anselin 1986). Following this, simulation was used to explore the issue of over- and under-specification of the weight matrix (Florax and Rey 1995), and the connectivity function used to define binary weight matrices (Kelejian and Robinson 1998). Kelejian and Robinson went one step further to quantify their test results by regressing the observed test rejection frequencies against the characteristics of the matrices used in their simulations. This technique was similarly applied by Farber et al. (2009) and again in Sect. 5 below.

Recently, simulation has been used to explore the impact of weight matrix topology on the network versions of SAR models. While simulation has seldom been used to investigate network topology and test strengths it has in two examples recently been used to study the impact of matrix density on estimation bias, a different but related issue. Mizruchi and Neuman (2008) found an increasing relationship between estimation bias of the dependence parameter and network density. In their conclusions they state that topological characteristics besides density are likely not relevant in determining bias, but they do suggest that due diligence is necessary, and that other classes of network structures (such as networks with different degree distribution functions) should nonetheless be investigated.

In an attempt to explain Mizruchi and Neuman's findings, Smith (2009) analytically investigated the relationship between estimation bias and network density. His argument is constructed around the behaviour of the SAR and SEM likelihood functions when the weight matrix is fully connected. In this case, the likelihood functions are unbounded as $\hat{\rho}$ decreases toward its minimum bound. It follows that networks within the neighbourhood of the fully connected one will inherit some of the properties of the degenerate case, namely, negative bias.

2.2 Network Topology

Network topology is a topic that has been widely investigated in the fields of mathematics, physics, and sociology. The geographies of networks has long been researched within the spatial sciences (Kansky 1963; Haggett and Chorly 1970) but the relationship between network topology and spatial econometric models is only now coming into focus. Recently, with the increase in size of real world networks that can be analyzed, statistical descriptions of network attributes have become necessary to obtain detailed information about connectivity that can no longer be obtained from visualization techniques (Newman 2003). A number of methods have been devised to gain a statistical description of large complex networks. Given such a statistical description, it is possible to investigate the effects of network topology on the behaviour of models such as those discussed in this chapter.

Among the different properties of networks discussed by Newman (2003, pp. 180–196), two measures of network topology have been singled out by recent research as important: degree distribution and clustering (often referred to as transitivity). For example, work pioneered by Barabasi regarding the degree distribution in many large complex networks has shown that networks follow scale-free or power-law distributions in empirical situations (e.g. Barabasi and Albert 1999). Simultaneously, the work of Watts and colleagues has demonstrated the important effects of clustering on network structure (e.g. Watts and Strogatz 1998).

The degree of a node in a network is the number of contacts to and from a node. In what follows, we consider only undirected networks, so that these are equivalent concepts. The degree distribution describes the frequency with which a randomly chosen node from a large network will have a given degree. Empirically, a summary measure of the degree distribution is mean degree z, defined as the average number of connections per unit of analysis.

Clustering measures the tendency of two nodes to have common neighbours. This property is sometimes described in the social networks literature as the probability that the friend of my friend is also my friend – in other words, it is a measure of transitivity in a network. Most work on clustering has focused on the small-world problem, that is, the tendency for networks with high clustering to simultaneously have short mean path length (i.e. the expected minimum path length between two randomly chosen nodes in a well-connected network). But clustering can have important affects on network topology in its own right, such as on the giant component size of a random network (Volz 2004). We use the definition of clustering proposed in Newman (2003). Accordingly, the clustering coefficient is given by the proportion of triads in a network out of those which could theoretically exist (a triad is a subset of three units, and the possible connections between them):

$$C = \frac{3N_{\Delta}}{N_3}$$

where N_{Δ} is the number of triads in the network and N_3 is the number of connected triplets of nodes. Note that in every triad there are three connected triplets.

2.3 Behaviour of the Likelihood Ratio Test When W is Dense

In this section, the likelihood functions for SAR and OLS are used to explain the observed negative relationship between matrix density and test strength reported by Farber et al. (2009). Through substitution, it is possible to derive a simple formula for the likelihood ratio test. Simulations are then used to illustrate the behaviour of the individual components comprising the test.

Consider the SAR model where **y** is a spatially lagged variable:

$$\mathbf{y} = \rho \mathbf{W}\mathbf{y} + \mathbf{X}\boldsymbol{\beta} + \boldsymbol{\varepsilon} \tag{3}$$

Páez et al. (2008) demonstrate that the OLS estimate for $\boldsymbol{\beta}$, $\hat{\boldsymbol{\beta}}_{OLS}$ is biased and:

$$\mathrm{E}\left[\hat{\boldsymbol{\beta}}_{OLS}\right] = (\mathbf{X}'\mathbf{X})^{-1}\mathbf{X}'(\mathbf{I} - \rho\mathbf{W})^{-1}\mathbf{X}\boldsymbol{\beta}. \tag{4}$$

The well known log-likelihood function for the linear regression model is:

$$\mathrm{L_{OLS}} = -\frac{n}{2}\ln 2\pi - \frac{n}{2}\ln \sigma^2_{OLS} - \frac{1}{2\sigma^2_{OLS}}\left(\mathbf{y} - \mathbf{X}\hat{\boldsymbol{\beta}}_{OLS}\right)'\left(\mathbf{y} - \mathbf{X}\hat{\boldsymbol{\beta}}_{OLS}\right). \tag{5}$$

Substituting σ^2_{OLS} with its conditional estimate

$$\sigma^2_{OLS} = \frac{1}{n}\left(\mathbf{y} - \mathbf{X}\hat{\boldsymbol{\beta}}_{OLS}\right)'\left(\mathbf{y} - \mathbf{X}\hat{\boldsymbol{\beta}}_{OLS}\right) \tag{6}$$

into equation (5) we obtain

$$\mathrm{L_{OLS}} = -\frac{n}{2}\ln 2\pi - \frac{n}{2} - \frac{n}{2}\ln\frac{1}{n} - \frac{n}{2}\ln\left[(\mathbf{y} - \mathbf{X}\hat{\boldsymbol{\beta}}_{OLS})'(\mathbf{y} - \mathbf{X}\hat{\boldsymbol{\beta}}_{OLS})\right]. \tag{7}$$

Similarly, the well known log-likelihood function of the SAR model is:

$$\begin{aligned}\mathrm{L_{SAR}} = &-\frac{n}{2}\ln 2\pi - \frac{n}{2}\sigma^2_{SAR} + \ln|\mathbf{I} - \hat{\rho}_{SAR}\mathbf{W}| \\ &-\frac{1}{2\sigma^2_{SAR}}\left((\mathbf{I} - \hat{\rho}_{SAR}\mathbf{W})\,\mathbf{y} - \mathbf{X}\hat{\boldsymbol{\beta}}_{SAR}\right)'\left((\mathbf{I} - \hat{\rho}_{SAR}\mathbf{W})\,\mathbf{y} - \mathbf{X}\hat{\boldsymbol{\beta}}_{SAR}\right).\end{aligned} \tag{8}$$

Since

$$\hat{\sigma}^2_{SAR} = \frac{1}{n}\left((\mathbf{I} - \hat{\rho}_{SAR}\mathbf{W})\,\mathbf{y} - \mathbf{X}\hat{\boldsymbol{\beta}}_{SAR}\right)'\left((\mathbf{I} - \hat{\rho}_{SAR}\mathbf{W})\,\mathbf{y} - \mathbf{X}\hat{\boldsymbol{\beta}}_{SAR}\right) \tag{9}$$

substituting (9) into (8) it is possible to obtain:

$$\begin{aligned}\mathrm{L_{SAR}} &= -\frac{n}{2}\ln 2\pi - \frac{n}{2} - \frac{n}{2}\ln\frac{1}{n} + \ln|\mathbf{I} - \hat{\rho}_{SAR}\mathbf{W}| \\ &\quad -\frac{\mathrm{n}}{2}\ln\left[\left((\mathbf{I} - \hat{\rho}_{SAR}\mathbf{W})\,\mathbf{y} - \mathbf{X}\hat{\boldsymbol{\beta}}_{SAR}\right)'\left((\mathbf{I} - \hat{\rho}_{SAR}\mathbf{W})\,\mathbf{y} - \mathbf{X}\hat{\boldsymbol{\beta}}_{SAR}\right)\right].\end{aligned} \tag{10}$$

Using (7) and (10) the likelihood-ratio test can be expressed as:

$$\begin{aligned}LR &= 2\left(\mathrm{L_{SAR}} - \mathrm{L_{OLS}}\right) \\ &= 2\ln|\mathbf{I} - \hat{\rho}_{SAR}\mathbf{W}| \\ &\quad -n\ln\left[\left((\mathbf{I} - \hat{\rho}_{SAR}\mathbf{W})\,\mathbf{y} - \mathbf{X}\hat{\boldsymbol{\beta}}_{SAR}\right)'\left((\mathbf{I} - \hat{\rho}_{SAR}\mathbf{W})\,\mathbf{y} - \mathbf{X}\hat{\boldsymbol{\beta}}_{SAR}\right)\right] \\ &\quad +n\ln\left[\left(\mathbf{y} - \mathbf{X}\hat{\boldsymbol{\beta}}_{OLS}\right)'\left(\mathbf{y} - \mathbf{X}\hat{\boldsymbol{\beta}}_{OLS}\right)\right].\end{aligned} \tag{11}$$

But (11) can be simply rewritten as

$$\mathrm{LR} = 2\ln|\mathbf{I} - \hat{\rho}_{SAR}\mathbf{W}| - n\ln\left(\hat{\boldsymbol{\varepsilon}}_{SAR}{}'\hat{\boldsymbol{\varepsilon}}_{SAR}\right) + n\ln\left(\hat{\boldsymbol{\varepsilon}}_{OLS}{}'\hat{\boldsymbol{\varepsilon}}_{OLS}\right). \tag{12}$$

It follows from (9) that

$$\mathrm{LR} = 2\ln|\mathbf{I} - \hat{\rho}_{SAR}\mathbf{W}| - n\ln\left(n\hat{\sigma}^2_{SAR}\right) + n\ln\left(n\hat{\sigma}^2_{OLS}\right) \tag{13}$$

Given the complex interactions embedded in each of the three terms in (13), and the difficulties associated with probing the effect of increasing density on the test, it is useful to proceed by illustrating the situation with a simple numerical experiment. In order to do this, twenty 100×100 binary symmetric matrices with zeros along the main diagonals were randomly generated while controlling for matrix density. Density, in this case, is the number of non-zero entries divided by 9,900, the maximum possible number of non-zero entries for a 100×100 weight matrix. Matrices were constructed for each level of density between 0.05 and 0.95 in increments of 0.05 and scaled by their largest eigenvalues to make the results comparable with Smith (2009) and Mizruchi and Neuman (2008). For each matrix, 1,000 SAR models were identified using exogenously determined coefficients and randomly drawn errors and covariates. The results of the numerical experiment are shown in Table 1. As expected, the likelihood ratio score declines with increasing density. It is apparent that the log-determinant term in the sixth column is negative but approaches zero as density increases. Thus, the true cause of the declining likelihood ratio is the decline of the third term toward the second (in the seventh and eighth columns of Table 1). In other words, as density increases the OLS residual variance approaches the residual variance for the SAR model, and the two terms effectively cancel each other out. Since the log-determinant

Table 1 Impact of matrix density on likelihood ratio

Matrix Density	Mean				1st LR Term[b]	2nd LR Term[c]	3rd LR Term[d]	LR
	$\hat{\rho}$[a] (0.3)	$\hat{\beta}_0$ (1.8)	$\hat{\beta}_1$ (1.2)	$\hat{\beta}_2$ (1.6)				
0.05	0.300	1.77	1.20	1.61	−1.26	459.52	600.84	140.06
0.10	0.300	1.81	1.20	1.60	−0.77	457.68	568.42	109.97
0.15	0.300	1.81	1.20	1.60	−0.56	457.61	544.95	86.78
0.20	0.297	1.85	1.20	1.60	−0.43	456.88	521.62	64.31
0.25	0.298	1.84	1.20	1.60	−0.36	456.40	514.20	57.44
0.30	0.299	1.79	1.20	1.61	−0.31	455.82	493.12	36.99
0.35	0.294	1.88	1.20	1.60	−0.26	455.55	487.05	31.24
0.40	0.295	1.89	1.20	1.60	−0.24	456.25	486.25	29.77
0.45	0.296	1.87	1.20	1.60	−0.21	456.13	482.41	26.07
0.50	0.292	1.92	1.20	1.60	−0.19	455.74	477.01	21.08
0.55	0.295	1.88	1.20	1.60	−0.18	455.65	471.40	15.57
0.60	0.292	1.93	1.20	1.60	−0.16	455.93	475.02	18.93
0.65	0.289	1.98	1.20	1.60	−0.15	456.38	468.65	12.12
0.70	0.293	1.94	1.20	1.60	−0.14	455.22	470.78	15.42
0.75	0.287	2.03	1.19	1.60	−0.13	456.00	464.95	8.82
0.80	0.269	2.32	1.20	1.60	−0.11	454.76	460.82	5.96
0.85	0.274	2.22	1.20	1.60	−0.11	455.85	461.84	5.89
0.90	0.237	2.85	1.20	1.60	−0.07	456.38	459.38	2.92
0.95	0.214	3.24	1.20	1.60	−0.06	456.32	458.59	2.22

[a]True values in parentheses
[b]$2\ln|I - \hat{\rho}_{SAR} W|$
[c]$n \ln n\hat{\sigma}^2_{SAR}$
[d]$n \ln n\hat{\sigma}^2_{OLS}$

term is negligible, the sum of the three terms approaches zero suggesting that the LR test will fail to reject the null hypothesis even in the presence of substantive autocorrelation.

Interestingly, the behaviour of the third term from (13) as seen in Table 1 suggests that the OLS estimate of the residual variance improves as network density increases. In this respect, the behaviour of the autocorrelation model is rather constant across the range of densities investigated but its relative advantage over the OLS model decreases with increasing density. This is an important subtlety to recognize since it attributes the behaviour of the likelihood-ratio test to the OLS estimates of variance and not to those for the SAR model.

We can use analysis to formalize the above finding and propose a potential cause for it. If $\hat{\sigma}^2_{OLS}$ is converging toward $\hat{\sigma}^2_{SAR}$ then from (9) and substituting (4) into (6) we see that this implies:

$$\left(\mathbf{y} - \mathbf{X}(\mathbf{X}'\mathbf{X})^{-1}\mathbf{X}'(\mathbf{I} - \rho\mathbf{W})^{-1}\mathbf{X}\mathbf{B}\right) \approx \left((\mathbf{I} - \hat{\rho}_{SAR}\mathbf{W})\mathbf{y} - \mathbf{X}\hat{\boldsymbol{\beta}}_{SAR}\right) \tag{14}$$

One case for (14) arises when $(\mathbf{I} - \rho\mathbf{W})$ and $(\mathbf{I} - \hat{\rho}_{SAR}\mathbf{W})$ simultaneously approach the identity matrix. Unfortunately, this cannot be the cause since $(\mathbf{I} - \rho\mathbf{W})$ certainly does not approach **I** with increasing matrix density. Currently it is not apparent why (14) holds for dense matrices and this analysis has indeed sprouted an interesting question to be answered in the future.

Another interesting question arises when considering that extreme negative estimation bias of the spatial parameter may incorrectly cause the likelihood ratio test to reject the null hypothesis even when no real autocorrelation is present. To test for this effect, the simulation experiment above was repeated for $\rho = 0$ and $\rho = -0.3$ (the table of results is omitted for brevity). In the first case, no Type I errors are committed despite similar levels of bias in estimating ρ and the regression constant at higher levels of matrix density as seen in Table 1. Furthermore, $\hat{\sigma}^2_{OLS}$ and $\hat{\sigma}^2_{SAR}$ are very similar across all matrix densities, resulting in consistent levels of the likelihood ratio. The fact that $(\mathbf{I} - \rho\mathbf{W})$ is equal to the identity matrix when $\rho = 0$ in conjunction with only slight levels of bias found in $\hat{\rho}_{SAR}$ is likely responsible for this result. For the second case, negative bias in $\hat{\rho}_{SAR}$ might be expected to increase the size of the likelihood-ratio. This however is not the case. Instead, as density increases, even though estimation inflates the magnitude of estimated autocorrelation, $\hat{\sigma}^2_{OLS}$ converges toward $\hat{\sigma}^2_{SAR}$ as before resulting in smaller and smaller likelihood ratios.

3 Experimental Design

The above section illustrates how the different components of the likelihood ratio test behave as network density increases toward its upper bound. In this section, the behaviour of the likelihood ratio test is explored further. Specifically, various additional topological properties are introduced into the simulation experiments, including network size, degree distribution function, the mean degree of connectivity, and the clustering coefficient. Additionally, the relationship between estimation bias and network density observed in Smith (2009) and Mizruchi et al. (2008) is revisited with respect to the larger set of topological properties discussed herein.

3.1 Simulating Networks with Tunable Degree Distribution and Clustering Coefficient

Given the two statistical descriptions of complex networks previously described (degree distribution and clustering), a natural problem arises as to how to create model networks which possess arbitrary combinations of these topologies. In order to generate the networks, in this chapter we employ the approach developed by Volz (2004) for generating random networks with any desired combination of degree distribution and clustering. Such networks allow us to explore a parameter space with potentially important implications for model behaviour. Readers seeking a more detailed description of the network generation process are encouraged to read Volz (2004).

For our application, networks are generated to exhibit a range of topologies, with both large and small levels of clustering, and degree distributions.

3.2 Monte Carlo Simulations

The Monte Carlo procedure undertaken commences with the creation of a simulated social network with known degree distribution function, clustering coefficient, and sample size. The domain of networks investigated herein can be described as the set of quadruplets:

$$\begin{aligned}
&Nets = \{(s, z, c, f)\}\\
&s \in \{100, 500, 1000\},\\
&z \in \{1.5, 3.5, 5.5, 7.5\},\\
&c \in \{0.2, 0.3, 0.4, 0.5, 0.6, 0.7\},\\
&f \in \{poisson, exponential\}
\end{aligned}$$

where s is the sample size, z is the mean of the degree distribution function with functional form f, and c is the clustering coefficient. In comparison to the wide range of densities explored in Sect. 2, the networks generated in the simulation study all have densities between 0.1% and 8%. This limits the scope of analysis to networks with densities comparable to geographic adjacency configurations (Farber and Páez 2009) and a possible direction for future research is to repeat this analysis on a range of densities approximating dense real-world networks and beyond into the theoretically achievable ranges of high density networks in Table 1. A further note about the range of variables tested is that different levels of mean degree imply different levels of network density depending on the total network size. For example when N = 100 nodes, a mean degree of 7.5 implies a network density of 0.076 versus 0.0076 for N = 1,000. Empirically we can imagine some types of networks, such as networks of immediate family relationships, which naturally decrease in density as the number of families in the network increases. In this respect, the mean degree can be thought of as an attribute of the linking function for the network which is invariant under network size. It is important to remark that for a given degree function the density will vary with different network sizes. At this point, the impact of scale effects can only be guessed at since the relationship between mean degree and density prevents the extrapolation of the results to topological properties beyond the range explicitly considered here.

Networks in this study were generated using the Poisson distribution,

$$p_k = z^k e^{-z} (k!)^{-1}\ , \text{ for } k \geq 0,$$

and the discretized exponential distribution:

$$\begin{aligned} p_k &= \int_k^{k+1} (1/\lambda)\, e^{-t/\lambda} dt \\ &= \left(1 - e^{-1/\lambda}\right) e^{-k/\lambda}, \text{ for } k \geq 0 \end{aligned}$$

where the mean of the distribution, z, equals $1/\lambda$.

According to Volz (2004), the Poisson and Exponential distributions are commonly employed in network simulations. For this reason, these two distributions are incorporated into his network generation algorithm. Previous work has considered only the Poisson distribution, and this choice can help clarify whether other distributions have similar impacts or not. If the selection of the distribution turns out to have an impact, that would provide evidence that the mean degree by itself may not be a sufficient indicator to assess potential loss of power in tests and bias in estimation, and that more complete descriptions of the distribution implicit in matrix **W** would be appropriate in empirical settings. Also, an avenue for further research is to consider other distributions such as power-law and log-normal distributions that are mentioned in the social networks literature. In total 144 networks were used in the experiments:

$$|Nets| = |s| \times |c| \times |z| \times |f| = 3 \times 4 \times 6 \times 2 = 144$$

The next step in the experiment is to generate regression data with known amounts of network autocorrelation for each of the synthetic networks. The weight matrices are row-standardized so as to constrain the identifiable range of the spatial dependence parameter to $[-1/\omega, 1]$ where ω is the largest positive eigenvalue of matrix **W**. In all, twelve levels of network dependence are investigated using the following levels of $\rho \in \{0, 0.01, 0.05, 0.10, 0.15, 0.20, 0.25, 0.30, 0.35, 0.40, 0.45, 0.50\}$. This range was selected because it reflects both the dominance in empirical settings of positive autocorrelation while also allowing for the testing of Type I errors. Moreover, likelihood ratio tests assume nearly perfect power when the true spatial dependence parameter is larger than 0.5 (Farber et al. 2009). The data generation process relies on random number draws to construct the error vector $\boldsymbol{\varepsilon} \sim N(0, 1)$ and the matrix of independent variables which includes a constant term and two random variables drawn from a uniform distribution over the range [2,5]. The dependent variable is then computed with known ρ (drawn from the set of possible values given above) and β, by solving (1) and (2) for **y** as follows:

$$\text{SAR Model}: \mathbf{y} = (\mathbf{I} - \rho\mathbf{W})^{-1}(\mathbf{XB} + \boldsymbol{\varepsilon})$$

$$\text{SEM Model}: \mathbf{y} = \mathbf{XB} + (\mathbf{I} - \rho\mathbf{W})^{-1}\boldsymbol{\varepsilon}$$

Following data generation, the model is estimated using LeSage's Spatial-Econometrics Toolbox for Matlab and all relevant results are stored for analysis

(LeSage 1999). Repeating this process 1,000 times for each combination of ρ and $\mathbf{W}$ provides us with reliable estimates of estimation bias and rejection frequencies of a variety of statistical tests of network dependence.

The tests considered herein are based on the principles of maximum likelihood and are used to assess the presence of network autocorrelation in the dependent variable (for the SAR model) or the error vector (in the case of the SEM model). Given either of these models, the null hypothesis of zero autocorrelation is $H_0 : \rho = 0$ while the alternative hypothesis is $H_a : \rho \neq 0$. Under the null condition both the SAR and SEM models are reduced to the standard linear regression model. The most frequently used tests of this type generally fall into three categories (likelihood ratio, Wald tests, and Lagrange multipliers). They have been shown to converge asymptotically, and previous experiments indicate that the differences between the tests within the domain of values used here are negligible (Farber et al. 2009). Thus, for sake of simplicity and brevity, only the likelihood ratio test is reported in this chapter.

The LR test statistic is χ^2 distributed with one degree of freedom so the null hypothesis of no network dependence is rejected at the 95% confidence level when $LR > 3.841$. The power of the test can thus be quantified as the test rejection frequency for each combination of weight matrix and exogenously determined level of autocorrelation.

4 Simulation Results

The Monte Carlo simulations provide measures of mean estimates for the regression parameters, estimation bias, and rejection frequency of the likelihood ratio test for each model and network/rho combination. Each of these vectors of results will be analyzed graphically or modeled statistically.

4.1 Likelihood Ratio Tests

To complement previous work on power of tests and network topology (Farber et al. 2009), the analysis is extended to networks generated with an exponential degree distribution function and to the SEM specification. Figure 1 shows the average likelihood ratio rejection frequencies for each combination of model and network type over different values of ρ. Generally, it can be seen that the LR test is much more powerful when testing for dependence in a SAR model compared to a SEM model (coinciding with previous findings (Anselin 1995)) and is more powerful given the exponential networks as compared to the Poisson networks. Interestingly, on average the test achieves an 80% rejection frequency at $\rho = 0.25$ for the SEM models and at values between 0.05 and 0.1 for the SAR models. The rejection frequencies used to produce Fig. 1 are for networks of varying sizes, degree

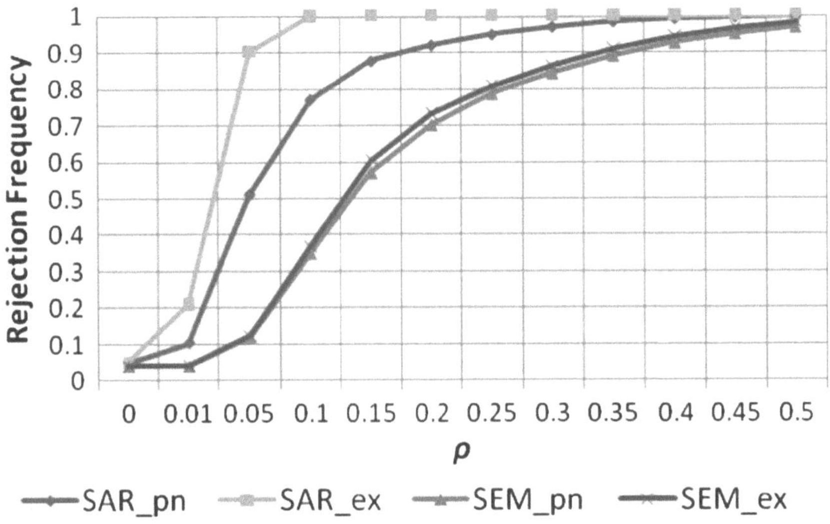

Fig. 1 LR test rejection frequency for difference levels of spatial dependence

distributions and clustering coefficients; therefore they mask all of the variation of test strength with respect to network topology and merely illustrate the most general of relationships between test strength, model specification and degree distribution function.

4.2 *Power of Test and Sample Size*

Figure 2 shows the difference in rejection frequency for networks with 100 versus 1,000 observations. Generally, the four curves have a similar shape indicating that sample size affects the rejection frequency most negatively in the middle of the autocorrelation range and most negligibly near $\rho = 0$ and $\rho = 0.5$. The figure suggests that the LR test for error correlation is the most severely affected by small sample size. Rejection rates for networks with 100 observations are up to 80 percentage points lower than for networks with 1,000 observations in the case of SEM while only up to 45 points lower for SAR models. The results also suggest that the SAR model with exponential networks is practically immune to the effects of small sample size for $\rho > 0.1$.

Figure 3 illustrates directly how sample size impacts the SEM rejection rate curves. For networks with 1,000 observations, the trajectory of the curve is steeply positive in the range of 0.05 and 0.15, indicating that the test becomes very powerful very quickly in the lower range of ρ. However, the rate of increase is less pronounced and smoother for the smaller networks with no sharp increases in test strength. Clearly practitioners using the SEM model on a small dataset should be weary of the limited test strength (in the lower range of simulated ρ's) and should

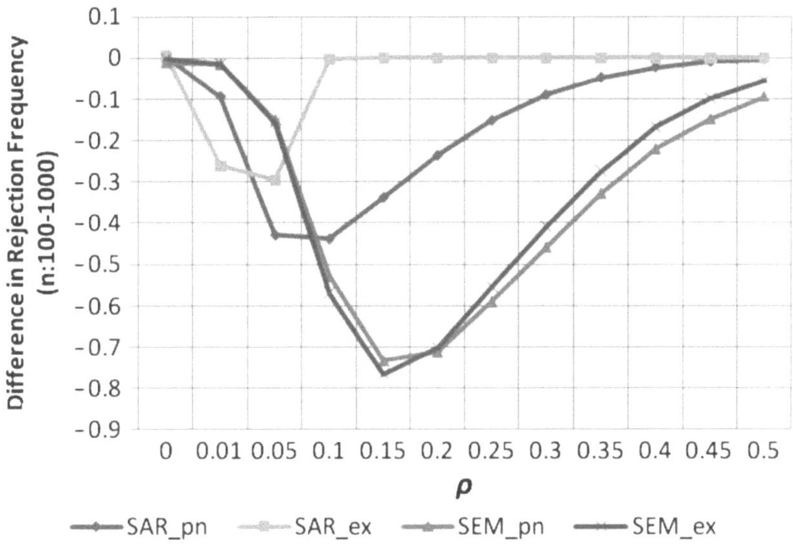

Fig. 2 The impact of sample size on rejection frequency

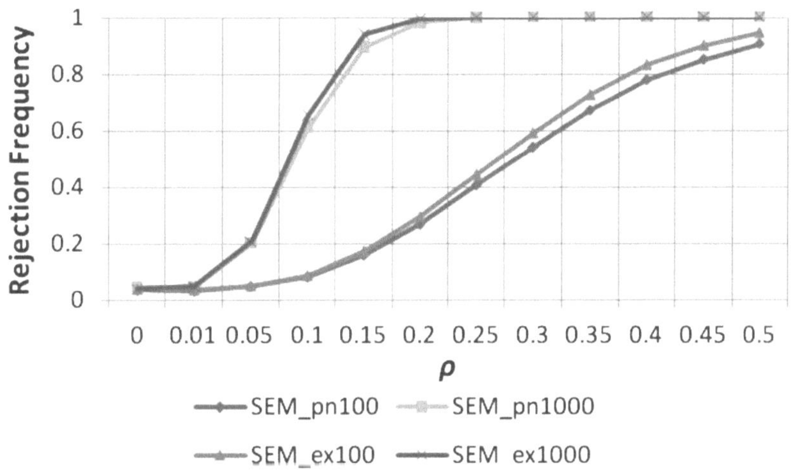

Fig. 3 Rejection frequency curves for two different sample sizes

not necessarily draw strong conclusions if their test for error autocorrelation fails to reject the null hypothesis.

4.3 Power of Test, Mean Degree and Sample Size

Previous results suggest that degree distribution is a significant factor in determining the power of tests (Farber et al. 2009). Moreover, the higher the ratio of mean

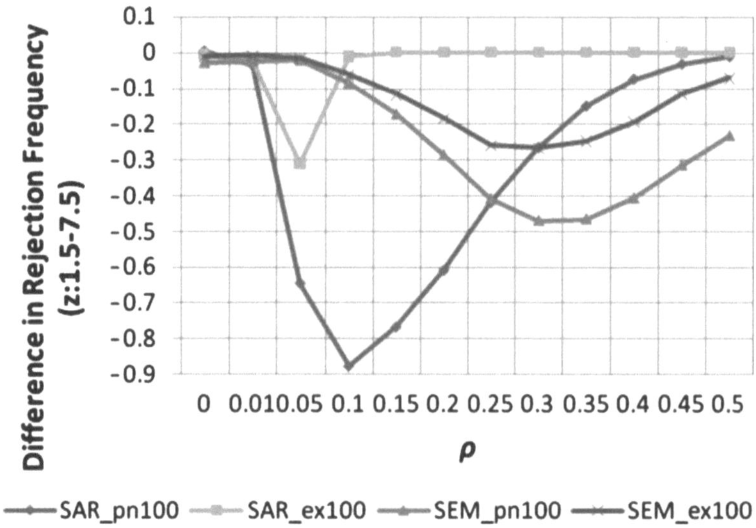

Fig. 4 The impact of mean degree on small networks

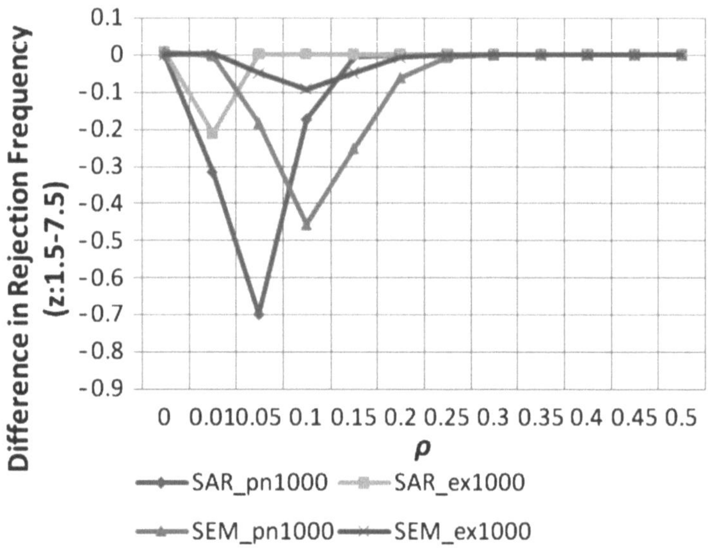

Fig. 5 The impact of mean degree on large networks

degree to sample size, the denser the weight matrix and the weaker the power. The key question is how much so. Figures 4 and 5 show the differences in test power between networks with the largest and smallest mean degree and sample size. Interestingly, the figures suggest that sample size and mean degree interact and impact test strength in a variety of ways. The first observation is that the power of the LR test for SAR models on networks with exponentially distributed degrees is quite robust to changes in mean degree except for $0.01 < \rho < 0.1$. The LR test for the SAR model using a Poisson distributed degree function appears to be the most sensitive to increases in mean degree, especially in the range of $0.01 < \rho < 0.3$ when the test is at least 30 and up to 90 percentage points weaker in more connected networks. The shapes of the curves for the SEM specifications differ substantially to those for the SAR specifications. While the SEM-Poisson combination produces slightly weaker LR strengths than the SEM-Exponential tests, the most striking observation is that both SEM curves are right-shifted indicating a relative immunity to mean degree in the lower ranges of ρ, and a vulnerability to degree in the higher investigated ranges of the spatial parameter ($0.2 < \rho < 0.5$).

The curves in Fig. 5 can be used to determine the degree to which increasing sample size mitigates the impact of mean degree on test strength. Since increasing sample size decreases weight matrix density for any given mean degree, we expect to find the curves in Fig. 5 to be flatter than those in Fig. 4. It is clearly seen that the effect of increasing sample size is an overall scale reduction in test weakness and a left-shift of the curves' minima. While the curves in Fig. 4 appear clustered by model specification (SEM versus SAR curves), those in Fig. 5 either appear clustered by degree distribution function (Poisson versus exponential) or not clustered at all; specifically, the test strengths on Poisson networks are more negatively impacted by mean degree than for exponential networks.

4.4 Power of Test and Clustering

Results from Farber et al. (2009) indicate that the clustering coefficient, a measure of overall transitivity, does not impact test strength for SAR models using a Poisson degree distribution function. As seen in Fig. 6, the impact of clustering is

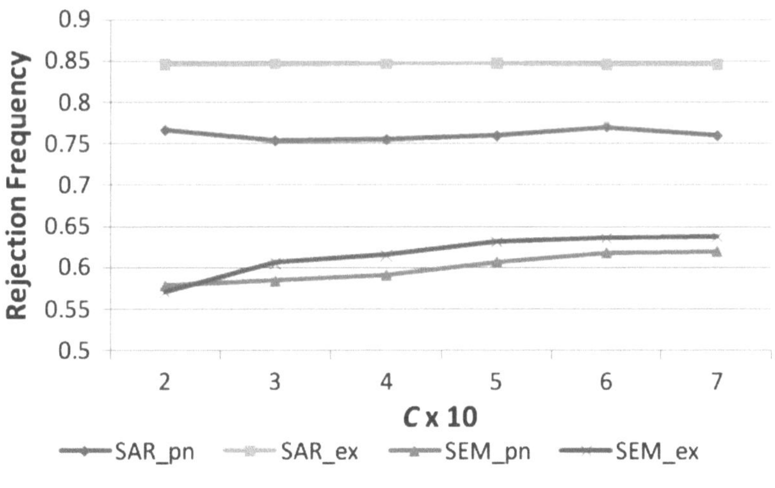

Fig. 6 The impact of clustering on rejection frequency

negligible on SAR models with both distribution function specifications. Whereas clustering does not have a discernable impact on tests for SAR dependence, there does seem to be a weak positive relationship between clustering and test strength for SEM dependence. To date there are no analytically driven results that would predict these simulation results, and these observations necessitate the investigation of the likelihood functions with respect to clustering before they can be fully understood.

4.5 Power of Test and Matrix Density

Recent work has identified matrix density to be an important determinant of estimation bias in spatial and network regression models (Smith 2009). Given a binary weight matrix **W** of size n x n, density is calculated as:

$$D = \frac{\sum_i \sum_j w_{ij}}{n(n-1)}$$

which is simply the sum of all the entries divided by the total number of possible entries.

Figure 7 shows the relationship between average test power and matrix density for each combination of model and degree distribution function. Each data point on

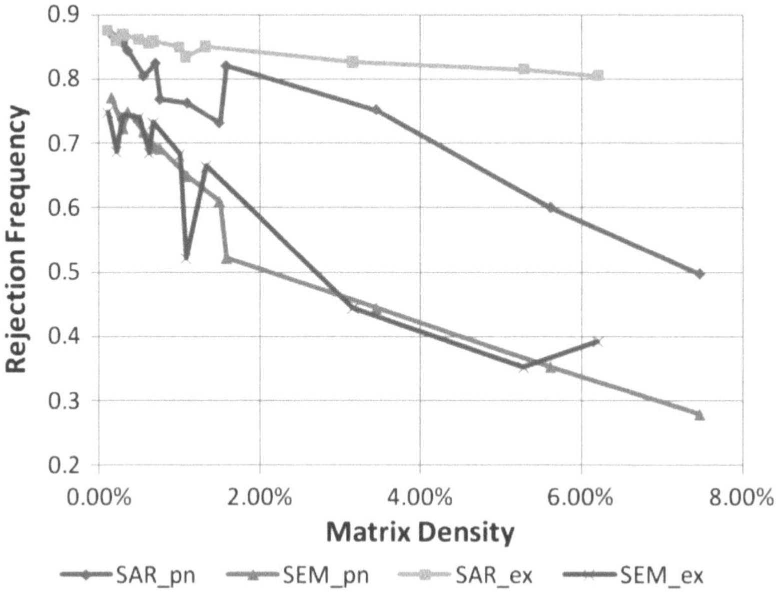

Fig. 7 The impact of matrix density on rejection frequency

the vertical axis is the LR rejection rate averaged over all ρ and C for each sample size and mean degree. The data points are clustered along the horizontal access due to the three disjoint sample-sizes used in the simulations. It appears that test strength generally declined with matrix density for all four data series. In support of the findings above, the LR tests are strongest for the SAR-exponential case and are quite insensitive to changes in matrix density. While the curve for SAR-Poisson is shifted upward from the SEM curves, the average rate of decline among the three curves appears to be equivalent. Notice that the exponential matrices contain fewer highly connected nodes in comparison to their Poisson distributed counterparts with similar mean degree and this may be one reason for improved LR test strengths for exponential network specifications.

4.6 Estimation Bias

Theoretically, one would expect the likelihood ratio tests to fail to reject the null hypothesis when the estimated ρ parameter is downward biased, as was shown to occur in dense networks by Smith (2009) and Mizruchi and Neuman (2008). However, neither Smith nor Mizruchi and Neuman investigate the possible impact of network structure (besides link density) on estimation bias. In this section we explore the simulation results to verify the previous findings regarding network density, and extend the frontier by investigating the impact of other topological properties.

In this context, the mean bias of the estimate for a given network and model specification is defined as:

$$\frac{\sum_r^R (\hat{\rho}_r - \rho)}{R},$$

where r is an iteration index, R is the total number of iterations (1,000 in our experiments), ρ is the true network dependence parameter, and $\hat{\rho}_r$ is the estimated dependence parameter in iteration r.

To begin, Fig. 8 displays the average mean bias for each level of ρ and combination of model/network specification. In concordance with previously published results, the amount of negative estimation bias increases with ρ (Mizruchi and Neuman 2008). This indicates that bias is a decreasing function of the spatial dependence parameter (at least in the range of parameters tested). Interestingly, the SEM estimates are far more biased than the SAR estimates which on average show a very weak negative bias. Theoretically, both the SEM and SAR estimates should be unbiased when using a properly specified weight matrix. In an effort to explore its possible causes, estimation bias is investigated with respect to the same topological properties used to explore LR test strength above.

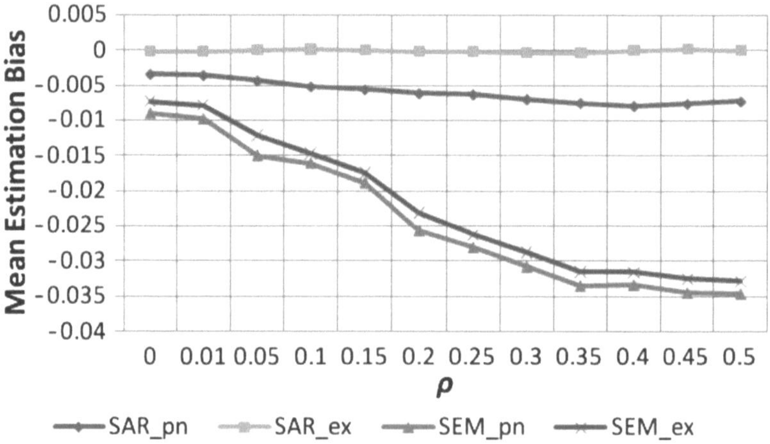

Fig. 8 Dependence parameter estimation bias for different levels of dependence

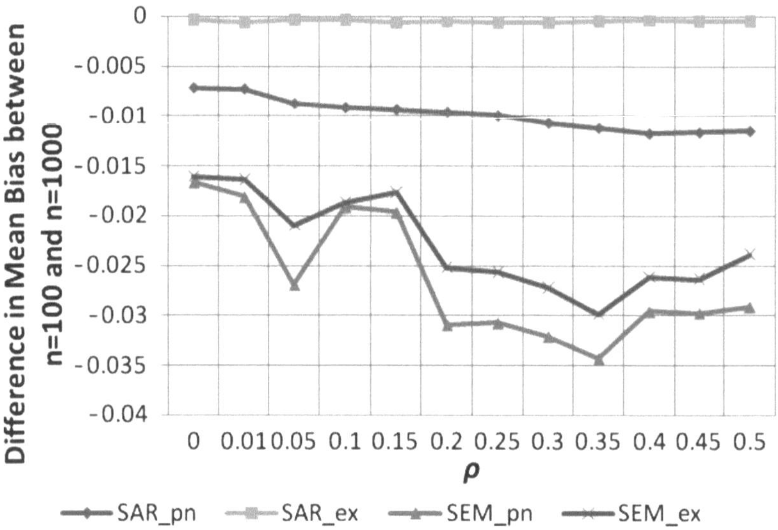

Fig. 9 The impact of sample size on dependence parameter estimation bias

4.7 Estimation Bias and Sample Size

The difference in estimation bias between the largest and smallest networks is displayed in Fig. 9. The figure shows that bias is not impacted by sample size for the SAR-exponential iterations. However, for the SAR-Poisson models the bias difference increases with ρ until 0.4 and then starts to decrease slightly. This indicates that as ρ increases the bias-reducing impact of increasing sample size gets stronger. This is true in general for both SEM specifications as well, however the inflection points occur slightly earlier at $\rho = 0.35$. The fact that increasing sample size reduces bias is not surprising since the strength of most statistical tests increases with sample size. Additionally however, bias is known to increase with density, which is inversely related to sample size. So in this specific case, it is difficult to separate and differentiate between the effects of increasing sample size and decreasing matrix density (see below for more). The fact that the impact of sample-size varies with ρ is difficult to explain and suggests that bias is functionally related to an interacting term containing sample-size and dependence.

4.8 Estimation Bias and Mean Degree Distribution

Figure 10 illustrates the impact of mean degree on estimation bias for the different levels of dependence. The trends for the different specifications are quite unique. As before, estimates for the SAR-exponential specification seem immune to degree distribution. On the other hand, the impact of high mean degree on the SAR-Poisson estimates grows more negative with increasing dependence up to $\rho = 0.45$ at which point the trend seems to reverse. The impact on SEM-exponential estimates decreases consistently with increasing dependence, while the impact on SEM-Poisson estimates is erratic for $\rho < 0.2$ and then decreases smoothly with increasing dependence. As before, it is not surprising that bias is negatively impacted by increasing mean degree; however the interaction between mean degree and dependence is puzzling.

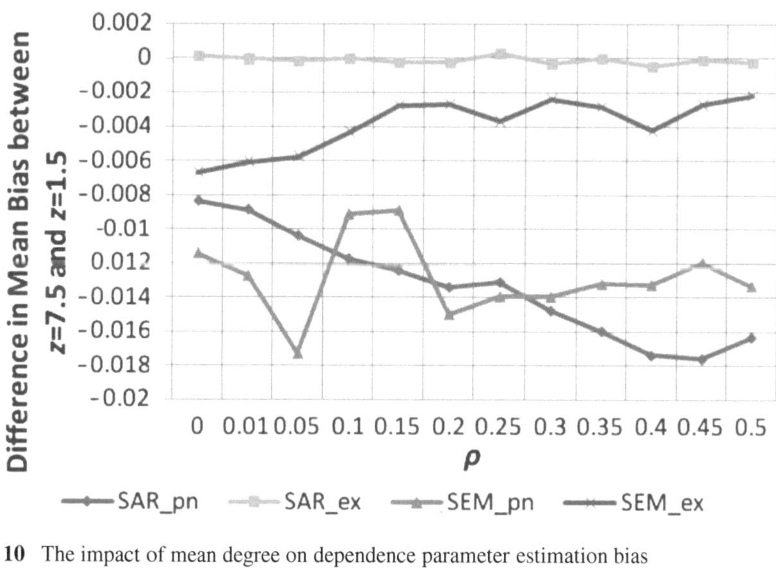

Fig. 10 The impact of mean degree on dependence parameter estimation bias

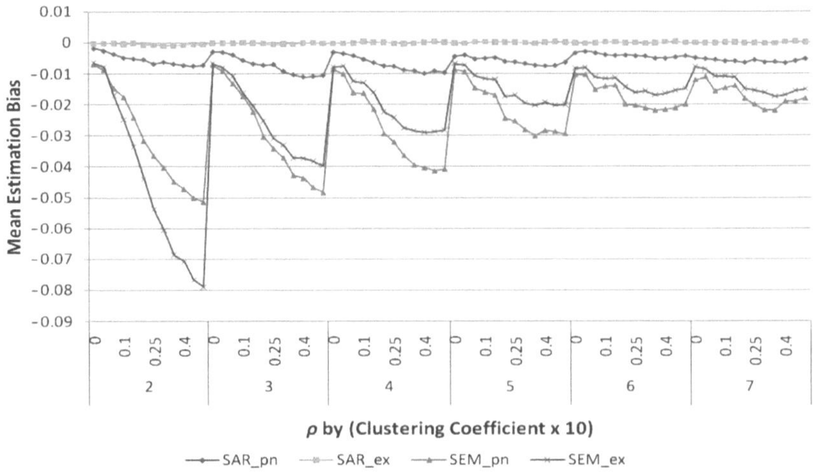

Fig. 11 The effect of clustering on dependence parameter estimation bias

4.9 Estimation Bias and Clustering Coefficient

Figure 11 shows the mean estimation bias for each combination of ρ and C, averaging over all sample sizes and mean degrees. The figure clearly shows that while bias increases with autocorrelation, the scale of bias decreases with increasing clustering. Moreover, the impact of clustering on estimation bias appears to be far more pronounced than that on LR test rejection above (Fig. 6), indicating that density is not necessarily the only relevant topological characteristic. While the impact of clustering seems to be strongly positive on both SEM specifications, it is only mildly effective and seemingly ineffective on the SAR-Poisson and SAR-exponential specifications respectively.

4.10 Estimation Bias and Matrix Density

The impact of matrix density on estimation bias is shown in Fig. 12. Each charted observation corresponds to the mean matrix density and estimation bias for each combination of sample size and degree distribution. This was done to simplify the chart especially since matrix density does not vary with ρ and only varies slightly with clustering. As expected, the chart confirms previous results by indicating that weight matrix density increases estimation bias, except in the case of SAR-exponential, which does not seem to be impacted by matrix density. While Mizruchi and Neuman's findings that network density introduces bias in the estimates of ρ are confirmed by the experiments herein, other topological properties also seem relevant to the bias discussion. Of course, if density causes bias, then so would degree distribution and sample size since they are functionally related

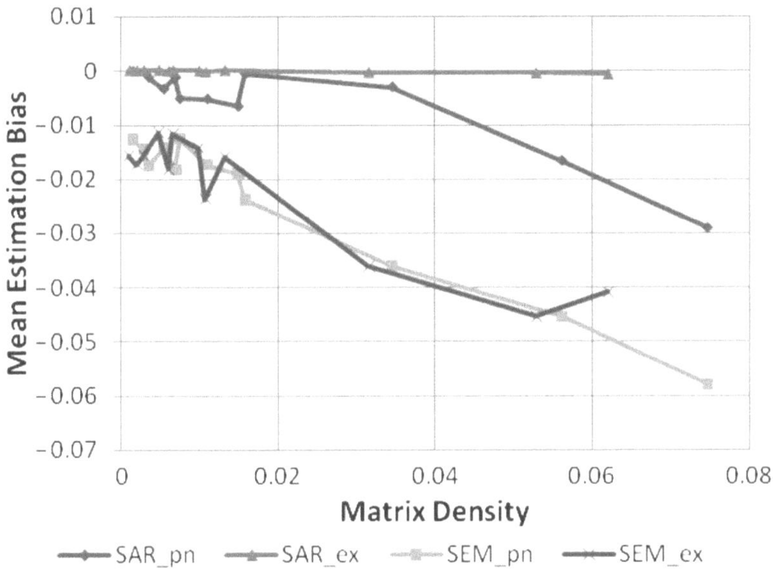

Fig. 12 The relationship between matrix density and estimation bias

properties. But the relationship between bias and network clustering displayed in Fig. 11 is unrelated to density and suggests that the causes of bias are related to topology in a more complex manner than previously suggested.

5 Regression Analysis

5.1 Logistic Regression for LR Test Results

Table 2 contains the results from a series of simple logistic regression models used to explore the relative influence of the various topological properties on LR rejection frequency. A separate regression is calibrated for each model type and degree distribution function in order to capture the heterogeneities between cases as observed in Sect. 4. In each case, the dependent variable is the frequency of LR test rejections over 1,000 trials. The purpose of the regressions is to organize the vast amount of simulation results into a parsimoniously defined functional relationship between LR rejection frequency and the topology of the networks. A linear model is a simple way to quantify and compare the impact of the various topological factors investigated herein. In order to allow for non-linear relationships, the topological characteristics of the networks, namely size, mean degree, and clustering, have been coded into dummy variables representing each factor level. While not displayed for the sake of brevity, several other model specifications were experimented with before settling with the ones in Table 2. In particular, sample-size and mean degree were replaced by the continuous measure of network density. Also, the dummy variables representing different levels of the clustering coefficient were replaced with the continuous measure of actual achieved clustering. In both cases, the signs and scales of the continuous variables were commensurate with the results in Table 2, however the model-fits did decrease slightly, and the imposition of the linearity constraint associated with the use of a single continuous variable was unfavourable. Furthermore, the

Table 2 Results of rejection frequency logistic regression

	SAR-Poisson		SAR-Exponential	
	b	*t*	*B*	*t*
Constant	-4.12	−166.1	-4.24	−165.5
Sample size				
$n = 100$			Reference	
$n = 500$	2.46	178.8	1.76	88.3
$n = 1000$	3.18	209.3	2.23	112.3
Mean degree				
$z = 1.5$			Reference	
$z = 3.5$	-1.15	−69.0	-0.26	−1.7E + 07
$z = 5.5$	-3.11	−171.1	-0.62	−35.2
$z = 7.5$	-4.06	−214.9	-0.80	−45.1
Clustering coefficient				
$c = 0.2$			Reference	
$c = 0.3$	-0.26	−15.2	0.03	2.2E + 06
$c = 0.4$	-0.23	−13.5	0.06	3.8E + 06
$c = 0.5$	-0.14	−8.2	0.09	6.0E + 06
$c = 0.6$	0.08	4.9	0.02	1.1E + 06
$c = 0.7$	-0.14	−8.2	0.01	8.7E + 05
Lag strength				
$\rho = 0.00$			Reference	
$\rho = 0.01$	0.95	41.3	1.70	86.2
$\rho = 0.05$	4.51	195.7	5.92	235.0
$\rho = 0.10$	6.50	251.4	10.54	96.0
$\rho = 0.15$	7.71	268.2	39.19	2168.9
$\rho = 0.20$	8.40	274.5	39.19	2168.9
$\rho = 0.25$	9.02	276.2	39.19	2168.9
$\rho = 0.30$	9.67	270.5	39.19	2168.9
$\rho = 0.35$	10.38	251.4	39.19	2168.9
$\rho = 0.40$	11.18	215.0	39.19	2168.9
$\rho = 0.45$	12.12	163.3	39.19	2168.9
$\rho = 0.50$	13.16	111.3	39.19	2168.9
Summary statistics				
Deviance	40949.0		9263.1	
Pseudo-R^2	0.9765		0.9938	
SSE	2.8605		0.5747	

(continued)

Table 2 (continued)

	SEM-Poisson		SEM-Exponential	
	B	*t*	*B*	*t*
Constant	-5.92	−242.8	-6.47	−258.3
Sample size				
$n = 100$				
$n = 500$	3.05	266.2	2.97	253.6
$n = 1000$	4.02	308.0	3.96	296.2
Mean degree				
$z = 1.5$				
$z = 3.5$	-0.6	−51.4	-0.36	−32.8
$z = 5.5$	-1.26	−112.4	-0.78	−70.3
$z = 7.5$	-1.81	−158.4	-0.73	−65.6
Clustering coefficient				
$c = 0.2$				
$c = 0.3$	0.07	5.4	0.43	32.9
$c = 0.4$	0.16	12.4	0.55	41.6
$c = 0.5$	0.36	27.3	0.76	57.2
$c = 0.6$	0.50	37.8	0.82	61.4
$c = 0.7$	0.52	39.5	0.84	62.6
Lag strength				
$\rho = 0.00$				
$\rho = 0.01$	-0.01	−0.2	0.05	1.8
$\rho = 0.05$	1.29	56.4	1.28	56.9
$\rho = 0.10$	3.06	141.2	3.06	143.1
$\rho = 0.15$	4.54	202.8	4.62	206.6
$\rho = 0.20$	5.58	237.9	5.67	240.6
$\rho = 0.25$	6.35	260.0	6.37	261.1
$\rho = 0.30$	6.94	275.0	6.93	275.6
$\rho = 0.35$	7.51	286.4	7.51	286.2
$\rho = 0.40$	8.07	293.0	8.13	289.3
$\rho = 0.45$	8.57	292.9	8.73	281.2
$\rho = 0.50$	9.10	284.7	9.39	258.8
Summary statistics				
Deviance	29271.0		32195.0	
Pseudo-R^2	0.9887		0.9815	
SSE	1.6064		2.192	

binary nature of the final set of variables displayed in Table 2 allows for the direct interpretation and comparison of coefficients. The unfortunate functional relationship, density= N× (mean degree)/(N × N–N), obviates the possibility to include these three variables simultaneously into a regression model without introducing high levels of multicolinearity. To this end, including the two terms (size and degree) was an appropriate way to model the rejection frequencies although it does inhibit the ability to estimate the impact of density directly.

All of the coefficients in all the models are significant with a very high degree of confidence as indicated by their associated t-values. It is noteworthy that such a high level of significance is achieved in part due to the extremely large sample size. One thousand repetitions for each of the 72 networks (per model and distribution function) and 12 levels of rho results in an overall sample size of 864,000 observations. Given such a large sample size it becomes necessary to use the t-values as a relative measure of significance between coefficients. In this way it becomes quite clear that the coefficients for clustering are the least reliable, and caution should be used when drawing conclusions based on them.[1] As an interesting side-note, the replacement of the clustering dummy variables with a single continuous variable produced a similarly low t-value for the clustering variable. Conversely, the single continuous density measure obtained an extremely high t-value when it replaced the sample size and mean degree variables in Table 2.

The coefficients themselves can be used to judge the relative influence of each variable on the rejection frequency. In all the models, the level of dependence obtains the highest regression coefficients. This confirms the above visual analysis whereby rejection frequency uniformly increased with the size of ρ. It is also important to observe the diminishing rate of increase of the regression coefficients as ρ increases. This indicates that the relationship between dependence level and LR rejection frequency is non-linear, an important finding that can be used in the future to validate analytical attempts at exploring the likelihood-ratio. Comparing the coefficients between the models, we observe that influence of dependence is stronger in the SAR models than in the SEM specifications.

Following dependence, sample size and degree distribution are the next most influential factors. Sample size seems to be a consistently strong positive influence on test-strength. Mean degree on the other hand has a negative influence, but its impact is for more pronounced in the Poisson cases, and especially in the SAR-Poisson case. It is difficult to determine exactly why this is, but the differences may be derived from the differences between the shapes of the distribution density functions. In particular, the Poisson distribution is more concentrated around its mean so most nodes obtain the mean number of connections. The exponential distribution is more dispersed and positively skewed, with most nodes obtaining a small number of connections, and some obtaining a very large number. The net effect is that for a given mean-degree and sample size, the simulated Poisson networks are more connected than the exponential ones. This might explain why mean-degree has a stronger influence on the Poisson networks than on the exponential ones. The models achieve a very high pseudo-R^2, an indication that a strong linear relationship exists between the observed and estimated rejection probabilities as seen in the scatterplots in Fig. 13. A second summary statistic, the sum of squared errors (SSE) is useful in comparing the strength of the models' fit. Considering that the

[1] The visualizations in Sect. 4 indicate that the LR test for SAR-exponential is extremely strong, even for quite small values of dependence. This would explain the extremely high coefficients on the dependence parameters, and perhaps the extremely significant but extremely small coefficients for clustering.

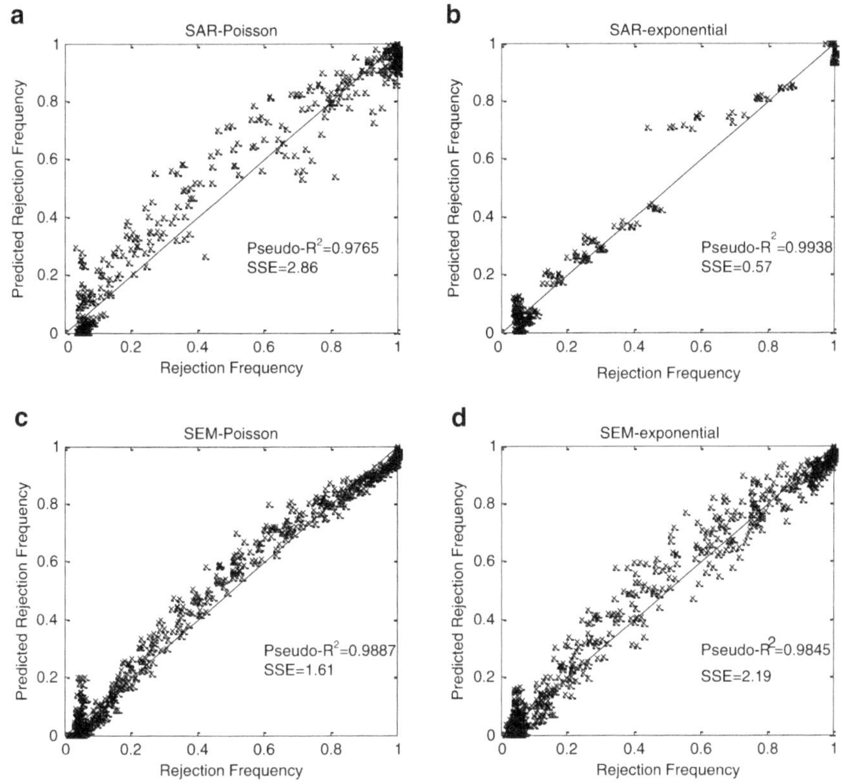

Fig. 13 Goodness of fit scatterplots

range of observed values for the four regressions is [0,1], the SSE indicates that the logistic model for SAR-exponential achieves the closest overall fit, followed by SEM-Poisson, SEM-Exponential, and SAR-Poisson. It is difficult to interpret why the models achieve the different levels of fit that they do, but it does appear that fit is related to the number of observed values at the extremes of the range [0,1]. The scatterplots also show evidence of overestimation near $LR = 0$ and $LR = 1$ and underestimation within the middle of the range. However, since the main focus of these logistic regressions is not predictive but explanatory, this pattern of over- and underestimation while noteworthy does not invalidate inferences drawn about topological characteristics from the regression coefficients.

6 Conclusions

This chapter expanded on previous work investigating the role of network topology on estimating network autocorrelation models and statistical tests of dependence. The work reported here benefits from recent research that includes advances in both analytical work and simulation studies. In this spirit, this chapter investigates the power of the likelihood ratio test for network dependence both analytically and with numerical simulations. The results of the two analyses are confirmatory in the sense that they are mutually supportive of the hypothesis regarding the negative relationship between test power and network density.

As part of the simulation study herein, the power of LR tests and estimation bias of SAR and SEM models were visualized with respect to three network sizes, two degree distribution functions, four levels of mean degree, six levels of clustering and twelve levels of autocorrelation used in the data generation process. General patterns of relationships between the factors were exposed through a series of visualizations and then organized in a logistic regression analysis.

For all model specifications, the level of network dependence indicated by ρ is the most significant factor in predicting the power of the LR test – and test strength is shown to increase with ρ. The visualizations illustrate how the rate of increase of test power with respect to ρ is non-linear and that tests for SAR models achieve higher power levels at lower levels of dependence as compared to SEM models. Similarly, tests on networks with exponential degree distributions are stronger than those with Poisson distributions, but this effect is far more pronounced in the case of SAR. A potential avenue for future research is to show analytically via an investigation of the LR statistics for SAR and SEM why ρ has a stronger positive effect on test strength in the SAR cases.

Second to the level of dependence are the combined effects of network size and mean degree. These factors while analyzed independently jointly define the concept of matrix density. Besides the SAR-exponential case which achieves almost perfect test strength in all of the simulations, the impacts of size and mean degree on test strength are very strong. For example, the SAR-Poisson test can be 90% stronger in networks with 1,000 nodes versus 100 and 70% stronger in loosely connected networks versus those with a mean degree of 7.5. When analyzed jointly, density has a very strong and negative impact on rejection frequency. While the range of network densities tested herein is quite small (roughly [0.01,0.075]) compared to other studies, the impact on test strength is very evident for all of the model formulations. Generally, network density more negatively impacts the SEM models with rejection frequencies ranging from 70 to 30% within the range of densities tested.

Third, the impact of clustering, while found to be significant in the logistic regression, is clearly not as influential as the other factors in determining test power. At this time, since clustering is unrelated to network density, there is still no analytical evidence that explains why clustering should have any impact on test power at all, so the particular pattern observed of increased power with increasing clustering is puzzling. We do know that highly clustered networks typically contain a higher frequency of isolated subgroups of nodes within the network and consequently a smaller giant component size. Investigating if this might be playing a role in the estimation of SAR or SEM models is a potentially rewarding future research avenue.

In addition to LR test power, in response to the recent evidence in the literature of estimation bias in models with dense weight matrices, we graphically illustrated the relationship between bias and the various topological properties of networks discussed above. Our findings are generally supportive of other research – primarily that bias increases with density – however the images in Sect. 4 clearly illustrate that this relationship is non-linear, and that the various topological characteristics interact to produce complex effects. Moreover, for the first time, clustering has been shown to reduce the level of bias in the estimate. Again, future research is required to further current understanding of this issue.

Finally, the last contribution of this chapter was to illustrate analytically why the strength of the LR test diminishes with increasing network density. The argument helps to identify the behaviour of the various terms which constitute the likelihood ratio with respect to changes in network density. Future work to solidify the argument into a formal proof and to extend the analysis to SEM specifications and specifications based on various levels of network clustering is needed.

The experiments in this chapter clearly illustrate that the topology of the network represented by matrix **W** used in autocorrelation models will impact the power of commonly used statistical tests and the accuracy of maximum-likelihood estimates. One of the general results found in the chapter is that bias and LR strength are worse for Poisson distributed networks than the exponential networks even when density is held constant. This may be attributed to the nature of the degree distributions. Of particular relevance to our study is that for a given mean (and therefore density), a higher frequency of low-degree nodes are generated in the exponential networks, suggesting that the distribution of the amount of network influence (i.e. the number of neighbours associated with each point) has some influence in determining how topology effects estimation bias and test strength.[2] Since the distribution function is found to be an active factor it is now clear that mean degree and network density are not the only relevant descriptors of matrix W and other commonly used degree-distributions, such as the power-law distribution, should be thoroughly investigated.

A second general finding in need of an explanation is the prevalence for more bias and weaker test strengths in SEM specifications as compared to SAR models. While neither SEM nor SAR estimates of $\boldsymbol{\beta}$ should be biased when the true W matrix is used, we know from (4) that OLS estimates for a SAR process are biased while those for SEM processes are not. This suggests that OLS achieves better estimation results for SEM processes. But, since the likelihood ratio is a function of the relative strength of the spatial model over OLS estimation, it follows that the LR test may generally be weaker in a SEM context as compared to a SAR context. A thorough analysis with the aid of some simulations will help solidify this argument in the future. Interestingly, while Smith (2009) shows that both SAR and SEM estimates of ρ are biased when **W** is dense, we still do not know why the SEM estimates appear to be more biased than SAR estimates and how this effects the power of the likelihood ratio test.

The implications of the findings regarding the likelihood ratio test in this research are twofold. First, practitioners utilizing network autocorrelation models in their research are advised to compare the topological characteristics of their networks

---

[2] We are grateful to the anonymous reviewer who brought this interpretation to our attention.

with those investigated herein. For if they find themselves using dense and less clustered networks they may incorrectly fail to observe significant autocorrelation even if the dependence does truly exist. Second, the chapter advances the theoretical knowledge of the factors influencing the likelihood ratio test since it is shown for the first time that it varies with respect to the degree of clustering and the degree distribution function specification in addition to sample-size and degree of connectivity. While the research in this chapter was focused on illustrating the *existence* of network topology effects, more research is needed in order to better understand *why* network composition, namely clustering and distribution function specification are related to both the likelihood ratio test and the bias in estimating the dependence parameter.

References

Anselin L (1986) Some further notes on spatial models and regional science. J Reg Sci 26:799–802
Anselin L (1988a) Lagrange multiplier test diagnostics for spatial dependence and spatial heterogeneity. Geogr Anal 20:1–17.
Anselin L (1988b) Spatial econometrics: methods and models. Kluwer, Dordrecht
Anselin L (2003) Spatial externalities, spatial multipliers, and spatial econometrics. Int Reg Sci Rev 26:153–166
Anselin L, Florax RJGM (1995) Small sample properties of tests for spatial dependence in regression models: some further results. In: Anselin L, Florax RJGM (eds) New directions in spatial econometrics. Springer, Berlin, pp 21–74
Anselin L, Rey S (1991) Properties of tests for spatial dependence in linear-regression models. Geogr Anal 23:112–131
Barabasi AL, Albert R (1999) Emergence of scaling in random networks. Science 286:509–512
Bartels CPA, Hordijk L (1977) Power of generalized moran contiguity coefficient in testing for spatial autocorrelation among regression disturbances. Reg Sci Urban Econ 7:83–101
Cliff A, Ord JK (1975) The choice of a test for spatial autocorrelation. In: Davis J, McCullagh M (eds) Display and analysis of spatial data. Wiley, Chichester
Cliff AD, Ord JK (1973) Spatial autocorrelation. Pion, London
Cliff, A. D. and Ord, J. K. (1981) Spatial Processes: models and applications. Pion, London
Cordy C, Griffith D (1993) Efficiency of least squares estimators in the presence of spatial autocorrelation. Commun Stat B 22:1161–1179
Dow MM, Burton ML, White DR (1982) Network auto-correlation – a simulation study of a foundational problem in regression and survey-research. Soc Networks 4:169–200
Farber S, Páez A, Volz E (2009) Topology and dependency tests in spatial and network autoregressive models. Geogr Anal 41:158–180
Florax RJGM, de Graaff T (2004) The performance of diagnostic tests for spatial dependence in linear regression models: a meta-analysis of simulation studies. In: Anselin L, Florax RJGM, Rey S (eds) Advances in spatial econometrics: methodology, tools and applications. Springer, Berlin, pp 29–65
Florax RJGM, Rey S (1995) The impact of misspecified spatial structure in linear regression models. In: Anselin L, Florax RJGM (eds) New directions in spatial econometrics, Springer, Berlin, pp 111–135
Haggett P, Chorly RJ (1970) Network analysis in geography. St. Martin's Press, New York
Haining R (1977) Model specification in stationary random fields. Geogr Anal 9:107–129
Haining R (1978) The moving average model for spatial interaction. Trans Inst Brit Geogr 3: 202–225

Kansky KJ (1963) Structure of transportation networks: relationships between network geometry and regional characteristics. Technical paper, University of Chicago

Kelejian HH, Robinson DP (1998) A suggested test for spatial autocorrelation and/or heteroskedasticity and corresponding Monte Carlo results. Reg Sci Urban Econ 28:389–417

Leenders RTAJ (2002) Modeling social influence through network autocorrelation: constructing the weight matrix. Soc Networks 24:21–47

LeSage JP (2009) Spatial econometrics toolbox. http://www.spatial-econometrics.com. 29 Sept. 2009

Mizruchi MS, Neuman EJ (2008) The effect of density on the level of bias in the network autocorrelation model. Soc Networks 30:190–200

Newman MEJ (2003) The structure and function of complex networks. SIAM Rev 45:167–256

Páez A, Scott DM (2007) Social influence on travel behavior: a simulation example of the decision to telecommute. Environ Plann 39:647–665

Páez A, Scott DM, Volz E (2008) Weight matrices for social influence: an investigation of measurement errors and model identification and estimation quality issues. Soc Network 30:309–317

Smith TE (2009) Estimation bias in spatial models with strongly connected weight matrices. Geogr Anal 41:307–332

Stetzer F (1982) Specifying weights in spatial forecasting models – the results of some experiments. Environ Plann 14:571–584

Volz E (2004) Random networks with tunable degree distribution and clustering. Phys Rev E 70:5056115–5056121

Watts DJ, Strogatz SH (1998) Collective dynamics of "small-world" networks. Nature 393: 440–442

Endogeneity in a Spatial Context: Properties of Estimators

Bernard Fingleton and Julie Le Gallo

1 Introduction

Endogeneity is a pervasive problem in applied econometrics, and this is no less true in spatial econometrics. However, while the appropriate treatment and estimation of the endogenous spatial lag has received a good deal of attention (Cliff and Ord 1981; Upton and Fingleton 1985; Anselin 1988, 2006), the analysis of the effects of other endogenous variables has been rather neglected so far.

Nevertheless, it is known that the consistent estimation of spatial lag models with additional endogenous variables is straightforward since it can be accomplished by two-stage least squares, with the lower orders of the spatial lags of the exogenous variables as instruments (see Anselin and Lozano-Gracia 2008; Dall'erba and Le Gallo 2008 for applications of this procedure). In addition, the case of endogenous variables and a spatial error process has been considered by Kelejian and Prucha (2004). Their paper generalizes the Kelejian and Prucha (1998) feasible generalized spatial two-stage least squares estimator to allow for additional endogenous variables on the right hand side when there is an explicit set of simultaneous equations. Kelejian and Prucha (2007) consider a general spatial regression model that allows for endogenous regressors, their spatial lags, as well as exogenous regressors, emphasizing that their model may, in particular, represent the *i*th equation of a simultaneous system of equations, but also mentioning its applicability to endogeneity in general. Fingleton and Le Gallo (2008a, b) develop the approach to consider endogeneity from various sources with either autoregressive or moving average error processes. However, there are certain specific aspects of spatial econometrics that lead to a somewhat different treatment of the endogeneity problem and its solution. In this chapter, we outline the problem in the spatial context, focusing on the relative impact of different sources of endogeneity. In particular, we focus on endogeneity and hence the inconsistency of the usual OLS estimators induced by

B. Fingleton (✉)
Department of Economics, Strathclyde University, 130 Rottenrow, Glasgow, Scotland G4 0GE, UK,
e-mail: bernard.fingleton@strath.ac.uk

A. Páez et al. (eds.), *Progress in Spatial Analysis*, Advances in Spatial Science,
DOI 10.1007/978-3-642-03326-1_4,

omitting a significant variable that should be in the regression model but which is unmeasured and hence is present in the residual. We also consider simultaneity and errors-in-variables.

The outline of the chapter is as follows. The next section describes the main sources of inconsistency considered in this chapter, namely omitted variables, simultaneity and measurement error. Also, we consider the particular case of omitted variables in a spatial context. Then, we perform the Monte-Carlo simulations aimed at analyzing the performance of a spatial Durbin model as a potential remedy for bias and inconsistency. The last section concludes.

2 Endogeneity and Spatial Econometric Models

We begin with a brief summary of the three sources of inconsistency considered in this chapter: simultaneity, omitted variable(s), and errors-in-variables (measurement error). We exclude the fourth source, namely the inclusion in a time series model of a lagged dependent variable as an explanatory variable where there is serial correlation in the disturbances. In each case, the source of the inconsistency is an inappropriate application of the usual OLS estimating equation. Consider for instance the case of a simple linear regression: $Y_i = b_0 + b_1 X_i + e_i$, where e_i is the error term with the usual properties. The OLS estimate of the coefficient b_1 is:

$$\hat{b}_1 = \frac{\sum\limits_{i=1}^{n} (X_i - \bar{X})(Y_i - \bar{Y})}{\sum\limits_{i=1}^{n} (X_i - \bar{X})^2} = \frac{Cov(X_i, Y_i)}{Var(X_i)} \tag{1}$$

where $\bar{X} = \left(\sum_{i=1}^{n} X_i\right)/n$, $\bar{Y} = \left(\sum_{i=1}^{n} Y_i\right)/n$ and n is the sample size.

It is easy to show that this leads to:

$$\hat{b}_1 = b_1 + \frac{Cov\,(X_i, e_i)}{Var\,(X_i)} \tag{2}$$

and that the consistency of the OLS estimator relies on the assumption that $Cov\,(X_i, e_i) = 0$, a condition that is violated under the four sources of endogeneity mentioned above. As a very simple example of simultaneity, consider the two variable system:

$$\begin{cases} Y_i = b_1 X_i + e_i \\ X_i = \gamma_1 Y_i + v_i \end{cases} \tag{3}$$

Simply estimating $Y_i = b_1 X_i + e_i$ by OLS leads to $Cov\,(X_i, e_i) \neq 0$, since:

$$Y_i = \frac{b_1 v_i + e_i}{1 - b_1 \gamma_1} \text{ and } X_i = \frac{\gamma_1 e_i + v_i}{1 - b_1 \gamma_1} \tag{4}$$

and hence X is correlated with e.

In the case of error-in-variables, suppose that the explanatory variable X_i is measured imprecisely by $\tilde{X}_i$ and we wish to estimate the true relationship $Y_i = b_1 X_i + e_i$. In fact, using $\tilde{X}_i$, the true relationship becomes:

$$Y_i = b_0 + b_1 \tilde{X}_i + \left[b_1 \left(X_i - \tilde{X}_i \right) + e_i \right] \tag{5}$$

since $b_1 \tilde{X}_i - b_1 \tilde{X}_i = 0$. Suppose we estimate $Y_i = b_0 + b_1 \tilde{X}_i + v_i$, then the error term $v_i = b_1 \left(X_i - \tilde{X}_i \right) + e_i$ contains the difference $\left(X_i - \tilde{X}_i \right)$. If $Cov\left(\tilde{X}_i, \left(X_i - \tilde{X}_i \right) \right) \neq 0$ then the OLS estimator $\hat{b}_1$ from $Y_i = b_0 + b_1 \tilde{X}_i + v_i$ is a biased and inconsistent estimator of the true parameter b_1 in $Y_i = b_0 + b_1 X_i + e_i$.

Finally, in the case of inconsistency arising from omitted variables, consider the true relationship $Y_i = b_1 X_i + b_2 Z + e_i$, but Z is omitted so that the estimated equation is $Y_i = b_1 X_i + v_i$, where $v_i = b_2 Z + e_i$. If $Cov(X, Z) \neq 0$, then $Cov(X, v) \neq 0$ and OLS is inconsistent.

One of the issues that has been raised in the recent literature has been whether an endogenous spatial lag needs necessarily be part of a spatial model. From a theoretical perspective, it has been acknowledged that the inclusion of a spatial lag is more generally appropriate than simply modeling spatial dependence via a spatial error process alone (see, for instance, Fingleton and López-Bazo (2006), in the context of growth econometrics and modeling of regional convergence).

Recently, Pace and LeSage (2008) and LeSage and Pace (2008) arrive at a similar conclusion from a purely statistical perspective. Conventionally, in time series analysis, the lagged endogenous variable is often used to capture the effects of omitted variables. Therefore, we envisage that the presence of the endogenous lag should help mitigate omitted variable bias in spatial regressions. They demonstrate that when one omits a variable which has spatial dependence because it is a spatial autoregressive process, and this variable is correlated with an included variable, then the resulting data generating process is the spatial Durbin model. The spatial Durbin is the model containing both endogenous and exogenous spatial lags, as given by Burridge (1981):

$$Y = \rho W Y + X b + W X \gamma + e \tag{6}$$

where Y is the $(n \times 1)$ vector of observations on the dependent variable; X is an $(n \times k)$ matrix of observations on k exogenous variables with b as the corresponding $(k \times 1)$ vector of parameters; e is the $(n \times 1)$ vector of $i.i.d.$ error terms; W is a $(n \times n)$ non-stochastic spatial weights matrix, with zeros on the main diagonal and non-negative values for $W_{ij}, i \neq j$. Conventionally, W is normalized so that rows sum to 1. In this case, the endogenous lag $\sum_{j=1}^{n} W_{ij} Y_j$ is the weighted average of Y_i in locations $j = 1, \ldots, n$ for which $W_{ij} > 0$. The parameter $\rho < 1$ quantifies the spatial dependence of Y on connected regions, as designated by the non-zero elements of W.

Consider the case of the model specification $Y_i = b_1 X_i + v_i$ in which the error term comprises an omitted variable, so that $v_i = Z_i$ where Z_i is spatially autocorrelated. We make an initial assumption that this autocorrelation is a spatial autoregressive process, so that, following LeSage and Pace (2008):

$$Z = \rho WZ + \varepsilon \tag{7}$$

and assume that $\varepsilon \sim iid\left(0, \sigma^2 I_n\right)$. If we estimate $Y = Xb_1 + \nu$, then we are subject to omitted variable inconsistency if this equation is estimated by OLS, provided that $Cov\,(X, Z) \neq 0$. The solution to this problem suggested by Pace and LeSage (2008) and LeSage and Pace (2008), is to eliminate the effect of the omitted variable by estimating a spatial Durbin model. This derives from an assumption that the correlation between X and Z causes correlation between X and $\varepsilon = (I - \rho W)\,Z$ and that this correlation is linear of the form:

$$\varepsilon = X\eta + \pi \tag{8}$$

We assume here that $\pi \sim iid\left(0, \sigma^2 I_n\right)$. It follows that:

$$\begin{aligned} Y &= Xb_1 + (I - \rho W)^{-1}\,(X\eta + \pi) \\ (I - \rho W)\,Y &= (I - \rho W)\,Xb_1 + (X\eta + \pi) \\ Y &= \rho WY + X\,(b_1 + \eta) - \rho WXb_1 + \pi \\ Y &= \rho WY + X\phi + WX\tilde{b} + \pi \end{aligned} \tag{9}$$

This indicates that although Z is omitted, provided W is known, unbiased estimates of the coefficients b_1, ρ in $Y = Xb_1 + \rho WZ + \varepsilon$ can be obtained by fitting this last equation in (9).

We next analyze the case where there is a spatial error process in addition to an omitted variable Z. This is a natural extension to what has already been considered and is important because we can never be sure that we have captured all of the spatial dependency by invoking specifically autoregressive variables such as Z. We prefer the less restrictive assumption that there might remain an unmodeled source of spatial dependence that can be captured by an error process. We consider here simply the autoregressive error process, so that the data generating process is given by:

$$\begin{aligned} Y &= Xb_1 + Zb_2 + e \\ Z &= \rho WZ + \varepsilon \\ e &= \lambda Me + \xi \end{aligned} \tag{10}$$

It is easy to show that this is exactly equivalent to:

$$Y = \rho WY + X\phi + WX\tilde{b} + \pi + (I - \rho W)\,(I - \lambda M)^{-1}\,\xi \tag{11}$$

We estimate (11) using two-stage least squares (2sls) and a spatial HAC (SHAC) estimation procedure (Kelejian and Prucha 2007). If the disturbances in (11) had been a simple parametric (AR or MA) error process, then for reasons of efficiency we might have preferred to treat it explicitly as such in estimation. However, SHAC provides consistent estimates of the error covariance matrix under rather general assumptions, accommodating various patterns of correlation and heteroscedasticity, including spatial ARMA(p,q) errors, and is appropriate in an IV context. In the wider context, going beyond (11) and extending to our other sources of endogeneity, namely simultaneity and measurement error as we do below, we exploit the generality of SHAC as a means of obtaining consistent covariance matrix estimates. In particular, Kelejian and Prucha 2007 assume that the $(n \times 1)$ disturbance vector e is generated as follows:

$$e = R\xi \tag{12}$$

where R is an $(n \times n)$ non-stochastic matrix whose elements are not known. The asymptotic distribution of the corresponding IV estimators implies the following variance-covariance matrix:

$$\Psi = n^{-1}\tilde{Z}'\Sigma\tilde{Z} \tag{13}$$

with $\Sigma = \left(\sigma_{ij}\right)$ denotes the variance-covariance matrix of e and $\tilde{Z}$ denotes a $(n \times f)$ full column rank matrix of instruments. Kelejian and Prucha (2007) show that the SHAC estimator for its (r,s)th element is:

$$\hat{\Psi}_{rs} = n^{-1}\sum_{i=1}^{n}\sum_{j=1}^{n}\tilde{z}_{ir}\tilde{z}_{js}\hat{e}_i\hat{e}_j K\left(d_{ij}^{*}/d_n\right) \tag{14}$$

where $\hat{e}_i$ is the IV residual for observation i; d_{ij} is the distance between unit i and unit j; d_n is the bandwidth and $K(.)$ is the Kernel function with the usual properties. In this chapter, we focus on the Parzen kernel as given by Andrews (1991), which is as follows:

$$K(x) = \begin{cases} 1 - 6x^2 + 6\left|x\right|^3 & \text{for} \quad 0 \le |x| \le 1/2 \\ 2\left(1 - |x|\right)^3 & \text{for} \quad 1/2 \le |x| \le 1 \\ 0 & \text{otherwise} \end{cases} \tag{15}$$

3 The Omitted Variable Case

As we have noted in the previous section, there is a specific circumstance in which we can estimate the augmented spatial Durbin model and consequently avoid omitted variable bias. While Z is unknown, what is required is precise knowledge of W which is part of the equation $Z = \rho WZ + \varepsilon$. In this section, we investigate the validity of these assertions using Monte-Carlo simulations aimed at analyzing the properties of estimators in augmented spatial Durbin models when there are omitted variables that are spatially autocorrelated.

We focus on two characteristics. First, the bias is defined as the median of the distribution minus the true value. Second, the RMSE, which gives equal weight to the two importation considerations for estimation: bias and dispersion. In theory, RMSE is the square root of the weighted average of the mean and variance. However, here we use the approximation given in Kelejian and Prucha (1999) and Kapoor et al. (2007):

$$RMSE = \left[bias^2 + \left[\frac{IQ}{1.35} \right]^2 \right]^{0.5} \tag{16}$$

where IQ is the difference between the 75 and 25% quantiles. This reduces to the standard RMSE statistic under a normal distribution, but will be more robust to outliers that may occasionally be generated by the Monte-Carlo replications.

In order to explore the bias and RMSE associated with our augmented spatial Durbin specification, as in (11), which includes the complex error process, we carry out various simulations.

The initial analysis is based on the following set-up. There are $n = 225$ square regions. We define W as a (15×15) matrix, with $W_{jk} = 1$ when regions j and k are contiguous (rook's definition) and $W_{jk} = 0$ otherwise. This matrix is subsequently standardized so that rows sum to 1. Moreover, we generate some artificial data for this landscape as follows: $\pi \sim N(0, 1)$, $X = 1 + x - y - 1.5x^2 + 5.5xy - 0.5y^2$, where x is the x-coordinate and y is the y-coordinate of each cell on the (15×15) matrix. This yields a spatially autocorrelated quadratic surface shown in Fig. 1a. We also assume $\rho = 0.5$ and initially assume that η, the level of correlation between X and Z, is equal to 0.5. Given these data, we generate $Z = (I - \rho W)^{-1} (X\eta + \pi)$. Finally, assuming $b_1 = 1$, $b_2 = 1$, $M = W$, and $e \sim N(0, 1)$, we generate the data using:

$$Y = b_1 X + b_2 Z + (I - \lambda M)^{-1} e \tag{17}$$

for $\lambda = 0.8$.

The augmented spatial Durbin (11) yields estimates of ϕ, $\tilde{b}$ and ρ, obtained via SHAC using the Parzen kernel with a cut-off distance on the lattice of 50 units. Since WY is endogenous, in addition to the exogenous variable X and its spatial lag WX, we divide the exogenous variable X into three groups, indexing the highest ranking values by +1, the middle ranked values by 0 and the lower ranked values by −1. This instrument is defined by analogy with the three-group method for measurement errors (Kennedy 2003) and has been used in a spatial framework by Fingleton (2003). In addition, we use the spatial lag of the three groups variable as an ancillary instrument. Repeating this process 100 times gives parameter estimate distributions. The estimates of the coefficients equal to $\rho = 0.5$, $\phi = b_1 + \eta = 1.5$ and $\tilde{b} = -\rho b_1 = -0.5$ for WY, X and WX respectively, are given in Fig. 1b–d.

Additional evidence regarding the bias and RMSE of the augmented spatial Durbin estimator on a (20×20) lattice, corresponding to $n = 400$ is provided by Table 1. In this case, we compute the bias using the median of the $\hat{\phi}$ distribution resulting from 500 replications. Given $Z = \rho WZ + \varepsilon$, we find that the bias is comparatively small under the augmented spatial Durbin specification. Table 1 indicates that there is evidently positive bias for $\lambda = -0.9, 0, 0.9$ and $\eta = 0.2, 0.5,$

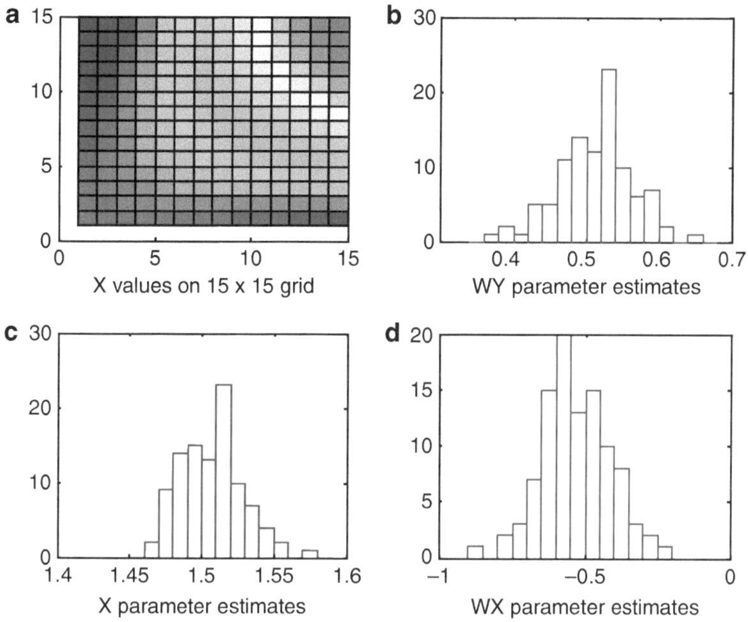

Fig. 1 Exogenous variable spatial distribution (**a**) and augmented spatial Durbin parameter distribution (**b**, **c** and **d**) resulting from Monte-Carlo simulations

Table 1 Spatial Durbin: 2sls-SHAC estimator bias and RMSE for b_1; omitted variable

λ	Bias			RMSE		
	$\eta = 0.9$	$\eta = 0.5$	$\eta = 0.2$	$\eta = 0.9$	$\eta = 0.5$	$\eta = 0.2$
−0.9	0.017615	0.023954	0.037268	0.106243	0.113733	0.101879
0	0.019026	0.014714	0.038249	0.110610	0.101569	0.100245
0.9	0.018510	0.027462	0.055672	0.104558	0.102497	0.102581

Table 2 OLS-SHAC estimator bias and RMSE for b_1; ignoring omitted variable

λ	Bias	RMSE
−0.9	0.397457	0.397463
0	0.397577	0.397582
0.9	0.397505	0.397511

0.9, but this is small compared to the bias, equal to $median(\hat{b}_1) - b_1$, produced by OLS-SHAC estimates of:

$$Y = b_1 X + (I - \lambda M)^{-1} e \tag{18}$$

that is, estimates obtained by ignoring the omitted variable Z from the data generating process. These estimates are given in Table 2.

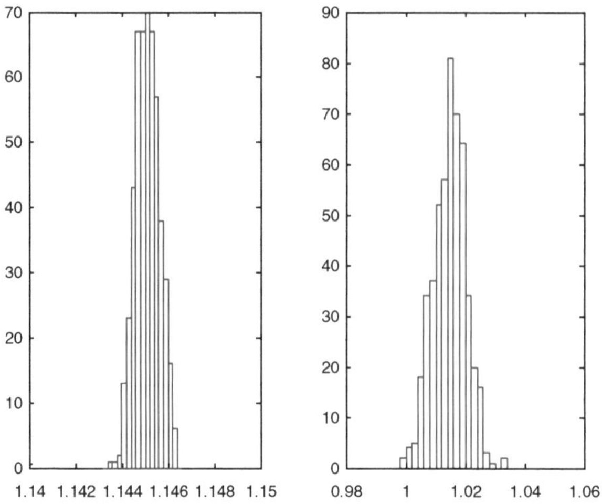

Fig. 2 Monte-Carlo distributions of the X parameter in (17) estimated by fitting (18) and (11)

Thus far we have assumed that $Z = (I - \rho W)^{-1} (X\eta + \pi)$ and shown that the augmented spatial Durbin (11) provides estimates of the coefficient on variable X that are evidently less biased than estimates using (18). Unfortunately the precise form of the W matrix may in practice be unknown, or the spatial pattern in Z may not be defined by an autoregressive process. Therefore, we next explore properties of the augmented spatial Durbin estimator given such an unknown omitted variable, comparing them with b_1 estimates provided by (18).

For that purpose, let $Z = 10 + 9.5x - 2y$, so that the correlation between X and Z is equal to 0.2726. Assume that the data are generated using (17) using in this case the parameters $b_1 = 1$, $b_2 = 1$, $\lambda = 0.8$ and $e \sim N(0, 1)$. Figure 2 on the left provides the distribution of 500 estimated b_1s from (18) obtained using OLS-SHAC for $n = 400$, based on the Parzen kernel with a cut-off distance on the lattice of 50 units. On the right, Fig. 2 shows the distribution of $\hat{\varphi}$ from (11) via 2sls-SHAC, again using the Parzen kernel with a cut-off distance on the lattice of 50 units, with the exogenous variable X and its spatial lag WX plus the three groups as instruments as above. It is again apparent that the estimates using (11) are relatively unbiased.

Table 3 summarizes the bias and RMSE for 500 replications of both (11) and (18) with data generated by (17). We use the same X, Z variables and parameters as were used to generate Fig. 2, but allow the parameter λ to vary, being equal to -0.9, 0 and 0.9.

Table 4 is generated exactly as Table 3, with the "single" difference that Z $=$ $10 + 3.5x - 20y$ or that Z $= 1 + x - y + 1.5x^2$. Consequently, the correlation between X and Z is equal to -0.6498 and to 0.37122 respectively. The bias is larger and negative, although again the bias and RMSE remain much smaller when estimation is via the augmented spatial Durbin than when the problem of an omitted variable is ignored and (18) is estimated.

Table 3 OLS-SHAC and 2sls-SHAC estimator bias and RMSE for b_1

λ	Bias		RMSE	
	Equation (11)	Equation (14)	Equation (11)	Equation (14)
−0.9	0.014535	0.145090	0.024364	0.145090
0	0.014671	0.145090	0.015515	0.145090
0.9	0.014038	0.144994	0.015188	0.144997

Table 4 OLS-SHAC and 2sls-SHAC estimator bias and RMSE for b_1

λ	Bias		RMSE	
	Equation (11)	Equation (14)	Equation (11)	Equation (14)
$Z = 10 + 3.5x - 20y$				
−0.9	−0.098457	−0.341260	0.102742	0.341260
0	−40.098083	−0.341264	0.098326	0.341264
0.9	−0.097348	−0.341189	0.097545	0.341191
$Z = 1 + x - y + 1.5x^2$				
−0.9	−0.01627	0.406744	0.029315	0.406744
0	−0.01538	0.406746	0.016545	0.406746
0.9	−0.01468	0.406848	0.015816	0.406850

Table 5 OLS-SHAC and 2sls-SHAC estimator bias and RMSE for b_1

λ	Bias		RMSE	
	Equation (11)	Equation (14)	Equation (11)	Equation (14)
	$b_1 = 1,\ b_2 = 1$			
−0.9	−0.01159	0.001201	0.013735	0.001202
0	−0.00019	0.001206	0.005985	0.001210
0.9	−0.00029	0.001245	0.005169	0.001602
	$b_1 = 0.5,\ b_2 = 100$			
−0.9	0.013744	0.119725	0.034811	0.119725
0	0.014417	0.119617	0.015985	0.119620
0.9	0.015044	0.119727	0.016075	0.119727

Consider next the case where Z is a dummy variable, generated so that Z equals 1 when X is greater than its mean, and zero otherwise, thus inducing correlation (0.8246) between Z and X. Table 5 displays the result in two cases. In the first case, $b_1 = 1, b_2 = 1$ while in the second case, $b_1 = 0.5, b_2 = 100$. Our results are the outcome of 500 replications. We use the Parzen kernel with a cut-off of 50 and the same instruments as before for 2sls-SHAC estimation. These results show that in the first case, while (11) produces a smaller bias, it is associated with a larger RMSE. In the second case, the bias and RMSE clearly favor estimation by (11).

In this section, we showed that under specific circumstances, the omission of a spatially autoregressive variable of the form $Z = \rho WZ + \varepsilon$ leads to only small bias and RMSE for the parameter estimate of correlated variable X when estimated via an augmented spatial Durbin model (11) using 2sls-HAC. This small bias and

RMSE is considered in comparison to what would be obtained by simply estimating the model without attempting to make any correction to allow for the absence of Z from the estimating equation. We then showed, from the limited number of simulations we have carried out, that the superiority of the augmented spatial Durbin model seems quite evident across a range of scenarios, even when the spatial pattern in omitted variable Z does not conform to the spatial autoregressive structure hitherto assumed. The estimates continue to support the notion that estimating the spatial Durbin is to an extent superior to doing nothing. In other words, while we cannot claim consistency or unbiasedness for our approach, the use of the spatial Durbin model as an estimator does seem, to some extent, to mollify the quite serious impact of omitted variable bias that otherwise would occur. Given this, we now go on to explore the impact of using (11) as an estimator when we have endogeneity due to the other causes outlined in the introduction, simultaneity and measurement errors.

4 Simultaneity and Measurement Errors

We begin our investigations with the case where endogeneity is due to system feedback. In order to analyze the properties of an augmented spatial Durbin model in this case, we consider a set-up that is similar to that of Anselin et al. (1997) and Fingleton and Le Gallo (2008a) where endogeneity is a result of feedback in a two-equation system. The two equations are as follows:

$$y_i = b_0 + b_1 x_{1i} + \gamma q_i + u_i \tag{19a}$$

$$q_i = \alpha_0 + \alpha_1 v_{1i} + \alpha_2 y_i + \xi_i \tag{19b}$$

where y_i and q_i are the endogenous variables for observation i; x_{1i} and v_{1i} are the exogenous variables for observation i; $b_0, b_1, \gamma, \alpha_0, \alpha_1, \alpha_2$ are unknown parameters to be estimated. We assume that the error terms are kept entirely separate by generating two innovations: $\xi_1 \sim iid\left(0, \sigma_1^2 I_n\right)$ and $\xi_2 \sim iid\left(0, \sigma_2^2 I_n\right)$, with $\sigma_1^2 = \sigma_2^2 = 1$, and then u and ξ as follows: $\xi = \xi_2$ and:

$$u_i = \lambda \sum_{j \neq i} w_{ij} u_j + \xi_i \tag{20}$$

or in matrix terms: $u = (I - \lambda W)^{-1} \xi$ with λ as the spatial autoregressive coefficient.

Given the set-up, the endogenous variables y and q must be generated using reduced forms:

$$y_i = \delta_0 + \delta_1 x_{1i} + \delta_2 v_{1i} + \delta_3 \xi_i + \frac{u_i}{\delta_4} \tag{21a}$$

$$q_i = \omega_0 + \omega_1 x_{1i} + \omega_2 v_{1i} + \omega_3 u_i + \frac{\xi_i}{\delta_4} \tag{21b}$$

with: $\delta_0 = (b_0 + \gamma\alpha_0)/\delta_4$; $\delta_1 = b_1/\delta_4$; $\delta_2 = \gamma\alpha_1/\delta_4$; $\delta_3 = \gamma/\delta_4$; $\delta_4 = 1 - \gamma\alpha_2$; $\omega_0 = (\alpha_0 + \alpha_2 b_0)/\delta_4$; $\omega_1 = \alpha_2 b_1/\delta_4$; $\omega_2 = \alpha_1/\delta_4$; $\omega_3 = \alpha_2/\delta_4$.

As previously, we assume that the spatial units are located on a square grid at locations $\{(r, s) : r, s = 0, 1, \dots m\}$. Therefore, the total number of units is $n = (m+1)^2$. The spatial weights matrix is a standardized rook-type matrix: two units are neighbors if their Euclidian distance is less than or equal to one. The coefficients $b_0, b_1, \alpha_0, \gamma, \alpha_0$ and α_1 are set to 1 and α_2 is set to 0.9. The spatial parameter λ takes on 3 values: −0.9, 0 and 0.9 and three lattice sizes have been specified: 121, 225 and 400.

Two cases have been considered relating to the method used to generate the variables. In the first case, x_1 and v_1 are generated from uniform distributions. In the second case, both x_1 and v_1 are spatially autocorrelated, i.e. they are generated as follows: $x_1 = (I - 0.5W)^{-1}\xi_3$ where $\xi_3 \sim iid\,(0, I_n)$, and $v_1 = (I - 0.5W)^{-1}\xi_4$ where $\xi_4 \sim iid\,(0, I_n)$. For each combination of spatial parameter λ and sample size, we perform 500 replications.

Table 6 displays the bias and RMSE for the parameter γ, which is the coefficient associated with the endogenous variable q, when OLS is applied to (19a) combined with SHAC estimation of the variance–covariance matrix. In the latter case, we use as previously a Parzen kernel, where $d_n = \left[n^{1/4}\right]$ and where $[z]$ denotes the nearest integer that is less than or equal to z.

We note that in all cases, the biases and RMSE's are symmetrically increasing when the absolute value of the spatial error coefficient is increasing. Also, they remain quite small and stable with increasing sample sizes. They are smaller when x_1 and v_1 are generated from a spatial autoregressive model than when they are generated from a uniform distribution.

Table 7 displays the bias and RMSE for the same parameter γ, associated to the endogenous variable q, when a spatial Durbin model is estimated, including x_1 and v_1 but also an endogenous spatial lag, together with spatial lags of x_1 and q. As previously, in addition to the exogenous variables and their spatial lags, we divide the exogenous variable x_1 into three groups, and use this three groups variable and its spatial lag as additional instruments. The model is then estimated using a combination of instrumental variables method and SHAC model for the variance–covariance matrix.

Table 6 OLS-SHAC estimator bias and RMSE for γ; simple model; simultaneity

λ	Bias			RMSE		
	$n = 121$	$n = 225$	$n = 400$	$n = 121$	$n = 225$	$n = 400$
	x_1 and v_1 generated from a uniform distribution					
−0.9	0.087024	0.085787	0.085239	0.08734	0.086000	0.08534
0	0.047353	0.047454	0.047604	0.04759	0.047600	0.04766
0.9	0.083356	0.084112	0.083921	0.08380	0.084320	0.08406
	x_1 and v_1 generated from a spatial autoregressive model					
−0.9	0.070423	0.068735	0.067791	0.071185	0.069092	0.068017
0	0.029910	0.029123	0.029564	0.030268	0.029306	0.029667
0.9	0.064828	0.066985	0.066628	0.065679	0.067432	0.066865

Table 7 IV-SHAC estimator bias and RMSE for γ; spatial Durbin model; simultaneity

λ	Bias			RMSE		
	$n = 121$	$n = 225$	$n = 400$	$n = 121$	$n = 225$	$n = 400$
	x_1 and v_1 generated from a uniform distribution					
−0.9	0.044564	0.044095	0.044706	0.047090	0.046300	0.046640
0	0.047116	0.047651	0.047406	0.047500	0.047830	0.047520
0.9	0.045475	0.045672	0.045775	0.048360	0.048150	0.047670
	x_1 and v_1 generated from a spatial autoregressive model					
−0.9	0.031398	0.029600	0.028885	0.037190	0.035469	0.036051
0	0.031167	0.031125	0.030771	0.031883	0.031636	0.031155
0.9	0.030216	0.030472	0.030284	0.031708	0.031653	0.031564

As previously, the biases and RMSEs are smaller when x_1 and v_1 are generated from a spatial autoregressive model than when they are generated from a uniform distribution. However, contrary to what we obtained when the model was estimated with OLS, they do not vary when the level of spatial autocorrelation varies. Moreover, they are lower than the values obtained when the simple model is estimated by OLS. This would indicate that, as in the omitted variable case, estimating a spatial Durbin model tends to decrease the extent of bias arising from system feedback.

The third source of endogeneity we analyze corresponds to the case where a measurement error affects one explanatory variable. In this case, assume that the data generating process corresponds to (19a) with spatial error autocorrelation as in (20). However, we assume that q_i is not observable. Instead, it is measured with errors (while x_1 is measured without errors) and we observe:

$$\tilde{q} = q + v \tag{22}$$

where v is a normally and independently distributed stochastic measurement error, which is independent of the explanatory variable x_1 and of the error term u.

Therefore, the Monte-Carlo simulation in this case runs as follows. With the same set-up for the parameter values for b_0, b_1, γ and λ, we generate the variables q and x_1 in two different ways: from a uniform distribution or as spatially autocorrelated variables. Then, y is generated using (19a) and a spatial autoregressive process for the error term. Finally, we generate v with $\sigma_v^2 = 0.1$ and compute $\tilde{q} = q + v$. Note that we have tried different values for the variance of the measurement error v. Indeed, as documented in the literature (Hausman 2001), it plays a significant role in the extent of the attenuation bias that arises from measurement error affecting an explanatory variable.

Table 8 displays the bias and RMSE for the parameter γ when OLS is applied to the following equation combined with SHAC estimation of the variance–covariance matrix with 500 replications:

$$y_i = b_0 + b_1 x_{1i} + \gamma \tilde{q}_i + u_i \tag{23}$$

Table 8 OLS-SHAC estimator bias and RMSE for γ; simple model; measurement error

λ	Bias			RMSE		
	$n = 121$	$n = 225$	$n = 400$	$n = 121$	$n = 225$	$n = 400$
	x_1 and q generated from a uniform distribution					
−0.9	−0.074931	−0.113590	−0.100580	0.634160	0.472750	0.380260
0	−0.104320	−0.119100	−0.098239	0.330620	0.254160	0.187990
0.9	−0.077574	−0.105580	−0.068526	0.646860	0.497910	0.326100
	x_1 and q generated from a spatial autoregressive model					
−0.9	0.005215	−0.006506	−0.0038462	0.119650	0.086081	0.067147
0	−0.007061	−0.005668	−0.006294	0.083374	0.059154	0.042928
0.9	−0.010263	−0.007889	−0.002871	0.247620	0.178490	0.132040

Table 9 IV-SHAC estimator bias and RMSE for γ; spatial Durbin model; measurement error

λ	Bias			RMSE		
	$n = 121$	$n = 225$	$n = 400$	$n = 121$	$n = 225$	$n = 400$
	x_1 and q generated from a uniform distribution					
−0.9	−0.079840	−0.089403	−0.117530	0.304830	0.264390	0.203220
0	−0.119570	−0.105440	−0.103350	0.357710	0.269830	0.216560
0.9	−0.110730	−0.116980	−0.103310	0.353990	0.271830	0.199030
	x_1 and q generated from a spatial autoregressive model					
−0.9	−0.008120	−0.012887	−0.007996	0.098675	0.071002	0.051835
0	−0.013136	−0.005938	−0.011747	0.097058	0.072987	0.054614
0.9	−0.011640	−0.008574	−0.016839	0.104920	0.073784	0.055478

Table 9 displays the bias and RMSE for the same parameter γ, when a spatial Durbin model is estimated using the instrumental variables method, including x_1 and $\tilde{q}$, an endogenous spatial lag, together with spatial lags of x_1 and $\tilde{q}$. The three-group variables associated with x_1 and $\tilde{q}$, together with their spatial lags are added as further instruments.

In this case, the effect of using a spatial Durbin model rather than a simple model to account for the effects of measurement errors primarily shows up in the RMSEs. Indeed, they are significantly lower in the latter case. Moreover, the sensitivity of the γ's RMSE to the level of spatial autocorrelation is, as in the simultaneity case, lower when using a spatial Durbin model. Finally, given the low variance attributed to the measurement error, the attenuation bias is here rather limited.

5 Conclusions

The presence of multiple endogenous variables on the right hand side of single equation spatial econometric models inevitably leads to 2sls, which is known to be a consistent estimator, although the practical application of 2sls often presents problems because of the difficulty of finding appropriate instruments. Given this, it is apt to consider alternative approaches to estimation as possible solutions to the problems caused by endogeneity.

In this chapter, we have focused on one particularly interesting case in the spatial context, which allows consistent estimation when endogeneity is induced by the omission of a (spatially autoregressive) variable. The approach uses the spatial Durbin model, which is the appropriate estimator given that the omitted variable is a spatially autoregressive process with known matrix W and is correlated with an exogenous variable. In the chapter we extend the data generating process (which includes the omitted variable) by allowing additional spatial dependence in the errors, and this leads to an augmented spatial Durbin model with a complex error process as a reasonably appropriate estimator, which we estimate using 2sls and SHAC. We explore the performance of the augmented spatial Durbin model relative to OLS-SHAC estimation of a specification minus the omitted variable. We then proceed to carry out Monte-Carlo simulations in which the omitted variable has different properties. We show that, for a limited range of simulations and compared with OLS-SHAC estimation of the omitted variable model, the spatial Durbin estimated by 2sls-SHAC remains superior in terms of bias and RMSE. We then proceed to the case where endogeneity is a consequence of simultaneity and errors in variables, and find that 2sls-SHAC estimation of the spatial Durbin model continues to provide superior estimates compared with ignoring the problem and estimating via OLS-SHAC a single equation model in which the endogeneity of one of the right hand side regressors (due to simultaneity or errors in variables) goes unacknowledged.

Finally, while we have shown some advantages associated with estimating the spatial Durbin, we do not claim that it is a consistent estimator; rather, we have shown that it appears to be better than simply ignoring the presence of endogeneity. The resulting estimates would appear to have less bias and a lower RMSE than they otherwise would have. At this point in time we do not know how using the spatial Durbin performs vis-à-vis spatial models with additional endogenous variables estimated via two-stage least squares, using as instruments the lower orders of the spatial lags of the exogenous variables. This is left for future research.

References

Andrews DWK (1991) Heteroskedasticity and autocorrelation consistent covariance matrix estimation. Econometrica 59:817–858

Anselin L (1988) Spatial econometrics: methods and models. Kluwer, Dordrecht

Anselin L (2006) Spatial econometrics. In: Mills TC, Patterson K (eds) Handbook of econometrics: volume 1, econometric theory. Palgrave MacMillan, Berlin, pp 901–969

Anselin L, Kelejian HH (1997) Testing for spatial error autocorrelation in the presence of endogenous regressors. Int Reg Sci Rev 20:153–182

Anselin L, Lozano-Gracia N (2008) Errors in variables and spatial effects in hedonic house price models of ambient air quality. Empir Econ 34:5–34

Burridge P. (1981) Testing for a common factor in a spatial autoregressive model. Environ Plann A 13:795–800
Cliff AD, Ord JK (1981) Spatial processes: models and applications. Pion, London
Dall'erba S, Le Gallo J (2008) Regional convergence and the impact of European structural funds over 1989–1999: a spatial econometric analysis. Pap Reg Sci 87:219–244
Fingleton B (2003) Increasing returns: evidence from local wage rates in Great Britain. Oxf Econ Pap 55:716–739
Fingleton B, Le Gallo J (2008a) Finite sample properties of estimators of spatial models with autoregressive, or moving average disturbances and system feedback. Ann Econ Stat, forthcoming
Fingleton B, Le Gallo J (2008b) Estimating spatial models with endogenous variables, a spatial lag and spatially dependant disturbances: finite sample properties. Pap Reg Sci 87:319–339
Fingleton B, López-Bazo E (2006) Empirical growth models with spatial effects. Pap Reg Sci 85:177–198
Hausman JA (2001) Mismeasured variables in econometric analysis: problems from the right and problems from the left. J Econ Perspect 15:57–68
Kapoor M, Kelejian HH, Prucha I (2007) Panel data models with spatially correlated error components. J Econom 140:97–130
Kelejian HH, Prucha IR (1998) A generalized spatial two-stage least squares procedure for estimating a spatial autoregressive model with autoregressive disturbances. J R Estate Finance Econ 17:99–121
Kelejian HH, Prucha IR (1999) A generalized moments estimator for the autoregressive parameter in a spatial model. Int Econ Rev 40:509–533
Kelejian HH, Prucha IR (2004) Estimation of simultaneous systems of spatially interrelated cross sectional equations. J econom 118:27–50
Kelejian HH, Prucha IR (2007) HAC estimation in a spatial framework. J Econom 140:131–154
Kennedy P (2003) A guide to econometrics, 5th edn. Blackwell, Oxford
LeSage J, Pace KP (2008) Spatial econometric modeling of origin-destination flows. J Reg Sci 48:941–967
Pace K, LeSage J (2008) Biases of OLS and spatial lags models in the presence of omitted variable and spatially dependent variables. In: Páez A, Le Gallo J, Buliung R, Dall'Erba S (eds) Progress in spatial analysis: theory and methods, and thematic applications. Springer, Berlin
Upton GJG, Fingleton B (1985) Spatial data analysis by example, vol 1. Wiley, Chichester

Dealing with Spatiotemporal Heterogeneity: The Generalized BME Model

Hwa-Lung Yu, George Christakos, and Patrick Bogaert

1 Introduction

Geographical studies involving natural systems and their attributes (e.g., environmental processes, land use parameters, human exposure indicators, disease variables, and financial indexes) often need to quantitatively assess spatiotemporal dependence and generate informative maps of the attributes across space-time. These are important, indeed, goals of spatiotemporal systems modelling and data analysis introduced in a modern statistical framework by Christakos (1990, 1991a,b, 1992). Subsequent works include Goodall and Mardia (1994), Haas (1995), Bogaert (1996), Christakos and Hristopulos (1998), and Kyriakidis and Journel (1999). Among the more recent developments one should notice the works of Serre et al. (2003), Kolovos et al. (2002, 2004), Douaik et al. (2004), Christakos et al. (2002, 2005), Stein (2005), Law et al. (2006), Porcu et al. (2006, 2008), Yu et al. (2007a–c), Renshaw et al. (2008), and Ruiz-Medina et al. (2008a,b).

In this chapter we present a spatiotemporal approach that is based on the fusion of two entities with separate goals and distinct conceptual structures: the generalized random field theory, on the one hand, and the epistematics knowledge synthesis framework, on the other. The entity resulting from this fusion is the *Generalized Bayesian Maximum Entropy* (*GBME*) approach that can be used in the spatiotemporal analysis and mapping of a wide variety of natural systems (physical, biological, social and cultural) with heterogeneous space-time patterns and dependence structures under condition of multi-sourced uncertainty.

H.-L. Yu (✉)
Department of Bioenvironmental Systems Engineering, National Taiwan University, 1, Section 4, Roosevelt Road, Taipei 10617, Taiwan, R.O.C.,
e-mail: hlyu@ntu.edu.tw

A. Páez et al. (eds.), *Progress in Spatial Analysis*, Advances in Spatial Science,
DOI 10.1007/978-3-642-03326-1_5,

2 Method

Epistemologically, GBME distinguishes between two major knowledge bases (KB):

1. The *general* or *core* KB (denoted by G) that includes: scientific laws and theories, mechanistic models, ecologic systems, population dynamics and social structures; theoretical space-time dependence models that are relevant to the system under investigation; and logical rules and reasoning principles of the human agents.
2. The *site-specific* or *specificatory* KB (denoted by S) that includes different sources associated with the particular system, such as: *hard* measurements characterized by a satisfactory level of accuracy (for all practical purposes); and *soft* data that include a non-negligible amount of uncertainty (secondary sources, imperfect observations, categorical data and fuzzy inputs).

In many applications of spatiotemporal data analysis under conditions of uncertainty, including temporal GIS (Christakos et al. 2002), one considers a spatiotemporal attribute $X(\boldsymbol{p}),\ \boldsymbol{p} = (\boldsymbol{s}, t)$, where the vector $\boldsymbol{s}$ denotes spatial location and the scalar t denotes time. Accordingly, the $\boldsymbol{p}_{map} = [\boldsymbol{p}_{hard}, \boldsymbol{p}_{soft}, \boldsymbol{p}_k]^T$ is a vector of space-time mapping points, which include hard data points ($\boldsymbol{p}_{hard}$), soft data points ($\boldsymbol{p}_{soft}$) and estimation points ($\boldsymbol{p}_k$).

Stochastic representation of an attribute $X(\boldsymbol{p})$ with heterogeneous space-time variation features is achieved by means of the powerful class of *generalized spatiotemporal random fields* (Christakos 1990, 1991a). Mathematically, one considers an operator Q that (a) transforms the original attribute $X(\boldsymbol{p})$ to a homogeneous/stationary field and (b) expresses the degree of departure from homogeneity and stationarity in terms of its corresponding orders ν and μ (which vary across composite space-time). The ν, μ give information about the mechanism underlying the attribute's space-time distribution.

A variety of mathematical Q-operators was examined by Christakos (1992) and Christakos and Hristopulos (1998). For example, the random vector $\boldsymbol{X} = [X(\boldsymbol{p}_1), X(\boldsymbol{p}_2), \ldots, X(\boldsymbol{p}_N)]^T$ can be decomposed as (Vyas and Christakos 1997):

$$\boldsymbol{X} = \mathbf{F}\beta + \Sigma, \tag{1}$$

where $\boldsymbol{F}$ is a matrix of space-time monomials with degrees ν, μ; the $\boldsymbol{\beta}$ is a vector of monomial coefficients; and $\Sigma = [\varepsilon(\boldsymbol{p}_1), \varepsilon(\boldsymbol{p}_2), \ldots, \varepsilon(\boldsymbol{p}_N)]^T$ is a random fluctuation vector. Then, the generalized random field operator can be expressed by:

$$Q = I - \mathbf{F}(\mathbf{F}^T\mathbf{F})^{-1}\mathbf{F}^T, \tag{2}$$

which removes the heterogeneous trend in the space-time variation of $X(\boldsymbol{p})$, i.e. $Q(\boldsymbol{F}\boldsymbol{\beta}) = 0$.

The ordinary covariance $c_X(\boldsymbol{p}, \boldsymbol{p}')$ of $X(\boldsymbol{p})$ is nonhomogeneous in space/nonstationary in time and can be decomposed as:

$$c_X(\boldsymbol{p}, \boldsymbol{p}') = \kappa_X(\boldsymbol{p} - \boldsymbol{p}') + P_{\nu/\mu}(\boldsymbol{p}, \boldsymbol{p}'), \tag{3}$$

where $\mathbf{p} - \mathbf{p}' = (\mathbf{s} - \mathbf{s}', t - t') = (r, \tau)$; $P_{\nu/\mu}$ is a polynomial function with spatial and temporal degrees ν and μ, respectively; and κ_X is the so-called generalized spatiotemporal covariance. There are several theoretical κ_X models (Christakos and Bogaert 1996; Christakos and Hristopulos 1998). In many applications an anisotropic space-time continuum is considered where the spatial variation is a function of spatial distance and the temporal variation is a function of time distance along with an anisotropy parameter α (Stein 1998). In these applications, a suitable generalized covariance model is as follows (Yu and Christakos 2009):

$$\kappa_X(\varsigma) = c\delta_\varsigma + \sum_{\rho=0}^{\max(\nu,\mu)} (-1)^{\rho+1} a_\rho \varsigma^{2\rho+1}, \tag{4}$$

where $\varsigma = \sqrt{r^2 + \alpha\,\tau^2}$, and the coefficients c and a_ρ should satisfy a set of the permissible conditions (Yu 2005; Yu et al. 2007a); the order of the anisotropic model is determined by the higher heterogeneity order, $\max(\nu, \mu)$. The covariance matrix estimation can be performed by means of the weighted least square technique (Christakos and Thesing 1993) or the maximum likelihood and the minimum variance unbiased quadratic estimation techniques (Kitanidis 1983; PardoIguzquiza 1997). An important feature of GBME is that only κ_X is required in spatiotemporal mapping. The class of generalized covariances is richer than that of the ordinary ones (Christakos 1991a).

In view of the above considerations, the GBME framework distinguishes between three main stages of spatiotemporal modelling and mapping that are described next.

2.1 Structural Stage

At this stage, GBME generates a probability density function (pdf), f_G, across space-time based on the available G-KB. In the case that $X(\boldsymbol{p})$ is a generalized random field, the G-KB may include the theoretical model κ_X and the heterogeneity orders ν, μ. The f_G model that satisfies the evolutionary principle of maximum expected epistemic information subject to G-KB is as follows (Christakos 2000):

$$f_G\left(\boldsymbol{\chi}_{map}\right) = A^{-1}\exp\left[-\frac{1}{2}\Theta\left(\boldsymbol{\chi}_{map}, \boldsymbol{\kappa}_{map}\right)\right] \sim N\left(\mathbf{0}, Q\boldsymbol{\kappa}_{map}Q^T\right) \tag{5}$$

where $\Theta\left(\boldsymbol{\chi}_{map}, \boldsymbol{\kappa}_{map}\right) = Q^T\left(\boldsymbol{\chi}_{map}\right)c_Q^{-1}\left(\boldsymbol{\kappa}_{map}\right)Q\left(\boldsymbol{\chi}_{map}\right)$; $c_Q\left(\boldsymbol{\kappa}_{map}\right) = \overline{Q\left(\boldsymbol{\chi}_{map}\right)Q^T\left(\boldsymbol{\chi}_{map}\right)}$, where the bar denotes stochastic expectation; $\boldsymbol{\chi}_{map}$ is a space-time realization of $X(\boldsymbol{p})$ associated with $\boldsymbol{p}_{map}$; $\boldsymbol{\kappa}_{map}$ is the matrix of the corresponding generalized covariances; N denotes the normal pdf; and A is a normalization coefficient. In light of some variate distribution results in Gupta and Nagar (2000), the pdf f_G in (5) can be simplified as follows (Yu 2005):

$$f_G\left(\boldsymbol{\chi}_{map}\right) \sim N\left(\mathbf{F}\boldsymbol{\beta}, \boldsymbol{\kappa}_{map}\right). \tag{6}$$

The generalized covariance model $\boldsymbol{\kappa}_{map}$ is conditionally positive definite, whereas $c_Q\left(\boldsymbol{\kappa}_{map}\right)$ is the "ordinary" covariance which is positive definite.

Despite the Gaussian shape of f_G in (6), the potentially non-positive definite character of $\boldsymbol{\kappa}_{map}$ can induce numerical instabilities when the f_G is directly calculated using the Gaussian formulation. Therefore, the calculation of f_G is based on (5) to assure numerical stability of the calculations. This involves the calculation of the spatiotemporal increments $Q\left(\boldsymbol{\chi}_{map}\right)$ at each local neighborhood which, in turn, depends on the unknown attribute value at the estimation point. In this study, the latter is assigned a uniform prior pdf to account for the non-informative state of uncertainty. If the $\boldsymbol{\kappa}_{map}$ is positive-definite, a more efficient numerical technique can be used by adopting the conditional Gaussian property (Gupta and Nagar, 2000). Let the matrix form of $\boldsymbol{\kappa}_{map}$ be:

$$\boldsymbol{\kappa}_{map} = \begin{bmatrix} K_{kk} & K_{kh} & K_{ks} \\ K_{hk} & K_{hh} & K_{hs} \\ K_{sk} & K_{sh} & K_{ss} \end{bmatrix} = \begin{bmatrix} K_{kk} & K_{kd} \\ K_{dk} & K_{dd} \end{bmatrix}, \tag{7}$$

where the subscripts denote covariances between various combinations of hard-soft-estimation points ($hs = \boldsymbol{p}_{hard}\text{-}\boldsymbol{p}_{soft}, hh = \boldsymbol{p}_{hard}\text{-}\boldsymbol{p}_{hard}, kk = \boldsymbol{p}_k\text{-}\boldsymbol{p}_k, kd = \boldsymbol{p}_{hard}\text{-}\boldsymbol{p}_{data}$, etc.). Equation (6) can lead to the following computationally efficient expression (Yu 2005):

$$f_G\left(\boldsymbol{\chi}_{map}\right) \sim N\left(M_{k|d}, K_{k|d}\right) N\left(M_{s|h}, K_{s|h}\right) N\left(M_h, K_{hh}\right), \tag{8}$$

where the M_h is a vector of attribute means at the data points; the $M_{k|d}$ and $M_{s|h}$ are vectors of conditional means at the estimation points given the data points, and at the soft data points given the hard data points, respectively; the K_{hh} is a vector of generalized covariances between the hard data points; and the $K_{k|d}$, and $K_{s|h}$ are vectors of conditional generalized covariances at the estimation points given the data points, and at the soft data points given the hard data points, respectively. These vectors can be calculated as follows (Yu and Christakos 2009):

$$\left.\begin{aligned} M_{k|d} &= M_k + K_{dk}\, K_{kk}^{-1} \left(\boldsymbol{\chi}_{data} - M_{data}\right) \\ M_{s|h} &= M_s + K_{sh}\, K_{hh}^{-1} \left(\boldsymbol{\chi}_{hard} - M_h\right) \\ K_{k|d} &= K_k - K_{dk}\, K_{kk}^{-1}\, K_{kd} \\ K_{s|h} &= K_{ss} - K_{sh}\, K_{hh}^{-1}\, K_{hs} \end{aligned}\right\}, \tag{9}$$

where M_h, M_s and M_k are the vectors of random field means at all hard data, soft data, and estimation points, respectively; as above, K_{hh}, K_{ss}, K_{kk}, K_{dk} and K_{sh} are vectors of generalized covariances between, the hard data points, the soft data points, the estimation points, the data and the estimation points, and the soft and the hard data points, respectively.

Table 1 Examples of S-KB

(a) Hard data (exact numerical values) χ_{hard} at points p_{hard}.	$Prob\left[x_{hard} = \chi_{hard}\right] = 1$
(b) Soft data in the form of intervals of possible values (uncertain data) are available at points p_{soft}.	$Prob\left[a < x_{soft} < b\right] = 1$, a and b are vectors of the lower and upper bounds of the intervals
(c) Soft data in the form of probability functions	$Prob\left[x_{soft} < u\right] = \int_{-\infty}^{u} d\chi_{soft}\, f_S\left(\chi_{soft}\right)$, f_S is a soft datum pdf

2.2 Specificatory Stage

At this stage, the case-specific knowledge available, S, is expressed into a form suitable for quantitative analysis. Common formulations of the S-KB include those depicted in Table 1. Case c is clearly a generalization of case b above; e.g., an interval datum is a pdf f_S uniformly distributed between the lower and upper bounds of the interval. Various other types of soft information that can be considered by the GBME technique are discussed in Yu and Christakos (2006), Kolovos et al. (2006) and Yu et al. (2007a).

2.3 Integration Stage

At this stage, the solution, f_G, of the structural stage above is updated using the evolutionary adaptation principle subject to S-KB (Christakos 2000, 2008), thus leading to the integration pdf:

$$f_K\left(\chi_k\right) = A^{-1} \int_D d\,\Xi_S\left(\chi_{\text{soft}}\right) f_G\left(\chi_{\text{map}}\right), \tag{10}$$

where A is a normalization constant; and the Ξ_S and D denote, respectively, a specificatory operator and the information range determined by the S-KB. The new pdf (10) describes the distribution of the $X(p)$ values at each estimation point p_k in light of the total knowledge base $K = G \cup S$. The pdf expression (10) is also known as the operational Bayesian conditional (to be distinguished from the standard Bayesian rule; see discussion in Christakos 2002).

For illustration, Table 2 presents some examples of S-KB together with the corresponding Ξ_S and D. The f_S denotes the probability function derived from S at the soft data points, $\mathbf{I}$ is a vector of $X(p)$ interval values at these points, and $\mathbf{I}_k$ is a vector of interval values at the estimation points themselves. The integration pdf (10), which is generally non-Gaussian, offers a complete stochastic characterization of $X(p)$ at each space-time point that integrates a wide variety of data (hard and soft) as well as the relevant core knowledge (κ_X, ν, μ etc.).

Table 2 Examples of soft data with integration domain $\boldsymbol{D}$ and operator Ξ_S – see, Equation (10)

S	D	Ξ_S
Interval	$\boldsymbol{I}$	χ_{soft}
Probabilistic	$\boldsymbol{I}$	$f_S\left(\chi_{soft}\right)$
Functional	$\boldsymbol{I} \cup \boldsymbol{I}_k$	$f_S\left(\chi_{soft}, \chi_k\right)$

From the integration pdf (10) we obtain various kinds of estimates across the space-time domain:

$$\left.\begin{array}{ll}\text{BMEmode} & \boldsymbol{\chi}_{\boldsymbol{k},\text{mode}} : \max_{\chi_k} f_K\left(\boldsymbol{\chi}_{\boldsymbol{k}}\right) \\ \text{BMEmean} & \boldsymbol{\chi}_{\boldsymbol{k},\text{mean}} : \int d\ \boldsymbol{\chi}_k\, \boldsymbol{\chi}_{\boldsymbol{k}} f_K\left(\boldsymbol{\chi}_{\boldsymbol{k}}\right)\end{array}\right\}, \tag{11}$$

The BMEmode provides the most likely value at the estimation point; the BMEmean minimizes the mean square estimation error.

Since the $\boldsymbol{\chi}_{\boldsymbol{k},\text{mean}}$ is used in the numerical experiment of the following section, we provide here some explicit expressions of the estimate:

$$\boldsymbol{\chi}_{\boldsymbol{k},\text{mean}} = A^{-1} \int_D d\boldsymbol{\chi}_k\, d\boldsymbol{\chi}_{soft}\, \boldsymbol{\chi}_k\, f_S\left(\boldsymbol{\chi}_{soft}\right) \exp\left[-\frac{1}{2}\Theta\left(\boldsymbol{\chi}_{map}, \boldsymbol{\kappa}_{map}\right)\right], \tag{12}$$

and the associated estimation error variance:

$$\begin{aligned}\sigma_k^2 = A^{-1} & \int_D d\boldsymbol{\chi}_k\, d\boldsymbol{\chi}_{soft}\, \left(\boldsymbol{\chi}_k - \boldsymbol{\chi}_{k,mean}\right)^2 f_S\left(\boldsymbol{\chi}_{soft}\right) \\ & \times \exp\left[-\frac{1}{2}\Theta\left(\boldsymbol{\chi}_{map}, \boldsymbol{\kappa}_{map}\right)\right],\end{aligned} \tag{13}$$

respectively.

Spatiotemporal GBME analysis and mapping possesses a number of attractive features (Yu and Christakos 2009). These are summarized in Table 3.

Below, we continue the presentation of spatiotemporal GBME analysis by means of numerical experimentation.

3 Numerical Experiments

Certain implementation features and performance indicators of the GBME approach can be investigated with the help of numerical experiments in controlled environments (Yu and Christakos 2009). The GBME approach is numerically compared with generalized kriging (GK), which is an advanced statistical regression technique of space-time estimation, also known as universal or intrinsic spatiotemporal kriging (Christakos 1990; Christakos and Raghu 1996).

Table 3 Summary of theoretical GBME properties

Its methodological underpinnings rely on evolutionary concepts of brain and behavioural sciences rather than on mechanistic schemes and technical recipes that lack cognitive reasoning substance.
Nonlinear estimators and non-Gaussian laws are automatically incorporated; i.e. no restrictive assumptions concerning estimator linearity and probabilistic normality are made.
It can study systems with heterogeneous space-time dependence patterns and synthesize various kinds of knowledge bases (core and site-specific) in a general and unified framework rather than in an *ad hoc* and arbitrary manner.
It can readily consider uncertain yet valuable information at the estimation (prediction) points themselves, when available.
It provides a sound space-time attribute characterization in terms of the complete predictive pdf at every point rather than just the first two estimation moments. In this way, more than one possibility can be available at each point, as far as estimation is concerned.
It derives several mainstream techniques (such as statistical regression, kriging and Gaussian process) as its special cases, a fact that amply demonstrates GBME's generalization power.

A spatiotemporal domain, $E_{s,t}$, was considered with dimensions $s_1 \times s_2 \times t \in [0,1] \times [0,1] \times [0,3]$ in suitable units, which includes $21 \times 21 \times 4$ grid nodes (i.e., the vector $\boldsymbol{p}_{map}$ has 1764 elements). A heterogeneous (spatially nonhomogeneous/temporally nonstationary) attribute $X(\boldsymbol{p}) = X(s_1, s_2, t)$ is simulated having a space-time dependent mean:

$$\overline{X(s_1, s_2, t)} = 4\cos(5s_1) + 2s_2 + 0.5\cos\left(\frac{1}{2}t\right); \tag{14}$$

and the corresponding fluctuation field $Y(s_1, s_2, t)$ has zero mean and space-time covariance:

$$c_Y(r, \tau) = c_0 \exp[-3r/a_r - 3\tau/a_\tau], \tag{15}$$

where $c_0 = 3$, $a_r = 0.5$ and $a_\tau = 3$ in suitable units. The separable covariance model was chosen since the random field generated using the model (15) exhibits a more complex pattern than other commonly used models (Gaussian or spherical models). This model refers to the fluctuation field (Y) and not the original space-time heterogeneous field (X). For illustration purposes, three $X(s_1, s_2, t)$-realizations are plotted in Fig. 1 (top row) at times $t = 0$, 1 and 2.

3.1 *Experiment 1*

A set of 30 hard data were drawn from the X (s_1, s_2, t) simulated field (Fig. 1, top row) at space-time points selected at random in the domain $E_{s,t}$ (i.e., the $\boldsymbol{p}_{hard}$ has 30 elements). Soft data in the form of uniform distributions were generated at ten randomly selected points ($\boldsymbol{p}_{soft}$ has ten elements; Fig. 2). Each soft datum has a mean equal to the closest simulated X $(s_1, s_2, t) = \chi$ value and a range $\chi \pm 0.5|\chi|$. In this experiment, the G-KB includes the theoretical model of space-time dependence

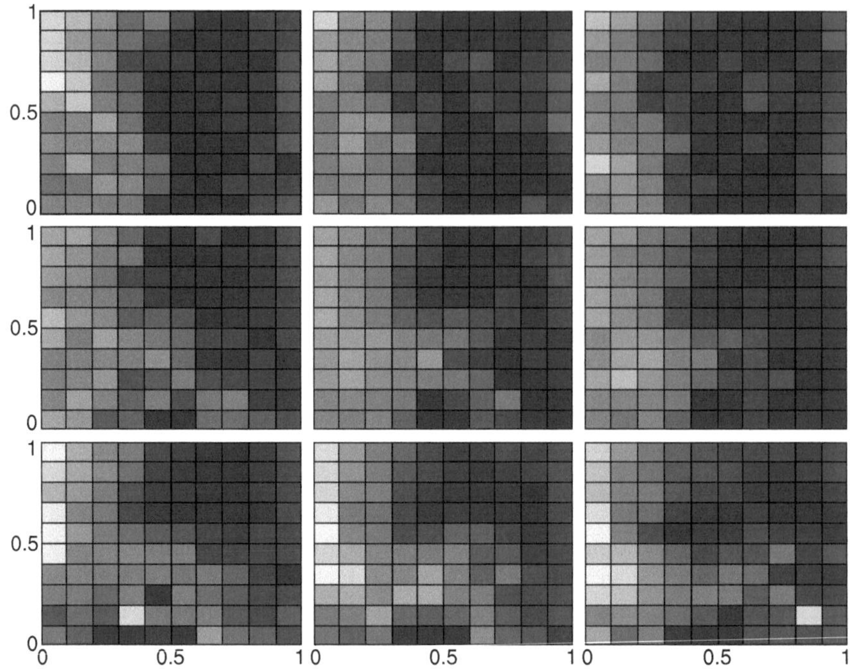

Fig. 1 Simulated random field realizations (*top row*); estimated field using GBME (*middle row*); and estimated field using GK (*bottom row*) at times $t = 0$ (*left column*), $t = 1$ (*middle column*), and $t = 2$ (*right column*)

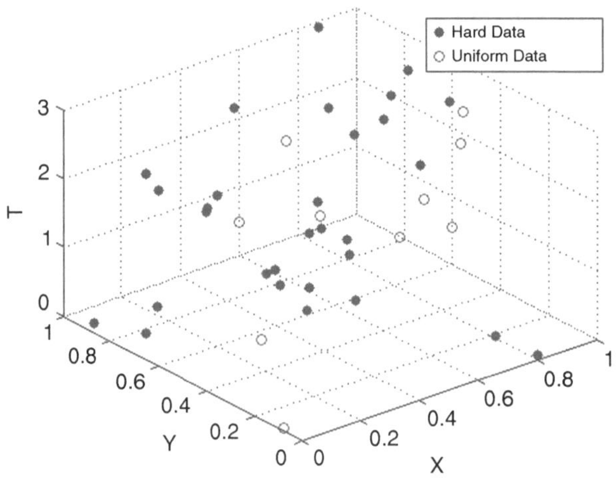

Fig. 2 Hard data (black circles), soft data in the form of uniform distributions (white circles), across space-time

in terms of ν, μ and κ_X. For illustration, Figs. 3 and 4 show maps of the heterogeneity orders at times $t = 0, 1, 2$. These maps offer information about relative trends in space-time. Spatial variation is more significant than temporal variation. Spatial and temporal trends are interrelated, since the geographical $X(s_1, s_2, t)$ distribution is also affected by temporal mechanisms. Not only the spatial order ν changes in space, but the temporal order μ varies in space, as well. These observations are in agreement with the spatiotemporal meantrend (14) of the simulated field. Due to its theoretical ability to rigorously incorporate various forms of S-KB in a unified theoretical framework, the GBME readily takes into consideration the 30 hard and the 10 soft data without any *ad hoc* modifications or computational schemes of questionable physical meaning and mathematical accuracy.

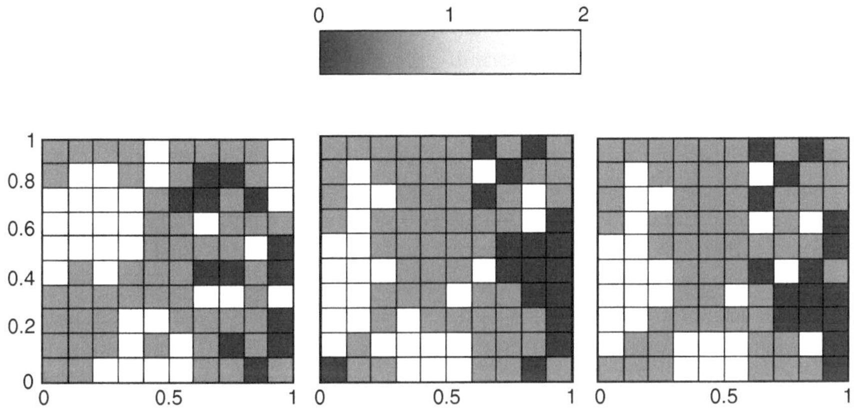

Fig. 3 Space-time distributions of the value of spatial order ν. (*Left*) $t = 0$, (*Middle*) $t = 1$, and (*Right*) $t = 2$

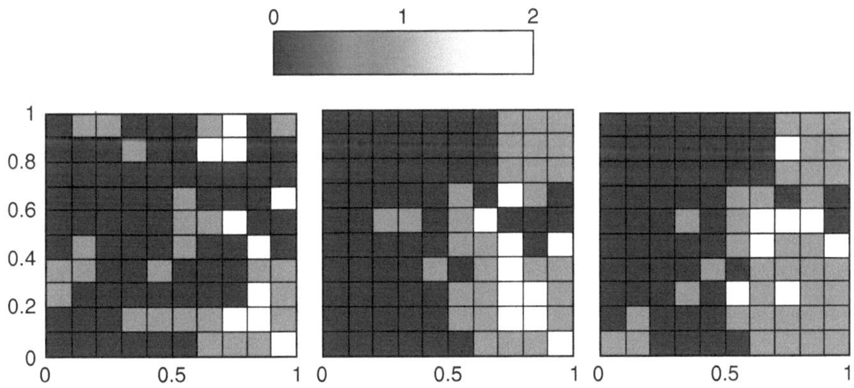

Fig. 4 Space-time distributions of the value of temporal order μ. (*Left*) $t = 0$, (*Middle*) $t = 1$, and (*Right*) $t = 2$

Next, by synthesizing the various core and case-specific knowledge sources, the GBME generated pdf f_K at all 484 grid points of the domain $E_{s,t}$, and from these pdf the space-time estimates $\hat{\chi}_{k,\text{mean}}$ were derived at each grid point. For illustration, in Fig. 1 (middle row) the estimated maps are plotted at times $t = 0, 1, 2$. These GBME maps provide an adequate representation of the simulated maps (Fig. 1, top row). Local space-time neighborhoods were conveniently used at every estimation point of the domain $E_{s,t}$. Figures 3 and 4 show that the space-time heterogeneity orders (ν, μ) vary between the different neighborhoods.

In view of its superior theoretical structure (Table 3), GBME should perform better than GK in numerical tests, as well (when such a comparison is possible and makes sense). To carry out a numerical comparison, space-time estimates were generated using GK. The GK makes certain restrictive assumptions concerning estimator linearity and probabilistic normality, and in this particular experiment GK readily processed only the 30 hard data (*ad hoc* modifications and approximations that allow GK to indirectly and partially account for the information provided by the soft data will be examined in the following numerical experiments).

GK estimates at times $t = 0, 1, 2$ are shown in Fig. 1, bottom row. Clearly, the generated GBME maps constitute a considerable improvement over the GK maps. To further support this conclusion, the histograms of the estimation errors ("simulated values-estimated values") for the two methods are plotted in Fig. 5. As was expected, GBME leads to more accurate estimates than GK (e.g., the GBME estimation errors are more closely concentrated around zero error than the GK estimations errors). GBME offers a complete characterization in terms of the non-Gaussian pdf f_K at each space-time point (different shapes are possible at different points), from which the estimates $\hat{\chi}_{k,\text{mean}}$ are calculated. The non-Gaussian shape implies that in

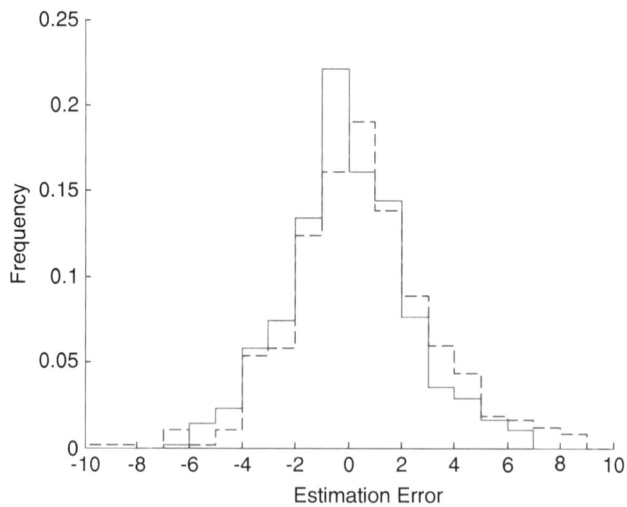

Fig. 5 Histograms of the estimation errors of the GBME (continuous line) and GK (dashed line) methods

some cases it may be possible to obtain better maps in terms of the estimates $\hat{\chi}_{k,\text{mode}}$ and $\hat{\chi}_{k,\text{median}}$, also readily calculated from the f_K. Such flexibility is not allowed by most mainstream methods, including GK and other statistical regression techniques.

Soft data generation as used in the present experiment may not be appropriate in real-world situations where the data amount is limited. Thus, soft data in the form of the uniform and the Gaussian probability laws were assumed, respectively, in the following Experiments 2 and 3. Moreover, in the GK context, instead of ignoring soft information, the mean values of the soft data are calculated across space-time and used as an additional set of hard data (more precisely, "hardened" soft data) in connection with the GK technique.

3.2 Experiment 2

As in Experiment 1, hard data points were drawn from the simulated field at 30 space-time points selected at random. This time, however, soft data in the form of uniform distributions $[\chi - 0.6\,|\chi|,\ \chi + 0.4\,|\chi|]$ were generated at ten randomly selected space-time points (Fig. 6) having a mean value different than the simulated field value at each point.

GBME's theoretical support readily and rigorously takes into consideration the 30 hard and the 10 soft data in a unified framework, with no need to employ any kind of approximation or computational trick. GBME generates estimates $\hat{\chi}_{k,\text{mean}}$ at the 484 mapping points in the domain $E_{s,t}$. The GK technique, on the other hand, while it can readily process the same 30 hard data as GBME, it cannot account for the information provided by the 10 soft data in a direct and theoretically unified

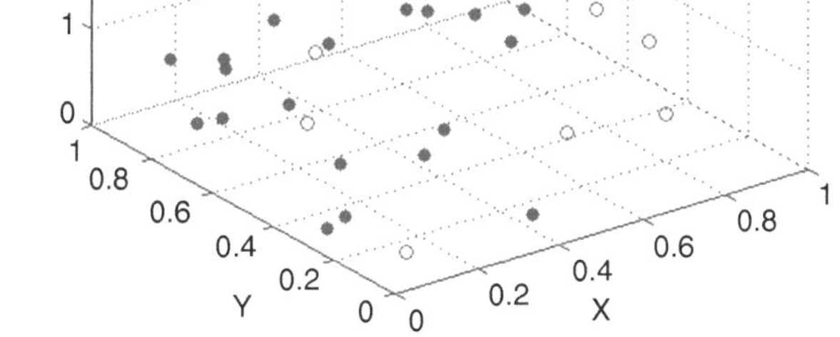

Fig. 6 Hard data (black circles) and uniform distributed data (white circles) across space-time

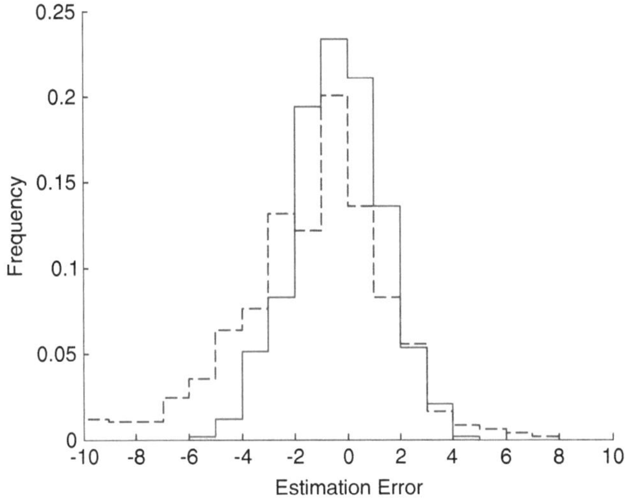

Fig. 7 Histograms of the estimation errors of the GBME (continuous line) and GK (dashed line) methods

manner. Instead, an *ad hoc* scheme was used that generated 10 more "hardened" soft data in terms of the expected $X(s_1, s_2, t)$ value at each soft datum point (i.e., the mean value was obtained from the uniform distribution at each point). Space-time estimates were then derived at the 484 mapping points in $E_{s,t}$ using GK. As is clearly demonstrated by the histograms of the estimation errors ("simulated field value-estimated value") for the two methods (Fig. 7), the GBME again leads to more accurate estimates than the GK method.

3.3 Experiment 3

As above, hard data points were drawn from the simulated field at 30 space-time points selected at random. Soft data in the form of Gaussian distributions were generated at ten randomly selected points with each mean equal to the local $\chi - 0.1\,|\chi|$ value and variance equal to three units (Fig. 8).

As before, the GBME's theoretical support readily takes into consideration the 30 hard and the 10 Gaussian data. GBME estimates $\hat{\chi}_{k,\text{mean}}$ were derived at all 484 mapping points in the space-time domain $E_{s,t}$.

The GK technique used the same 30 hard data, but its mathematical apparatus could only indirectly and incompletely account for the soft information in terms of the mean value of the Gaussian distribution at each soft data point, thus generating ten more ("hardended" soft data) values treated as hard data. Space-time GK estimates were subsequently obtained, and the histograms of the estimation

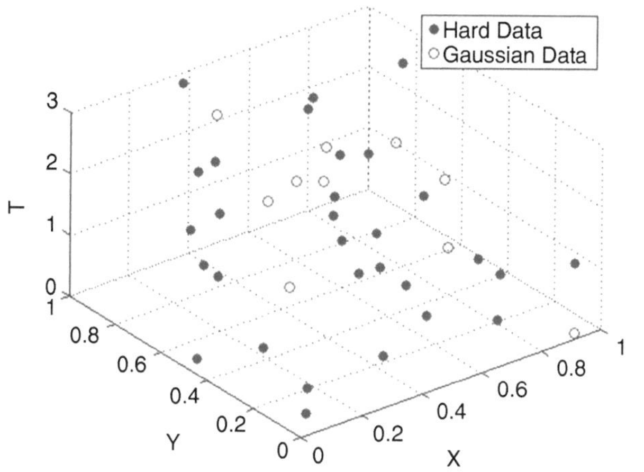

Fig. 8 Hard data (black circles), and Gaussian-distributed data (white circles) across space-time

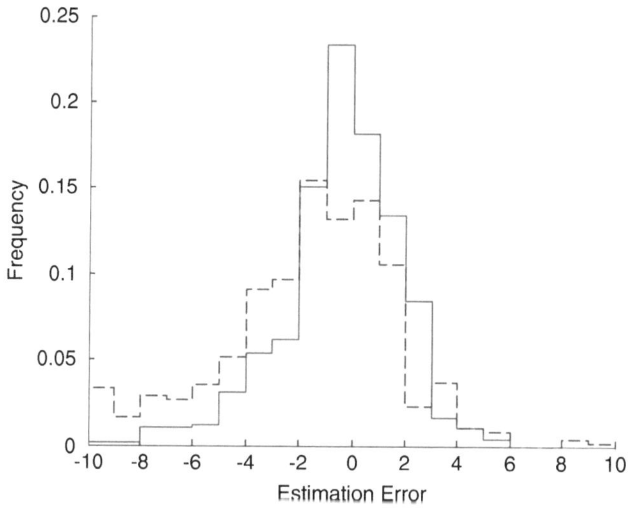

Fig. 9 Histograms of the estimation errors of the GBME (continuous line) and GK (dashed line) methods

errors for the two methods were plotted in Fig. 9. Under the current experimental circumstances, the spatiotemporal GBME analysis once more showed a superior performance compared to the GK technique. In addition, some real-world case studies are discussed in Yu and Christakos (2009).

4 Discussion

Due to its strong theoretical support and computational efficiency, GBME is a more accurate and realistic approach of spatiotemporal data analysis than many mainstream quantitative geography methods. Numerical comparisons of the conventional GK technique vs. the GBME approach were made in this chapter. When comparing GBME vs. GK, one should keep in mind that GBME readily considers non-Gaussian distributions (e.g., Eq. 10), whereas GK is implicitly restricted by the Gaussian assumption (Kitanidis and Shen 1996); and that GBME is generally a non-linear estimator, whereas the GK is limited to linear estimators.

Although many statisticians prefer to routinely use the mean value of the probability distribution at each (soft data) point as the "hardened" value to be used in space-time estimation (e.g., GK), there are often sound reasons for making a different choice in accordance with the epistemic context of the situation. Selecting different values from the distribution at different points (say, mean, highest, lowest, most probable and most improbable values), rather than insisting on the mean value at all points, provides sufficient flexibility and may be a more realistic approach. For example, a spatiotemporal analysis that, on occasion, takes into account low-probability values, may turn out to be very informative, since, when these values occur, they can be highly consequential (as practitioners of the financial markets can testify, it does not matter how rare an event is if its occurrence is too costly to bear).

The experimental GBME performance was shown to be generally superior to that of GK. This is due to the more general theoretical structure of GBME vs. GK and the different ways the two methods incorporate soft information. The GBME accomplishes the spatiotemporal analysis tasks in a theoretically general and unified manner (e.g., accounts for complete information provided by soft data), whereas the GK results are highly influenced by its restrictive assumptions and the occasional tricks used to "harden" the available uncertain information. Also, if the soft data mean value is not exactly the same as the true value, GK can produce biased estimation results (Figs. 7 and 9). On the other hand, even when the soft data mean (or mode) does not coincide with the true value, GBME analysis will still be more accurate and less biased than GK. This is very valuable in practical applications where a significant amount of inaccurate or highly uncertain data is considered.

5 Conclusions

A GBME approach was discussed that can be used in the case of natural systems and attributes with spatiotemporal heterogeneities. On theoretical grounds the GBME approach has certain important advantages compared to mainstream methods (including statistical regression, standard Bayesian, Gaussian process and geostatistical kriging techniques). The computational GBME properties were also investigated by means of numerical experiments under controlled conditions in which GBME was compared to the GK technique (which is considered the most general and powerful among the mainstream geostatistics techniques):

- The estimation domain of the GBME is much larger than that of many mainstream techniques, since the former includes non-linear estimators and non-Gaussian probability laws. Statistical regression, on the other hand, is often based on some kind of linearity and Gaussian assumptions (this is the case of geostatistical kriging, statistical regression and Gaussian process techniques).
- Instead of formulating a "modified" GK matrix to incorporate the effect of soft data in a mathematically approximate and physically questionable manner, the GBME analysis accounts for the uncertain information of site-specific datasets in a rigorous and unified manner.
- GBME is based on a generalized stochastic theory especially developed for space-time heterogeneous random fields and does not employ any *ad hoc* assumptions and approximations (e.g., no deterministic trend is arbitrarily defined and substracted from the original data, as in certain statistical regression techniques).

GBME analysis – together with other spatiotemporal analysis, modelling and mapping techniques – can be found in the *SEKS-GUI* software library (Spatiotemporal Epistematics Knowledge Synthesis-Graphical User Interface). This library is available at the following webpages:

Geography Department, San Diego State University (California, USA):
http://geography.sdsu.edu/Research/Projects/SEKS-GUI/SEKS-GUI.html
and:
Department of Bioenvironmental Systems Engineering, National Taiwan University (Taipei, Taiwan):
http://homepage.ntu.edu.tw/~hlyu/software/SEKSGUI/SEKSHome.html
See, also, Kolovos et al. (2006) and Yu et al. (2007a).

Acknowledgements Partial support for this work was provided by a grant from the California Air Resources Board, USA (Grant No. 55245A).

References

Bogaert B (1996) Comparison of kriging techniques in a space-time context. Math Geol 28:73–86
Christakos G (1990) Random Field modelling and its applications in stochastic data processing. PhD Thesis, Applied Sciences Division, Harvard University, Cambridge, MA
Christakos G (1991a) On certain classes of spatiotemporal random fields with application to space-time data processing. IEEE T Syst Man Cyb 21:861–875
Christakos G (1991b) Some applications of the BME concept in geostatistics. Fund Theories Phy: 215–229, Kluwer, Amsterdam, The Netherlands
Christakos G (1992) Random field models in earth sciences. Academic, San Diego, CA. Also, Dover Publ. 2005
Christakos G (2000) Modern spatiotemporal geostatistics. Oxford University Press, New York, NY
Christakos G (2002) On the assimilation of uncertain physical knowledge bases: Bayesian and non-Bayesian techniques. Adv Water Resour 25:1257–1274

Christakos G (2008) Bayesian maximum entropy. In: Kanevski M (ed.) Advanced mapping of environmental data: geostatistics, machine learning, and Bayesian maximum entropy. Wiley, New York, NY, pp 247–306

Christakos G, Bogaert P (1996) Spatiotemporal analysis of spring water ion processes derived from measurements at the Dyle Basin in Belgium. IEEE Trans Geosci Remote Sens 34:626–642

Christakos G, Bogaert P, Serre ML (2002) Temporal GIS. Springer, New York, with CD-ROM

Christakos G, Hristopulos DT (1998) Spatiotemporal environmental health modelling: a Tractatus Stochasticus. Kluwer, Boston

Christakos G, Olea RA, Serre ML, Yu HL, Wang LL (2005) Interdisciplinary public health reasoning and epidemic modelling: the case of black death. Springer, New York

Christakos G, Raghu VR (1996) Dynamic stochastic estimation of physical variables. Mathematical Geology 28:341–365

Christakos G, Thesing GA (1993) The Intrinsic Random-Field Model in the Study of Sulfate Deposition Processes. Atmos Environ Gen Top 27:1521–1540

Douaik A, van Meirvenne M, Toth T, Serre ML (2004) Space-time mapping of soil salinity using probabilistic BME. Stoch Environ Res Risk Assess 18:219–227

Goodall C, Mardia KV (1994) Challenges in multivariate spatio-temporal modeling. In: Proceedings of the XVIIth International Biometric Conference, 1–17, Hamilton, Ontario, Canada, 8–12 August 1994

Gupta AK, Nagar DK (2000) Matrix variate distributions. Chapman & Hall, Boca Raton, FL

Haas TC (1995) Local prediction of spatio-temporal process with an application to wet sulfate deposition. J Am Stat Assoc 90:1189–1199

Kolovos A, Christakos G, Serre ML, Miller CT (2002) Computational BME solution of a stochastic advection-reaction equation in the light of site-specific information. Water Resour Res 38:1318–1334

Kitanidis PK (1983) Statistical estimation of polynomial generalized covariance functions and hydrologic applications. Water Resour Res 19:909–921

Kitanidis PK, Shen KF (1996) Geostatistical interpolation of chemical concentration. Adv Water Resour 19:369–378

Kolovos A, Christakos G, Hristopulos DT, Serre ML (2004) Methods for generating non-separable spatiotemporal covariance models with potential environmental applications. Adv Water Resour 27:815–830

Kolovos A, Yu HL, Christakos G (2006) SEKS-GUI v.0.6 user manual. Department of Geography, San Diego State University, San Diego, CA

Kyriakidis PC, Journel AG (1999) Geostatistical space-time models: a review. Math Geol 31:651–684

Law DC, Bernstein K, Serre ML, Schumacher CM, Leone PA, Zenilman WC Miller, Rompalo AM (2006) Modelling an early syphilis outbreak through space and time using the bayesian maximum entropy approach. Ann Epidemiol 16:797–804

PardoIguzquiza E (1997) GCINFE: a computer program for inference of polynomial generalized covariance functions. Comput Geosci 23:163–174

Porcu E, Gregori P, Mateu J (2006) Nonseparable stationary anisotropic space-time covariance functions. Stoch Environ Res Risk Assess 21:113–122

Porcu E, Mateu J, Saura F (2008) New classes of covariance and spectral density functions for spatio-temporal modelling. Stoch Environ Res Risk Assess 22:65–79

Renshaw E, Comas C, Mateu J (2008) Analysis of forest thinning strategies through the development of space-time growth-interaction simulation models. Stoch Environ Res Risk Assess. doi:10.1007/s00477–008–0214-x

Ruiz-Medina MD, Angulo JM, Anh VV (2008a) Spatiotemporal statistical analysis of influenza mortality risk in the State of California during the period 1997–2001. Stoch Environ Res Risk Assess 22:15–25

Ruiz-Medina MD, Angulo JM, Anh VV (2008b) Multifractality in space–time statistical models. Stoch Environ Res Risk Assess 22:81–86

Serre ML, Kolovos A, Christakos G, Modis K (2003) An application of the holistochastic human exposure methodology to naturally occurring Arsenic in Bangladesh drinking water. Risk Anal 23:515–528

Stein A (1998) Analysis of space-time variability in agriculture and the environment with geostatistics. Statistica Neerlandica 52:18–41

Stein ML (2005) Space-time covariance functions. J Am Stat Assoc 100:310–321

Vyas VM, Christakos G (1997) Spatiotemporal analysis and mapping of sulfate deposition data over eastern USA. Atmos Environ 31:3623–3633

Yu HL (2005) Development and implementation of knowledge synthesis methods for stochastic natural systems. PhD Thesis, University of North Carolina at Chapel Hill, Department of Environmental Sciences and Engineering, Chapel Hill, NC, USA

Yu HL, Christakos G (2006) Spatiotemporal modelling and mapping of the bubonic plague epidemic in India. Int J Health Geogr 5 (Internet Journal)

Yu HL, Kolovos A, Christakos G, Chen JC, Warmerdam S, Dev B (2007a) Interactive spatiotemporal modelling of health systems: The SEKS-GUI framework. J Stoch Environ Res Risk Assess – Special Issue on "Medical Geography as a Science of Interdisciplinary Knowledge Synthesis under Conditions of Uncertainty," Griffith DA, Christakos G (eds) 21:555–572

Yu HL, Christakos G, Modis K, Papantonopoulos G (2007b) A composite solution method for physical equations and its application in the Nea Kessani geothermal field (Greece). J Geophys Res B Solid Earth Planets 112:B06104. doi:10.1029/2006JB004900

Yu HL, Christakos G (2009) Generalized BME processing and imaging of heterogeneous space-time data. Working paper available from the authors

Local Estimation of Spatial Autocorrelation Processes

Fernando López, Jesús Mur, and Ana Angulo

1 Introduction

The difficulties caused by the lack of stability in the parameters of an econometric model are well known: biased and inconsistent estimators, misleading tests and, in general, wrong inference. Their importance explains the attention that the literature has dedicated to the problem. The first formal test of parameter stability is that of Chow (1960), which considers only one break point, known a priori, under the assumption of constant variances. Dufour (1982) extends the discussion to the case of multiple regimes and Phillips and Ploberger (1994) and Rossi (2005) place it in a context of model selection. Simultaneously, Quandt (1960) started another line of research in which the break point is unknown and the variance can change. The CUSUM test, based on recursive residuals (Brown et al. 1975), the various methods for endogenizing the choice of the break point (as in Banerjee et al. 1992), and the extension to multiple structural changes in a system of equations (Qu and Perron 2007) are natural proposals in this line. Other more peculiar approaches include the tests for continuous parameter variation (Hansen 1996), the Markov switching regression (García and Perron 1996) and the Bayesian approaches (e.g., Salazar 1982; Zivot and Phillips and Ploberger 1994; Koop and Potter 2007).

The discussion quickly took on a spatial context with the work of Casetti (1972, 1991), in which a parametric approach predominates. In fact, Casetti proposes explicitly modeling how the break in the parameters is produced through the so-called "*contextual*" variables. In the nineties, there was a great leap forwards when concern about the "*pockets of local nonstationarity*," characteristic of the literature dedicated to the LISA (Getis and Ord 1992; Anselin 1995) coincided with the development of nonparametric procedures for analyzing spatial data (McMillen 1996; McMillen and McDonald 1997). The best-known approach in this line is what Brunsdon et al. (1996) call Geographically Weighted Regressions (GWR in what

F. López (✉)
Department of Quantitative Methods and Computing, Technical University of Cartagena,
Paseo Alfonso XIII, 50 Cartagena 30203, Spain,
e-mail: fernando.lopez@upct.es

A. Páez et al. (eds.), *Progress in Spatial Analysis*, Advances in Spatial Science,
DOI 10.1007/978-3-642-03326-1_6,

follows), whose immediate precursor are the Locally Weighted Regressions (LWR from now on) proposed in the seminal papers of Cleveland (1979) and Cleveland and Devlin (1988). In all these papers, interest shifts from the general to the local. Some have a merely descriptive objective, as in the case of the LISA, while others adopt more ambitious proposals, like the LWR and the GWR.

The convenience of local approaches is clear when the heterogeneity of the data is very high and escapes the control of the model or when the appropriate functional form is doubtful, in which case it is recommendable to use generic and flexible specifications. In a spatial context, the LWR or GWR estimation has also been used to correct the problems of spatial correlation that come from an inadequate treatment of the spatial heterogeneity in the data (Páez et al. 2002a,b).

The question that we wish to deal with is in the same line, although changing the focus slightly. We analyze what happens when the heterogeneity that we observe in the data is a consequence of the instability in the mechanisms of spatial dependence that act in the model. This is the idea developed by Rietveld and Wintershoven (1998) referring to a "*border effect*" that operates between regions which are geographically contiguous but separated by an international frontier (an idea later taken up by Lacombe 2004; Ertur et al. 2006). As pointed out by Brunsdom et al. (1998, p. 958) the "*alternative conjecture.... that in some areas this spatial influence is more marked than in others*" may be preferable in cases of great heterogeneity. Pace and Lesage (2004) focus the discussion on the estimation of models with symptoms of instability of this type for which they propose a recursive maximum-likelihood algorithm, called SALE (Spatial Autoregressive Local Estimation), which Ertur et al. (2007) convert into the BSALE by introducing Bayesian criteria.

Our intention is to better understand the problem of instability in the spatial dependence mechanisms. Furthermore, we will analyze how the local estimation algorithms work when applied to autoregressive spatial structures. In Sect. 2, we look at the problem in greater detail. Using a ML approach, we develop a preliminary test of parameter stability in the parameter of spatial dependence. The remaining sections of the chapter deal with a Monte Carlo exercise which constitutes the main part of our work. The design of the exercise appears in Sect. 3. In Sect. 4, we analyze the consequences of a break of this type on the most common diagnostic measures in applied research. In Sect. 5, we examine the behavior of the SALE estimation under different conditions. In Sect. 6, we consider possible solutions to the problem of how to identify the regions that are affected by the problem of structural breaks. Finally, Sect. 7 presents our conclusions and the lines that remain open for further research.

2 Instability in Parameters of Spatial Autocorrelation

Our point of departure will be the case of an econometric cross-sectional model with spatial dependence of a substantive nature (the Spatial Lag Model, SLM in what follows), specified under the assumption of stability:

$$y = \rho \mathbf{W} y + x\beta + \varepsilon, \quad \varepsilon \sim N\left(0, \sigma^2 I\right) \tag{1}$$

where y is the (R×1) vector of the observations of the endogenous variable, x is an (R×k) matrix of observations of the k explanatory variables, ϵ is a random vector of error terms and **W** is an (R×R) exogenous weighting matrix. Finally, ρ is a parameter of spatial interaction and β a (k×1) vector of coefficients.

The model of (1) has been specified under the assumption of homogeneity, which may not hold in some circumstances. As indicated by McMillen (2004, p. 232): "*spatial relationships are typically more complicated. Statistical tests based on simple functional forms often reveal that coefficients vary over space*"; in other words, a certain heterogeneity that is omitted from the base model often persists in the data. The solutions proposed in the literature try to give more flexibility to the specification of (1) by acting either on the systematic part of the equation or on the error term. In the first case, an intermediate alternative is to group the regions into clubs that share the same vector of parameters. This solution is common in the analysis of regional convergence (Baumol 1986; Quah 1986), where each vector of parameters identifies a stationary state. However, in more extreme cases, this option is not enough (because the equation is a poor approximation, because the geographical effects are very important, etc.) and it will be necessary to make use of nonparametric methods like the LWR or the GWR. The introduction of heteroskedasticity into the error term, by groups or at the individual level, is another way of dealing with problems of heterogeneity (Anselin 1988b). All this discussion is well documented in the literature and we will not go into further details here.

Our interest lies in the assumption that parameter ρ is constant in (1). Habitually, this is a *maintained* hypothesis which can cause problems if it is imposed unduly. Brunsdom et al. (1996, p. 1962) find "*that the level of homeownerships exhibits a different level of clustering (or autocorrelation) in different areas*" (in the housing market of Tyne and Wear in northeast England), while Parent and Riou (2005, p. 767) conclude that the process of the spatial diffusion of knowledge in Europe presents "*an increasing spatial dependence as we move from the core of Europe to the peripheral regions.*" Di Giacinto (2003), Pace and Lesage (2004), Ertur et al. (2007) and Mur et al. (2008) also find evidence of instability in the mechanisms of spatial dependence that they introduce into their respective applications.

Going back to the SLM of (1), a simple way of giving greater flexibility to the autoregressive part of the model is to add several spatial lags of the endogenous variable to the right-hand side of the equation. Each lag, as in Huang (1984), is associated with an interaction parameter and a specific weighting matrix:

$$y = \rho_1 \mathbf{W}_1 y + \rho_2 \mathbf{W}_2 y + \cdots + \rho_p \mathbf{W}_p y + x\beta + \varepsilon, \quad \varepsilon \sim N\left(0, \sigma^2 I\right) \tag{2}$$

The matrices $\mathbf{W}_j$ should be linearly independent for the model to be identified and, habitually, they are specified in decreasing order of proximity. Nevertheless, they can also be used to alter the intensity with which certain points are related to the others. The equation that corresponds to this proposal is:

$$y = \rho \mathbf{W} y + \gamma \mathbf{W}^{*} y + x\beta + \varepsilon, \quad \varepsilon \sim N\left(0, \sigma^{2} I\right) \tag{3}$$

where **W** is the common weighting matrix associated with an *overall* level of dependence, as measured by the parameter ρ. We use the matrix $\mathbf{W}^*$ to intervene in the regions with peculiarities, measured by the parameter γ. The matrix $\mathbf{W}^*$ must reproduce a certain part of matrix **W** for the resulting variable, $\mathbf{W}^*$y, to be able to be treated as a dummy variable of the multiplicative type with respect to **W**y. With small changes, the same idea appears in the works of Rietveld and Wintershoven (1998), Lacombe (2004), Ertur et al. (2006), and Mur et al. (2008).

The generalization of (3) leads us to model (4) where the capacity of each region to interact with its surroundings is a circumstance specific to each region:

$$\left.\begin{aligned} y &= \rho \mathbf{H} \mathbf{W} y + x\beta + \varepsilon, \quad \varepsilon \sim N\left(0, \sigma^{2} I\right) \\ \mathbf{H} &= \operatorname{diag}\left\{h\left(Z_{r}\alpha\right); r = 1, 2, \ldots, R\right\} \\ \alpha &= \left[\alpha_{0}\ \alpha^{*}\right]' \quad Z_{r} = \left[1\ z_{r}\right] \\ h\left(Z_{r}\alpha\right) &< \infty; h\left(\alpha_{0}\right) = \kappa < \infty \end{aligned}\right\} \tag{4}$$

with vectors α and Z_r of order (2×1). As before, the hypothesis is that there is a basic level of dependence for all the regions, associated with parameter ρ. If this parameter is zero, the conclusion is that there is no cross-sectional dependence in the sample and the discussion finishes at this point. Only if coefficient ρ is different from zero, is there any point in asking whether it is, furthermore, constant over space. In (4), we propose that the intensity of the dependencies evolves depending on a certain variable.[1] It is not necessary for the function h[-] to be known, although it must be finite, continuous and stable over space. In these conditions, the fundamental piece of information is the indicator of heterogeneity, variable z, whose variability generates instability in the measures of spatial dependence.

For example, in the paper of Parent and Riou (2005), the stock of R&D infrastructures is an important factor for explaining the different capacity of interaction of a region with its neighbors in the question of technological innovations. If z_r is a regional indicator of this type of infrastructures, a reasonable option may be to introduce a logistic function into h[-]:

$$h\left(z_{r}\alpha\right) = \frac{e^{\alpha_{0}+\alpha_{1} z_{r}}}{1 + e^{\alpha_{0}+\alpha_{1} z_{r}}}$$

so that its capacity for interaction improves if $\alpha_1 > 0$ and worsens if $\alpha_1 < 0$. Fisher and Stirböck (2006), examining the hypothesis of convergence between the European regions, find evidence of a club structure between the center of the continent and the periphery. If these differences also affected the mechanisms of spatial

[1] For simplicity, we suppose that this capacity of interaction only depends on one factor, although the discussion can be generalized to the case of p variables.

dependence, it would be sufficient to specify the function h[-] as:

$$h\left(z_r\alpha\right) = \alpha_0\left(1 + \alpha_1 d_r\right)$$

where d_r is a binary variable that takes the value 1 if the region belongs to the center and zero otherwise.

The estimation of model (4), assuming normality, can be resolved by Maximum Likelihood (ML) methods. We can write the log-likelihood function (Anselin 1988a):

$$l\left(y;\theta\right) = -\frac{R}{2}\ln\left(2\pi\right) - \frac{R}{2}\ln\sigma^2 + \ln\left|\mathbf{A}\right| - \frac{\left(\mathbf{A}y - x\beta\right)'\left(\mathbf{A}y - x\beta\right)}{2\sigma^2} \qquad (5)$$

where $\mathbf{A} = \mathbf{I} - \rho\mathbf{HW}$ and there are $k+p+2$ parameters: $\theta = \left[\rho, \beta, \alpha, \sigma^2\right]'$, p being the number of parameters included in α. Assuming that the break responds to only one variable, the number of parameters is $k + 3$ and the score, highly nonlinear, is:

$$\left.\begin{aligned}
\frac{\partial l}{\partial\beta} &= \frac{1}{\sigma^2}\left(\mathbf{A}y - x\beta\right)' x \\
\frac{\partial l}{\partial\rho} &= -\mathrm{tr}\mathbf{A}^{-1}\mathbf{HW} + \frac{\left(\mathbf{A}y - x\beta\right)'\mathbf{HW}y}{\sigma^2} \\
\frac{\partial l}{\partial\alpha} &= -\rho\mathrm{tr}\mathbf{A}^{-1}\mathbf{H}_1\mathbf{ZW} + \rho\frac{\left(\mathbf{A}y - x\beta\right)'\mathbf{H}_1\mathbf{ZW}y}{2\sigma^2} \\
\frac{\partial l}{\partial\sigma^2} &= -\frac{N}{2\sigma^2} + \frac{\left(\mathbf{A}y - x\beta\right)'\left(\mathbf{A}y - x\beta\right)}{2\sigma^4}
\end{aligned}\right\} \qquad (6)$$

Matrices $\mathbf{H}_1$ and $\mathbf{Z}$ come from the differential of $\mathbf{H}$, $\frac{\partial\mathbf{H}}{\partial\alpha} = \mathbf{H}_1{}^{\mathbf{Z}}$, where:

$$\mathbf{H}_1 = \begin{bmatrix} \frac{\partial h\left(z_1'\alpha\right)}{\partial\left(z_1'\alpha\right)} & 0 & \cdots & 0 \\ 0 & \frac{\partial h\left(z_2'\alpha\right)}{\partial\left(z_2'\alpha\right)} & \cdots & 0 \\ \cdots & \cdots & \cdots & \cdots \\ 0 & 0 & \cdots & \frac{\partial h\left(z_R'\alpha\right)}{\partial\left(z_R'\alpha\right)} \end{bmatrix}; \mathbf{Z} = \begin{bmatrix} z_1 & 0 & 0 & \cdots & 0 \\ 0 & z_2 & 0 & \cdots & 0 \\ 0 & 0 & z_3 & \cdots & 0 \\ \cdots & \cdots & \cdots & \cdots & \cdots \\ 0 & 0 & 0 & \cdots & z_R \end{bmatrix} \qquad (7)$$

If we have all the information about the break (characteristics of the function h[-], of the indicator z, etc.), the system of (6) can be resolved using, for example, numerical methods. Otherwise, the previous exercise is a merely theoretical artifice.

An interesting aspect of this approach is that it allows us to resolve tests of stability in the mechanism of spatial dependence. In accordance with the specification of (4), our interest focuses on:

$$\left.\begin{aligned} H_0 &: \alpha^* = 0 \\ H_A &: \alpha^* \neq 0 \end{aligned}\right\} \qquad (8)$$

The null hypothesis implies that there is no break in the coefficient of spatial dependence, in relation to the variable used, while the alternative points towards a certain mechanism of break in which the indicator z intervenes. The test is conditioned upon the indicator used in clear analogy with respect to the heteroskedasticity test of Breusch and Pagan (1979). The Lagrange Multiplier is relatively simple to obtain because we only need the estimation of the model under the null; that is, of the SLM model of (1). The expression of the statistic is the following:

$$\Rightarrow \mathrm{LM}_{\mathrm{Break}}^{\mathrm{SLM}} = \left[\frac{\left(\mathrm{tr}\mathbf{A}^{-1}\mathbf{ZW} - \frac{\varepsilon'\mathbf{ZW}\mathrm{y}}{2\sigma^2} \right)^2}{\mathrm{I}_{11}^{*} - \mathrm{I}_{12}^{*}\mathrm{V}(\varphi)\,\mathrm{I}_{21}^{*}} \right] \tag{9}$$

In this expression, ϵ is the vector of ML residuals while the terms of the denominator come from the information matrix of the model of (4), under the null of (8). Specifically:

$$\begin{aligned}
&\mathrm{I}_{11}^{*} = \mathrm{I}_{\alpha\alpha}; \quad \mathrm{I}_{12}^{*} = \left[\mathrm{I}_{\alpha\rho}\ \mathrm{I}_{\alpha\sigma^2}\ \mathrm{I}_{\alpha\beta} \right]; \quad \mathrm{I}_{21}^{*} = \mathrm{I}_{12}^{*\,\prime} \\
&\rightarrow \mathrm{I}_{\alpha\alpha} = \left[\mathrm{tr}\mathbf{A}'^{-1} \left(\mathbf{ZW'}_{\mathbf{A}}{}^{-1}\mathbf{ZW} + \mathbf{W'}_{\mathbf{Z}}{}^{2}\mathbf{W}_{\mathbf{A}}{}^{-1} \right) + \frac{\beta'\mathrm{x'}_{\mathbf{A}}{}'^{-1}\mathbf{W}_{\mathbf{Z}}'{}^{2}\mathbf{W}_{\mathbf{A}}{}^{-1}\mathrm{x}\beta}{\sigma^2} \right] \\
&\rightarrow \mathrm{I}_{\alpha\rho} = \left[2\mathrm{tr}\mathbf{A}^{-1}\mathbf{ZWA}^{-1}\mathbf{W} + \frac{\beta'\mathrm{x'}_{\mathbf{A}}{}'^{-1}\mathbf{W'ZW}_{\mathbf{A}}{}^{-1}\mathrm{x}\beta}{\sigma^2} \right] \\
&\rightarrow \mathrm{I}_{\alpha\sigma^2} = \frac{\mathrm{tr}\mathbf{ZW}_{\mathbf{A}}{}^{-1}}{\sigma^2} \\
&\rightarrow \mathrm{I}_{\beta\alpha} = \frac{\beta'\mathrm{x'}_{\mathbf{A}}{}'^{-1}\mathbf{W'Z}\mathrm{x}}{\sigma^2} \\
&\rightarrow \mathrm{V}(\varphi) = \begin{bmatrix} \mathrm{I}_{\rho\rho} & \mathrm{I}_{\rho\sigma^2} & \mathrm{I}_{\rho\beta} \\ \mathrm{i}_{\rho\sigma^2} & \mathrm{I}_{\sigma^2\sigma^2} & 0 \\ \mathrm{I}_{\rho\beta} & 0 & \mathrm{I}_{\beta\beta} \end{bmatrix}^{-1}
\end{aligned} \tag{10}$$

φ is the vector of parameters of the SLM, under the assumption of homogeneity, $\varphi = \left[\rho, \beta, \sigma^2\right]'$, and $\mathrm{V}(\varphi)$ its covariance matrix.

In the discussion above we have adopted a parametric perspective to the problem of the lack of instability, which has allowed us to resolve an exercise of ML inference with full information. If we introduce uncertainty into this scenario, either because the function that relates the endogenous variable to the regressors of (1) is unknown or because we lack information about the characteristics of the break that affects the mechanisms of spatial dependence, the situation becomes more favorable to the application of nonparametric methods. In the following sections, we examine more deeply the advantages and disadvantages of local algorithms, like the SALE, through a Monte Carlo exercise.

3 Main Characteristics of the Monte Carlo Experiment

Multiple factors, from the purely economic to the statistical or geographical, intervene in a structural break. In this contribution, we have decided to focus on the basic aspects of the problem by designing a simple Monte Carlo experiment. To begin with, the model simulated only includes one exogenous variable, along with the constant and the spatial lag of the endogenous variable. In matrix notation:

$$\left.\begin{array}{l} y = \rho \mathbf{W} y + X\beta + \varepsilon \\ \varepsilon \sim N\left(0, \sigma^2 I\right) \end{array}\right\} \tag{11}$$

where $X = [1, x]$ and $\beta = [\beta_0, \beta_1]$. The data of variable x come from a N(0,1) distribution, as do those of ϵ (that is, $\sigma^2 = 1$ in 11); obviously, both distributions are independent. In all cases, β_0 has been set to 2 and β_1 to 3.[2] We have used regular lattice systems of orders (7×7) and (20×20). This means that the sample sizes are 49 and 400 observations, respectively. The weighting matrix has been specified accordingly, first of a binary type using a contiguity criterion and queen-type movements. Afterwards, the resulting matrix has been row-standardized in the usual way.

We have introduced a break of the discrete type, with only two values in the coefficient of spatial autocorrelation, so that each parameter acts in one part of the space:

$$\left.\begin{array}{l} y = H\mathbf{W} y + X\beta + \varepsilon \\ \varepsilon \sim N\left(0, \sigma^2 I\right) \end{array}\right\}$$

$$H = \text{diag}\{h_r; r = 1, 2, \ldots, R\} \quad \text{where } h_r = \begin{cases} \rho_a & \text{if } r \in \text{Periphery (resp. West)} \\ \rho_b & \text{if } r \in \text{Center (resp. East)} \end{cases} \tag{12}$$

We defined two spatial regimes which correspond to an East–West and a Center–Periphery structure, as shown in Fig. 1. The color, white or gray, identifies the cells included in each regime.

We have used several combinations of values for the parameters of spatial autocorrelation, as shown in Table 1. Case 0 is the control case and includes a medium level of spatial interaction, 0.5, which is homogeneous for the whole lattice. The other cases cover different situations. In Case 1, there are very important discrepancies in the values of the spatial interaction coefficients whereas they are small in the other two cases. In Case 3, there is a high level of spatial dependence and Case 2 is closer to the control case, with a medium level of spatial dependence. Finally, each configuration has been repeated 1,000 times.

[2] If ρ were zero in (11), the R^2 coefficient of the corresponding OLS regression should oscillate around 0.9.

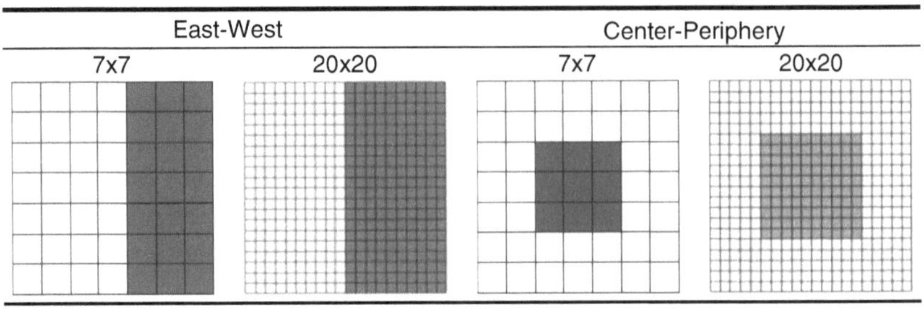

Fig. 1 Spatial regimes used in the experiment

Table 1 Coefficients used in the simulation

	ρ_a	ρ_b
Case 0	0.5	0.5
Case 1	0.1	0.9
Case 2	0.4	0.6
Case 3	0.8	0.9

4 Diagnostic Measures and Maximum Likelihood Estimation When There Is Instability in the Mechanisms of Spatial Interaction

We think it interesting, firstly, to examine what happens to the usual diagnostic measures, obtained from the Ordinary Least Squares (OLS) estimation, when we introduce a structural break into the parameter of spatial autocorrelation. That is, having estimated the static version of the model of (11), $y = X\beta + u$, we check the results for symptoms of misspecification. Specifically, we will test for the hypothesis of normality, using the test of Jarque-Bera (JB), homoskedasticity, by means of the Breusch-Pagan (BP) and White (WH) tests, and linearity, through the RESET test up to order 3 (Greene 2003). We also include the usual tests of spatial dependence, namely, Moran's I, the LMERR, the LMEL, the LMLAG, the LMLE and the SARMA tests (Florax and de Graaff 2004). The results corresponding to the mean and standard deviation for each of these tests, together with the percentage of rejections of the respective null hypothesis at the 5% level of significance, appear in Tables 2 and 3.

There are several results worth highlighting. For example, the strong signs of non-normality that we find in the OLS residuals, in spite of the normality of the random term of the DGP. The distortion suffered by the JB test is the consequence of a certain bimodality of the data induced by the existence of two spatial regimes in the DGP. As is well known, the null hypothesis of the JB test is that the skewness of the OLS residuals is zero and the kurtosis is 3. The break in the mechanism of spatial dependence gives rise to OLS residuals which are clearly non-symmetrical with respect to the origin and that tend to be leptokurtic. The impact of the bimodality gets stronger as the difference between the coefficients that operate in the two zones

Table 2 Diagnostic statistics in the static model. No spatial effects. Lattice: 7 × 7[a]

		I	LM LAG	LM ERR	LMEL	LMLE	SAR MA	JB	BP	WH	RE SET
Case 0	Mean	2.71	6.88	25.22	0.86	19.21	26.08	1.57	0.98	2.03	1.02
	Std	1.03	4.95	7.66	1.13	6.06	7.46	2.08	1.56	2.02	1.07
	% Rej.	76.6	99.9	67.8	3.1	99.4	99.8	2.9	4.5	4.3	5.2
East–West regime											
Case 1	Mean	7.86	54.94	55.05	4.00	4.11	59.05	6.64	0.97	2.60	1.16
	Std	0.52	6.98	10.12	4.80	4.70	6.68	3.26	1.51	2.42	1.15
	% Rej.	100.0	100.0	100.0	36.1	37.3	100.0	46.3	5.6	9.2	9.0
Case 2	Mean	3.81	13.79	27.61	1.20	15.01	28.81	2.23	1.02	2.03	1.10
	Std	1.32	8.71	10.07	1.74	6.53	10.09	3.57	1.50	2.04	1.21
	% Rej.	90.3	99.8	87.1	7.3	96.1	99.5	6.0	5.8	5.3	7.8
Case 3	Mean	6.78	41.44	55.04	0.75	14.35	55.79	3.16	0.99	2.27	1.19
	Std	1.07	12.09	10.32	1.13	7.27	10.14	2.80	1.39	2.11	1.33
	% Rej.	100.0	100.0	100.0	3.1	94.2	100.0	9.5	4.3	5.0	8.6
Center–Periphery regime											
Case 1	Mean	4.37	17.60	15.55	4.48	2.43	20.03	18.12	1.66	2.05	1.07
	Std	1.22	7.62	8.66	3.62	2.83	8.17	12.85	2.28	2.27	1.30
	% Rej.	93.5	89.8	92.2	50.4	22.1	92.5	84.5	12.4	5.7	6.4
Case 2	Mean	2.65	6.72	20.26	0.86	14.40	21.13	1.89	0.96	1.95	1.09
	Std	1.08	4.97	7.39	1.22	6.08	7.28	2.81	1.37	1.98	1.18
	% Rej.	73.6	99.7	65.4	4.3	97.2	98.9	5.9	4.8	4.0	7.6
Case 3	Mean	5.34	25.28	46.89	0.54	22.16	47.44	1.74	1.02	2.14	1.07
	Std	0.85	7.85	8.41	0.69	6.00	8.36	1.74	1.47	1.98	1.15
	% Rej.	99.9	100.0	99.8	0.3	99.7	100.0	2.7	5.8	4.7	5.9

[a]% Rej.: Percentage of rejection at the 5% level of significance

of the lattice increases. The heteroskedasticity tests also react to the break in parameter ρ although only when the sample size is large (lattice of 20 × 20) and the difference between the two coefficients of autocorrelation is high. Moreover, the East–West regime has a greater tendency to show problems of heteroskedasticity than the Center–Periphery regime. Lastly, the RESET test does not detect significant problems.

The results tend to confirm the lack of cross-sectional independence in the data. Moran's I, as well as the raw Lagrange Multipliers (LMERR and LMLAG), almost always reject their null hypotheses. The average value of these statistics increases with the size of the lattice, much more rapidly than their standard deviation. For the largest sample, Table 3, the raw Multipliers take very high values. Another aspect to underline is the behavior of the robust Multipliers (LMEL and LMLE). The model that we are simulating is of the SLM type, so we should accept the null hypothesis of the LMEL test and reject that of the LMLE. This only occurs in the control case or when the difference between the coefficients of autocorrelation is low (cases 2 and 3) and, simultaneously, the sample size is small (7 × 7 lattice). The break tends to confuse the two tests so that the LMEL rejects the null hypothesis more than it

Table 3 Diagnostics statistics in the static model. No spatial effects. Lattice: 20×20[a]

		I	LM LAG	LM ERR	LMEL	LMLE	SAR MA	JB	BP	WH	RESET
Case 0	Mean	8.47	71.19	222.4	0.89	152.1	223.2	2.24	1.20	2.22	0.99
	Std	1.12	18.55	25.86	1.19	17.82	26.07	2.49	1.71	2.47	1.06
	% Rej.	100.0	100.0	100.0	3.6	100.0	100.0	7.2	7.4	6.4	5.4
East–West regime											
Case 1	Mean	25.99	665.6	665.3	16.45	16.13	681.7	42.14	4.41	12.18	0.98
	Std	0.28	14.25	21.04	10.57	12.70	12.72	5.30	3.16	5.79	0.93
	% Rej.	100.0	100.0	100.0	92.5	83.5	100.0	100.0	47.1	86.6	3.9
Case 2	Mean	13.03	168.4	268.5	15.64	115.7	284.1	9.24	2.00	2.90	1.04
	Std	1.39	35.48	37.76	7.57	19.89	40.23	8.86	2.85	3.06	1.02
	% Rej.	100.0	100.0	100.0	97.5	100.0	100.0	56.1	16.%	13.6	5.7
Case 3	Mean	23.42	540.9	612.6	6.72	78.46	619.3	22.56	2.00	3.43	1.03
	Std	0.84	38.11	37.84	4.92	19.20	34.50	21.78	2.83	3.33	1.00
	% Rej.	100.0	100.0	100.0	66.0	100.0	100.0	97.0	16.1	18.7	5.6
Center–Periphery regime											
Case 1	Mean	24.35	584.5	540.3	52.79	8.54	593.0	214.5	5.34	3.96	1.06
	Std	0.50	23.70	48.12	24.70	8.55	26.62	70.16	5.77	3.41	1.20
	% Rej.	100.0	100.0	100.0	99.9	61.4	100.0	100.0	48.3	22.2	6.2
Case 2	Mean	10.74	115.3	211.8	10.20	106.6	222.0	12.72	1.68	2.32	0.98
	Std	1.62	34.24	34.87	6.98	18.53	38.44	14.42	2.50	2.38	1.01
	% Rej.	100.0	100.0	100.0	79.7	100.0	100.0	61.6	13.5	7.3	4.9
Case 3	Mean	21.06	438.2	542.4	5.89	110.1	548.2	33.08	1.85	2.55	1.06
	Std	1.25	50.83	42.27	3.72	21.20	40.89	32.35	2.73	2.44	1.06
	% Rej.	100.0	100.0	100.0	66.5	100.0	100.0	84.7	14.4	8.9	5.5

[a]% Rej.: Percentage of rejection at the 5% level of significance

should while the LMLE seems to be downwardly biased. The sample size does not correct these anomalies; on the contrary, it accentuates them.

If there is a break in ρ, the malfunctioning of two robust Multipliers complicates the selection of the right model. It could be shown that the suggestion of Florax et al. (2003) in relation to the model that should be specified in situations where both robust (and raw) Multipliers are significant: "*If both tests are significant, estimate the specification pointed to by the more significant of the two tests. For example, if LMLAG > LMERR then estimate the spatial lag model (...). If LMLAG < LMERR then estimate the spatial error model*" (p. 561), results in a clear bias towards the spatial error model, SEM. This situation contrasts very sharply to what happens in the control case, where we tend to select the spatial lag model, SLM, on most occasions.

In Table 4, we present some details of the performance of the LM^{SLM}_{Break} test introduced in Sect. 2. Specifically, the table shows the percentage of rejections of the null hypothesis of stability in the parameter ρ. We also include the results of the Chow test of parameter stability and of the Rao Score test of no cross-sectional correlation in the residuals of an SLM (Anselin and Bera 1998), applied in both cases to the

Table 4 Testing the SLM, under the hypothesis of stability[a]

		7 × 7			20 × 20		
		$RS_{\lambda\vert\rho}$	LM^{SLM}_{Break}	CHOW	$RS_{\lambda\vert\rho}$	LM^{SLM}_{Break}	CHOW
Case 0	Mean	1.01	1.03	2.27	0.95	1.06	2.11
	Std	1.41	1.46	2.62	1.31	1.52	2.14
	% Rej.	5.0	4.7	7.3	4.5	6.7	6.5
East–West regime							
Case 1	Mean	4.13	15.66	40.75	4.46	33.20	108.09
	Std	3.89	7.20	22.92	5.33	6.28	27.22
	% Rej.	42.3%	100.0%	99.6%	40.5%	100.0%	100.0%
Case 2	Mean	1.61	8.02	9.38	8.84	56.84	51.61
	Std	2.19	5.11	7.65	7.40	12.92	16.36
	% Rej.	12.4%	77.0%	60.1%	70.2%	100.0%	100.0%
Case 3	Mean	3.05	12.23	16.30	11.29	51.91	68.93
	Std	3.64	6.96	11.40	9.97	20.89	26.80
	% Rej.	29.1	93.1	83.9	74.0	100.0	100.0
Center–Periphery regime							
Case 1	Mean	6.72	54.33	79.36	14.99	220.81	219.36
	Std	4.42	17.74	42.91	9.87	74.05	58.24
	% Rej.	70.6	99.7	98.0	91.0	100.0	100.0
Case 2	Mean	1.25	7.46	8.07	6.82	70.48	48.51
	Std	1.80	6.44	7.09	5.37	18.79	15.69
	% Rej.	7.3	63.5	52.1	64.8	100.0	100.0
Case 3	Mean	1.80	11.58	14.58	15.36	110.64	88.37
	Std	2.18	6.77	9.94	9.46	29.89	23.63
	% Rej.	14.3	85.1	80.5	90.6	100.0	100.0

[a]% Rej.: Percentage of rejection at the 5% level of significance

model of (11). The null hypothesis of the first test (Anselin 1990) is that the vector of parameters β is constant, assuming stability in the parameter ρ. Under the null hypothesis, the Chow test is distributed, asymptotically, as a χ^2(k), the order of β being k. The Rao Score test, $RS_{\lambda|\rho}$, is a $\chi^2(1)$ under the null.

The results of Table 4 indicate that the SLM of (11) will not, probably, pass the check. That is, there is a fairly high probability of finding symptoms of cross-sectional correlation in the residuals of the model and, almost certainly, traces of a structural break will be observed. It should not be forgotten that, although the error term of the DGP is a white noise, (11) is misspecified due to the lack of stability in the coefficient ρ. This misspecification will produce ML residuals with a strong structure of cross-sectional dependence, which will increase differences in the autoregressive coefficients. Furthermore, the LM^{SLM}_{Break} and CHOW tests coincide in their rejection of their respective null hypotheses of stability, the first correctly but the second spuriously. If we reverse the starting point, introducing a break into the vector β of the DGP while the parameter ρ remains stable, the situation does not change substantially. The three statistics tend to reject their respective null hypothesis with a higher probability as the sample size increases or as the break becomes more serious.

The solution is to re-specify the model of (11) in order to identify the origin of the instability in the ML estimations. To do so, we first introduce a break into the vector β and replicate the tests of misspecification; then, we do the same but introducing the break into the coefficient ρ. If the cause of the instability is in the vector β, the first enlarged model, but not the second, should pass the check, and vice versa.[3]

5 Local Estimation in the Cases of Stability and Instability in the DGP

Briefly stated, the local estimation technique consists of fitting individual regressions to selected points in the sample, with more weight assigned to observations that are closer to the point of interest (McMillen 1996). Repeating this exercise for every point in the sample, we can construct estimation surfaces in order to discuss the nonstationarity of each parameter in the model. The concept of "*closeness*" is flexible and must be adapted to the objectives of the study. Moreover, the distribution of the weights among the neighboring observations with respect to point r is determined by the kernel function (Cressie 1991). In the case of the GWR, this is a decreasing function of the distance between the points, and the bandwidth determines how rapidly the weights decline with distance. We decided to use a rectangular or uniform kernel with a fixed bandwidth of m for every point. This means that the m nearest neighbors will receive a weight of one, and the other points zero.

In our case we have to resolve the local estimation of an SLM for which it is not advisable to use the OLS algorithm. Following the example of Brunsdon et al. (1998) and of Pace and Lesage (2004), we will obtain the local estimators from the ML estimation of the local model:

$$\mathrm{y}_{\mathrm{r}}^{(\mathrm{m})} = \rho_{\mathrm{r}}^{(\mathrm{m})}\mathbf{W}_{\mathbf{r}}^{(\mathbf{m})}\mathrm{y}_{\mathrm{r}}^{(\mathrm{m})} + \mathrm{x}_{\mathrm{r}}^{(\mathrm{m})}\beta_{\mathrm{r}}^{(\mathrm{m})} + \mathrm{u}_{\mathrm{r}}^{(\mathrm{m})};\, \mathrm{u}_{\mathrm{r}}^{(\mathrm{m})} \sim \mathrm{N}\left(0, \sigma_{\mathrm{r,m}}^{2}\mathrm{I}_{\mathrm{m}}\right) \tag{13}$$

The indexes r and m mean that the data correspond to the local system defined by m elements around point r. Therefore, $\mathrm{y}_{\mathrm{r}}^{(\mathrm{m})} = (\mathrm{y}_{\mathrm{r}}, \mathrm{y}_{\mathrm{i}_1}, \mathrm{y}_{\mathrm{i}_2}, \ldots, \mathrm{y}_{\mathrm{i}_{\mathrm{m}-1}})$ where $\mathrm{i}_{\mathrm{k}} \in \mathrm{N}(\mathrm{r})$, being N(r) the bundle of indexes of the m-1 neighbours nearest to the point r. The same criterion is used to define $\mathrm{x}_{\mathrm{r}}^{(\mathrm{m})}$. Matrix $\mathbf{W}_{\mathrm{r}}^{(\mathrm{m})}$ refers to the weighting matrix obtained for this local system, defined with the same connectivity criteria that are used to obtain the global W matrix, specified following standard criteria. Finally $\rho_{\mathrm{r}}^{(\mathrm{m})}$, $\beta_{\mathrm{r}}^{(\mathrm{m})}$ and $\rho^{2}{}_{\mathrm{r,m}}$ are the local parameters of interest. This is what we call the *Zoom* estimation (different to the SALE algorithm of Pace and Lesage (2004), in that, in each local system, we use the matrix $\mathbf{W}_{\mathrm{r}}^{(\mathrm{m})}$ specific for the local network

[3] The tables of estimated power for the cases just described have not been included for reasons of length but are available from the authors upon request.

around point r). We refer to m as the *Zoom size* (equivalent to *window size* in nonparametric literature).

The aim of the present section is to study the behavior of the *Zoom* algorithm in different situations. Specifically, we want to know the reaction of this algorithm when the DGP that has intervened in the generation of the data is stable. In the second part, we use data that have been generated in a context of instability. Throughout this section, we pay particular attention to the estimation of the parameter ρ.

5.1 The Zoom Estimation When the DGP Is Stable

If the mechanism of spatial dependence is the same for all the space, the best option is to apply ML using all the sample information available. Asymptotically, the ML estimators are unbiased, consistent and efficient. The local estimation restricts the quantity of information used at each estimation point, depending on the kernel function. This means that it makes no sense to speak about consistency and efficiency: the local estimators will be biased and inconsistent.[4] Therefore, neither the Law of Large Numbers nor the Central Limit Theorem (Davidson 2000) are applicable. In spite of these comments, we think that it is interesting to examine what type of local estimations we obtain for the parameter ρ and its sensitivity to changes in the *Zoom* size.

This part of the Monte Carlo experiment maintains the design of the previous section but introduces the restriction of stability in the parameter ρ, for which we use three different values (0.1, 0.5 and 0.8). With respect to the value of m, we decided to use small *Zooms* with a maximum size of half the sampling space. This means that for the 7×7 lattice we have considered *Zooms* of sizes 9, 16 and 25 around each cell and for the 20×20 lattice 9, 16, 25 and 100.

The main results appear on the contour plots of Figs. 2 and 3. Each line on these plots links points at which the mean of the local estimation of the parameter is the same, after resolving 1,000 iterations (we have identical information about the other parameters of the model, which are omitted for reasons of space).

We have also obtained several indices with which to measure the bias of the *Zoom* estimation of the parameter ρ. The first is the Average Global Bias:

$$\mathrm{S}_{\rho} = \frac{1}{\mathrm{R}} \sum_{\mathrm{r}=1}^{\mathrm{R}} (\bar{\rho}_{\mathrm{r}}^{(\mathrm{m})} - \rho) \quad \text{with} \quad \bar{\rho}_{\mathrm{r}}^{(\mathrm{m})} = \frac{1}{1000} \sum_{\mathrm{k}=1}^{1000} \rho_{\mathrm{r}}^{(\mathrm{m})} \quad \mathrm{r} = 1, \ldots, \mathrm{R} \tag{14}$$

where $\rho_{\mathrm{r}}^{(\mathrm{m})}$ is the local estimation at point r considering a *Zoom* of size m. The second is the usual Mean Squared Error:

[4] If the function were linear and the model were well specified, the local estimators would be unbiased, as is the case with the LWR or GWR estimation.

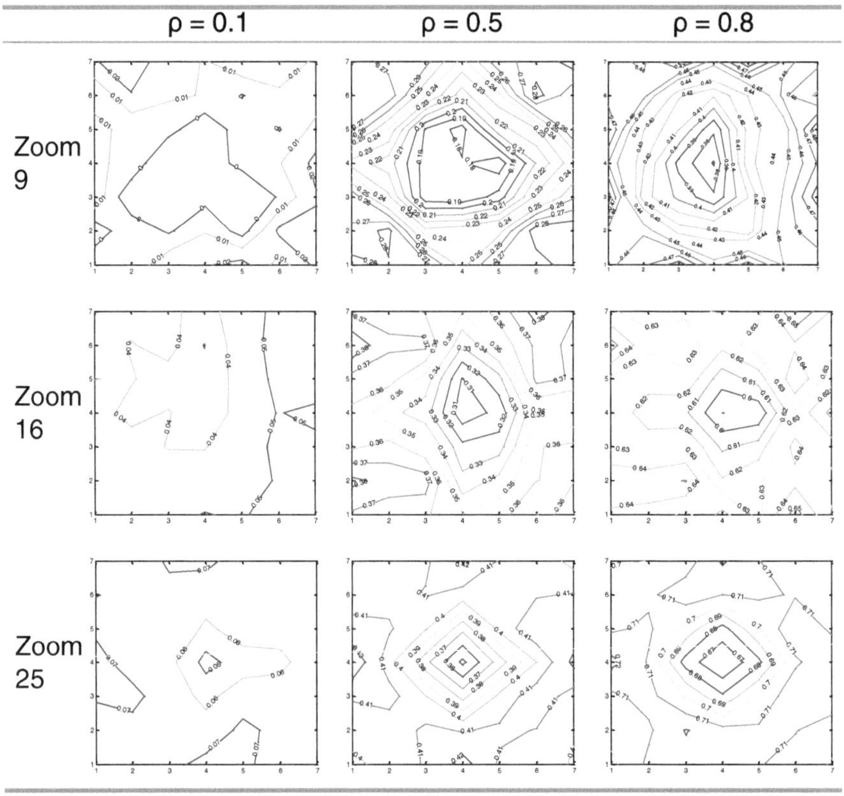

Fig. 2 Spatial distribution of ρ_r. Lattice $7 \times 7^{(*)}$

$$\mathrm{MSE}_\rho = \frac{1}{\mathrm{R}} \sum_{r=1}^{\mathrm{R}} \left(\bar{\rho}_r^{(m)} - \rho \right)^2 \tag{15}$$

Table 5 presents the results of these indices, together with some other statistics like the mean, maximum and minimum values for each series of values of $\rho_r^{(m)}$.

As we said earlier, the ML local estimators of an SLM are biased because of the nonlinearity of the algorithm,[5] which leads to a tendency to underestimate the parameter ρ. The index S_ρ is always negative, indicating that the average estimation of $\rho_r^{(m)}$ is systematically smaller than the true value of ρ. Moreover, in all the cases, the maximum value obtained from the local algorithm is smaller than the true value of ρ. The size of the bias, measured by the ECM, grows when the ratio m/R becomes smaller and diminishes when the size of the *Zoom* increases.

[5] An additional source of bias is that we are not using the original, global, **W** weighting matrix but specify the corresponding local weighting matrix, $\mathbf{W}_r^{(m)}$ for each local system of estimation.

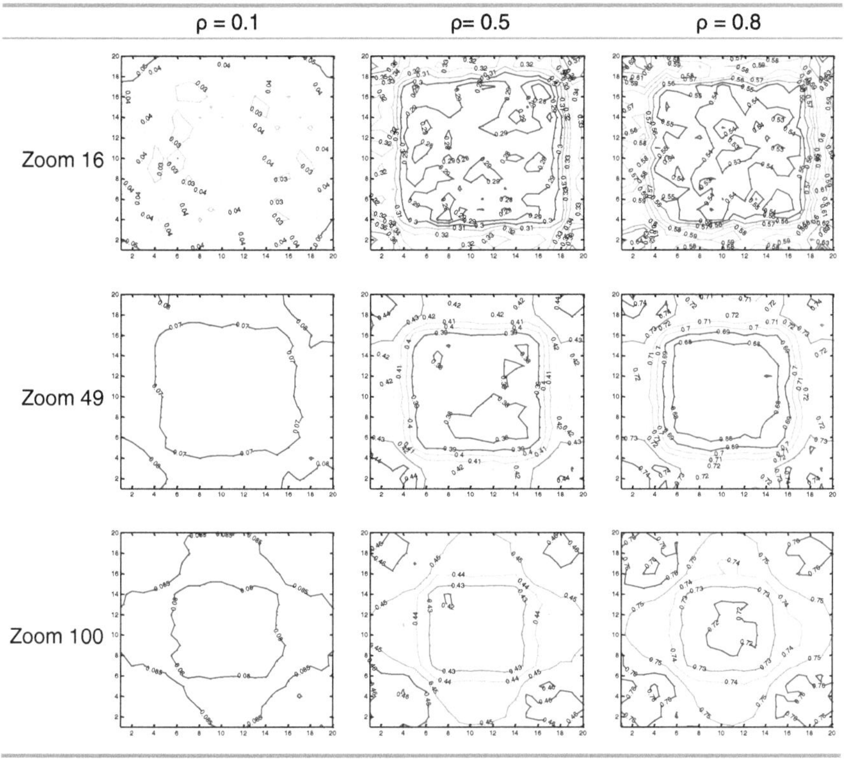

Fig. 3 Spatial distribution of ρ_r. Lattice 20 × 20

Table 5 Zoom estimation under the null hypothesis. Some descriptive statistics

		7 × 7			20 × 20		
		0.1	0.5	0.8	0.1	0.5	0.8
Zoom 16	$\bar{\rho}_r^{(m)}$	0.045	0.355	0.629	0.038	0.308	0.568
	S_ρ	−0.055	−0.145	−0.171	−0.062	−0.192	−0.232
	ECM_ρ	0.003	0.021	0.029	0.004	0.037	0.055
Zoom 25	$\bar{\rho}_r^{(m)}$	0.066	0.404	0.706	0.055	0.353	0.634
	S_ρ	−0.034	−0.096	−0.094	−0.045	−0.147	−0.166
Zoom 49	ECM_ρ	0.001	0.009	0.009	0.002	0.022	0.028
	$\bar{\rho}_r^{(m)}$	–	–	–	0.072	0.410	0.709
	S_ρ	–	–	–	−0.028	−0.090	−0.091
	ECM_ρ	–	–	–	0.001	0.009	0.009

Another interesting aspect is the stability observed in the spatial distribution of the estimations of $\rho_r^{(m)}$. The shape of the contour plots tends to repeat itself in the different cases, independently of the lattice used. As we increase the size of the *Zoom*, the dispersion of the local estimations is significantly reduced. In spite of this stability, it is easy to recognize a kind of *doughnut effect* in all the figures: the contour plots tend to drop at the center of the lattice. The lowest zones of the estimations, in all cases, are in the central part of the lattice while, as we move out from the center, the average of the local estimations increases slightly. This *doughnut effect* shows irregularities when the *Zoom* is small but they tend to disappear for large *Zooms*.

5.2 The Zoom Estimation When the DGP Is Not Stable

Now we focus on the behavior of the *Zoom* algorithm when the DGP is not stable due to instability in coefficient ρ. We expect that the instability in this parameter will have some impact on the local estimation algorithm, helping us to identify the characteristics of the structural break.

In this case, we have extended the Monte Carlo exercise using series of data generated with the DGP of (11) but introducing the structural break described in Fig. 1 and Table 1. The break contains only two regimes of parameters. Table 6 presents a summary of the main results obtained from this part of the simulation, including the minimum, the maximum and the mean value of the estimates of $\rho_r^{(m)}$ after 1,000 draws. We present the results for only two *Zoom* sizes, 16 and 25 nearest neighbors.

The range of variation of the local estimates depends, mainly, on the values of the real parameters used in the DGP. In any case, this dispersion is higher than that observed in the case of stability in Table 5. Moreover, the tendency of the ML algorithm to underestimate the parameter ρ disappears in this case and we can find estimates below and above the corresponding local true parameters.

Apparently, the local estimations are more precise when the regimes are not mixed, as occurs in the East–West structure. The use of large *Zooms* tends to smooth out the discrepancies between the local estimations, which has positive

Table 6 Zoom estimation when the DGP is unstable in ρ^a. Descriptive statistics

		East–West regime			Center–Periphery regime		
		Case 1	Case 2	Case 3	Case 1	Case 2	Case 3
7 × 7							
Zoom 16	Min	0.026	0.276	0.676	0.265	0.175	0.334
	Max	0.966	0.927	0.991	0.880	0.845	0.932
	Mean	0.607	0.629	0.883	0.711	0.685	0.824
Zoom 25	Min	0.086	0.375	0.863	0.626	0.528	0.687
	Max	0.984	0.949	0.998	0.910	0.871	0.952
	Mean	0.834	0.831	0.966	0.864	0.820	0.917
20 × 20							
Zoom 16	Min	0.003	0.206	0.542	−0.151	0.025	0.328
	Max	0.974	0.927	0.994	0.745	0.587	0.741
	Mean	0.485	0.398	0.744	0.218	0.247	0.547
Zoom 25	Min	0.052	0.159	0.596	0.005	0.264	0.690
	Max	0.963	0.984	0.998	0.988	0.947	0.998
	Mean	0.501	0.401	0.788	0.622	0.523	0.869

[a] The values of the parameters associated to each case are as follows: Case 1 ($\rho_a = 0.1$; $\rho_b = 0.9$), Case 2 ($\rho_a = 0.4$; $\rho_b = 0.6$), Case 3 ($\rho_a = 0.8$; $\rho_b = 0.9$)

consequences (the contour plots are more structured) but also some negative ones (the transition from one regime to the other is more diffuse). As was expected, the size of the lattice has only a minor impact.

The spatial distributions of the local estimations for the different cases simulated appear in the contour plots of Figs. 4 and 5. These figures confirm that the *Zoom* algorithm produces useful information to discuss whether there is a problem of instability in our model.

As we have already stated, the local estimation seems to work better in structures of an East–West type (when the regimes tend to be separate), and the size of the *Zoom* appears to be more important than the size of the sample. In fact, this technique is more efficient when the size of the *Zoom* is small. In this case, the discrepancies between the local estimations are greater and their spatial distribution adjusts better to the spatial distribution of the regimes in the parameters. As we increase the size of the *Zoom*, the general appearance of the contour plots is not so sharp.

We also find a *doughnut effect* in this case. If there is Center–Periphery type regime in which the parameter that intervenes in the central region of the lattice is higher than that which acts on the Periphery, we find an unexpected fall in the local estimations corresponding to the central zone. If the break is of the East–West type, with a higher value in the East, the local estimations corresponding to this part of the lattice tend to fall as we move to the right. In Fig. 6, we show two contour plots obtained from the (20×20) lattice with a *Zoom* size of 16, which are representative of the two cases. The anomaly is less evident when the spatial break is of the East–West type because the transition from one regime to the other is quick and the fall produced as we move towards the right is less steep.

6 A Proposal to Identify Spatial Regimes in the Parameter of Spatial Interaction

If, as we suggested in the previous section, local estimation may be a useful tool for detecting situations of instability in the parameter ρ, the problem now is how to identify the regions that belong to each regime of parameters. This situation is nothing new in the literature and we can find interesting proposals there (see, e.g., Tsionas 2000; Bloom et al. 2003; LeGallo et al. 2003; Ertur et al. 2006; Fisher and Stirböck, 2006; LeGallo and Dall'erba 2006; Ramajo et al. 2008; or Battisti and Di Vaio 2007). In general, these papers make use of some indicator of local spatial dependence like the G_i statistic of Getis–Ord (Getis and Ord 1992) or employ techniques based on the mixing of distribution functions (for mixture densities, see Titterington et al. 1985; McLachlan and Peel 2000).

The strategy that we propose is based on the use of two descriptive techniques. First we use the *Zoom* algorithm to obtain the local estimation of the parameter ρ and, with the results, we carry out a cluster analysis. The idea is to use this procedure to detect zones in which a mechanism of spatial dependence, different to the rest of the space, seems to be acting. The final decision as to whether there are

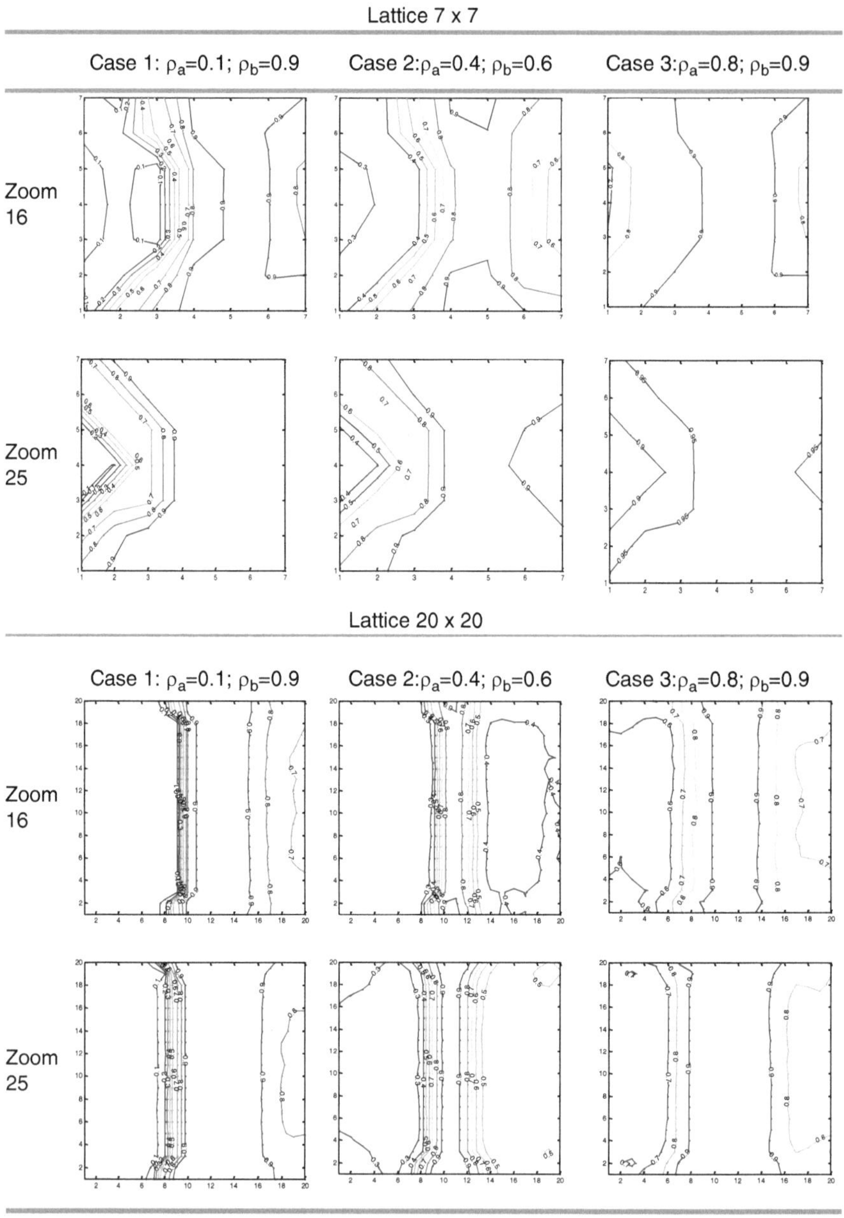

Fig. 4 Spatial distribution of ρ_r under the break. East–West structure

significant differences or not in these mechanisms must be taken, obviously, using formal statistical criteria such as the test LM_{Break}^{SLM}, or some adjusted version of it.

In this section, we are going to compare four strategies for obtaining clusters. The first two allow us to form as many groups as necessary while the other two only

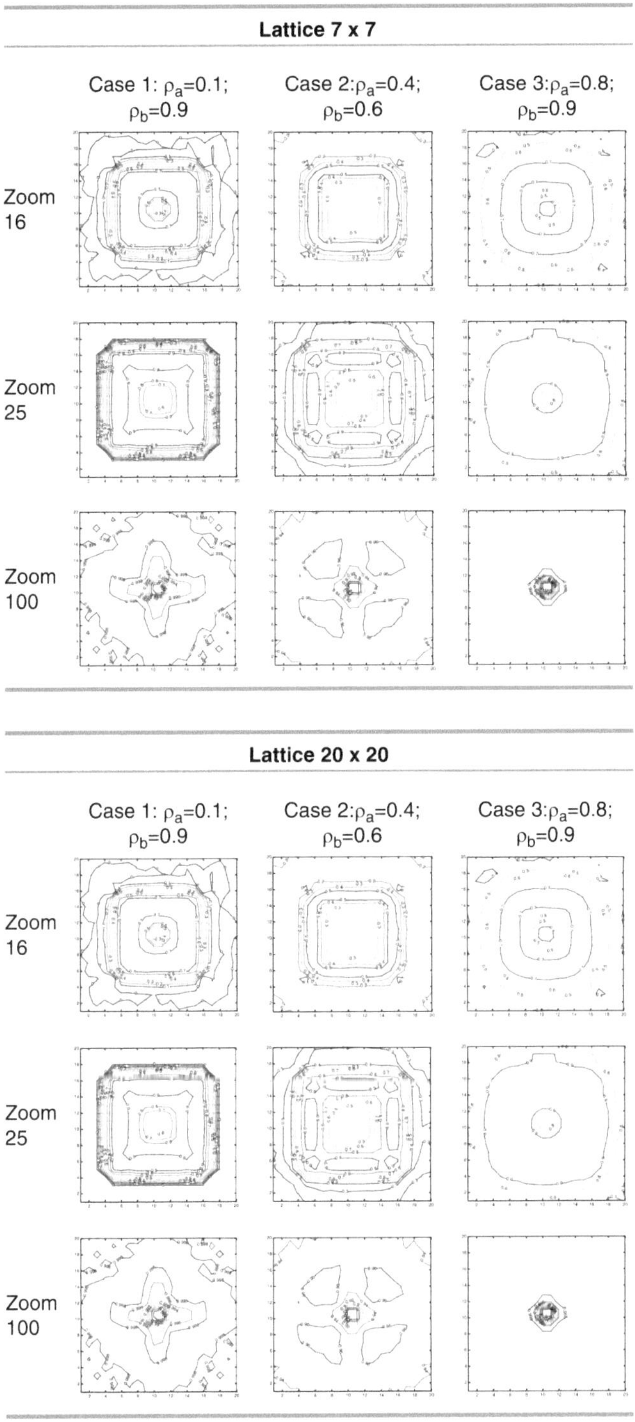

Fig. 5 Spatial distribution of ρ_r under the break. Center–Periphery structure

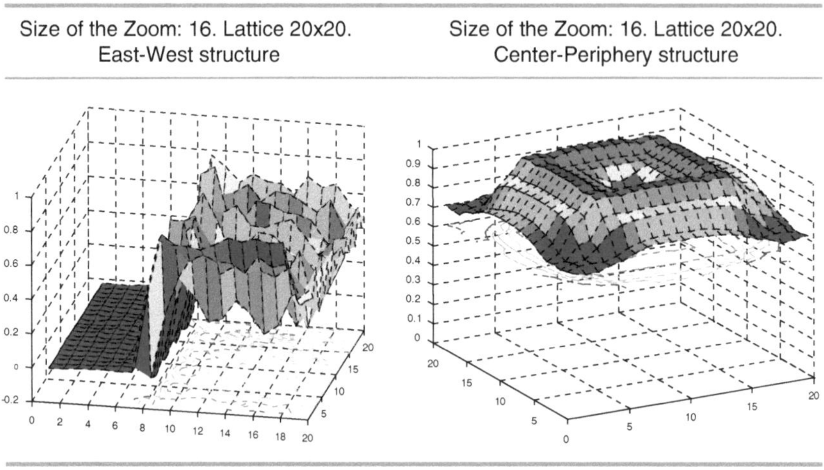

Fig. 6 The *doughnut effect* and the *Zoom* estimation

allow the formation of two clusters. The first is the algorithm known as k-means (Seber 1984) in which, on the basis of two randomly selected cells, others are added with the aim of minimizing a measure of internal distance between the values of $\rho_r^{(m)}$. The algorithm has been repeated 10 times starting with different points, selecting the solution that offers the lowest internal distance, so that the final solution does not depend on the initial points. The second, called Gaussian Mixture Models (GM), uses the Expectation Maximization (EM) algorithm to identify the regimes (McLachlan and Peel 2000). This method has also been used to identify clusters of regions of homogeneous behavior in the presence of spatial dependence (Tsionas 2000; Bloom et al. 2003; Battisti and Di Vaio 2007). It is adequate when we suspect that there is a mixture of several populations in the data, each of them coming from a Gaussian distribution with different first and second order moments. As before, the EM algorithm has been repeated 10 times in each case and we have selected the solution that maximizes the likelihood of the solution.

We also present the results of two other techniques that only allow us to tackle bipolarity. The first is well known in the literature and is based on the sign of the standardized G_i statistic obtained from the local estimations of the coefficient ρ (LeGallo et al. 2003; Ertur et al. 2006; Fisher and Stirböck 2006; LeGallo and Dall'erba 2006; Ramajo et al. 2008). Lastly, we also include a very simple classification rule based on the 10% trimmed mean (tme): we assign the regions whose value of $\rho_r^{(m)}$ is lower than the trimmed mean to one cluster and the others to the other cluster.

Table 7 evaluates the effectiveness of these four techniques through the average percentage of cells correctly classified after the 1,000 iterations. The results allow us to draw important conclusions. For example, the percentage of correctly classified cells improves in the four criteria as the difference between ρ_a and ρ_b increases. The size of the *Zoom* is another important factor. The worst results are obtained when

Table 7 Percentage of cells correctly classified

		Center–Periphery				East–West			
		k_m	GM	G_i	tme	k_m	GM	G_i	tme
		Case 1: $\rho_a = 0.1$; $\rho_b = 0.9$							
Zoom 9	20 × 20	0.815	0.950	0.842	0.790	0.703	0.688	0.755	0.715
Zoom 16		0.815	0.955	0.842	0.790	0.914	0.882	0.933	0.911
Zoom 25		0.767	0.792	0.766	0.766	0.882	0.842	0.897	0.879
Zoom 49		0.380	0.692	0.478	0.641	0.636	0.570	0.669	0.625
Zoom 9	7 × 7	0.664	0.416	0.565	0.672	0.775	0.794	0.801	0.848
Zoom 16		0.353	0.355	0.196	0.506	0.751	0.828	0.808	0.805
Zoom 25		0.154	0.235	0.183	0.303	0.589	0.745	0.720	0.666
		Case 2: $\rho_a = 0.4$; $\rho_b = 0.6$							
Zoom 9	20 × 20	0.531	0.410	0.605	0.594	0.525	0.513	0.545	0.527
Zoom 16		0.528	0.409	0.603	0.592	0.664	0.547	0.729	0.679
Zoom 25		0.724	0.744	0.689	0.716	0.591	0.524	0.732	0.671
Zoom 49		0.499	0.638	0.465	0.584	0.283	0.316	0.325	0.282
Zoom 9	7 × 7	0.457	0.443	0.489	0.554	0.573	0.559	0.619	0.637
Zoom 16		0.332	0.335	0.234	0.464	0.672	0.695	0.702	0.712
Zoom 25		0.205	0.202	0.169	0.379	0.602	0.669	0.683	0.596
		Case 3: $\rho_a = 0.8$; $\rho_b = 0.9$							
Zoom 9	20 × 20	0.399	0.490	0.522	0.548	0.475	0.466	0.457	0.463
Zoom 16		0.423	0.546	0.519	0.548	0.599	0.645	0.637	0.644
Zoom 25		0.430	0.492	0.381	0.603	0.635	0.633	0.643	0.641
Zoom 49		0.473	0.630	0.253	0.583	0.414	0.352	0.495	0.363
Zoom 9	7 × 7	0.349	0.314	0.356	0.541	0.469	0.470	0.488	0.499
Zoom 16		0.255	0.259	0.183	0.411	0.544	0.527	0.620	0.466
Zoom 25		0.111	0.141	0.183	0.313	0.639	0.642	0.649	0.430

the size is large. A bandwidth of between 9 and 16 seems to be the most adequate to identify the composition of the clusters, no matter what happens with the other factors (sample size, structure, etc.). The size of the lattice is also of some importance, though much less, in the number of correct classifications (the percentage falls as the size of the lattice decreases). The difference is inappreciable with respect to the spatial regime of the break.

There is no criterion that clearly dominates over the others when constituting the clusters. The k-means presents the worst results in most cases. That based on the trimmed mean works well with small sample sizes (7 × 7 lattice) and also when the distance between the values of ρ is small. The differences between the other two methods (GM and G_i) are difficult to appreciate, especially with medium and large (20 × 20) samples. Nevertheless, the GM method has the advantage of being more general, its statistical basis is more solid and it allows the constitution of two or more regimes.

7 Conclusions

The use of models which include rigid, uniform spatial structures for the whole set of regions has evolved, in recent decades, towards more flexible specifications. In this process, the concept of local estimates appears very important. Our suggestion is that this trend must be extended to models in which there are mechanisms of spatial dependence, in spite of the computational cost implied.

In this work, we have presented a wide Monte Carlo exercise dedicated specifically to checking the possibilities of the local estimation techniques. The main conclusions that we have obtained from this experiment are the following:

- Local estimation may be a useful technique since it has the potential to provide accurate estimates even when the true model is not known. However, it should be used with caution because, in general, we will perceive symptoms of heterogeneity in the local estimates with and without spatial breaks in the parameter of spatial dependence.
- In relation to the above, it is important to have some statistical test that confirms/rejects the existence of spatial breaks.
- The size of the *Zoom* (the bandwidth in kernel literature) appears to be one of the most important aspects in the performance of the local estimation algorithm. This technique seems to work better with very small *Zoom* sizes.
- The problem of identifying the regions that belong to the different regimes is not simple. A geographical cluster analysis is the possibility that we have explored in this chapter and, although the results are not thoroughly disappointing, we are sure that its performance could be improved.

Acknowledgements This work has been carried out with the financial support of project SEJ2006–02328/ECON of the Ministerio de Ciencia y Tecnología of the Reino de España.

References

Anselin L (1988a) Spatial econometrics: methods and models. Kluwer, Dordrecht

Anselin L (1988b) Lagrange multiplier test diagnostics for spatial dependence and spatial heterogeneity. Geogr Anal 20:1–17

Anselin L (1990) Spatial dependence and spatial structural instability in applied regression analysis. J Reg Sci 30:185–207

Anselin L (1995) Local indicators of spatial association. Geogr Anal 27:93–115

Anselin L, Bera A (1998) Spatial dependence in linear regression models with an introduction to spatial econometrics. In: Ullah A, Giles D (eds) Handbook of applied economic statistics. Marcel Dekker, New York, pp 237–289

Banerjee A, Lumsdaine R, Stock J (1992) Recursive and sequential tests of the unit root and trend break hypotheses: theory and international evidence. J Bus Econ Stat 10:271–287

Battisti M, Di Vaio G (2007) A spatially filtered mixture of P-convergence regressions for EU regions, 1980–2002. Empir Econ 34:105–121

Baumol W (1986) Productivity growth, convergence, and welfare: what the long-run data show. Am Econ Rev 76:1072–1085

Bloom D, Canning D, Sevilla J (2003) Geography and poverty traps. J Econ Growth 8:355–378
Breusch T, Pagan A (1979) A simple test for heteroscedasticity and random coefficient variation. Econometrica 47:1287–1294
Brown R, Durbin J, Evans J (1975) Techniques for testing the constancy of regression relationships over time. J Roy Stat Soc B 37:149–192
Brunsdom C, Fotheringham S, Charlton M (1996) Geographically weighted regression: a method for exploring spatial nonstationarity. Geogr Anal 28:281–298
Brunsdom C, Fotheringham S, Charlton M (1998) Spatial nonstationarity and autoregresive models. Environ Plann A 30:957–973
Casetti E (1972) Generating models by the expansion method. application to geographical research. Geogr Anal 4:81–91
Casetti E (1991) The investigation of parameter drift by expanded regressions: generalities and a family planning example. Environ Plann A 23:1045–1051
Chow G (1960) Tests of equality between sets of coefficients in two linear regressions. Econometrica 28:591–605
Cleveland W (1979) Robust locally weighted regression and smoothing scatterplots. J Am Stat Assoc 74:829–836
Cleveland W, Devlin S (1988) Locally weighted regression: an approach to regression analysis by local fitting. J Am Stat Assoc 83:596–610
Cressie N (1991) Statistics for spatial data. Wiley, New York
Davidson J (2000) Econometric theory. Blackwell, New York
Di Giacinto V (2003) Differential regional effects of monetary policy: a geographical SVAR approach. Int Reg Sci Rev 7:313–341
Dufour J (1982) Recursive stability analysis of linear regression relationships. J Econometrics 19:21–76
Ertur C, LeGallo J, Baumond C (2006) The regional convergence process, 1980–1995: do spatial regimes and spatial dependence matter? Int Regional Sci Rev 29:3–34
Ertur C, LeGallo J, LeSage J (2007) Local versus global convergence in Europe: a Bayesian spatial econometric approach. Rev Reg Stud 37:82–108
Fisher M, Stirböck C (2006) Pan-European regional income growth and club-convergence. Insights from a spatial econometric perspective. Ann Reg Sci 40:693–721
Florax R, de Graaff T (2004) The performance of diagnostics tests for spatial dependence in linear regression models: a meta-analysis of simulation studies. In: Anselin L, Florax R, Rey S (eds) Advances in spatial econometrics: methodology, tools and applications. Springer, Berlin, pp 29–65
Florax R, Folmer H, Rey S (2003) Specification searches in spatial econometrics: the relevance of Hendry's methodology. Reg Sci Urban Econ 33:557–579
García R, Perron P (1996) An analysis of the real interest rate under regime shifts. Rev Econ Stat 78:111–125
Getis A, Ord J (1992) The analysis of spatial association by use of distance statistics. Geog Anal 24:189–206
Greene W (2003) Econometric analysis. Prentice Hall, New York
Hansen B (1996) Inference when a nuisance parameter is not identified under the null hypothesis. Econometrica 64:413–430
Huang J (1984) The autoregressive moving average model for spatial analysis. Aust J Stat 26:169–178
Koop G, Potter S (2007) Estimation and forecasting in models with multiple breaks. Rev Econ Stud 74:763–789
Lacombe D (2004) Does econometric methodology matter? An analysis of public policy using spatial econometric techniques. Geogr Anal 36:105–118
LeGallo J, Dall'erba S (2006) Evaluating the temporal and spatial heterogeneity of the European convergence process, 1980–1999. J Reg Sci 46:269–288

LeGallo J, Ertur C, Baumont C (2003) A spatial econometric analysis of convergence across European regions, 1980–1995. In: Fingleton B (ed) European regional growth. Springer, Berlin, pp 99–130
McLachlan G, Peel D (2000) Finite mixture models. Wiley, New York
McMillen D (1996) One hundred fifty years of land values in chicago: a nonparametric approach. J Urban Econ 40:100–124
McMillen D (2004) Employment densities, spatial autocorrelation, and subcenters in large metropolitan areas. J Reg Sci 44:225–244
McMillen D, McDonald J (1997) A nonparametric analysis of employment density in a policentric city. J Reg Sci 37:591–612
Mur J, López F, Angulo A (2008) Symptoms of instability in models of spatial dependence. Geogr Anal 40:189–211
Pace K, Lesage J (2004) Spatial autoregressive local estimation. In: Getis A, Mur J, Zoller H (eds) Spatial econometrics and spatial statistics. Palgrave, London, pp 31–51
Páez A, Uchida T, Miyamoto K (2002a) A general framework for estimation and inference of geographically weighted regression models: 1: location-specific kernel bandwidth and a test for locational heterogeneity. Environ Plann A 34:733–754
Páez A, Uchida T, Miyamoto K (2002b) A general framework for estimation and inference of geographically weighted regression models: 2: spatial association and model specification tests. Environ Plann A 34:883–904
Parent O, Riou S (2005) Bayesian analysis of knowledge: spillovers in European regions. J Reg Sci 45:747–775
Phillips P, Ploberger W (1994) Posterior odds testing for a unit root with data-based model selection. Economet Theor 10:774–808
Quah D (1986) Regional convergence clusters across Europe. Eur Econ Rev 40:951–958
Quandt R (1960) Tests of the hypothesis that a linear regression system obeys two separate regimes. J Am Stat Assoc 55:324–330
Qu Z, Perron P (2007) Estimating and testing structural changes in multivariate regressions. Econometrica 75:459–502
Ramajo J, Márquez M, Hewings G, Salinas M (2008) Spatial heterogeneity and interregional spillovers in the European Union: do cohesion policies encourage convergence across regions? Eur Econ Rev 52:551–567
Rietveld P, Wintershoven H (1998) Border effects and spatial autocorrelation in the supply of network infrastructure. Paper Reg Sci 77:265–276
Rossi B (2005) Optimal tests for nested model selection with underlying parameter instability. Econ Theor 21:962–990
Salazar D (1982) Structural changes in time series models. J Econometrics 19:147–163
Seber G (1984) Multivariate observations. Wiley, New York
Titterington D, Smith A, Makov U (1985) Statistical analysis of finite mixture distributions. Wiley, New York
Tsionas E (2000) Regional growth and convergence: evidence from the United States. Reg Stud 34:231–238
Zivot E, Phillips P (1994) A Bayesian analysis of trend determination in economic time series. Economet Rev 13:291–336

Part II
Spatial Analysis of Land Use and Transportation Systems

"Seeing Is Believing": Exploring Opportunities for the Visualization of Activity–Travel and Land Use Processes in Space–Time

Ron N. Buliung and Catherine Morency

1 Introduction

The study of the relationship between activity–travel behaviour and the development of city-regions is a matter of great concern among researchers and urban planners. Much of the current debate focuses on understanding and influencing the relationship between transportation and land use systems, with a view to achieving economic, sustainability, and quality of life policy objectives. The essence of the transport-land use link is that the development of "new" or the presence of "old" transport infrastructure (e.g., road, rail, etc.) increases the relative accessibility and hence attractiveness of place, giving rise to several possible outcomes including: the enhancement of economic growth and spatial interaction. The economic benefits that materialize in this context, however, have been the subject of debate (Black 2001).

Accessibility effects have also become prominent in policy-based discourse and research focused on the efficacy of urban design as a mechanism for reducing transports' negative externalities. Researchers have set out to test the conventional wisdom that placing and mixing the "things" people want to or have to do, close to where people "want" to or "have to" live or work, will facilitate reductions in automobile use, energy consumption, and environmental emissions (e.g., Buliung and Kanaroglou 2006b; Cervero and Kockelman 1997; Crane 2000). The results appear to be somewhat inconsistent, with context specific evidence suggesting that the relationship between transport and land use tends to vary from person to person, and place to place.

Understanding how transportation and land use processes evolve both independently and jointly in space–time, and how these systems influence one another, is a complex endeavour that can arguably be enhanced through the development and application of tools for graphical analysis and data visualization. In the current

R.N. Buliung (✉)
Department of Geography, University of Toronto Mississauga, 3359 Mississauga Road North, Mississauga, ON L5L 1C6, Canada,
e-mail: ron.buliung@utoronto.ca

A. Páez et al. (eds.), *Progress in Spatial Analysis*, Advances in Spatial Science,
DOI 10.1007/978-3-642-03326-1_7,

context, where transportation and land use data are becoming more abundant, available in an increasing variety of formats, and collected using a wide range of instruments, the opportunity exists to usefully increase stakeholder and institutional awareness of the possible policy and programme relevant applications of public data. Moreover, the availability of proprietary and open software environments, coupled with implementation of innovative approaches for data collection, presents fresh opportunities for the construction of spatiotemporal knowledge of transport and land use processes. Opportunities now exist to match the impressive and occasionally controversial visual qualities of transport and land use systems with comparatively impressive and instructive experiments with data visualization.

The goal of this chapter is to make the case for engaging more fully with the exploratory analysis, visualization, and ultimately the communication of various spatial, temporal and demographic qualities of transportation and land use systems and processes. Using examples drawn primarily from the Greater Montreal Area (GMA) and the Greater Toronto Area (GTA), Canada, the chapter examines how numerous and appropriate visualization techniques and tools can be used, often in a complementary way, to clarify the spatial and temporal qualities of transport and land use processes.

The chapter is organized into five sections. First, some discussions on the role of visualization in the spatial analysis process are reported. Then, a general framework for analysing, understanding and observing the urban system is presented, along with a description of the two city-regions that provide the contextual setting for the chapter. A third section focuses on examples of the visualization of selected outcomes of activity–travel processes. Attention then turns toward visualizing the land-use system and development processes. The chapter concludes with summary observations concerning the visualization examples presented throughout the chapter.

2 The "Art and Science" of Visualization

The greatest value of a picture is when it forces us to notice what we never expected to see.

(Tukey 1977)

Data visualization and graphical analysis have historically featured prominently across disciplines centred on the construction of knowledge of natural processes and human activities (e.g., CSISS 2008; Tufte 2001). Recent examples, including the ground-breaking 3D rendering, visualization, and analysis of the Mona Lisa (Borgeat et al. 2007), and Time Magazine's, "One Day in America" exposé on commuting within major US cities (Time 2007) suggest that visualization is coming of age, becoming a more regular part of the scientific and human experience than ever before.

The aim of scientific or data visualization is to increase human understanding of complex processes through the creation and viewing of imagery constructed from data (Gahegan 2000; Hearnshaw and Unwin 1994). To draw a parallel with "the arts," visualization can be compared to an anamorphosis interpreter wherein the act of visualization makes use of specialized devices (e.g., computer programs, statistical tools, GIS, interactive spreadsheets), or compels the viewer to occupy a specific perspective (spatial, temporal, or social feature), with a view to reconstituting the "original" for the purpose of developing a clearer understanding of "process."

From a scientific or quasi-scientific perspective, visualization can be viewed as an exploratory exercise and, when used interactively by modellers, can be applied to support hypothesis generation and case study analyses (Buliung and Kanaroglou 2006a; Buliung et al. 2008; Gahegan 2000). In other words, visualization can make use of available data to engage in the a priori exploration and development of hypotheses that may be formally and more rigorously tested later. When visualization tools are rigorously constructed, they can also prevent data from being misused or misinterpreted by non-specialists, particularly when the analyst implements constraints that prevent non-specialist development of erroneous relationships in n-dimensional space that violate statistical or other scientific principles.

Visualization can also be conceptualized as a communication process. If done well, i.e., attention is given to achieving clarity, precision, and efficiency, visualization can facilitate the representation of processes using structures and schematics which are less abstract, making it easier for non-specialists to understand complex human and/or physical processes. The communication motivation also provides opportunities for the democratization of specialized knowledge through the engagement of lay-audiences in "conversations" about complex phenomena. Examples of this sort can be taken from the participatory planning literature, where stakeholders, professional planners, and other decision makers simultaneously engage with the visualization process to shape planning decisions and outcomes (e.g., Al-Kodmany 1999, 2002; Lewis and Sheppard 2006; Tress and Tress 2003).

The case for visualization has been argued by many, with perhaps the most salient and earliest endorsements emerging from John Tukey's influential work, published in 1977, "Exploratory Data Analysis". The concept of EDA (Exploratory Data Analysis) initiated many discussions with respect to graphical methods and tools for data processing and analysis. The aim of EDA is to facilitate the identification of patterns in data using graphical, visual, and numerical methods. EDA is typically more descriptive (and intuitive) than formal.

While EDA methods facilitate pattern identification within a dataset, they do not explicitly integrate the spatial quality of data that describe geographical processes. The development and widespread availability of GIS then made inevitable the emergence of exploratory techniques specifically dedicated to spatial data (Anselin 1995). Exploratory spatial data analysis (ESDA) is the extension of EDA to spatial data, integrating additional techniques to detect spatial patterns and formulate and test hypothesis based on the geography of processes (Haining et al. 2000).

As a "geographical" extension of visualization, *geovisualization* maintains an important role in spatial analysis and ESDA (Haining et al. 2000; Wise et al. 1999). Geovisualization integrates spatial and non-spatial theory, methods, and technologies to facilitate exploration, analysis, synthesis, and communication of geographical processes and data (MacEachren and Kraak 2001). Haining et al. (2000), argue that the focus on "the visual" is justifiable because: (1) the power of modern graphical interfaces means that graphics are no longer a way of simply presenting results in the form of maps or graphs, but a tool for the extraction of information from data; and (2) graphical, exploratory methods are felt to be more intuitive for non-specialists when compared with numerical spatial methods, enabling broader participation in the scientific process.

Haining's position arguably favours the horizontal construction of scientific knowledge, a model that is useful in certain contexts (e.g., participatory planning), and potentially unworkable in other cases (i.e., instances where the testing of "theory" and development of "laws" occasionally relies on the application of more sophisticated computationally intensive or "classical" approaches to hypothesis testing). With this caveat in mind, the opportunity to enhance our understanding of processes in space–time through geovisualization provides adequate justification for the practice.

2.1 A Brief Note on "Tools"

The "integrative" aspect of geovisualization, and the goal of simplifying the complexity of spatial processes and data is largely facilitated by the efforts of a relatively small number of spatial scientists, and private agencies, who engage in the development of software for ESDA. These initiatives have involved various platforms and philosophies for development and distribution. While the advancement of geographic data analysis capabilities within the proprietary domain are widely known (e.g., ESRI, MapInfo), less attention, beyond the academic domain, has been given to the advancement of Free and/or Open Source (OS) projects. In this chapter, examples have been developed using both proprietary (e.g., ESRI's ArcGIS, Microsoft Excel), and Open Source (e.g., the R Language and its libraries) software.

Open source (OS) projects are generally distinguishable from their proprietary counterparts because they typically include data and code in the distribution. Moreover, the range of OS licensing options enhances the process of collaborative, transparent, incremental, and rapid application development. Current OS activities related to spatial analysis and geovisualization include GRASS, THUBAN, SAGA, the *spatstat*, *spdep*, and *aspace* packages for R (Baddley and Turner 2005; Bivand 2006; Buliung and Remmel 2008), the R-Grass interface (Bivand and Neteler 2000), STARS (Rey and Janikas 2006), Terralib, and Geovista (Takatsuka and Gahegan 2002). Open Source organizations have also developed to support knowledge exchange and the distribution of OS GIS and spatial analysis software (e.g., OSGeo.org, opensourcegis.org, FreeGIS.org, MapTools.org).

Spatial analysis freeware has also become available, with GeoDA (Anselin et al. 2006) and CrimeStat (Levine 2006), for example, offering a wide range of capabilities for the visualization and analysis of spatial processes. ESDA toolkits in general have also become increasingly interactive over time, providing opportunities to link planar and non-planar views of data in both static and dynamic ways (e.g., Anselin 1995, 2000; Anselin et al. 2006; Buliung and Kanaroglou 2006a; Kwan 2000).

The evolution of Tukey's EDA toward the widely distributed ESDA solutions available today presents researchers interested in the relationship between urban policy and urban change with an opportunity to engage in innovative spatial science. Using examples drawn from two of Canada's largest city-regions, the GTA, and the GMA, the remainder of the chapter demonstrates the application of geovisualization to the study of transportation and land use processes. Where possible (i.e., similar data are available), comparative exploratory analyses have been undertaken. The overall goal of the chapter, however, is not to comparatively describe the geography of transport and land use across the city-regions, but to illustrate, using secondary data from these places, the utility of visualization for enhancing our understanding of urban spatial processes.

3 Geovisualizing Transportation and Land Use Processes

Understanding interactions between activity–travel and land use processes is a complex task. For some time, researchers have engaged in the empirical study of the reciprocal influences shared across transportation and land use systems. The literature suggests that the magnitude and direction of the effects tend to vary from place to place, and across segments of the population (Badoe and Miller 2000; Buliung and Kanaroglou 2006b). Moreover, the influences arguably operate at different geographical and temporal scales. With a view to advancing current thinking on the behavioural and physical aspects of transport and land use interactions, Morency (2004) proposed a conceptualization of the urban system based on ten critical dimensions (Fig. 1). The framework embodies much of the theoretical and empirical data on the subject of activity–travel behaviour developed since Jones (1979).

Morency's framework specifies that observed urban travel behaviours proceed from and contribute to the structuring of housing, transportation and activity functions. Activity–travel behaviours are intimately linked to individuals and households and to space–time decisions regarding residential location and activity participation. Formalising the urban system using ten comprehensive dimensions can arguably draw the attention of decision makers toward specific interactions. Enhancing this conceptual experience with dynamic, theoretically grounded, data-based views of how key agents and processes evolve in space–time can also clarify urban issues for various stakeholders. The visualization exercise can therefore become part of stakeholder conversations regarding policy and planning issues (i.e., Al-Kodmany 1999). Using this framework as a conceptual background, the remainder of the chapter presents examples of the geovisualization of urban processes.

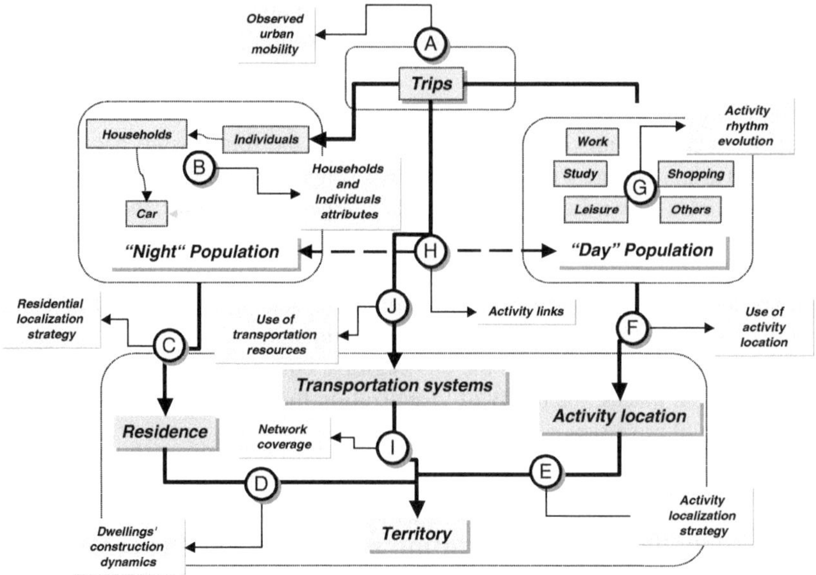

Fig. 1 Critical dimensions and interactions between activity-travel and land-use systems

3.1 Regional Context and Data Sources

The geovisualization examples have been constructed using data drawn from two of Canada's largest city-regions, the GMA and the GTA (Fig. 2). The GTA is Canada's largest metropolitan region, with a population exceeding five million. During the last decade, the population of the suburban regional municipalities has grown at a faster rate than the City of Toronto, historically, the economic and cultural hub of the region. The GMA is the largest urban area in the province of Quebec, with 3.5 million inhabitants distributed across 5,500 km^2. Over the last 15 years, and across both study areas, the population has become more geographically dispersed, household size has decreased, and car ownership has been on the rise.

Large-scale travel surveys (trip diary) are conducted regularly in both city-regions and used to support academic and practitioner research and planning activities. Around 5% of households are surveyed in the GTA as part of the Transportation Tomorrow Survey, while roughly 5% of the population are interviewed, in the GMA. Both surveys are conducted approximately every 5 years; data in Montreal have been collected since 1970, while the TTS was first launched in 1986. The surveys facilitate construction of large datasets containing demographic data on households and individuals, as well as spatiotemporal details regarding daily trips.

Other secondary datasets are available in both regions to document additional transport-related issues. For example, data from other types of surveys or observation systems can be used to study additional dimensions of transportation and land use systems. Transaction datasets from carsharing organizations, retail databases, parking inventories, car ownership files, and GPS traces from onboard systems can

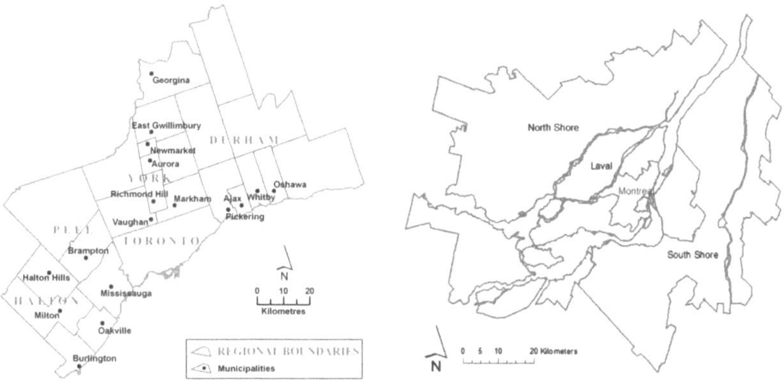

Fig. 2 The Greater Toronto Area (GTA) and Greater Montreal Area (GMA)

provide insights to behavioural processes that are typically not observed through the sort of large scale, legacy travel surveys conducted in both regions. At the same time, developing large representative samples for regional scale analysis and modelling is arguably difficult using the sort of "near" real-time data collection enabled by new integrative mobility technologies.

4 Activity–Travel Processes

Activity–Travel processes (e.g., planning/scheduling) can give rise to the movement of objects, such as people, families, friends, cars, or information in space–time. For some time, and in the presence of increasingly available space–time activity microdata, and GIS tools, Transportation Geographers have been developing approaches to manage and visualize activity–travel data within a spatiotemporal context (e.g., Buliung and Kanaroglou 2006a; Kwan 2000; Shaw and Wang 2000; Shaw et al. 2008). In this section, examples of both the implicit and explicit processing of the location of objects over time are presented. For instance, the 24-h monitoring of people example *explicitly* incorporates space–time in the visualization experiment, while accumulation profiles (i.e., a running count of activity–travel objects in space–time) *implicitly* reference the spatial and temporal dimensions of travel demand.

4.1 People and Cars in Space and Time

Population size and density are often included in travel behaviour models to explain systematic variation in travel demand (e.g., transit share, kilometres travelled, trip generation). Metrics of this sort are also used as benchmarks for understanding the extent to which cities and regions are moving toward planning goals, such as the intensification of land use. Typically, these metrics are examined at a single point in time. However, such a static measurement approach does not reflect the potentially time-varying use of space by people or other moving objects (e.g., cars). While pursuing the study of the time-varying use of systems and space is not new (e.g., Civic Transportation Committee 1915; Goodchild and Janelle 1984; Janelle and Goodchild 1983), the availability of GIS, and activity–travel micro-data presents new opportunities to communicate to stakeholders how cities and regions are used in space and time (e.g., Time 2007).

Travel survey data as detailed as those available in both Montreal and Toronto enable exploration of temporal variation in the use of space by people and households during a typical weekday. It is possible to follow moving objects in space–time, using time of departure for every trip, estimated travel times, and the precise location of trip ends. The 24-h record of travel demand permits evaluation of potentially more useful indicators of spatial occupancy, in comparison to classical static measures. In the following set of examples, travel data were processed for both the GMA and GTA to explore spatiotemporal patterns of mobility. First, data were processed to document the daily mobility of residents across the two city-regions, to assess the importance of regional sub-areas in providing activity opportunities (e.g., central business district, other centres), and to expose any day–night differential in the occupancy of local environments. Additionally, the reported examples were developed using separate "tools" to illustrate the utility of both proprietary and Open Source solutions for this sort of data exploration. Second, data were processed, for the GMA only, to classify boroughs (geopolitical areas) according to their attractiveness with respect to activity provision and parking supply.

With respect to the first application, and for the GMA only, an animation of the activity of respondents, on a 24-h basis, was produced by allocating time-stamped trip-ends to 1 km^2 grid cells in a grid-based tessellation (raster) covering the study area. Six snapshots of the daily mobility of people over space are presented as 3D maps in Fig. 3. Each cell is projected from the plane by a factor equivalent to the population (survey respondents) density in every cell at a specific hour of a typical weekday. In addition to being clearly understood by non-specialists, this visualization gathers all of the information required to compute dynamic densities over time. For example, the data suggest that, at 12h00, the population density is 80 times higher in the central business district (up to 13,000 people per km^2 for the most active cells) compared with what is observed in the evening, a level of activity that is never reached in outer areas.

This experiment appears to follow the conclusions of other scholars concerning the continued significance of the Montreal CBD as a place of work (Shearmur and Coffey 2002), while simultaneously highlighting other "busy" sub-areas. At the unit level, the approach facilitates estimation of the inequality between day and night densities. This visualization approach is currently in use by Montreal's Metropolitan Agency of Transportation to summarise daily patterns of mobility in space–time. Interestingly, the visualization approach is also used to communicate to local constituencies, the value-added proposition of the travel survey data collection programme.

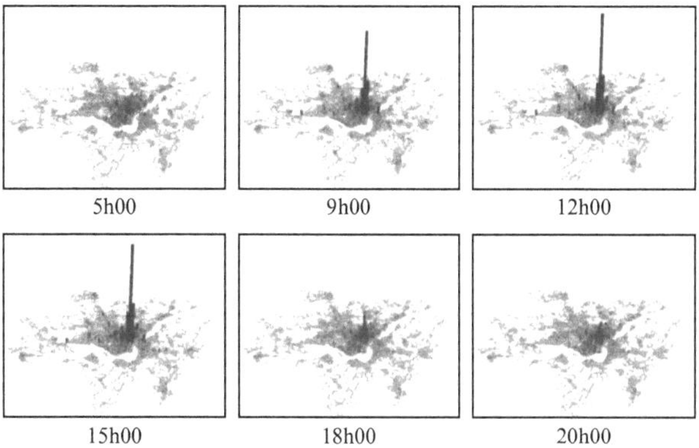

Fig. 3 Chronology of the spatial location of the mobile population during an average weekday in the GMA (1998)

A similar example has been developed for the GTA using data from the 2001 Transportation Tomorrow Survey (Fig. 4). While the Montreal example was developed using the ArcScene extension to ESRI's ArcGIS proprietary software, the GTA example has been developed using the *scatterplot3d* library from the Open Source R language for statistical and graphical analysis (see www.r-project.org). In this case, a 3D scatterplot (e.g., lollipop plot) has been developed by projecting traffic zone centroids from the plane by a factor equivalent to the trip density (number of trip-ends/traffic zone size in km^2) associated with each zone for a prescribed time interval.

The time interval indicated on each panel refers to the start-time of trips ending in each traffic zone. The series conveys an initial sketch of the trip density for all purposes (conducted by any mode) across the study area during the early to mid-morning, mid-day, and late afternoon to evening time periods. The lighter grey points represent the location of those traffic zones in the top 5%, with respect to trip density, for the prescribed time interval. Of course, the travel data can be explored further by examining trips by purpose, mode, and/or market segment.

The print-form visualization (Fig. 4) sheds light on the time-varying nature of travel demand in the GTA, and the overwhelming intensity of travel to the City of Toronto during a typical weekday. A more complete picture of the GTAs regional economy emerges when those zones with the highest travel densities are removed from the exploration (Fig. 5), decentralized sub-centres are clearly more visible. Moreover, both figures convey the time-varying intensity of use of those sub-centres located to the east and west of Toronto, immediately adjacent to Lake Ontario.

Scholars have noted the unique differences across Montreal and Toronto with respect to the spatial structure of the regional economy (Shearmur and Coffey 2002). In contrast to the centralization of economic productivity in Montreal, empirical work has shown that the GTA has typically followed "US" development patterns characterized by a decline in the share of jobs located in the CBD, and the consolidation of growth in secondary suburban centres. The identification of decentralized

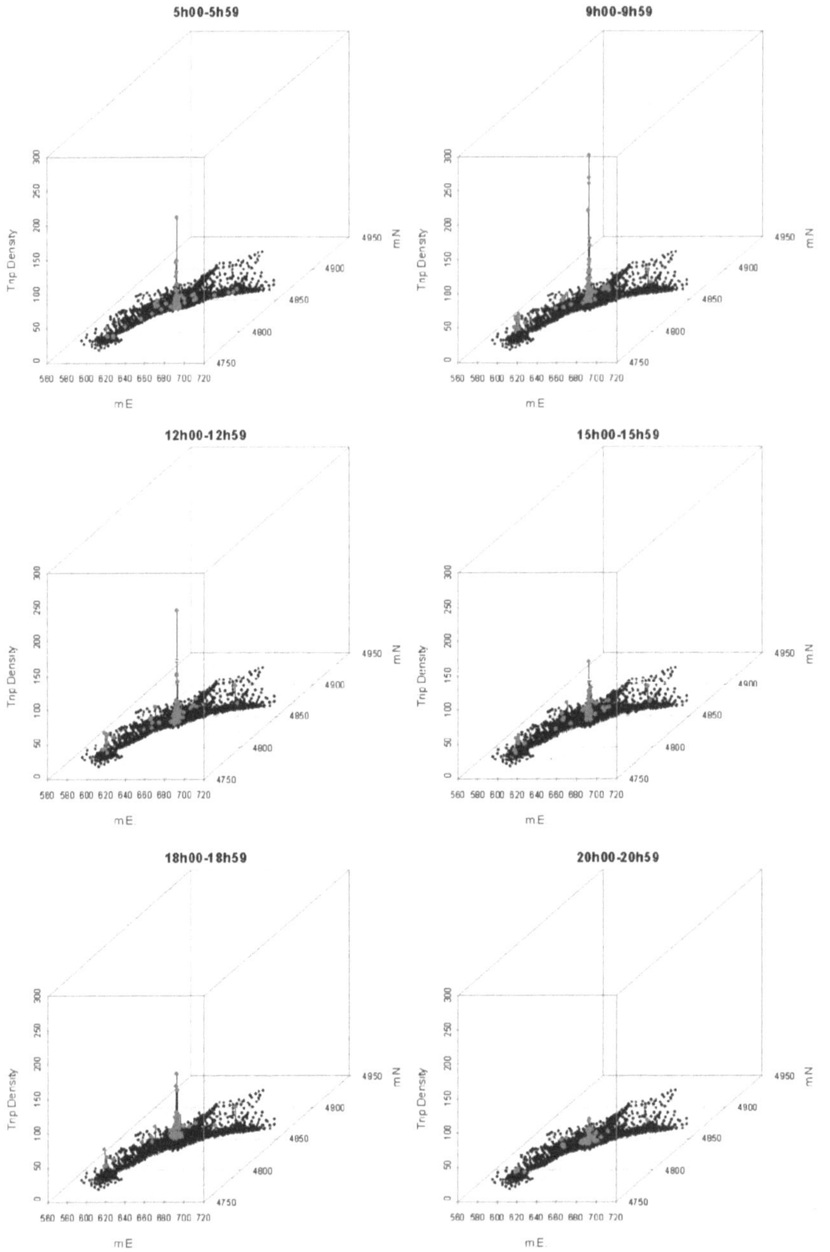

Fig. 4 Chronology of the spatial location of the mobile population during a typical weekday in the GTA & Hamilton (2001)

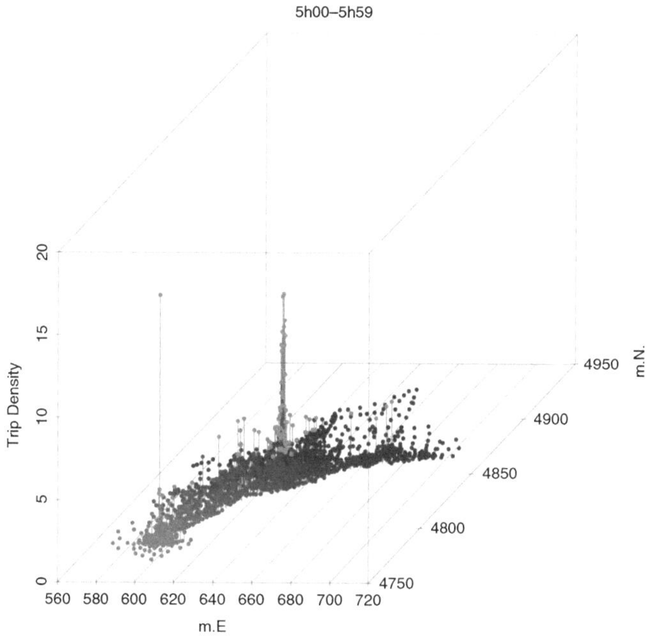

Fig. 5 GTA trip density excluding high density CBD traffic zones (2001 TTS)

traffic zones in the top 5% with respect to trip density (Figs. 4 and 5) provides additional evidence of a polycentric economy in the GTA.

Overall, however, while the total share of regional employment located in the central area of both city-regions has declined during the post-war era (Heisz and LaRochelle-Côté 2005) – the city centres in both the GTA and GMA continue to be used more intensively than other places across the regions. Notably, the Montreal and Toronto visualization experiments highlight the continued power and identity of the central business districts of two of Canada's oldest and largest cities within their broader regional economies.

The second application of the GMA travel survey data explores the attractiveness of geopolitical sub-areas with respect to activity participation and parking supply. People and car accumulation profiles (e.g., PAP and CAP) are developed to examine variation in the accumulation of these potentially mobile objects across space, for a specified time period. Since all the attributes of the moving object travel with it (household and people attributes), it is also possible to describe the set of objects simultaneously located in a specific zone using these attributes. Holding the destination constant, and collecting trips originating in other districts, clarifies the use of space by the non-resident users of the fixed destination. Moreover, the approach translates the "abstract" origin-destination flow matrix into a coherent graphic (e.g., Tufte 2001), facilitating analysis and discussion of the spatiality of travel demand using familiar geo-political constructs (e.g., neighbourhoods, boroughs, etc.).

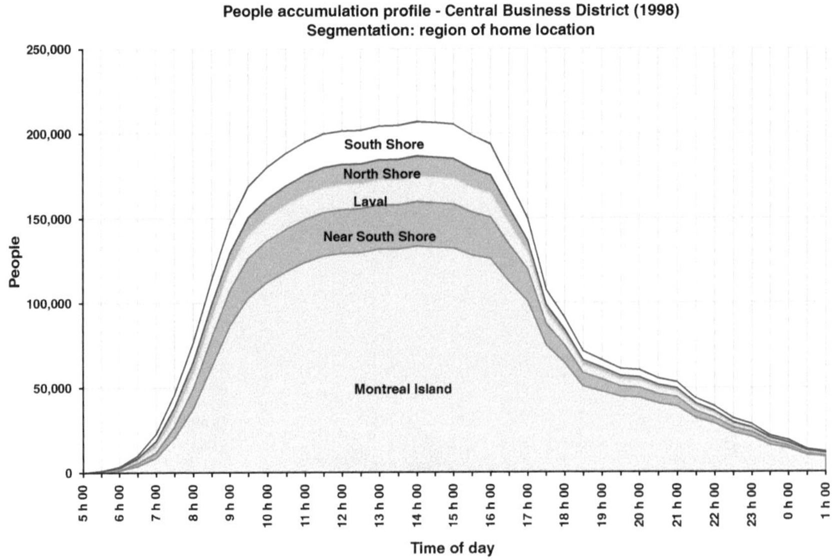

Fig. 6 People accumulation profile in the Central Business District (GMA) segmented by region of home location (1998)

As an example of the PAP approach, Fig. 6 reveals, at each time of the day, the extent to which non-resident users (i.e., individuals living in five large areas linked to specific transportation planning agencies and urban planning strategies) benefit from activity destinations located in the central part of the GMA. For example, around 15h00, there are more than 200,000 people located in this area, with more than 35% not being residents of the Montreal Island (hence not directly contributing to the local fiscal burden). For policy makers and planners, the approach facilitates visual exploration of the tensions between the use of space by central city and suburban residents. Measuring the consumption of centralized infrastructure by non-residents helps assess the scale of the benefits non-residents receive by accessing opportunities located in the CBD. Evidence of this sort can inform action directed at distributing the costs associated with the provision of centralized activities and infrastructure equitably across all users of the urban system.

The availability of subsidized parking influences mode choice toward solo-driving, particularly for the journey to work (Downs 2004; Willson and Shoup 1990). Moreover, the opportunity cost of providing parking spaces in urban markets with high land values is arguably very high. Understanding spatiotemporal trends in the demand for parking can inform the development of parking availability and taxation strategies aimed at adjusting the use of private and public parking resources, or shifting demand toward "sustainable" alternatives. Despite the relevance of parking to transportation planning, few public agencies have access to parking inventory data and when they do, the data are often incomplete (private/public parking) and limited to capacity (no information on the use of the parking spaces) (Morency et al. 2006).

The last two regional travel surveys conducted in the GMA asked questions regarding the type of parking space available at trip destinations. Responses include data on the physical type of parking (street, outdoor, indoor) and cost (free, paid, employer subsidized). These data have been used to construct car accumulation profiles (CAP) that demonstrate the availability and use of parking in space–time. CAPs were first developed for Montreal's Metropolitan Agency of Transportation as a means for having standardized and sector-based statistics on parking availability and use (Morency and Trépanier 2008). For example, the data indicate that the Montreal CBD has the highest car density, highest proportion of immobile cars (cars owned by residents that are not used), lowest internal use index (proportion of the parking spaces × hours used by residents), and highest emptying factor (ratio between the number of cars at 12h00 and the number of cars at 4h00) during a typical weekday. Following publication of the Montreal Transportation Plan, boroughs were asked to develop specific projects based on their respective transportation issues. In this context, CAP figures (e.g., Fig. 7) were produced along with maps illustrating the spatial distribution of vehicles parked in an area throughout a typical day (Fig. 8).

4.2 *Spatio-Demographic Travel Indicators*

Many city-regions are undergoing or are expected to undergo dramatic demographic changes that will hold important implications for urban planning, policy, and service provision (e.g., transportation, health, etc.). Profound changes are expected to arise as Canada's population "ages in place" or elsewhere (Health Canada 2002). Transportation planners and engineers should therefore be prepared to meet the needs of a population increasingly composed of non-working individuals, many of whom will experience progressively reduced mobility as they age. In addition to heavy cultural trends that have occurred in recent years, age and life cycle also influence the travel behaviours of individuals (Morency and Chapleau 2008). In this context, it is critical that planners be aware of demographic trends and the links between demographic change and travel behaviours.

The transportation-oriented age pyramid is introduced here, as a new approach to operationalize the spatiodemographic visualization of travel demand. The example shown in Fig. 9 was developed to provide comprehensive data on the demography of transportation to local planning authorities in Montreal (Morency 2004). Similar to the PAP and CAP experiments, standard spreadsheet functions are used to facilitate interactive geodemographic visualization of transportation processes (Chapleau and Morency 2005). What is new and relevant in this visualization tool is the possibility to simultaneously observe demographic change and travel demand in space–time. Combining the well-known age-pyramid with regional travel survey and geopolitical boundary data provides an intuitive visual framework for understanding interactions between people, where they live, and how they use the transportation system. For example, Fig. 9 presents the 1987 and 1998 age pyramids for the population of central Montreal. Each cohort is segmented according to transit

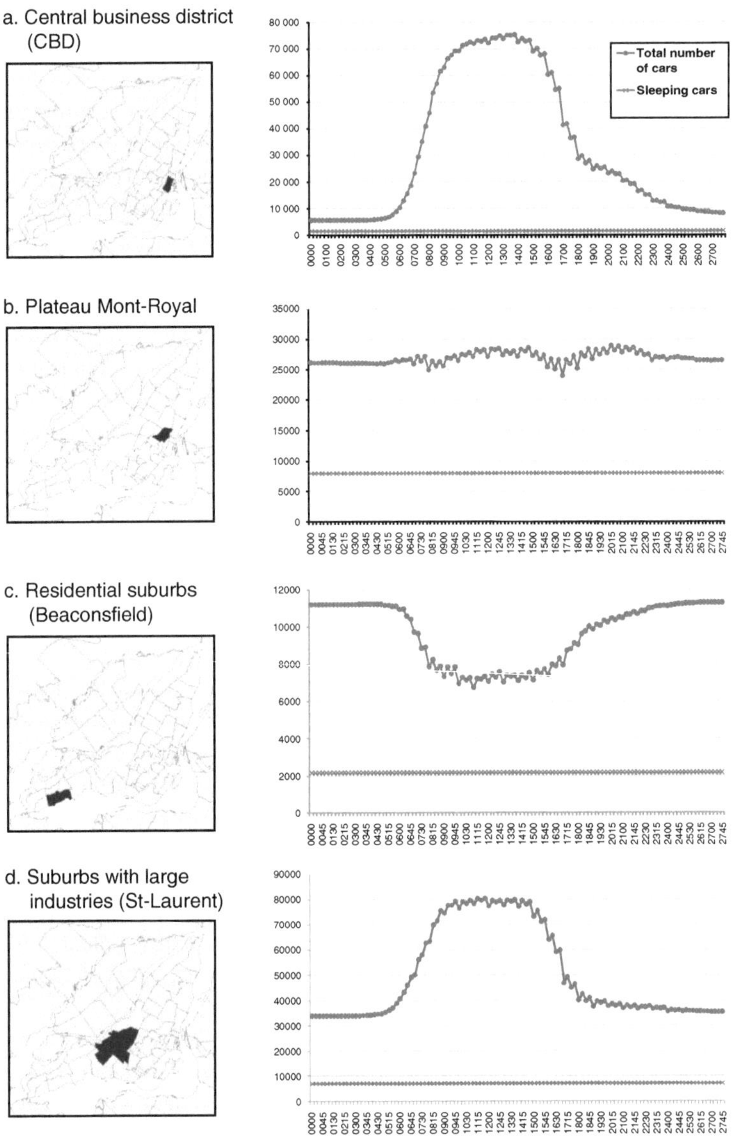

Fig. 7 2003 Car accumulation profile (CAP), four districts (x: time of day, y: number of cars)

use: people who have used transit at least once during the day are classified as transit users, all other mobile respondents as non-transit users, and the residual population as non-movers for the observation period. The figure demonstrates the evolution of the age pyramid over time and the decline in transit use across all cohorts. Other activity–travel dimensions – i.e., duration, type, or kilometres travelled by mode, can also be visualized using a similar approach.

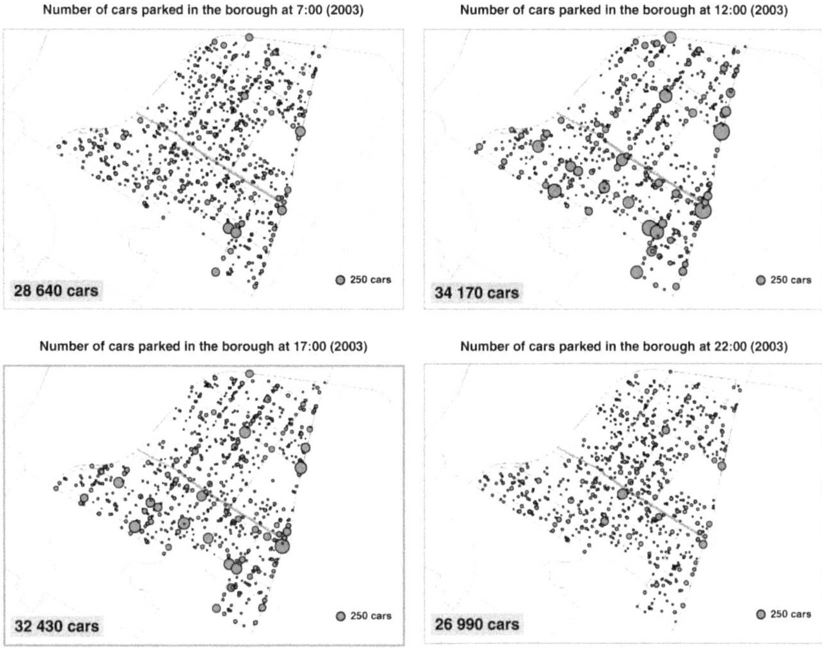

Fig. 8 Monitoring of the number of cars parked in a specific area during a typical weekday

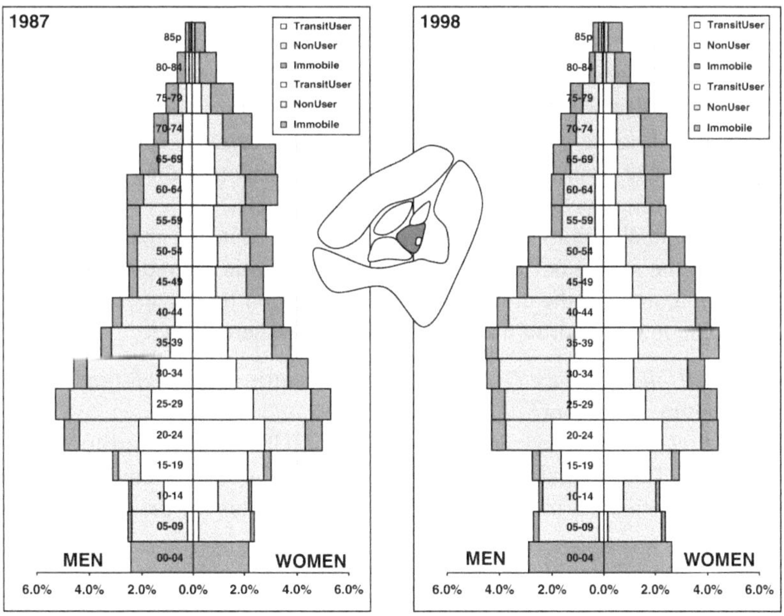

Fig. 9 Demographic structure with segmentation related to transit use (1987 & 1998 OD surveys), central Montreal

5 Land Use Processes

Land-use processes refer to the evolution, in space and time, of the objects that define how an area is structured, and to a certain extent, how space is used by various behavioural agents (e.g., individuals, firms, households). Land uses provide the built-environment foundation for activities and travel that occur in the "physical" city-region. Land-use processes, however, can also include the development of infrastructure that facilitate activities in Cyberspace and/or virtual worlds (e.g., Buliung 2007; Dodge and Kitchin 2001). In this chapter, focus is given to the geovisualization of land use in the physical city-region. The examples focus on the conceptualization of transportation as a "type" of land use, and the evolution of commercial development in space–time.

5.1 Transportation Network Coverage

The location of development, and consumers, relative to a city's central business district (CBD) is often used to describe patterns of development and settlement, and as a predictor of travel behaviour. This monocentric conceptualization of the spatial structure of the city-region has changed somewhat because many cities have evolved away from Von Thünen's isolated state (Anas et al. 1998; Shearmur and Coffey 2002). Nevertheless, distance to the CBD remains a variable of interest to scholars and practitioners interested in the dispersion and expansion of economic activity across space (e.g., Bonnafous and Tabourin 1998; Peguy 2002; Scheou 1998).

The radioconcentric conceptualization of space can be adjusted, however, with a view to understanding the space occupied by the transportation system by trading Euclidean distance with network distance, and Euclidean space with network space. This sort of abstraction can be valuable for understanding the coverage of transportation infrastructure (e.g., networks, stations, etc.).

Four conceptualizations of urban space are advanced in this chapter, and used to describe: (a) the spatial deviation of the city-region from the radioconcentric conceptualization, and (b) the coverage of the city-region by transportation facilities (Fig. 10). The four models of urban space considered here include:

1. *Isotropic Uniform Space (IUS)*: the radioconcentric space measured around an urban centre, of area πr^2
2. *Urban Area Space (UAS)*: an area equal to the IUS excluding territories beyond the extent of the spatial planning district (i.e., the space covered by a regional origin-by-destination travel survey, or the spatial extent of the developed urban area)
3. *Transportation Network Space (TNS)*: the surface covered by a passenger transportation network, estimated using a uniform buffer (e.g., 100 m) applied to either side of a network's road segments. The area of this space, for any radial distance r from the CBD, is given by the sum of buffered segments

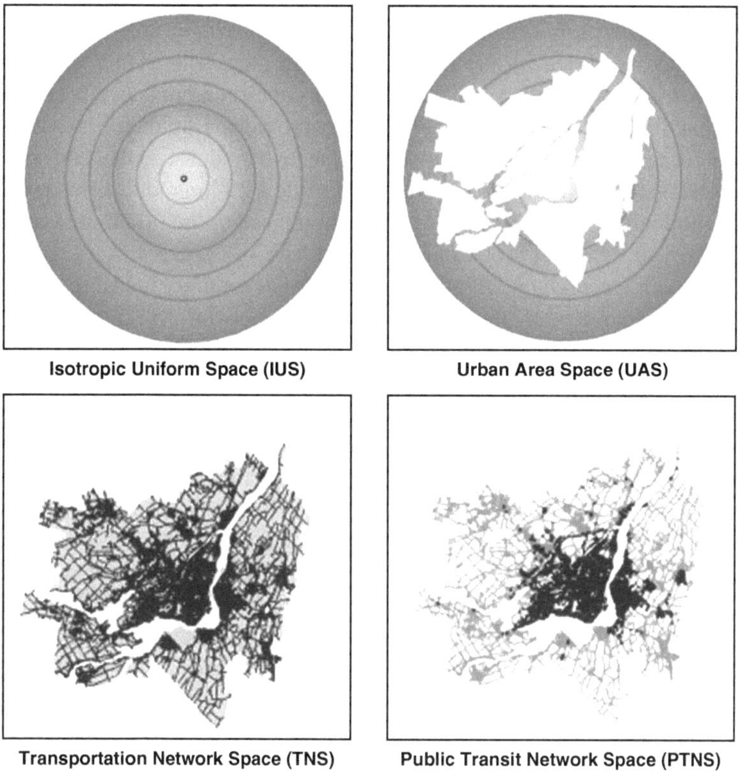

Fig. 10 Geopolitical and network based conceptualizations of urban areas

4. *Public Transit Network Space (PTNS)*: This space is a subset of the TNS since the public transit network is generally superimposed over the road network, even in the case of heavy transit (subway, rail) where stations are necessarily linked to the road network. This measure of transit network coverage is more complex because the level of supply fluctuates throughout the day, the week and the season. In this example, the PTNS is estimated by the application of an accessibility buffer (500 m) around every bus stop, subway and rail station, notwithstanding the level of service. Only the part of this accessibility buffer located inside the TNS is considered. The 500-m limit assures the coverage of the space accessible by foot, and avoids the inclusion of undeveloped areas (around heavy transit stations).

Within the current framework, and for a given radial distance r from the CBD, it is always the case that: $IUS \geq UAS \geq TNS \geq PTNS$. Relationships between these spatial abstractions are further developed here into indices that hold value for measuring the proportion of an urban area occupied by: (a) the urban area as defined by its geopolital extent; (b) the road transportation network; and (c) the passenger transit system. Following the introduction of the indices, two examples are developed for the GMA.

The *Spatial Discontinuity Index (SDI)* measures the proportion of the theoretical isotropic space occupied by the developed urban area. The SDI is expressed as: $SDI(r) = UAS(r)/IUS(r)$, where r is simply the radial distance used to define the isotropic limit of the estimation. The index provides a summary measure of the deviation of an urban area from the radioconcentric model. For the GMA, the city of Montreal occupies almost 90% of the IUS inside a 20-km radius. This proportion declines gradually beyond 30 km. At the regional scale, the urban area represents less than 40% of the IUS (> 60 km radius from CBD).

The *Network Occupancy Index (NOI)* provides a summary measure of the share of an urban area covered by the road transportation network: $NOI(r) = TNS(r)/UAS(r)$. In the GMA, approximately 37.5% of the urban area is covered by the transportation network. As expected, the *NOI* increases to roughly 90% inside a 10 km radius from the CBD. The *NOI* can be used to enhance understanding of geographical variation in the supply of road transportation facilities. For example, Fig. 11 shows the spatial distribution of the NOI estimated for 100 traffic analysis zones. As expected, the result suggests a decline in the use of space for road transportation with distance from the Montreal CBD.

Lastly, the *Transit Network Occupancy Index (TNOI)* expresses the proportion of the TNS allocated to operational transit services: $TNOI(r) = PTNS(r)/TNS(r)$. As Fig. 11 suggests, the supply of transit infrastructure declines steadily with distance from the Montreal CBD. At the GMA scale, transit covers roughly 40% of the transportation network. In the core area, the transit network practically covers the entire transportation network, while the supply of transit infrastructure appears to be unequivocally negligible in the suburban districts. These indicators provide a synthetic profile of network coverage at the metropolitan scale by mode, complementing classical measures of network accessibility (i.e., population and activity locations within a specified distance of transit facilities).

The spatial conceptualization and measurement of transportation network coverage described here can facilitate the development of unique spatial perspectives on the allocation of urban land to the transportation system. The relationship between the value of the various indices and proximity to the CBD are not entirely unexpected due to the historical and contemporary significance of central Montreal to the regional economy. Future work with these measures should, however, examine their sensitivity to several well known geographical estimation problems. The boundary delineation problem, and the scale and zoning effects associated with the modifiable areal unit problem (MAUP) are expected to influence the estimation results (Openshaw and Taylor 1979).

While an exhaustive and specific diagnosis of the sensitivity of these metrics to geographical estimation problems is beyond the scope of this chapter, there is some value in the hypothetical exploration of the issues. With respect to the boundary delineation problem, the SDI will exhibit sensitivity to the researcher's conceptualization and measurement of the built-up or developed UAS. The dynamic disequilibria of land use processes at the "edge," and the adaptive re-use or wholesale transformation of industrial spaces within the "urban area," complicates the task

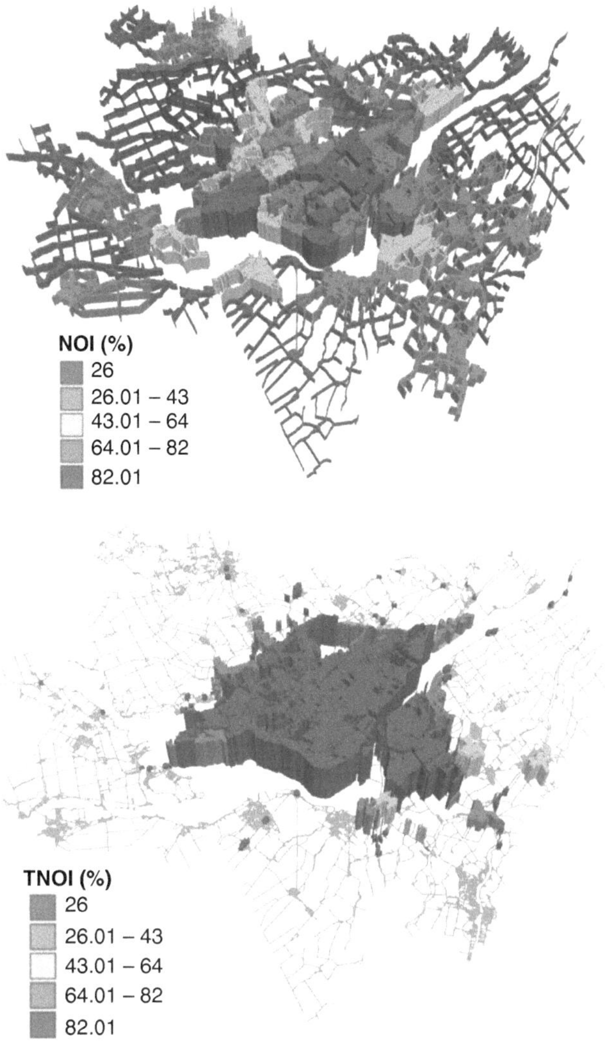

Fig. 11 Network Occupancy Index (top) and Transit Network Occupancy Index (bottom) estimated for 100 traffic analysis zones

of fixing the urban area to a discrete contiguous space. As a result, and given an IUS with a fixed radius, r, the value of the SDI will increase with the size of the UAS.

The scale component of the modifiable areal unit problem refers to the inconsistency or variation in results across scales of analysis, while the zoning effect pertains to changes in results manifest in the permutation of zone boundaries given a fixed analytical scale (Openshaw and Taylor 1979). Notably, the indicators of urban area and network coverage described above are sensitive to the specification of the scale parameter (r). Specifically, and as described in the text, estimates of urban area (SDI) and network (NOI, TNOI) coverage have been shown to decline with distance from the Montreal CBD.

This is not necessarily an undesirable property of these metrics because the scale dependent variability of the estimates is precisely what the analyst is seeking to describe (i.e., the change in network coverage with distance from a major centre). Perhaps more problematic, and related to the scale effect, is the identification of the specific intervals at which the indicators should be estimated, with a view to effectively communicating the geography of urban area and network coverage. Moreover, and specifically related to the indicators of network coverage, additional attention should be given to the specification of buffer-size (i.e., the proportion of a traffic zone covered by the network will increase with buffer-size).

With regard to zoning effects, the graphical display and estimation of the NOI and TNOI shown in Fig. 11 is expected to change with the permutation of zone boundaries. Assuming a fixed scale of analysis (i.e., the number of zones is held constant), the proportion of each zone covered by transportation facilities is expected to change with an adjustment to the geography of zone boundaries. Speculatively, and for Montreal, a similar understanding of spatial variation in network coverage at the global scale (e.g., across the GMA) is expected to materialize (i.e., the concentration of transit close to the CBD), despite local variation in the results. The next section changes course, demonstrating the use of spatial statistics for describing the evolution of commercial development processes in space–time.

5.2 Commercial Development in Space–Time

The most significant structural change in North America's retail economy during the last decade or more has been the introduction of new formats (Jones and Doucet 2000). In particular, the introduction of the "big-box" store – i.e., a store that is typically three times larger than other facilities in the same category, has fundamentally altered the geography of retailing, from the way consumers interact with retail opportunities, to the pace and spatial patterning of development. What follows is an experiment designed to demonstrate the utility of geovisualization as an approach to exploring regional variation in the intensity of *power centre* retail development (i.e., three or more big-box retailers with a shared parking lot) across the GTA during the period 1996–2005. In this context, geovisualization can be used as a mechanism to quantify spatiotemporal patterns of growth, with a view to improving our understanding of the geography of commercial development processes.

Data for the series of examples have been drawn from the Centre for the Study of Commercial Activity (CSCA) retail databases, Ryerson University, Toronto. The CSCA databases provide a comprehensive national inventory of Canadian retailing. Since the early 1990s, these data have been used to trace the evolution of new retail formats in the Canadian retail economy. Each record contains information on the location, type of business, size, and type of location (enclosed mall, big-box stores).

In this case, a database has been extracted to reflect the location of retail power centres across the GTA for each year covering the period 1996–2005. Each power centre is modelled as a planar coordinate pair (x_i, y_i), with a weight, w_i, attached to reflect the total retail square footage of each power centre location (i.e., the sum of the retail square footage of the individual retail locations included in the power centre).

Two spatial analytic approaches are demonstrated, the first approach involves the use of kernel estimation, a method developed to obtain smooth probability estimates from univariate or multivariate data (Bailey and Gatrell 1995). The kernel approach has been used elsewhere to describe spatial variation in various types of point events recorded in *cross-sectional surveys*. Examples include, firm locations (Maoh and Kanaroglou 2007), the incidence of disease (Bailey and Gatrell 1995), and the spatial and spatiotemporal variation in activity–travel behaviour (Buliung 2001; Kwan 2000). The second approach applies a centrographic statistic, weighted mean centre, to explore the spatial expansion of big-box retail capacity over time (Bachi 1963; Ebdon 1988).

The weighted bivariate kernel density used in the first example can be expressed as:

$$\hat{f}(x, y) = \sum_{i=1}^{n} K\left\{[w_i \cdot I_i] \cdot \frac{1}{2\pi\sigma^2} e^{\left[(x-\mu_x)^2+(y-\mu_y)^2\right]/2\sigma^2}\right\} \tag{1}$$

where $K\{\}$ is a bivariate probability density function referred to as the kernel, w_i is a weight attached to each power centre i (retail square footage), I_i is the intensity of the spatial point process at each observed power centre (i.e., the mean number of retail locations per unit area), x, y are planar x and y coordinates representing the location of each power centre, and σ is a scale parameter or bandwidth (specified in measurement units).

There are several decisions left to the analyst when applying kernel estimation. First, the kernel function $K\{\}$ can take one of several possible forms (e.g., Gaussian, quartic, triangular), and second, a decision needs to be taken regarding the value for the scale parameter, σ (Bailey and Gatrell 1995; Levine 2006). The choice regarding $K\{\}$ is typically guided by the application context. For example, regional scale analyses are potentially better suited to the application of the Gaussian distribution because the function returns estimates of spatial intensity for every tessellated location across the study area (Levine 2006).

With respect to the scale parameter, the degree of smoothing is influenced by the specification of σ, with larger values providing a smoother estimate of spatial intensity. While empirical approaches to identify an appropriate bandwidth have been introduced to the literature (Bailey and Gatrell 1995; Levine 2006; Rowlingson and Diggle 1993), selection can also arise from a quasi-empirical approach where the analyst qualitatively evaluates successive interpolations. A decision regarding the scale parameter will be influenced by the degree to which a particular value of σ

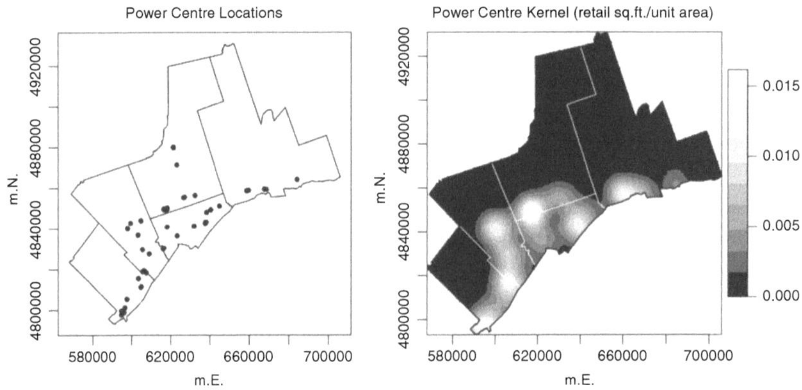

Fig. 12 Weighted Gaussian bivariate kernel estimation

provides an instructive view of the spatial process under examination. It is this latter approach that has been adopted in this case.

A demonstrative example of the input and result of the kernel estimation process is shown in Fig. 12 ($\sigma = 4{,}500\,m$). A bivariate kernel estimate has been created for the spatial pattern of weighted (retail square footage) power centre retail locations identified for the year 2000. The estimation was achieved by inputting the weighted power centre locations to a weighted Gaussian bivariate kernel function implemented in the *spatstat* library for R (see Baddley and Turner 2005). Regional trends in power centre retailing become apparent, as overlapping retail events (weighted point locations) are transformed into a smoothed surface of retail development intensity. The resulting surface provides a cross-sectional view of the multiple foci of power retailing in the GTA. The data indicate greater intensity of power retail development in the outer suburbs.

The power centre data from the CSCA database have been organized into a discrete time-series reflecting the spatial pattern of operational power centres during each year for the period 1996–2005. The procedure outlined above has been independently applied to each power centre distribution ($\sigma = 4{,}500$ m).

Notably, and while it is not possible to demonstrate in the print medium, the panels have also been organized into an animation, established to illustrate the spatiotemporal evolution of power retail development (Buliung and Hernandez 2007). A uniform scale and gradient are applied to all panels based on the range of estimated values for the year 2005. This approach facilitates visual comparison across the years included in the analysis.

The visualization effectively communicates the suburban focus of power retail development during the 10-year period. Overall, the data suggest that (a) power retailing has emerged primarily as a suburban phenomena; (b) that there is regional variation in the intensity of this sort of retail activity; (c) that specific regional nodes have materialized during the last decade or more; (d) that established nodes within the inner and outer-suburbs have, in many cases, continued to expand over time; and (e) that there is one sub-area in particular, at the north-western edge of the City of

Toronto, that appears to have the largest share of overall power retail capacity in the region.

Centrographic analysis has also been conducted to demonstrate an alternative and perhaps more general approach to describe the geographical trend in power centre growth. Centrographic statistics include bivariate summary measures such as the mean centre and standard distance (Bachi 1963). The mean centre of a spatial point pattern is essentially a bivariate extension of the univariate mean (Bachi 1963). When spatial point data are available in a time series, estimation and geovisualization of the mean centre across a set of prescribed time intervals can provide insight into the geographical migration of a process through time. For example, the US Census Bureau has used the mean centre to illustrate the westward movement of the population of the US between 1790 and 2000 (US Census Bureau 2007). Applying a weight to the mean centre has the advantage of drawing the bivariate mean toward point events with large weights.

The weighted mean centre for the spatial point pattern of power centre locations, from a particular year in the longitudinal power centre database can be expresses as:

$$\bar{x}_t = \frac{\sum_{i=1}^{n} w_{i,t} x_{i,t}}{\sum_{i=1}^{n} w_{i,t}}, \quad \bar{y}_t = \frac{\sum_{i=1}^{n} w_{i,t} y_{i,t}}{\sum_{i=1}^{n} w_{i,t}} \tag{2}$$

where $(\bar{x}_t, \bar{y}_t)$ is a 1×2 coordinate vector defining the location of the weighted mean centre in Cartesian space for time t (e.g., one of the years during the period 1996–2005), $x_{i,t}$ and $y_{i,t}$ are planar coordinates describing the geographical location of each power centre destination i during year t, and $w_{i,t}$ is a weight variable (i.e., total retail square footage of power centre i in year t).

The result of estimating the weighted mean centre (WMC) for each power centre distribution during the period 1996–2005 is shown in panel A of Fig. 13. Panel B shows the same series of WMCs displayed at a scale defined by the minimum and maximum coordinate locations of the WMC points in the series. In addition, each WMC in panel B has been scaled to reflect the change in retail square footage within the overall retail system from one year to the next. This scale factor, δ, is expressed as:

$$\delta = \frac{\left[\sum_{i=1}^{n} S_{i,t} - \sum_{i=1}^{n} S_{i,t-1}\right]}{\sum_{i=1}^{n} S_{i,t-1}} \cdot 10 \tag{3}$$

where $S_{i,t}$ represents the total retail square footage of a power centre i in the time series during year t, and $S_{i,t-1}$, is the total retail square footage of a power centre i during the previous year. All WMCs in the series have been connected by a hatched line (WMC Path) to aid in the visual interpretation of the migration of the power centre development process during the 10-year observation period.

Complementing the results of the kernel estimation, the evidence suggests a period of overall expansion of power retail capacity measured in terms of retail

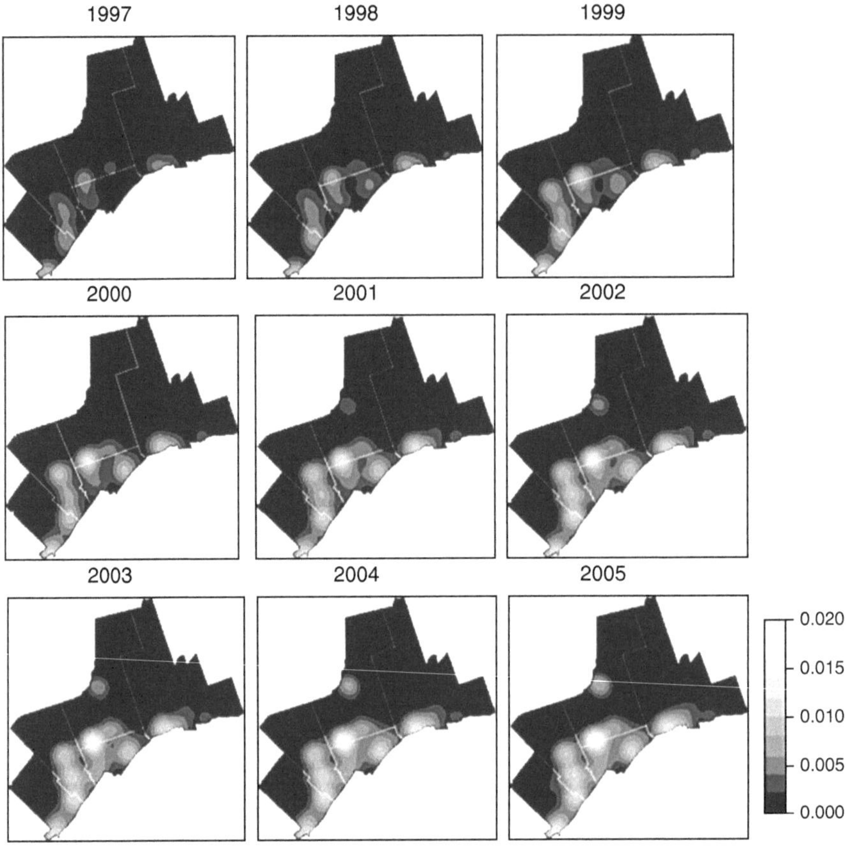

Fig. 13 Geovisualization of power retail capacity in the Greater Toronto Area (1997–2005)

square footage. With respect to the geography of the development process, the WMC path suggests the presence, in time, of periods of outward expansion and horizontal infill (i.e., take note of the growth between 1998 and 1999, versus the changing location of the WMC estimates from 1999 to 2001). The period of relatively rapid growth in power centre capacity that occurred during the late 1990s, a time period in which the study area experienced strong economic growth, appears to be followed more recently by positive, but marginal capacity increases.

The development of power retail capacity within the GTA has been explored in this section using bivariate kernel estimation and centrographic statistics. The bivariate kernel provides a more intuitive visual sketch of retail change for both specialist and non-specialist audiences than the centrographic analysis. The geovisualization of retail development presented in Fig. 14 arguably satisfies many of the criteria for graphical excellence outlined by Tufte (2001). The viewer is encouraged to think about "process" over "method," the data are presented in a uniform, coherent manner, the example serves a clear purpose, a relatively large volume of data are

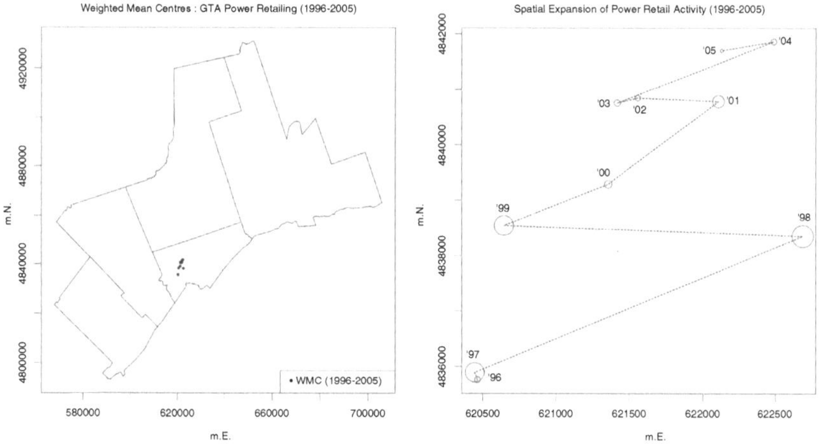

Fig. 14 Centrographic estimation and geovisualization of power centre expansion

displayed in a relatively small space, and the data are nested within statistical and verbal descriptions of method and process.

6 Conclusion

The *geovisualization* process is integrative; combining theory, methods, and technology to facilitate the construction of knowledge about processes through the identification of patterns in spatial data. If executed thoughtfully, geovisualization can be used to effectively communicate "ideas" to heterogeneous audiences comprised of individuals with and without specialized knowledge about the processes under investigation. Arguably, the practice is most effective when it serves as "part" of a conversation, and when it simultaneously reduces the complexity or abstraction of the original information, revealing fundamental qualities of spatial phenomena.

In this chapter, attention has been given to the geovisualization of transportation and land use processes. Most examples were developed with a view to exposing the dynamic behaviours of urban agents (e.g., individuals), objects (e.g., cars), and systems (e.g., transportation networks, commercial activities). The examples demonstrate how intrinsically abstract, geographically situated data structures (e.g., origin-destination matrices, retail databases) can be transformed into intuitive spatiotemporal constructs that shed light on the use and evolution of urban systems in space and time. The various examples also illustrate how visualization tools can assist in unravelling complex phenomena and translate, into a universal language, the most important outputs and trends sometimes hidden within the original data.

Many of the examples, particularly those drawn from the GMA (e.g., accumulation profiles, transportation-oriented age pyramids), have been used to communicate the spatiotemporal and demographic qualities of the city-region's transportation and land use system to professional planners, engineers, and other community stakeholders. The examples from the GTA have yet to be applied in a similar manner, and have been constructed with a view to visualizing what is intuitively understood about the demand for travel in the GTA, and to add-value to research aimed at understanding commercial development processes. In this regard, the chapter included examples covering the continuum from scientific visualization to public realm communication (MacEachren 1995).

A key ingredient facilitating the sort of exploratory and visual analyses described in this chapter is the availability of the necessary expertise and resources to gather and interpret the appropriate data. While the outcome of geovisualization can be graphical results intended for a lay audience, it goes without saying that the analyst should have a rich, theoretically grounded understanding of the processes being conveyed. In addition, and reflecting upon the various epistemological traditions that permeate the social sciences, engineering, and natural sciences, "seeing" should more appropriately be viewed as "part" of, or one approach for constructing or testing "beliefs" about urban processes. Geovisualization is simultaneously an "art and science," a technology enabled, data-driven practice that can arguably add value to inquiry, no matter the lens.

Acknowledgements The authors wish to thank the anonymous reviewers for their contributions to this manuscript. The first author wishes to thank Dr. Tony Hernandez at the Centre for the Study of Commercial Activity, Ryerson University for providing access to the retail opportunities data. The second author extends her gratitude to the transport authorities by whom the large-scale surveys, mainly Household Origin-Destination surveys, are conducted. Those surveys authorize the continuation of a travel behaviour observational and analytical culture in the GMA: STM, RTL, STL, AMT and MTQ. Both authors acknowledge support from the Natural Sciences and Engineering Research Council of Canada (NSERC).

References

Al-Kodmany K (1999) Using visualization techniques for enhancing public participation in planning and design: process, implementation, and evaluation. Landsc Urban Plann 45:37–45

Al-Kodmany K (2002) Visualization tools and methods in community planning: from freehand sketches to virtual reality. J Plann Lit 17:189–211

Anas A, Arnott R, Small KA (1998) Urban spatial structure. J Econ Lit 36:1426–1464

Anselin L (1995) Local indicators of spatial association – LISA. Geogr Anal 27:93–115

Anselin L (2000) Computing environments for spatial data analysis. J Geogr Syst 2:201–220

Anselin L, Syabri I, Kho Y (2006) GeoDa: an introduction to spatial data analysis. Geogr Anal 38:5–22

Bachi R (1963) Standard distance measures and related methods for spatial analysis. Pap Reg Sci Assoc 10:83–132

Baddley A, Turner R (2005) Spatstat: an R package for analyzing spatial point patterns. J Stat Software 12:1–42

Badoe DA, Miller EJ (2000) Transportation-land use interaction: empirical findings in North America, and their implications for modeling. Transport Res D 5:235–263

Bailey TC, Gatrell AC (1995) Interactive spatial data analysis. Addison-Wesley-Longman, Cambridge

Bivand RS (2006) Implementing spatial data analysis software tools in R. Geogr Anal 38:23–40

Bivand RS, Neteler M (2000) Open source geocomputation: using the R data analysis language integrated with GRASS GIS and PostgreSQL data base systems. Presented at GeoComputation 2000, University of Greenwhich, Kent

Black WR (2001) An unpopular essay on transportation. J Transport Geogr 9:1–11

Bonnafous A, Tabourin E (1998) Modélisation de l'évolution des densités urbaines, in Données Urbaines n°2, ed. Anthropos, mai 98: 273–285

Borgeat L, Godin G, Massicotte P, Poirier G, Blais F, Beraldin J-A (2007) Visualizing and analysing the Mona Lisa. Real-time interaction with complex models, IEEE Computer Society, Nov./Dec. 2007

Buliung RN (2001) Spatiotemporal patterns of employment and non-work activities in Portland, Oregon. In: Proceedings of the 2001 ESRI International User Conference, San Diego, California

Buliung RN (2007) Broadband technology and metropolitan sustainability: an interpretive review. Report submitted to The Ministry of Government Services, Province of Ontario, Broadband Research Initiative. Available at: http://kmdi.utoronto.ca/broadband/publications/default.html, 29 Sept. 2009

Buliung RN, Hernandez T (2007) The growth and change of retail opportunity in the Greater Toronto Area. Paper presented at the 54th Annual Meeting of the North American Regional Science Association International, Savannah, GA, November 2007

Buliung RN, Kanaroglou PS (2006a) A GIS toolkit for exploring geographies of household activity/travel behavior. J Transport Geogr 14:35–51

Buliung RN, Kanaroglou PS (2006b) Urban form and household activity-travel behaviour. Growth Change 37:174–201

Buliung RN, Remmel TK (2008) Open source, spatial analysis, and activity-travel behaviour research: capabilities of the aspace package. J Geogr Syst 10:191–216

Buliung RN, Roorda MJ, Remmel T (2008) Exploring spatial variety in patterns of activity-travel behaviour: initial results from the Toronto Travel-Activity Panel Survey (TTAPS). Transportation 35:697–722

Center for Spatially Integrated Social Sciences (CSISS) (2008) CSISS classics. Available at: http://www.csiss.org/classics/. Accessed Mar 2008

Cervero R, Kockelman KM (1997) Travel demand and the 3ds: density, diversity, and design. Transport Res D 2:199–219

Chapleau R, Morency C (2005) Dynamic spatial analysis of urban travel survey data using GIS. Twenty-Fifth Annual ESRI International User Conference, San Diego, California

Civic Transportation Committee [map] (1915) Scale not given. "Diagram Showing Home-ward Passenger Movement During The Evening Rush Period Mid-week Conditions 4–30 To 7–30 P.M." University of Toronto Data, Map & Geographic Information Systems Centre. http://prod.library.utoronto.ca:8090/maplib/digital/rushhour.jpg. 24 Sept. 2008

Crane R (2000) The influence of urban form on travel: an interpretive review. J Plann Lit 15:3–23

Dodge M, Kitchin, R (2001) Mapping cyberspace. Routledge, New York

Downs, A. (2004) Still stuck in traffic. Brookings Institute, Washington

Ebdon D (1988) Statistics in Geography, 2nd edn. Blackwell, Oxford

Gahegan M (2000) The case for inductive and visual techniques in the analysis of spatial data. J Geogr Syst 2:77–83

Goodchild, MF, Janelle, DG (1984) The city around the clock: space-time patterns of urban ecological structure. Environ Plann A 16:807–820

Haining R, Wise S (1997) Exploratory spatial data analysis, NCGIA Core Curriculum in GIScience. Available at: http://www.ncgia.ucsb.edu/giscc/units/u128/u128.html, Mar. 2008

Haining R, Wise S, Ma J (1998) Exploratory spatial data analysis in a geographical information system environment. The Statistician 47:457–469

Haining R, Wise S, Ma J (2000) Designing and implementing software for spatial statistical analysis in a GIS environment. J Geogr Syst 2:257–286

Haining R, Wise S, Signoretta P (2000) Providing scientific visualization for spatial data analysis: criteria and an assessment of SAGE. J Geogr Syst 2:121–140

Health Canada (2002) Canada's aging population. (Cat. H39–608/2002E). Minister of Public Works and Government Services, Ottawa, Canada

Hearnshaw HM, Unwin D (1994) Visualization in geographical information systems. JohnWiley, Chichester

Heisz A., LaRochelle-Côté S (2005) Work and commuting in census metropolitan areas, 1996–2001. (Catalogue No. 89–613-MIE). Ottawa, Statistics Canada

Janelle DG, Goodchild M (1983) Diurnal patterns of social group distributions in a Canadian city. Econ Geogr 59:403–425

Jones K, Doucet M (2000) Big-box retailing and the urban retail structure: the case of the Toronto area. J Retailing Consum Serv 7:233–247

Jones, P.M. (1979) New approaches to understanding travel behavior: the human activity approach. In: Hensher DA, Stopher PR (eds) Behavioural travel modelling, Redwood Burn Ltd, London, pp 55–80

Kwan, MP (2000) Interactive geovisualization of activity-travel patterns using three-dimensional geographical information systems: a methodological exploration with a large data set. Transport Res C 8:185–203

Levine N (2006) Crime mapping and the Crimestat program. Geogr Anal 38:41–56

Lewis JL, Sheppard SRJ (2006) Culture and communication: can landscape visualization improve forest management consultation with indigenous communities? Landsc Urban Plann 77:291–313

MacEachren AM (1995) How maps work: representation, visualization, and design. The Guilford Press, New York.

MacEachren AM, Kraak M-J (2001) Research challenges in geovisualization. Cartography and Geographic Information Science 28:3–12

Maoh H, Kanaroglou PS (2007) Geographic clustering of firms and urban form: a multivariate analysis. J Geogr Syst 9:29–52

Morency C (2004) Contributions à la modélisation totalement désagrégée des interactions entre mobilité urbaine et dynamiques spatiales, Thèse de doctorat, École Polytechnique de Montréal

Morency C, Chapleau R (2008) Age and its relation with home location, household structure and travel behaviors: 15 years of observation. Paper presented at the 86th Annual Meeting of the Transportation Research Board, Washington, DC

Morency Catherine, Trépanier Martin, Characterizing parking spaces using travel survey data, CIRRELT, CIRRELT-2008-15, 2008

Morency C, Saubion B, Trépanier M (2006) Evaluating the use of parking spaces in strategic urban areas using travel survey data. Paper presented at the 2006 North American Meetings of the Regional Science Association International 53rd Annual Conference, Toronto

Openshaw S, Taylor PJ (1979) A million or so correlation coefficients: three experiments on the modifiable areal unit problem. In: Wrigley N (ed) Statistical applications in the spatial sciences. Pion, London, pp 127–144

Peguy P-Y (2002) Analyse économique des configurations urbaines et de leur étalement, Thèse pour le Doctorat en Sciences Économiques, mention Économie des Transports, Université Lumière Lyon 2, Faculté de Sciences Économiques et de Gestion

Rey SJ, Janikas MV (2006) STARS: space-time analysis of regional systems. Geogr Anal 38:67–86

Rowlingson BS, Diggle PJ (1993) Splancs: spatial point pattern analysis code in s-plus. Comput Geosci 19:627–655

Scheou, B (1998) L'estimation de la population totale à un niveau communal: utilisation du modèle de René Bussière, Document de travail no 98/01, http://web.mrash.fr/let/francais/indexpub.htm, Dec. 2002

Shaw S-L, Wang D (2000) Handling disaggregate spatiotemporal travel data in GIS. GeoInformatica 4:161–117

Shaw S-L, Bombom LS, Wu H (2008) A space-time GIS approach to exploring large individual-based spatiotemporal datasets. Transactions in GIS 12:425–441

Shearmur R, Coffey WJ (2002) A tale of four cities: intrametropolitan employment distribution in Toronto, Montreal, Vancouver, and Ottawa–Hull, 1981–1996. Environ Plann A 34:575–598
Takatsuka M, Gahegan M (2002) GeoVISTA studio: a codeless visual programming environment for geoscientific data analysis and visualization. Comput Geosci 28:1131–1144
Time (2007) One day in America. Available at: http://www.time.com/time/2007/america_numbers/commuting.html, Mar. 2008
Tress B, Tress G (2003) Scenario visualisation for participatory landscape planning – a study from Denmark. Landsc Urban Plann 64:161–178
Tufte ER (2001) The visual display of quantitative information, 2nd edn. Graphics Press, Cheshire
Tukey J (1977) Exploratory data analysis, Addison-Wesley, Cambridge
US Census Bureau (2007) Mean centre of the population of the United States: 1790 to 2000. Available at: http://www.census/gov/geo/www/cenpop/cntpop2k.html. Accessed Dec 2007
Willson RW, Shoup DC (1990) Parking subsidies and travel choices: assessing the evidence. Transportation 17:141–157
Wise SM, Haining RP, Signoretta P (1999) Scientific visualization and the exploratory analysis of area-based data. Environ Plann A 31:1825–1838

Pattern-Based Evaluation of Peri-Urban Development in Delaware County, Ohio, USA: Roads, Zoning and Spatial Externalities

Darla K. Munroe

1 Introduction

As urban areas continue to disperse and decentralize, new urban growth is increasingly occurring in peri-urban or rural areas beyond the suburban fringe, but within commuting distance of metropolitan areas. This trend is referred to in a variety of ways, including urban expansion, urban dispersion, or peri-urbanization. Many communities are concerned with seemingly uncontrolled urban sprawl and expansion into peri-urban areas for a variety of reasons, including the fiscal, environmental and social impacts associated with urban land-use change. Urbanization can alter major biogeochemical cycles, add or remove species, and have drastic effects on habitat (Vitousek et al. 1997), particularly when such development is low-density and scattered (Theobald 2004). Urban decentralization can also decimate the inner-city tax base (Downs 1999). Growth at the urban fringe, or in the rural portions of metropolitan counties, has greatly increased, and is of significantly lower density than the surrounding urbanized areas and clusters (Heimlich and Anderson, 2001). In Ohio, low-density development outside urbanized areas has increased from 58 to 72% of total land area between 1970 and 2000 (Partridge and Clark 2008).

Explanations for fragmented urban development patterns include effective decreases in development costs due to improvements in roads and highways (Anas et al. 1998), flight-from-blight processes as medium-higher income residents flee perceived urban ills such as higher crime and lower quality schools in search of greater stability in outlying areas (Mieskowski and Mills 1993; Downs 1999), avoidance of spatial externalities such as congestion (Irwin 2002), uncoordinated local land use policy where jurisdictions compete for new growth (Carruthers and Ulfarsson 2002; Byun and Esparza 2005), and the effects of developer behavior (Byun and Esparza 2005). All of these explanations could be paired against the so-called "natural evolution" hypothesis (Mieskowski and Mills 1993) that states that urban

D.K. Munroe
Department of Geography, The Ohio State University, 154 N. Oval Mall, Columbus, OH 43210, USA,
e-mail: munroe.9@osu.edu

A. Páez et al. (eds.), *Progress in Spatial Analysis*, Advances in Spatial Science,
DOI 10.1007/978-3-642-03326-1_8,

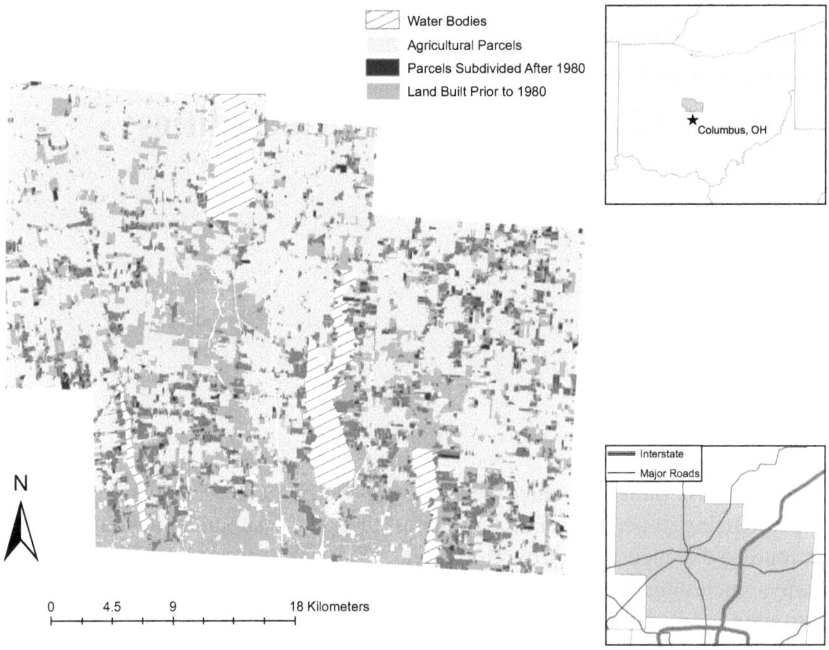

Fig. 1 Study area

growth and decentralization are inevitable in an era of rising average incomes and falling agricultural prices, which effectively reduce the opportunity cost of land development at the urban fringe.

This chapter reports on a study conducted to examine patterns of land conversion and development within one of the fastest growing counties in the U.S., Delaware County, located near Columbus, Ohio (Fig. 1). According to the U.S. Census, the population of Delaware County was 66,929 in 1990 and 109,989 in 2000, a 64% increase. Residential land conversion is continually unfolding, and the location of past changes influence future conversion (An and Brown 2008). Explaining peri-urban development, and designing policy requires careful attention to (1) the inter-temporal nature of urban growth; (2) the spatially heterogeneous landscape upon which such growth is ultimately distributed; and (4) how overall development patterns reflect such processes in the aggregate. Several techniques are used to study the growth and change of Delaware County in time and space. First, a descriptive analysis of the spatial pattern of urban change was undertaken to quantify the amount of urban decentralization that has occurred concomitant with a qualitatively new pattern of growth. Then, a complementary log–log survival model of land conversion was estimated to explain the overall impact of various spatial factors on the timing of development. Finally, parameters from the survival models were used to simulate predicted patterns of growth under three scenarios. In these scenarios, hypothetical development risk was estimated holding the following factors constant (i.e., as if they had not influenced the pattern): (1) proximity to roads, (2) maximum density policies, and (4) the percentage of development within 1/2 to 3/4 miles of each parcel, to examine how each of these factors influenced the overall configuration of exurban development.

The remainder of the chapter is organized as follows. First, a description of the study area is given. Then, the conceptual model underlying the empirical analysis is explained. Third, the methods are described, including the implementation of a complementary log–log survival model to estimate the probability of land conversion between 1988 and 2003, and a landscape pattern analysis of Delaware County. The discussion of the results is followed by a summary of the paper and concluding remarks.

2 Study Area and Data

Delaware County is located within the Columbus, Ohio, Metropolitan Statistical Area (MSA). Its county seat, the city of Delaware, lies just 27 miles (44 km) from downtown Columbus (Fig. 1) and 12 miles (20 km) from the county's southern boundary. Since 1995, growth within the Columbus MSA was significantly higher than in any other metropolitan area in Ohio, especially when compared with its outlying counties, such as Delaware (Partridge et al. 2007). Despite dramatic residential development, Delaware County remains agriculturally important. Though the county is generally flat, with much prime agricultural land, there is some variation in topographical relief, and more importantly, there are several water reservoirs throughout (Hite et al. 2003), with recreational opportunities including water sports and hiking trails.

Parcel data and records of land conversion between 1988 and 2003 were obtained from the Delaware County Auditor. The year 1988 reflects a period just before development accelerated and became a public concern in Columbus. For example, the impact of "urban sprawl" on local farmers was first discussed in the early 1990s (see Steel, 1992). Data obtained from the tax assessor included parcel boundaries, zoning delineation, tax district and the year a new structure was built. Because the focus of this paper is land-use conversion, a "peri-urban field" was defined according to Clark et al. (2005). This field includes only those parcels outside portions of the county designated by the U.S. Census in 2003 to be "urbanized" or "urban clusters" (i.e., leaving parcels in rural areas at the urban fringe). It is possible that certain areas classified as urbanized in 2003 would have been agricultural in 1988, effectively underbounding the true peri-urban field. However, a more conservative representation of peri-urbia may be preferable to including observations within the higher density portions of the county.

The subdivision of an agricultural parcel in the process of conversion to residential development is the fundamental process of interest here. Records of the original agricultural parcels are not directly available, however. From the database of parcels developed in rural areas, individual parcels developed as one subdivision were determined. Out of 69,467 total parcels developed since 1980, 51,157 parcels or roughly 74% were developed within subdivisions. The remaining 18,310 parcels (26%) were recorded as developed individually, not as part of a subdivision. In order to represent the subdivision of an agricultural parcel as the key theoretical process, parcels developed within one subdivision were aggregated to one observation, and parcels developed individually were kept as unique observations.[1] Within the peri-urban field, those agricultural parcels that were either converted to residential uses or remained in agricultural use from 1988 to 2003 summed to 7,245 observations.

The attribute "year built" is used as a proxy for the year the original agricultural parcel was subdivided because the date of sale of the undeveloped parcel was unavailable. Year built can be problematic because if a structure on an existing residential parcel is torn down and rebuilt, "year built" reflects the year of this new construction. This problem is less likely in the study area because the peri-urban field lies outside of urbanized areas and urban clusters where most such redevelopment occurs, and the greatest amount of residential land conversion has happened during the time interval under observation (1988–2003). One remaining confounding factor that cannot be observed is whether agricultural land is bought by a third party and held for speculative value until some further time point. Data on such speculative purchase of land is difficult to obtain. If speculative purchases occur randomly within a county, this effect would not confound the identification of other key influences.

3 Processes Underlying Peri-Urban Development

3.1 Timing of Development

In order to examine the processes underlying land conversion at the urban fringe, economists have traditionally focused on explaining how rural land uses, such as agriculture, come to be replaced by other urban uses, such as residential. Such explanations generally compare estimated economic returns from agriculture with those of urban uses. Agricultural returns may be relatively easy to observe, given the underlying productivity of a particular parcel and producer price indices. Changes in urban returns, or urban land rent, may be harder to quantify, as conversion is the local realization of urban-level growth processes (due to income growth, population growth, or both (Munroe and York 2003)). Therefore, urban growth pressure, which

[1] There are two possible measurement errors that could be induced by this method: (1) multiple agricultural parcels were simultaneously converted within one subdivision; and (2) one agricultural parcel was subdivided to create more than one of the developed parcels not part of a subdivision. To the degree that the optimal development timing process considered the full spatial extent of the eventual subdivisions, problem (1) should not cause bias in the model coefficients. Problem (2) could potentially cause bias, but of unknown direction; because estimation was run on a spatially stratified sample, this potential bias was likely minimized.

in turn increases the opportunity cost of agricultural production at the urban fringe, can be conceptualized as a latent variable that is not directly observable, but may be related to timing.

Capozza and Helsley (1989) developed the classic model of development timing by postulating that land should be converted to a residential use when the expected annualized value of residential benefits is equal to foregone agricultural rents plus conversion costs. Irwin (2002) adapted this model to yearly time steps, which is not unreasonable given an expected time lag between the sale, subdivision and subsequent development of an agricultural parcel. The value of development of an agricultural parcel i, at the optimal time t^* can be represented as:

$$V\left(i,t^*\right) - \sum_{t=0}^{\infty} A\left(i,t^* + t\right)\delta^t, \tag{1}$$

where V is net returns from the sale of the parcel i less conversion costs, A is foregone agricultural returns from farming an additional time step t, and δ is the discount rate of the landowner, i.e., the amount by which she discounts the value of the sale of the parcel in the future (Irwin 2002). Because land value is rising over time, if the agricultural landowner is to maximize returns from the sale of a given parcel that is unique in its particular attributes, it may be worthwhile to delay development of the parcel until some future date when net returns are at their highest. Irwin (2002) state that if returns are positive (i.e., exceed agricultural revenues), and the landowner will receive more at time t^* than $t^* + 1$, the owner will develop at the smallest t (i.e., at the earliest possible time) satisfying the inequality:

$$\frac{V\left(i,T+1\right) - \{V\left(i,T\right) - A\left(i,T\right)\}}{V\left(i,T\right) - A\left(i,T\right)} < r, \tag{2}$$

where the ratio of additional returns from waiting another year to develop, over the returns in developing during the current time period, is less than the interest rate. Thus, the analyst must pay attention (a) to the fact that parcels are heterogeneous in terms of their underlying productivity for agriculture or respective market value for development, based on spatial attributes, and (b) to the temporally dependent nature of the development process; the underlying risk of development is higher at certain times than others. How these two factors come together in a particular setting will influence the total amount and spatial configuration of new residential land.

3.2 Spatial Influences

There are myriad spatial factors, operating at various scales, influencing urban land-use change and urban form (Verburg et al. 2004). Urban land-use conversion may result from federal policies such as mortgage interest deductions, regional population and employment trends, accessibility and available transportation to local community public goods and amenities (Zhang 2001). It is generally assumed that development is more likely to occur in areas where conversion costs arising from topographical variations are lowest. Ironically, the soil that tends to be the best for agricultural production is also prime for development. The ruggedness of the terrain can also increase conversion costs, though households may prefer areas with such relief (Lake et al. 2000).

Any number of land-use policies, such as development taxes and zoning, can affect the timing of development, as well as the amount and location of development. Hite et al. (2003) has emphasized that local variations in tax rates and public services matter a great deal when examining development trends in the peri-urban landscape. The timing and the amount of development may be affected because of increased conversion costs, as well as the opportunity costs induced by restrictive development policies limiting land conversion. Interestingly, the spatial location of development can be affected both directly and indirectly by policy. First, the spatial pattern that development takes can directly relate to new costs generated by policy, particularly if such policies are heterogeneous in space. For example, minimum lot sizes may be higher in outlying areas, causing development in those areas to be less densely distributed than it would be without the policy. There can also be indirect spatial effects; restrictions on development in one location, all things equal, could lead to increased development in a neighboring location. For example Esparza and Carruthers (2000) showed that land-use density requirements led to the "leapfrogging" of development in Arizona.

Prior research (Irwin et al. 2003; Irwin 2002, 2004; Munroe 2007) has demonstrated that spatial externality effects can influence returns to land use. These influences are not generated by the individual land owner, but are a cost or a benefit accruing to a land owner. For instance, Irwin (2002) demonstrated that the estimated pattern of development was significantly more fragmented along the Baltimore – Washington, D.C. corridor, all other factors held equal, due to an observed pattern of avoiding previously densely developed areas. Land conversion in the neighborhood of a particular parcel can also alter the timing of development for neighboring parcels. Access to opportunities such as commercial land, and the desire to avoid industrial land uses can also be important (Munroe 2007). Finally, proximity to valuable natural amenities has been shown to be an important part of returns to land use, often making a significant contribution to land value (Irwin 2002; Campbell and Munroe 2007).

One can assume that urban conversion is likely to be irreversible: once a parcel has been subdivided and built-up, it is unlikely that it will return to its agricultural state. Therefore, it is reasonable to assume that there is a spatiotemporal path dependence in how development unfolds (An and Brown 2008). Two implications of development timing models have particular relevance for understanding the spatial pattern of land conversion. First, there must be net benefits to clearing, less conversion costs; and second, it may be optimal to delay conversion until some time in the future when net benefits are higher. Thus, though *on average* more desirable land is likely to be converted, *at the margins* the most desirable land may be held for future conversion. For example, a plausible scenario may be that early on, there is limited demand for newly converted land in peri-urban regions, and marginally productive agricultural areas would be the first to drop out. As development pressure increases, developers may “upgrade” their offered product by selecting those areas most likely to command the highest premium, and other factors may become more important. Spatially, this implies that areas proximate to valuable natural amenities, for example, could be developed later because the market price per acre for these parcels will be higher.

4 Methods

Statistical land-use models can be useful for summarizing available information regarding past changes, in order to investigate underlying processes or predict where new change is likely to occur (Veldkamp and Lambin 2001; Munroe and Müller 2007). Statistical models are often less useful in providing insights on the development process when the process itself is not temporally stationary (An and Brown 2008), though there is a growing literature engaging with the issue of the timing of land-use change more centrally (Irwin 2002, 2004; Vance and Geoghegan 2002; Hite et al. 2003; An and Brown 2008). In this chapter, both survival models and landscape pattern analysis approaches were used to study patterns of land conversion and development within Delaware County, Ohio. The analysis described in the remainder of the chapter proceeds as follows: (1) descriptive analysis was conducted of the spatial pattern of development within the study area between 1988 and 2003; (2) the association between spatial factors and the observed timing of development was then examined using multivariate survival models; (4) the sensitivity of the results to variation in key processes was examined by holding several development and policy-based factors constant and re-estimating development probabilities; and (5) the likely aggregate spatial effect of these factors on the overall pattern of development (e.g., dispersed, compact) is considered.

4.1 Landscape Pattern Analysis of Development 1988–2003

Three measures of landscape pattern were used to examine variations in the configuration of recent residential development in Delaware County: the number of patches (e.g., the number of clusters of contiguous parcels that share a common boundary), the landscape shape index (LSI), and Euclidean nearest neighbor distance between noncontiguous (isolated) patches of developed land. Each of these approaches is briefly described below.

In order to quantify landscape fragmentation (e.g., urban land conversion that affects the configuration and connectivity of remaining agricultural land in a nonlinear way), it is useful to characterize new developments in terms of the relative complexity of the landscape, as some function of the edge to area ratio. In other

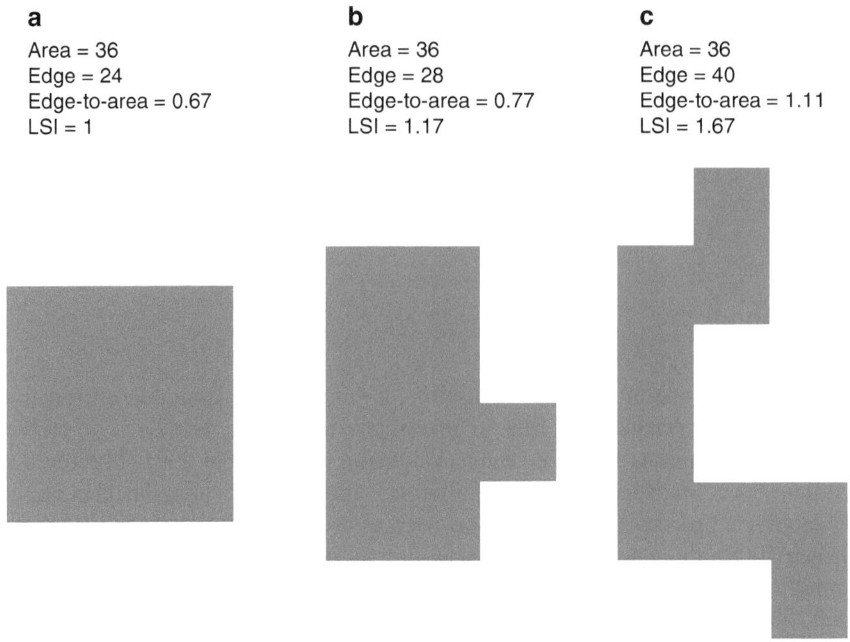

Fig. 2 Graphical illustration of variations in edge-to-area ratio and the corresponding landscape shape index (LSI). (**a**) A square patch made up of nine individual squares of dimension 2×2. (**b**) A non-square patch made up of the same nine individual squares, arranged less squarely. (**c**) A non-square patch made up of nine individual squares, arranged nearly linearly

words, more compact development has a smaller impact on the configuration of remaining agricultural land. The LSI is a desirable measure of the amount of edge in the landscape because it controls for the fact that patches may vary in size. Moreover, the index does not penalize for a large patch with a correspondingly large edge length. The LSI is defined as follows:

$$LSI = \frac{E}{\min E}, \tag{3}$$

where E represents the total edge in the landscape (e.g., the sum of the perimeter of patch boundaries) divided by the minimum total length of edge that would be possible if the landscape were a single patch (e.g., all residential parcels were contiguous) (McGarigal et al. 2002). Thus, it is an area-weighted measure of edge density. The LSI ranges from one (most compact) to infinity (the most edge-to-area). As the value of the LSI increases, the more fragmented the distribution of individual patches of development (Fig. 2). The Euclidean nearest neighbor distance (or ENN) is a measure of the Euclidean distance from a patch to the nearest neighboring patch of the same cover type (in meters) from edge to edge. The survival analysis and simulation study described in the following sections provides complementary insight to the factors driving the patterns of development described through the application of these landscape ecology metrics.

4.2 *Survival Models*

Timing-of-development models, coupled with information regarding the spatial heterogeneity of the land surface, have proven exceedingly useful in explaining where and when development is likely to happen. These models are typically specified to explain the point in time when a particular parcel is developed as a function of the spatial variation in several key attributes of residential property (e.g., proximity to roads and natural features, surrounding land uses, and land-use policy). Statistically, the sort of conceptual model of development timing described earlier (e.g., Capozza and Helsley 1989) can be implemented using survival analysis. Within a land use change context, survival models allow the researcher to estimate the conditional probability of conversion, given that a particular parcel has remained undeveloped for some duration. Certain factors, related to the suitability of land for conversion, neighborhood variables, and access to amenities, employment and commercial activities, will amplify or dampen the risk of conversion. Some of these factors are static, and some change over time. In land-use change analyses, there are several examples of such models (Irwin 2002, 2004; Vance and Geoghegan 2002; An and Brown 2008) investigating agricultural land conversion, or the clearing of forest for agricultural uses.

For this analysis, a complementary log–log formulation of a hazard model was implemented, following McCullagh (1980). The standard hazard function (Cox 1972) defines the risk of failure at time t conditional on the value of the covariates for each observation, x:

$$\lambda(t; x) = \lambda_0(t) \exp(\beta x) \tag{4}$$

where λ_0 is the hazard function when $x = 0$, and β is a set of parameters to be estimated that indicate the effect of the covariates on the likelihood that a particular agricultural parcel will be developed, compared to the overall rate of conversion $\lambda_0(t)$. From the hazard function, we can derive the survival function (S)

$$\lambda(t) = -d \log[S(t)] / dt, \tag{5}$$

where $S(t)$ is the probability of survival (e.g., an agricultural parcel remaining in agriculture) up to time t. Following from (4) and (5):

$$S(t + \Delta t) / S(t) = \exp[-\exp(\alpha_t + \beta_k x_k)], \tag{6}$$

where the left hand side of (6) expresses the conditional probability that observation k will remain in agriculture beyond t given that it has not been developed up until t.

The coefficient α_t represents the log of the integral of the hazard function over total duration t. Finally, if the intervals $(t, t + \Delta t)$ are indexed by i, increasing with time, the model can expressed in the complementary log–log formulation:

$$log[-log\,(1 - p_i)] = exp(\alpha_i - \beta_k x_k), \tag{7}$$

where p_i is the probability of the developed occurring in time period i conditional on conversion not happening prior to i (Abbott 1985).

The complementary log–log formulation of the hazard model is appealing due to several properties. These models allow for the specification of temporally varying covariates (Allison 1982), and allow for a specification of time that is not entirely continuous; i.e., that the event is only specific to the year in which occurred (Vance and Geoghegan 2002). This specification is useful here because the construction of a residential parcel is measured in the tax asessor's data at yearly intervals, a coarse representation of the underlying temporal dynamics of land conversion (An and Brown 2008).

Survival models can indicate how specific factors influence the likelihood of an individual parcel's conversion from an agricultural to a residential land use. Because urban growth pressure varies over time, conversion of agricultural parcels is more likely to occur at certain points in time than others (Irwin 2002). Therefore, it is empirically (and theoretically) challenging to separate the influence of spatial and non-spatial parcel attributes from the overall average affect of development pressure, which is not directly measurable. A well-specified survival model allows the researcher to separate the influence of spatial factors from the question of timing; ex-post specification testing and evaluation of model fit are important to check this assumption (An and Brown 2008).

4.3 Survival Model Estimation

Each parcel was coded as agricultural (0) or residential (1) for each yearly time step, depending on its development history. Use of the full dataset in estimation was not computationally feasible; the model was estimated using a subset of the data. A 30% spatially stratified sample[2] was drawn (2,331 parcels) and the model parameters were used to estimate the probability of conversion for all the observations (a total of out of a total 7,245 parcels). A White-Huber correction for underlying heteroskedasticity in the model residuals was applied. The estimated coefficients were also used to generate predicted values for each of the different scenarios in the simulation analysis described later in the paper. Following the estimation of the survival

[2] Fitting a model using only a sample of the data increases the risk that the estimated parameters are specific to the observations selected. Three distinct samples were initially drawn, and neither coefficients nor standard errors varied significantly. In addition, tests of leverage, i.e., for multivariate outliers, were also not significant.

model, the parcel layers were transformed into grids with a cell size of 10,000 ft^2 (or approximately 929 m^2). This was done to examine whether the conversion of agricultural land to an urban use occurred more often within more compact (discrete blocks of converted land) or fragmented (individual, isolated parcels) parts of the study area. Proximate cells of agriculture or urban land were aggregated into one "patch" if they were contiguous, including diagonal neighbors. The software package Fragstats (McGarigal et al. 2002) was then used to evaluate landscape configuration.

Independent variables were included in the specification of the survival model to estimate the average effect, independent of time period, of relevant spatial factors in contributing to the risk of development. To capture the effect of accessibility to economic and cultural opportunities on the likelihood of development, network distance to the core area of the Metropolitan Statistical Area (Columbus, capital city of Ohio) as well as the micropolitan city of Delaware were included. Euclidean distance to the nearest major highway and interstates (from the 2000 Census TIGER line files) was also included. Historical datasets regarding roads were not available; however, the limited access highways were in existence through the duration of the study period. To capture the effect of the spatially varying terrain on the risk of development, the estimated percentage of prime farmland within each parcel and the maximum slope (in degrees) were also included.

Several potential externality influences were also controlled for. First, because there are a number of lakes and reservoirs with recreational opportunities within the county, Euclidean distance to the nearest water body was calculated. It is also assumed that surrounding residential development can either increase or decrease development risk. The presence of existing development may reduce an undeveloped parcel's development risk, if negative externalities such as congestion and avoidance of neighboring development are present. On the other hand, more locally, there may be positive spillover effects from existing development due to common amenities within a neighborhood or otherwise unmeasured spatial autocorrelation (Irwin 2002). In order to specify the size of the neighborhood where spillovers are likely to occur, Fleming (2000) suggests the use of geostatistical techniques to determine the effective distance of positive or negative spillovers. In this research, two neighborhoods were used: the proportion of previously developed land within a 1/2 mile radius of each parcel, and between 1/2 and 3/4 miles of each parcel the year before conversion at each time step. Semivariogram analysis (Cressie 1993), conducted using the Geostatistical Analyst in ArcGIS, indicated that the observed spatial autocorrelation in the year built variable peaked at approximately 2,320 ft (1/2 mile = 2,640 ft); negative spatial autocorrelation was evident beyond this distance, dropping off to insignificance at approximately 4,000 ft (3/4 mile = 3,960 ft).

To study the spatial effects of variation in land-use policy, data on the maximum allowable density, according to the relevant township plan, was recorded for each parcel. Within Delaware County, there are 25 separate townships or municipalities that have the ability to set their own allowable densities, measured as the maximum number of dwelling units allowed per acre. Within the county, these densities range

Table 1 Landscape pattern analysis, 1988–2003

	% Developed area[a]	Contiguous patches	Shape index	Nearest neighbor distance, m
1980	22.75	2,252	65.91	0.1264
1990	26.04	1,924	63.49	0.1279
1995	28.75	1,636	61.01	0.1304
2000	32.10	1261	56.84	0.1420
2003	32.90	1,126	54.06	0.1507
% change, 1988–2003	44.65	−50.00	−17.98	19.23

[a]Includes all developed parcels (residential, commercial and industrial).

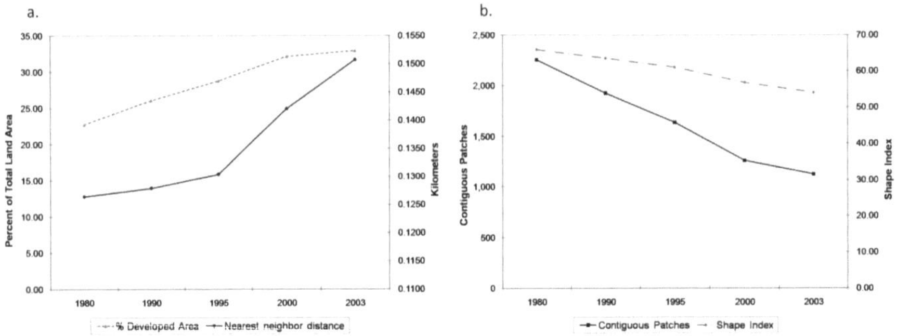

Fig. 3 Landscape pattern analysis of Delaware County, 1988–2003. (**a**) Percent developed area (of total land) and Euclidean nearest neighbor distance edge-to-edge between contiguous parcels (km). (**b**) The number of patches (contiguous parcels sharing a common boundary) and the landscape shape index (higher = greater proportional edge in the landscape)

from as low as 1.0 unit per two acres to 9.09 units per acre (or from nearly 125 to 2,000 units per km^2). All variables were tested for deviations from normality, especially those calculated as proportions, and transformed as necessary. All variables except the maximum allowable density and the proportions of surrounding developed land were transformed using a natural logarithm.

5 Results

The presentation of the results begins with a descriptive analysis of development pattern. Then, exploratory data analysis was conducted, comparing residential and agricultural parcels. Third, the results from the complementary log–log survival models are presented. Finally, the results of the simulation analysis are discussed. Table 1 reports the results of the landscape pattern analysis of the entirety of developed land (defined as the total acreage contained within the sum of commercial, industrial, residential and developed public lands parcels) in Delaware County at various intervals between 1988 and 2003. Fig. 3 contains a graphical depiction of these changes. Out of a total land area of $1,146\,km^2$, nearly 23% was developed in 1980, whereas in 2003, 33% of this area was developed, an increase of nearly 45%. Surprisingly, the number of contiguous patches decreased by 50% and the shape index (as a relative measure of total edge in the landscape) decreased by nearly 18%. Both of these trends in the metrics indicate that across the landscape, development was more compact in 2003 than it was in 1988. At the same time, the mean Euclidean nearest neighbor distance across contiguous patches of developed parcels increased by nearly 20%. Therefore, there appears to be interesting changes in development pattern unfolding in this county. Total developed area has increased dramatically in this 15-year period, and much of this development has occurred in non-contiguous clusters outside small municipalities. These distinct patches of development have become, on average, more dispersed than they were at the beginning of this period, though in the aggregate, the configuration of contiguous patches of development is slightly more compact than before. Visual inspection indicates that initial developments were more scattered, and a significant amount of infill occurred over time, corresponding in part to an increase in peri-urban subdivisions, as more development occurred along arterial roads. To the degree to which newer parcels are more likely to share common boundaries with existing parcels, this observed pattern is consistent with infill.

Table 2 contains descriptive statistics for all peri-urban parcels, broken up into agricultural (censored observations, remaining unconverted by 2003) and residential parcels. There were fewer than 30 industrial parcels developed in this time period. Overtime, there was some development of commercial space across the county, but the year of development for commercial parcels was not available; thus, these areas were not included in the analysis. The median values for these variables differed significantly between the two groups according to a nonparametric Mann–Whitney U-test ($p < 0.01$), except for the maximum allowable density. Distance to Columbus, interstate access points and water, and the percentage share of prime farmland are all higher on agricultural parcels. Slope, distance to Delaware (county seat), and the proportion of developed land within 1/2 and 3/4 miles were all, on average, higher on converted parcels.

Table 3 presents the estimated coefficients for the complementary log–log model associating the previously described factors with the risk of urban conversion between 1988 and 2003. A significant increase in the instantaneous risk of development (implying that an increase in the relevant covariate would make development of a particular parcel more likely at all time periods) was associated with slope, distance to water and the proportion of developed land within 1/2 mile of the parcel. A significant decrease in the instantaneous risk of development was associated with increasing density per acre (higher maximum density zoning made conversion less likely, all things equal) and with the proportion of developed land farther than 1/2 mile, but less than 3/4 miles away. Because the coefficients of the hazard model are difficult to interpret with regard to the change in probability of conversion, they can be transformed by exponentiation into "relative risk ratios," which are bounded between 0 and infinity. To examine the average contribution (separate from time) of each of these independent variables on the probability of conversion,

Table 2 Descriptive statistics, peri-urban agricultural parcels, and parcels developed, 1988–2003

	Agricultural parcels				
	Mean	Median	Std. Dev.	Min.	Max.
Lot size, acres	37.49	28.25	37.42	0.01	509.25
Slope, degrees	0.30	0.15	0.50	0.00	10.15
Dist. to Columbus	22.38	22.56	5.10	10.47	33.96
Dist. to Delaware	8.96	8.48	4.23	1.11	19.99
Dist. to nearest major road	2.43	1.96	1.90	0.00	9.21
Dist. to nearest interstate	7.50	6.56	4.73	0.00	19.62
Percent prime farmland	80.47	82	16.51	58.00	99.00
Dist. to water	2.04	1.74	1.32	0.00	6.02
Maximum density (units per acre)	3.45	2	2.25	0.50	9.09
Proportion developed within 1/2 mile, 1988	0.27	0.23	0.21	0.00	0.95
Proportion developed within 1/2 mile, 2003	0.33	0.21	0.24	0.00	0.99
Proportion developed between 1/2 and 3/4 miles, 1988	0.27	0.29	0.20	0.00	0.93
Proportion developed between 1/2 and 3/4 miles, 2003	0.35	0.31	0.23	0.00	0.95
Year built	n/a		n/a	n/a	n/a
N	4,389				
	Residential parcels developed 1988–2003				
Lot size, acres	5.99	5.01	6.14	0.11	100.67
Slope, degrees	3.61	2.49	3.08	0.00	19.54
Dist. to Columbus	20.45	20.46	4.65	10.15	33.86
Dist. to Delaware	9.52	8.66	4.39	1.00	20.06
Dist. to nearest major road	2.45	2.15	1.58	0.00	9.17
Dist. to nearest interstate	6.04	5.60	4.09	0.06	19.28
Percent prime farmland	76.92	79	15.94	30.50	99.00
Dist. to water	1.85	1.51	1.40	0.00	6.20
Maximum density (units per acre)	3.53	2	2.17	0.50	9.09
Proportion developed within 1/2 mile, 1988	0.35	0.29	0.19	0.00	0.95
Proportion developed within 1/2 mile, 2003	0.51	0.33	0.21	0.01	0.99
Proportion developed between 1/2 and 3/4 miles, 1988	0.32	0.40	0.18	0.00	0.93
Proportion developed between 1/2 and 3/4 miles, 2003	0.42	0.51	0.21	0.00	0.97
Year built	1996.13	1997	4.58	1988	2003
N	2,856				

Table 3 Results of complementary log–log model of urban conversion, 1988–2003

30% sample n = 2,331	Beta	Std. Error	Prob.	Relative risk ratio	% change in prob. with unit change in X
Intercept	−7.5687	2.4451	0.00	0.0005	
Distance to Columbus	0.1339	0.1842	0.47	1.1433	14.33
Distance to Delaware	−0.0713	0.0806	0.38	0.9312	−6.88
Slope	1.3283	0.0396	0.00	3.7746	277.46
Distance to major road	0.0542	0.0401	0.18	1.0557	5.57
Distance to water	0.136	0.038	0.00	1.1457	14.57
Density per acre	−0.155	0.0469	0.00	0.8564	−14.36
Proportion developed within 1/2 mile	3.1603	0.2486	0.00	23.5777	2, 257.77
Proportion developed between 1/2 and 3/4 mile away	−1.8805	0.2913	0.00	0.1525	−84.75
Log likelihood = $-3,630.74$ $\rho^2 = 0.375$					

we can subtract one and multiply by 100 to derive percentages. For example, a one unit increase in slope, all things equal, increases the likelihood of conversion by 277%. Because of multicollinearity, the influence of interstate accessibility and prime farmland were dropped from the analysis. The effect of distance to roads was not significant; this coefficient had a comparatively large standard error (e.g., development risk was likely high both close to and far from roads). One issue with the influence of accessibility is that temporally varying information for these variables is not available, including such factors as road widening or other improvements, which may be significant in the aggregate.[3] Likewise, development was more likely close to Delaware and far from Columbus, but these effects were not significant.

Regarding the temporally varying covariates (e.g., the map of surrounding land use that was updated in each time period), there are interesting results to report. Conversion probability was very highly positively associated with the proportion of developed land within 1/2 mile of a parcel, while at the same time negatively associated with the proportion of development greater than 1/2 mile but less than 3/4 mile away. This finding supports the interpretation of the landscape pattern analysis: that considerable infill happens within small-to-medium "clusters" of development over time, but these clusters are relatively spread out. It is important to note, however that unmeasured influences associated with local development risk may be captured by the smaller neighborhood density variable, particularly if these influences tend to be spatially correlated (Irwin and Bockstael 2004).

[3] However, road improvements are unlikely to be independent of development risk: roads are more likely to be widened where prior growth has occurred.

6 Simulations

To connect three suggested drivers of dispersed development (roads, inconsistent policy and spatial externalities) to the observed pattern of change in Delaware, three simulations were conducted: (1) no maximum density; i.e., observations were treated as if there were no differences in zoning density requirements across the county; (2) no negative spatial externality effect; i.e., the "avoidance" effect of development within 1/2 to 3/4 miles was dropped; and (4) no effect of distance to roads. Table 4 presents the results of this analysis. In each case, the described predictor variable (e.g., density, externalities, roads) was held constant across all parcels, and new predicted probabilities were generated. Following Irwin (2002) it was assumed that the same number of parcels (2,856) would be developed in each scenario; thus, each time the 2,856 parcels with the highest estimated instantaneous risk were coded as developed, whereas the remaining 4,380 parcels were coded as agricultural.

These models were evaluated on the basis of overall fit against the observed parcels as well as their ability to replicate the observed landscape pattern. First, the peri-urban pattern generated by the base model was compared against the observed data. To evaluate predictive accuracy, the Kappa statistic was employed; this statistic is commonly used in the land-use modeling community (Pontius 2002). The Kappa statistic compares the ratio of predicted to observed land-use changes (in this case, conversion), adjusted by the expected number of predicted outcomes that may occur by chance. For example, if we have two outcomes (agricultural, residential) that are observed with equal frequency, a model that randomly assigns observations to one or the other category could be expected to obtain 50% accuracy. A Kappa value above 0.60 is said to have substantial agreement (Landis and Koch 1977). Pattern outcomes were compared using total area developed (summing the acreage of converted parcels), the number of contiguous patches, the LSI, and the mean estimated Euclidean nearest neighbor distance between isolated patches of development.

The base model yielded a Kappa statistic of 0.83, which implies an excellent overall fit. Regarding the various landscape metrics, all of the metrics estimated on the base model were within 6% of the actual values, except for total developed

Table 4 Landscape pattern analysis of actual and predicted development patterns

Landscape pattern analysis	Total developed area, sq. km	Contiguous patches	Landscape shape index	Euclidean nearest neighbor distance, km	Kappa statistic
Observed conversion pattern	0.1552	1,599	53.72	18.08	n/a
Base model	0.2095	1,526	50.26	17.48	0.8255
Simulations					
1. No maximum density	0.2295	1,462	49.88	16.52	0.7226
2. No negative spatial externality	0.2677	1,358	46.84	16.33	0.6087
3. No roads	0.2264	1,467	49.95	16.45	0.7295

parcels, which was predicted to be about 35% higher than actual. This result is not surprising, however, due to the fact that agricultural parcels are much larger on average than residential ones, so commission errors (false positives) bear a large penalty for this metric. Across the various simulations, the least "correct" simulation in terms of the Kappa statistic was the one omitting negative spatial externalities. This simulation had the highest predicted developed area, the fewest number of contiguous patches, and the smallest LSI and Euclidean nearest neighbor distance. This result is completely intuitive because if there is no penalty on the instantaneous risk of development due to surrounding development, we would expect the resulting development pattern to be much more compact than with this effect. Conversely, with the omission of a spatially varying maximum density requirement, there appears to be fewer contiguous patches, a lower shape index and lower nearest neighbor distance than the base model (with roughly an 8% difference overall in these patterns). Finally, though the distance to roads variable was not significant, omitting its effect resulted in small differences in pattern relative to the base, roughly consistent with the maximum density effect. It is interesting to note that omitting maximum density and roads both decreased the number of patches, the shape index and the nearest neighbor distance compared to the base, but less so than the negative spatial externality effect. Overall prediction in these two scenarios varied between 7 and 10% from the base model.

7 Discussion

It is possible to have some confidence that the estimated model separates temporal trends in development from the average spatial influences. On the whole, the estimated results conform to theory. One notable exception is that distance from water bodies was positively associated with development risk: indicating that at any given time, areas *farther* from water are more desirable. This counterintuitive fact can be explained by the fact that recent subdivisions have been built very close to these water bodies and are among the most expensive real estate in the county. Moreover, due to additional costs of construction because of hydrological constraints in these areas, it was relatively expensive to build in these locations. Given that urban growth pressure rises over time before peaking, and the Capozza and Helsley (1989) model posits that decision-makers are optimizing returns from development over time, it could be that it was not financially lucrative to build in these areas until recently.

The impact of slope on development can be complicated, and the average effect on development risk could be either negative or positive. Steeper slopes are likely to increase construction costs, so development risk could be lower in areas of steep slope. On the other hand, steeper areas are the least valuable for agriculture, and thus have a low opportunity cost. In addition, steeper slopes can provide a scenic view, which may make these areas more desirable (Cavailhès et al. 2006). Because increases in slope in Delaware were also associated with higher instantaneous risk of development, it is likely that the marginal agricultural value and/or the aesthetic value of steeper slopes increased the development risk.

Regarding the effect of the other covariates, it is surprising that the measures of accessibility were largely insignificant, though the simulation analysis indicates that the effect of access to roads is perhaps not unimportant, just highly variable. Because of the well-documented endogeneity between roads and development (Anas et al. 1998; Irwin and Bockstael 2001), a statistical correction for this endogeneity may be helpful.

There is also evidence to suggest that variation in maximum density zoning affects development timing. Areas with higher maximum density are less attractive for developers, because land is more scarce and perhaps less desirable to consumers. The simulation analysis indicates that there are likely important indirect effects related to these policies: growth would be more compact in the county as a whole if density requirements were equal everywhere. Finally, the finding of variation in development risk according to the proportion of development within two non-overlapping spatial neighborhoods was very interesting. Within 1/2 mile, a greater proportion of prior development very significantly increased development risk: this effect was of higher magnitude than any other. However, greater than 1/2 mile and less than 3/4 mile, development was much less likely. Therefore, there appears to be a balance: small neighborhoods, or neighborhood subdivisions are the norm in this area, but relatively isolated areas are preferable, perhaps related to a premium residents are willing to pay for surrounding open space (Irwin 2002). If such avoidance of neighborhood externalities is truly this important as to greatly influence overall development patterns as the simulations suggest, policy makers concerned with fragmented development should address this effect. One way to do so would be to increase plans for mixed-use development, perhaps by including improved, permanent open space for neighborhood residents. Particularly because the environmental impacts of increasing urban decentralization may be nonlinear, fine-scale variations in low-density urban form warrant more attention (Theobald 2004).

8 Conclusion

Given recent advances in spatial analytical tools and technology, coupled with newly available spatial data, increasingly sophisticated models of the development process are possible (Irwin et al. 2003). New insights include the recognition that urban development processes can be spatially heterogeneous, depending on how land conversion pressures are allocated on a variable landscape (Irwin and Bockstael 2004). In addition to predicting the location of land-use changes, land-use models that provide insights regarding landscape level patterns (or spatial configuration) can be useful for planning and policy (Parker and Meretsky 2004). This research indicates that inconsistent local land-use policy and more growth in lower density areas both contribute to the emergence of a peri-urban landscape that was more disperse in 2003 than in 1988. Furthermore, the findings suggest that development in Delaware County has not simply been a "natural evolution" of the Columbus metropolitan area.

Statistical land-use models have traditionally been useful for exploring the *where* and the *why* of land-use change, but it has been analytically more difficult to focus on the *when* (An and Brown 2008). The overarching objective of this analysis was to explore the relative influence of various spatial factors on the observed pattern of peri-urban residential development in Delaware County, given that such development likely exhibits spatiotemporal path-dependency. The estimated complementary log–log model allows the researcher to ask, given that a certain level of development is expected at a certain point in time, where is this development likely to be distributed? Thus, we can abstract from the temporally dependent nature of the development process to explore how various influences amplify or dampen development risk over space. Then, in turn, the collective pattern, both the amount and configuration of residential development, in the county can be shown to vary depending on how urban growth pressure touches down on the spatially heterogeneous landscape, and how past conversion continues to influence future patterns.

Overall, the results indicated that though development has become more far-flung or dispersed overtime (i.e., increasingly permeating into the far reaches of the county), such development is more compact than it was in 1988. Thus, in terms of fiscal impact, such as service provision, the trend of increasing dispersion of residential clusters may be partially offset by the increasingly locally clustered nature of new development. Because increased maximum density appears to reduce development within a particular township, increased cooperation among townships in setting and enforcing land-use policies may curb some of the decentralization pressure. Finally, the results were consistent with the assumption that more scenic areas or bucolic landscapes remain a development pull, which might imply that suburban jurisdictions may need to think carefully about open space preservation and other such efforts to reduce the flight of their constituents to farther outlying areas.

References

Abbott RD (1985) Logistic regression in survival analysis. Am J Epidemiol 121:465–471

Allison PD (1982) Discrete-time methods for the analysis of event histories. In: Leinhardt S (ed) Sociological methods and research. Jossey-Bass, San Francisco, pp 61–98

Anas A, Arnott, R Small KA (1998) Urban spatial structure. J Econ Lit 36:1426–1464

An L, Brown DG (2008) Survival analysis in land change science: integrating with GIScience to address temporal complexities. Ann Assoc Am Geogr 98:323–344

Byun P, Esparza AX (2005) A revisionist model of suburbanization and sprawl: the role of political fragmentation, growth control, and spillovers. J Plann Educ Res 24:252–264

Campbell Jr, HS, Munroe DK (2007) Greenways and greenbacks: the impact of the Catawba regional trail on property values in Charlotte, North Carolina. SE Geogr 47:118–137

Capozza DR, Helsley RW (1989) The fundamentals of land prices and urban growth. J of Urban Econ 26(3):295–306

Carruthers J, Ulfarsson, G (2002) Fragmentation and sprawl: evidence from interregional analysis. Growth Change 33:312–340

Cavailhès J, Brossard T, Foltête J-C, Hilal M, Joly D, Torneux F-P, Tritz C, Wavresky P (2006) Seeing and being seen: a GIS-based hedonic price valuation of landscape. Presented at the 1ère Recontre du Longement, Marseille, Octobre 2006

Clark JK, McChesney R, Munroe DK, Irwin EG (2005) Spatial characteristics of exurban settlement pattern in the US. Paper prepared for the 52nd Annual North American Meetings of the Regional Science Association Las Vegas, NV, November 2005
Cox D (1972) Regression models and life tables. J Roy Stat Soc B 34:187–220
Cressie N (1993) Statistics for spatial data. Wiley, New York, NY
Downs A (1999) Some realities about sprawl and urban decline. Housing Policy Debate 10:955–974
Esparza AX, Carruthers JI (2000) Land use planning and exurbanization in the rural mountain West. J Plann Educ Res 20:23–36
Fleming M (2000) Spatial statistics and econometrics for models in fisheries economics: discussion. Am J Agr Econ 82:1207–1209
Heimlich R, Anderson WD (2001) Development at the urban fringe and beyond: impacts on agriculture and rural land. Agricultural Economic Report No 803, United States Department of Agriculture, Washington, DC
Hite D, Sohngen B, Templeton J (2003) Zoning, development timing, and agricultural land use at the suburban fringe: a competing risks approach. Agr Res Econ Rev 32:145–157
Irwin EG (2002) The effects of open space on residential property values. Land Econ 78:465–480
Irwin EG, Bockstael NE (2002) Interacting agents, spatial externalities and the evolution of residential land use patterns. J Econ Geogr 2:31–54
Irwin EG, Bockstael NE (2004) Endogenous spatial externalities: empirical evidence and implications for the evolution of exurban residential land use patterns. In: Anselin L, Florax RJGM, Rey SJ (eds) Advances in spatial econometrics: methodology, tools and applications. Springer, Berlin Heidelberg New York pp 359–380
Irwin EG, Geoghegan J (2001) Theory, data, methods: development spatially explicit economic models of land use change. Agr Ecosyst Environ 85:7–24
Irwin EG, Bell KP, Geoghegan J (2003) Modeling and managing urban growth at the rural-urban fringe: a parcel-level model of residential land use change. Agr Res Econ Rev 32:83–102
Lake IR, Lovett AA, Bateman IJ, Day B (2000) Using GIS and large-scale digital data to implement hedonic pricing studies. Int J Geogr Inform Sci 14:521–541
Landis JR, Koch GG (1977) The measurement of observer agreement for categorical data. Biometrics 33:159–174
McCullagh P (1980) Regression models for ordinal data. J Roy Stat Soc B 42:109–142
McGarigal K, Cushman SA, Neel MC, Ene E (2002) FRAGSTATS: spatial pattern analysis program for categorical maps. Computer software program produced by the authors at the University of Massachusetts, Amherst
Mieskowski P, Mills E (1993) The causes of metropolitan suburbanization. J Econ Perspect 7:135–147
Munroe DK (2007) Exploring the determinants of spatial pattern in residential land markets: amenities and disamenities in Charlotte, NC, USA. Environ Plann B 34:336–354
Munroe DK, Müller D (2007) Issues in spatially explicit statistical land-use/cover change (LUCC) models: examples from western Honduras and the Central Highlands of Vietnam. Land Use Pol 24:521–530
Munroe DK, York AM (2003) Jobs, houses, and trees: changing regional structure, local land-use patterns, and forest cover in southern Indiana. Growth and Change 34:299–320
Parker DC, Meretsky V (2004) Measuring pattern outcomes in an agent-based model of edge-effect externalities using spatial metrics. Agr Ecosyst Environ 101:233–250
Partridge MD, Clark JK (2008) Our joint future: rural-urban interdependence in 21st Century Ohio. White Paper Prepared for the Brookings Institution, Greater Ohio
Partridge MD, Sharp JS, Clark JK (2007) Growth and change: population change in Ohio and its rural-urban interface. The Exurban Change Project and Swank Program in Rural-Urban Policy, Summary Report May 2007
Pontius Jr. RG (2002) Statistical methods to partition effects of quantity and location during comparison of categorical maps at multiple resolutions. Photogramm Eng Rem Sens 68:1041–1049

Steel S (1992) Farm markets mix entertainment, produce to lure customers. Columbus Dispatch, pp 2D

Theobald DV (2004) Placing exurban land-use change in a human modification framework. Front Ecol Environ 2:139–144

Vance C, Geoghegan J (2002) Temporal and spatial modeling of tropical deforestation: a survival analysis linking satellite and household survey data. Agr Econ 27:317–332

Veldkamp A, Lambin EF (2001) Predicting land-use change. Agr Ecosyst Environ 85:1–6

Verburg PH, van Eck JRR, de Nijs TCM, Dijst MJ, Schot P (2004) Determinants of land-use change patterns in the Netherlands. Environ Plann B 31:125–150

Vitousek PM, Mooney HA, Lubchenco J, Melillo JM (1997) Human domination of earth's ecosystems. Science 277:494–499

Zhang T (2001) Community features and urban sprawl: the case of the Chicago metropolitan region. Land Use Pol 18:221–232

Demand for Open Space and Urban Sprawl: The Case of Knox County, Tennessee

Seong-Hoon Cho, Dayton M. Lambert, Roland K. Roberts, and Seung Gyu Kim

1 Introduction

Urban sprawl is often blamed for causing negative environmental effects from unsustainable land consumption and increased traffic congestion. While there is no generally accepted definition of urban sprawl, the process is well-described as the expansion of urban development into rural areas surrounding major cities, and the leapfrogging of development beyond the city's outer boundary into smaller rural settlements (Hanham and Spiker 2005). Many studies have pointed toward the lifestyle choices of the economically affluent society for the rapid growth of urban sprawl (Brueckner 2000; Carruthers and Ulfarsson 2002; Frumkin 2002; Gordon and Richardson 1998, 2000, 2001a,b; Krieger 2005; Nechyba and Walsh 2004; Skaburskis 2000; Stone and Gibbins 2002). These lifestyle choices include preferences for larger homes and lot sizes, low density housing, mobility afforded by private vehicles, and the demand for open space. This kind of growth has raised concern about the potential negative impacts, especially the loss of benefits provided by farmland and open space, and higher costs of infrastructure and community services. Concerns about the negative consequences of urban sprawl have led local policymakers and nongovernmental activists to turn to urban and suburban open space conservation as potential mechanisms to counter urban sprawl.

One example of these mechanisms includes "smart growth" policies. Smart growth policies are development initiatives that protect open space and farmland, revitalize communities, keep housing affordable, and provide more transportation

[1]International City/County Management Association (2007) has laid out 100 policies and guidelines for communities to realize smart growth. The mechanisms include zoning, building design, transfer of development rights (TDRs), purchase of development rights (PDRs), multimodal transportation systems, and land value taxation. We do not address mechanisms other than land value taxation in this chapter.

S.-H. Cho (✉)
Department of Agricultural Economics, University of Tennessee, 321 Morgan Hall, Knoxville, TN 37996-4511, USA,
e-mail: scho9@utk.edu

A. Páez et al. (eds.), *Progress in Spatial Analysis*, Advances in Spatial Science, 171
DOI 10.1007/978-3-642-03326-1_9,

choices (International City/County Management Association 2008).[1] Local governments have incorporated "smart growth" principles designed to encourage compact development and preserve open space to curtail urban sprawl (Tracy 2003). Compact development is a key component of most smart growth policies. A large body of planning literature has addressed a variety of local strategies that are grouped under the rubric of "smart growth" (e.g., Blakely 1994; Daniels 2001; Handy 2005; Weitz 1999).

Local and regional governments have incorporated smart growth principles to stimulate demand for private and public open space. Some communities with commitments to stimulate demand for open space through smart growth directives continue to struggle with policy implementation (Cho and Roberts 2007). Stimulating demand for open space is challenging because little is known about the factors influencing it (Bates and Santerre 2001). Consequently, clearly defined policy tools to stimulate demand for open space are lacking. Understanding the structure of demand for open space is crucial to planners as they place more emphasis on smart growth policies to stimulate demand for open space. More specifically, the sensitivity of demand to factors closely associated with urban sprawl, i.e., income, house and lot size, housing density, and price of open space, needs to be examined.

Studies have estimated the willingness to pay for open space using contingent valuation methods (Blaine et al. 2003; Breffle et al. 1998; Rosenberger and Walsh 1997; Sorg et al. 1985; Stevens 1990; Tyrväinen and Väänänen 1998). However, Flores and Carson (1997) explain that demand for public or environmental goods is not necessarily the same as the willingness to pay for these goods. In addition, Cummings and Taylor (1999) show that willingness to pay estimates are often subject to hypothetical bias when derived from contingent valuation methods. Bates and Santerre (2001) estimated the local public demand for open space using two-stage least squares endogeneizing the price of open space. While their study was the first attempt to estimate demand for open space, the model had limitations. First, the price and income elasticities of demand for open space were assumed to be constant across communities. This assumption disregards possible spatial heterogeneity with respect to estimates of the elasticities. Spatial heterogeneity refers to variation in some condition or measure from one geographic area to another (Cho et al. 2006), which also has adverse effects on the properties of least squares estimators (Anselin 1988). Second, spatial dependencies causing spatial autocorrelation were not considered in their study. Spatial dependencies between cross-sectional units may lead to biased, inefficient, or inconsistent estimates.[2]

---

[2] There are two distinct forms of spatial dependence, namely spatial error dependence and spatial lag dependence. Spatial error dependence occurs where the dependence pertains to the error terms, whilst spatial lag dependence occurs where the dependence pertains to the dependent variable. Spatial error dependence is likely to yield inefficient, but unbiased and consistent estimates of standard errors while spatial lag dependence likely yields biased and inconsistent estimates arising form the endogenous nature of the lagged dependent variable (Anselin 1988). In this analysis, we only address the issue of spatial error dependence.

In this chapter, a geographically weighted regression (GWR), modified to attend to problems arising from spatial error dependencies, is used to estimate the demand for open space with cross-sectional data from Knox County, Tennessee. GWR allows local elasticities of demand for open space to be measured and mapped. The maps of GWR parameter estimates in particular may help policy makers or planners in developing location-specific smart growth policies to stimulate demand for open space.

In the next section, we develop the empirical model that uses a two-step procedure to estimate the demand for open space. Section 3 describes the study area and data. The empirical results are discussed in Sect. 4. Policy implications with respect to the findings and general conclusions are provided in the final section of the chapter.

2 Empirical Model

The following two-step procedure was applied to estimate the demand for open space. In the first step, the marginal implicit price of open space was estimated with a hedonic price model at the parcel level using GWR corrected for spatial error autocorrelation. GWR allows coefficients to vary across space by way of a moving window regression (Brunsdon et al. 1996).[3] In step two, an open-space demand equation was estimated using the marginal implicit price of open space estimated in the first step as a proxy for the price of open space. As in the first step, this demand relationship was estimated at the parcel level using GWR corrected for spatial error autocorrelation.

2.1 Step 1 – Estimation of the Marginal Implicit Price of Open Space

Because demand for open space is largely determined by the housing market, which in turn is a function of demand for open space within a reasonable neighborhood, i.e., a 1.0-mile radius (circular shaped buffer), a system of simultaneous equations was estimated to represent the demand for open space and the hedonic housing price (Geoghegan et al. 2003; Irwin and Bockstael 2001; Walsh 2007). Following Irwin and Bockstael (2001), the instrumental variables (IV) estimation approach was used to account for open-space endogeneity,

$$\ln p_i = \alpha x_i + \beta \ln \hat{o}_i + \varepsilon_i \tag{1}$$

$$\ln o_i = \gamma \chi_i + \eta_i \tag{2}$$

[3] GWR is a computationally intensive modeling approach. Each model run for this study took approximately 72 h to complete.

where $\ln p_i$ is the natural log of the value of house i; x_i is a vector of factors determining the value of house i; $\ln o_i$ is the natural log of open space in the vicinity of house i; $\hat{o}_i$ is the predicted value from (2); χ_i is a vector of instruments that are correlated with $\ln o_i$ and uncorrelated with ε_i; and (ε_i, η_i) are a random disturbances with expected values of zero and unknown variances. The instruments used in (2) are identified in Table 1.

The GWR hedonic model with spatially autocorrelated disturbances is:

$$\ln p_i = \sum_k \beta_k (u_i, v_i)\, \hat{x}_{ik} + \varepsilon_i, \quad \varepsilon_i = \lambda \sum_{j=1, j \neq i}^{n} w_{ij}\, \varepsilon_j + \xi_i,$$
$$\xi_i \sim iid\left(0, \sigma^2\right) \tag{3}$$

where $\hat{x}_{ik}$ is a vector of exogenous variables, including the predicted value of $\ln \hat{o}_i$; (u_i, v_i) denotes the coordinates of the ith location in the housing market; $\beta_k (u_i, v_i)$ represents the local parameters associated with house i; w_{ij} is an element of an m by n spatial weighting matrix between points i and j; and λ is a spatial error autoregressive parameter.

The specification in (3) allows a continuous surface of parameter values with spatially autocorrelated disturbances, and measurements taken at certain points denote the spatial heterogeneity of the surface (Fotheringham et al. 2002). Previous studies have found that a log transformation of the distance and area explanatory variables generally performs better than a simple linear functional form, as the log transformation captures the declining effects of these distance variables (Bin and Polasky 2004; Iwata et al. 2000; Mahan et al. 2000). Thus, a natural log transformation of the distance and area-related variables is used in this study.

Given estimation of (3), GWR residuals are tested for spatial error autocorrelation using a Lagrange Multiplier (LM) test (Anselin 1988). A row-standardized inverse distance matrix was used to test the hypothesis of spatial error independence. Rejection of the null hypothesis suggests a GWR-spatial autoregressive error model (GWR-SEM) as a way to address spatial heterogeneity and spatial error autocorrelation. The GWR-SEM combines well-founded methods typically used in conventional spatial econometric analyses, i.e., the Cochran–Orcutt method of filtering dependent and explanatory variables to address spatial error autocorrelation (Anselin 1988), with local regression techniques in a parametric framework. The filtering mechanism $[(\mathbf{I} - \lambda \mathbf{W})]$ partials out spatial error autocorrelation associated with the explanatory and dependent variables while estimating local coefficients. It helps to envision GWR as running n parametric regressions at n locations to control spatial heterogeneity, and then testing whether the residuals generated by these local regressions are spatially correlated. If the hypothesis of no spatial autocorrelation is rejected, conventional methods are applied to filter the dependent and explanatory variables (e.g., Anselin 1988, p. 183), and the GWR model is estimated again using the transformed variables.

A convenient procedure to estimate λ is Kelejian and Prucha's (1998) general moments approach, based on the set of GWR residuals. Given determination of λ, the closed form solution to (3) is:

Table 1 Variable names, definitions, and descriptive statistics

Variable (Unit)	Definition	Mean	Std. Dev.
Dependent variable			
Housing price ($)	Sale price adjusted to 2000 by the housing price index	129,610.227	95,460.498
Variables closely associated with urban sprawl			
Income[a] ($)	Median household income	51,505.871	20,940.122
Finished area[a] (feet^2)	Total finished square footage of house	1,929.689	975.633
Lot size[a] (feet^2)	Total parcel square footage	25895.720	69956.690
Housing density[a] (houses per acre)	Housing density for census-block group	1.105	0.927
Open space $(10^3 \times \text{feet}^2)$	Area of open space within a buffer of 1.0 mile drawn around each house sale transaction	53,822.711	15,490.449
Price of open space ($)	Marginal implicit price of increasing additional 10,000 ft^2 of open space within 1.0-mile buffer (assuming individual housing price and open-space area)	47.618	38.610
Structural variables			
Age[a] (year)	Year house was built subtracted from 2006	29.207	21.733
Brick[a]	Dummy variable for brick siding (1 if brick, 0 otherwise)	0.254	0.435
Pool[a]	Dummy variable for swimming pool (1 if pool, 0 otherwise)	0.055	0.229
Garage[a]	Dummy variable for garage (1 if garage, 0 otherwise)	0.635	0.481
Bedroom[a]	Number of bedrooms in house	3.068	0.647
Stories[a]	Height of house in number of stories	1.340	0.474
Fireplace[a]	Number of fireplaces in house	0.729	0.575
Quality of construction[a]	Dummy variable for quality of construction (1 if excellent, very good and good, 0 otherwise)	0.352	0.478
Condition of structure[a]	Dummy variable for condition of structure (1 if excellent, very good and good, 0 otherwise)	0.734	0.442
Distance variables			
Distance to CBD[a] (feet)	Distance to the central business district	44,552.592	20,713.081
Distance to greenway[a] (feet)	Distance to nearest greenway	7,886.866	5,573.062

(continued)

Table 1 (continued)

Variable (Unit)	Definition	Mean	Std. Dev.
Distance to railroad[a] (feet)	Distance to nearest railroad	6,978.618	5,463.655
Distance to sidewalk[a] (feet)	Distance to nearest sidewalk	3,060.270	4,229.282
Distance to park[a] (feet)	Distance to nearest park	8,652.930	5,556.530
Park size[a] ($feet^2$)	Size of nearest park	1,454.759	5,094.984
Distance to golf course[a] (feet)	Distance to nearest golf course	10,680.078	4,942.615
Distance to water body[a] (feet)	Dist. to nearest stream, lake, river, or other water body	8,440.579	5,884.047
Size of water body[a] (1,000 $feet^2$)	Size of nearest water body	19,632.026	39,026.745
High school district dummy variables (1 if in School District)			
Doyle[a]	Dummy variable for Doyle High School District	0.077	0.266
Bearden[a]	Dummy variable for Bearden High School District	0.157	0.363
Carter[a]	Dummy variable for Carter High School District	0.027	0.161
Central[a]	Dummy variable for Central High School District	0.092	0.290
Fulton[a]	Dummy variable for Fulton High School District	0.053	0.224
Gibbs[a]	Dummy variable for Gibbs High School District	0.055	0.228
Halls[a]	Dummy variable for Halls High School District	0.057	0.231
Karns[a]	Dummy variable for Karns High School District	0.147	0.354
Powell[a]	Dummy variable for Powell High School District	0.065	0.247
Farragut[a]	Dummy variable for Farragut High School District	0.148	0.355
Austin[a]	Dummy variable for Austin High School District	0.014	0.116
Census block-group variables			
Vacancy rate[a] (ratio)	Vacancy rate for census-block group (2000)	0.063	0.031
Unemployment rate[a] (ratio)	Unemployment rate for census-block group (2000)	0.037	0.029
Travel time to work[a] (min)	Average travel time to work for census-block group (2000)	22.519	3.314
Other variables			
Knoxville[a]	Dummy variable for City of Knoxville (1 if Knoxville, 0 otherwise)	0.343	0.475

(*continued*)

Table 1 (continued)

Variable (Unit)	Definition	Mean	Std. Dev.
Flood[a]	Dummy variable for 500-year floodplain (1 if in stream protection area, 0 otherwise)	0.010	0.097
Interface[a]	Dummy variable for rural–urban interface (1 if in census block of mixed rural–urban housing, 0 otherwise)	0.223	0.417
Urban growth area[a]	Dummy variable for urban growth area (1 if in urban growth area, 0 otherwise)	0.083	0.276
Planned growth area[a]	Dummy variable for planned growth area (1 if in planned growth area, 0 otherwise)	0.431	0.495
Season[a]	Dummy variable for season of sale (1 if April through September, 0 otherwise)	0.559	0.497
Prime interest rate[a]	Average prime interest rate less average inflation rate	4.267	2.104

[a]Indicates instrumental variables used in the first step estimation

$$\hat{\boldsymbol{\beta}}(u_i, v_i) = (\mathbf{X}'(\mathbf{I} - \lambda\mathbf{W})'\mathbf{A}(\mathbf{I} - \lambda\mathbf{W})\mathbf{X})^{-1}\mathbf{X}'(\mathbf{I} - \lambda\mathbf{W})'\mathbf{A}(\mathbf{I} - \lambda\mathbf{W})\mathbf{P} \tag{4}$$

which is analogous to the GLS estimator in the spatial econometric literature, $\boldsymbol{\beta}_{SEM} = (\mathbf{X}'(\mathbf{I}\text{–}\lambda\mathbf{W})'\boldsymbol{\Omega}(\mathbf{I}\text{–}\lambda\mathbf{W})\mathbf{X})^{-1}\mathbf{X}'(\mathbf{I}\text{–}\lambda\mathbf{W})'\boldsymbol{\Omega}(\mathbf{I}\text{–}\lambda\mathbf{W})\boldsymbol{y}$, where $\boldsymbol{\Omega}$ is a an n by n diagonal matrix with a set of weights corresponding with each observation, except that it generates i sets of local parameters. The n by n matrix $\mathbf{A}$ (which is a function of u_i and v_i) addresses spatial heterogeneity, with diagonal elements identifying the location of other houses relative to house i and zeros in off-diagonal positions (Fotheringham et al. 2002). Houses near house i have more influence in the estimation of the parameters associated with house i than other houses located farther away.

When $\lambda = 0$, (4) generates the usual GWR estimates. Pseudo-standard errors for the i sets of regression parameters are based on the covariance matrix (cov):

$$cov(\hat{\boldsymbol{\beta}}(u_i, v_i)) = \sigma_i^2(\mathbf{X}'(\mathbf{I} - \lambda\mathbf{W})'\mathbf{A}(\mathbf{I} - \lambda\mathbf{W})\mathbf{X})^{-1} \tag{5}$$

where $\sigma_i^2 = \boldsymbol{e}'(\mathbf{I} - \lambda\mathbf{W})'\mathbf{A}(\mathbf{I} - \lambda\mathbf{W})\boldsymbol{e}/(q-k)$ is the variance associated with the ith regression point (Fotheringham et al. 2002).[4] Statistical significance of the estimates from the GWR-SEM at the ith regression point is evaluated with the Pseudo-t tests

[4] Those standard errors do not take into consideration the first stage estimation. Further studies will consider a covariance matrix adjusted for the first stage regression.

derived from the Pseudo-standard errors of the location-specific covariance matrices. Based on the GWR-SEM, the marginal implicit price of an additional $10,000\ \mathrm{ft}^2$ of open space is estimated.

2.2 Step 2 – Open-Space Demand Estimation

The demand for open space is estimated using the marginal implicit price of open space estimated in the first step as a proxy for the price of open space. The demand equation for open space in the GWR framework is:

$$\ln o_i = \zeta(u_i, v_i) \ln \hat{p}_i + \sum_k \alpha_k(u_i, v_i) x_{ik} + \upsilon_i, \quad k = 1, \ldots, m^5 \qquad (6)$$

where $\ln \hat{p}_i$ is the natural log of the estimated marginal implicit price of open space for house i, and x_{ik} is the kth of m variables determining the demand of open space for house i. The x_{ik} includes variables closely associated with urban sprawl (e.g., income, house and lot size, and housing density), structural attributes of the house, census-block group variables (e.g., vacancy rate, unemployment rate, and travel time to work), distance measures to amenities (e.g., lakes, parks) or disamenities (e.g., railroads), school districts, and other spatial dummy variables (e.g., urban growth area and planned growth area) (see Table 1 for the complete list). The statistical significance of the local estimates at the ith regression point is evaluated with t-tests derived from the standard errors of the location-specific covariance matrices.

Another concern in regression models with many explanatory variables is multicollinearity, which occurs when two (or more) independent variables are linearly related. Multicollinearity may inflate estimates of standard errors, rendering hypothesis testing inconclusive. Multicollinearity can be detected by variance inflation factors (VIF) (Maddala 1992). VIFs are a scaled version of the multiple correlation coefficients between a variable and the rest of the independent variables (Maddala 1983). There is no clear guideline for how large the VIF must be to reflect serious multicollinearity, but a rule of thumb is that multicollinearity may be a problem if the VIF for an independent variable is greater than ten (Gujarati 1995). The VIFs were lower than ten for all but three variables, namely dummy variables differentiating the rural–urban interface (22), the City of Knoxville (12), and the Bearden high school district (11) in the demand for open space equation. In general, multicollinearity does not appear to be too great a concern because many of the location-specific coefficients were significantly different from zero at the 5% level.[6]

[5] Covariance of $\hat{p}_i$ is not adjusted for first stage regression.

[6] If the VIF is large but the coefficient is significant, multicollinearity is not a problem with respect to the estimation of the standard errors. If a coefficient is significant using a weak t-test caused by collinearity (inflated standard error), it would be significant using the stronger t-test associated with the lack of collinearity (inflated standard error).

Nevertheless, those three variables with high VIFs were not excluded for lack of sufficient justification.

3 Study Area and Data

Knox County, Tennessee was chosen as a case study for this research because (1) Knoxville is the eighth most sprawling U.S. metropolitan region (Ewing et al. 2002), and (2) the area consists of both rapid and slow regions of housing growth. Knox County is located in East Tennessee, one of the three "Grand Divisions" in the state. The City of Knoxville is the county seat of Knox County. Knoxville comprises 101 miles^2 of the 526 miles^2 within Knox County. Total populations of Knoxville and the Knoxville Metropolitan Area were 173,890 and 655,400 in 2000, respectively (US Census Bureau 2002). The University of Tennessee and the headquarters of Tennessee Valley Authority (TVA) are near downtown Knoxville, and the US Department of Energy's Oak Ridge National Laboratory is 15 miles northwest of Knoxville. These institutions are the major employers of the area. Maryville is located approximately 14 miles southwest of Knoxville and it is home to ALCOA, the largest producer of aluminum in the United States. Farragut, a bedroom community, is located along the edge of the western end of Knox County (see Fig. 1). The Smoky Mountains, the most-visited National Park in the United States, and a large quantity of lake acreage (17 miles^2 of water bodies) developed by the TVA are on Knoxville's doorstep.

It is important to note that push/pull factors of the geography surrounding the study area were not modeled because data were not available. However, to our knowledge, no other hedonic studies have successfully addressed this issue. Admittedly, these omitted factors may cause some estimates to be biased. But understanding this context beforehand aids in the interpretation of patterns generated by mapped coefficients. It is also important to note that the results of this study may not be representative of other urban areas. The data set does not represent most typical urban areas, and because of the local amenities and job opportunities, Knox County may be more of an outlier case compared to other rapidly growing metropolitan areas. Nevertheless, the methods used in this case study can be applied to other urban areas where similar data exist.

This research used five GIS data sets: individual parcel data, satellite imagery data, census-block group data, boundary data, and environmental feature data. The individual parcel data, i.e., sales price, lot size, and structural information, were obtained from the Knoxville, Knox County, Knoxville Utilities Board Geographic Information System (KGIS 2009), and the Knox County Tax Assessor's Office. Data were used for single-family home sales transactions between 1998 and 2002 in Knox County, Tennessee. A total of 22,704 single-family home sales were recorded during this period. Of the 22,704 houses sold, 15,500 were randomly selected for this analysis. County officials suggested that sales prices below $40,000 were probably gifts, donations, or inheritances, and would therefore not reflect true market value.

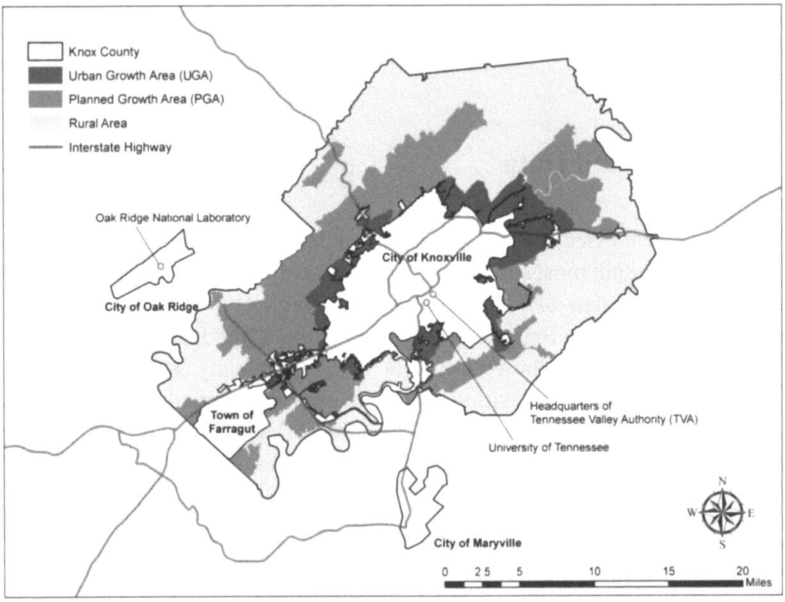

Fig. 1 Study area

Officials also suggested that parcel records less than 1,000 ft^2 might be misinformation. Therefore, parcels smaller than 1,000 ft^2 were eliminated from the sample data. There were 15,335 observations after eliminating these outliers. Selecting a random sample of sales transactions saved time in running the GWR. Prices were converted to 2000 (year) dollars to account for real estate market fluctuations in the Knoxville metro region. This adjustment was made using the annual housing price index for the Knoxville metro statistical area obtained from the Office of Federal Housing Enterprise Oversight (OFHEO 2006).

Land cover information was derived from Landsat 7 imagery for 2001. The classified national land cover database from the multi resolution land characteristics consortium (NLCD 2001) includes the GIS map used in the analysis to identify open space in the study area. There are 21 land cover classifications in the NLCD 2001 database. Of the 21 classified land covers, 11 classifications were considered as open space in our study.[7] The open-space classification was loosely based on the definition of "open area" or "open space" in Sect. 239-y of the General Municipal Law (Open space inventory 1999).[8]

[7] The 11 classifications include developed open space, barren land (rock/sand/clay), deciduous forest, evergreen forest, mixed forest, shrub/scrub, grassland/herbaceous, pasture/hay, cultivated crops, woody wetlands, and emergent herbaceous wetlands.

[8] Section 239-y defines "open area" as any area characterized by natural beauty or, whose existing openness, natural condition or present state of use, if preserved, would enhance the present or potential value of abutting or surrounding development or would offer substantial conformance

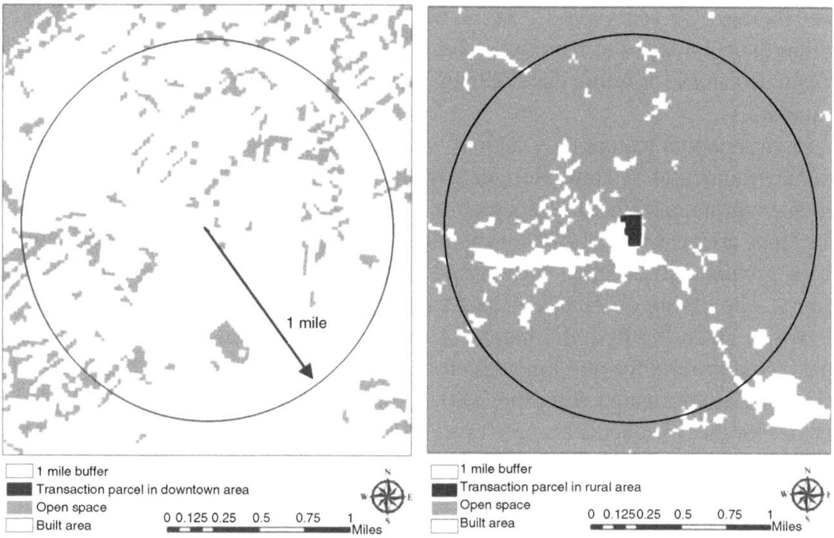

Fig. 2 Transaction parcel with surrounding open space and 1.0-mile buffer

To define the open-space demand for individual households, the space in the 11 open-space classifications was aggregated within a 1.0-mile radius (buffer) of each housing sales transaction (see Fig. 2). Buffer sizes found in the literature were not consistent, resulting in different estimates of open space value (McConnell and Walls 2005). For example, Geoghegan et al. (2003) used two buffers, a 100-m radius around the property and a 1,600-m radius. Acharya and Bennett (2001) also used a 1,600-m buffer. Nelson et al. (2004) used 0.1-mile, 0.25-mile, and 1.0-mile buffers and Irwin (2002) used a 400-m buffer. Lichtenberg et al. (2007) used buffers of 0.5, 1, and 2 miles. Although buffer sizes are arbitrarily chosen without using a systematic framework, a 1-mile buffer was chosen for this study because the 1-mile distance is what can be enjoyed within an easy walk assuming sidewalks or uncongested roads.

The boundary data, i.e., high school districts and jurisdiction and growth boundaries, were obtained from the Knoxville Knox County Metropolitan Planning Commission (KGIS 2009). Three classifications of land, i.e., rural areas, urban growth area (UGA), and planned growth area (PGA), and jurisdiction boundaries are used to capture the effects of regional core boundaries, as well as, inner and outer suburb boundaries. The rural areas include land to be preserved for farming, recreation, and other non-urban uses. The UGA is reasonably compact but adequate to accommodate the entire city's expected growth for the next 20 years. The PGAs are large enough to accommodate urban growth expected to occur in unincorporated areas

with the planning objectives of the municipality or would maintain or enhance the conservation of natural or scenic resources.

over the next 20 years (MPC 2001). Most current residential development exists within the boundaries of Knoxville and Farragut while the UGA and PGA serve as designated areas for future development. Farragut and UGA also function as suburb boundaries.

Environmental feature data such as water bodies and golf courses were found in the Environmental Systems Research Institute Data and Maps 2004 (ESRI 2004). Other environmental feature data such as railroads were acquired from KGIS (2009). The study area consists of 234 census-block groups. Information from these census-block groups was assigned to houses located within the boundaries of the block groups. The timing of the census and sales records did not match except for 2000. However, given the periodic nature of census taking, census data for 2000 were considered proxies for real time data for 1998, 1999, 2001, and 2002. By the same token, variables created from the 2001 national land cover database were used as proxies for the other years because open space was not expected to change appreciably during the study period. Detailed statistics for individual variables are reported in Table 1.

4 Empirical Results

The overall performance of the hedonic price and open-space demand equations estimated with GWR, GWR-SEM, and OLS are compared in Table 2. The OLS model is called the "global model" hereafter, in contrast to the GWR models (GWR and GWR-SEM). The spatial error Lagrange Multiplier (LM) statistic for the GWR model is 82% lower than the LM statistic for the global model, and the GWR-SEM model reduces the spatial LM statistic by 96% compared with GWR. In the open-space demand equation, the spatial LM statistic for the GWR model is 39% lower than for the global model, and the GWR-SEM model further reduces the spatial LM statistic by 4% compared with GWR. Nevertheless, the null hypothesis of no spatial error autocorrelation is still rejected in the hedonic and open-space demand equations using the GWR-SEM estimation method. Spatial error autocorrelation remains in both equations. Although the local models significantly mitigate spatial autocorrelation in both equations, they do not completely eliminate it and, thus, the statistical results must be interpreted with caution. As a result, the GWR-SEM model can be viewed as a complement to the global model rather than an alternative to it.

In the hedonic model, the Akaike Information Criterion (AIC) for the GWR-SEM model is 3,045, lower than for the global model (4,655), and slightly lower than for the GWR model (3,502). The error sum of squares for the GWR-SEM model is 1,066, lower than for the global model (1,206) and slightly lower than for the GWR model (1,085). The global F-test comparing the global and local models confirms that the GWR and GWR-SEM models outperform the global model. Given these diagnostics, estimates from the GWR-SEM specification are used to calculate

Table 2 Comparison of performance among OLS, GWR, and GWR-SEM

Statistic	Hedonic price (Dependent variable = ln (Housing price))			Open-space demand (Dependent variable = ln (open-space area within 1-mile buffer)		
	OLS	GWR	GWR-SEM	OLS	GWR	GWR-SEM
Error sum of squares	1,206	1,085	1,066	490	220	77
Global F test		11.05	21.59		187.48	21.59
Spatial error LM test[a]	2,786	497	18	189,902	115,905	111,263
AIC	4,655	3,502	3,045	−8,948	−21,189	1,130

[a]Critical value for LM test at 0.01% is 15.14 (1 degree of freedom)

marginal implicit prices of open space and create maps. The marginal implicit prices are mapped to visually highlight their spatial variations.

In the open-space demand equation, the corrected AIC for the GWR model is −21,189, lower than for the global (−8,948) and the GWR-SEM (1,130) models. The error sum of squares for the GWR-SEM model is 77, lower than for the global (490) and the GWR (220) models. The global F-test comparing the global and local models confirms that the GWR and GWR-SEM models outperform the global model. The overall fit of the GWR-SEM model is better than the GWR model, and the GWR-SEM model more effectively accounts for spatial error autocorrelation. Given these diagnostics, the estimates of the demand for open space are discussed based on the GWR-SEM estimates.

The results of the global hedonic price equation and the open-space demand equation are presented in Table 3. The estimates from the local model (GWR-SEM) are too numerous to show in Table 3. Instead, the coefficients for open space in the hedonic model and the coefficients for the variables closely associated with urban sprawl, i.e., income, house and lot size, housing density, and price of open space that are significant at the level of 5% are mapped in the Figs. 3–8.

The positive coefficient for open space in the global hedonic model indicates that households place significant value on more open space in the area surrounding their houses. An additional $10,000\,\text{ft}^2$ of open space within a 1.0-mile buffer adds $42 to the value of a house, other things constant. The estimated marginal implicit prices for open space for individual households, that are significant at the 5% level in the hedonic GWR-SEM model, are mapped in Fig. 3. The map indicates that open space significantly influences housing prices in the entire study area and the amenity values of open space increase from the east toward west Knox County. The open space area within a 1-mile buffer varies inversely with price and directly with income. According to the regression results of the global open-space demand equation, the price elasticity of open-space demand is −0.07. The income elasticity of open-space demand in the global model is 0.07. Both elasticities are significant at the 1% level. The results imply that the demand curve for open space is downward sloping on average and open space is a normal good in the Knoxville area; demand for open space increases with household income.

Table 3 Parameter global estimates of global (OLS) models

Variable	Dependent Variable = ln (house price)		Dependent Variable = ln (open space)	
	Coefficient	Std. Error	Coefficient	Std. Error
Intercept	3.830***	1.274	14.392***	0.124
Variables closely associated with urban sprawl				
Income			0.070***	0.004
ln (Finished area)	0.545***	0.009	0.031***	0.005
ln (Lot size)	0.049***	0.005	0.026***	0.002
Housing density			−0.029***	0.002
ln (Open space)	0.021**	0.086		
ln (Price of open space)			−0.070***	0.004
Structural variables				
Age	−0.004***	0.000	−0.002***	0.000
Brick	0.073***	0.006		
Pool	0.060***	0.010		
Garage	0.091***	0.006		
Bedroom	0.016***	0.005		
Stories	0.096***	0.007		
Fireplace	0.042***	0.005		
Quality of construction	0.168***	0.007		
Condition of structure	0.098***	0.006		
Census block-group variables				
Vacancy rate	−0.079	0.094	0.006	0.059
Unemployment rate	−0.059	0.147	−1.116***	0.061
Travel time to work	0.000	0.001	0.008***	0.001
Distance variables				
ln (Dist. to CBD)	−0.044	0.032	0.283***	0.009
ln (Dist. to greenway)	−0.027***	0.004	0.026***	0.002
ln (Dist. to railroad)	0.002	0.003	0.020***	0.002
ln (Dist. to sidewalk)	−0.019***	0.004	0.027***	0.002
ln (Dist. to park)	−0.002	0.005	−0.039***	0.002
ln (Park size)	0.017***	0.004	−0.021***	0.002
ln (Dist. to golf course)	−0.004	0.007	−0.056***	0.003
ln (Dist. to water body)	−0.036***	0.003	−0.007***	0.002
ln (Size of water body)	0.005***	0.001	−0.004***	0.001
High school district dummy variables				
Doyle	−0.248***	0.057	0.548***	0.010
Bearden	−0.046**	0.021	−0.152***	0.008
Carter	−0.233***	0.028	0.141***	0.013
Central	−0.100***	0.017	0.135***	0.008
Fulton	−0.039	0.024	−0.214***	0.010
Gibbs	−0.202***	0.025	0.145***	0.011
Halls	−0.127***	0.023	0.117***	0.011
Karns	−0.080***	0.013	−0.031***	0.008

(*continued*)

Table 3 (continued)

Variable	Dependent Variable = ln (house price)		Dependent Variable = ln (open space)	
	Coefficient	Std. Error	Coefficient	Std. Error
Powell	−0.110***	0.020	0.082***	0.010
Farragut	−0.079***	0.030	−0.251***	0.011
Austin	−0.230***	0.027	0.016	0.016
Other variables				
Knoxville	−0.056***	0.013	0.049***	0.009
Flood	−0.017***	0.024	−0.009	0.015
Interface	0.001	0.009	0.002	0.006
Urban growth area	−0.021	0.014	0.077***	0.008
Planned growth area	0.006	0.011	−0.067***	0.006
Season	0.024***	0.005		
Prime CPI	0.003***	0.001	0.000	0.001
Adjusted R^2	0.732	0.730		

***, **, and * indicate statistical significance at the 1%, 5%, and 10% levels respectively. Sample size is 15,335 and the optimal number of neighbors is 6,080

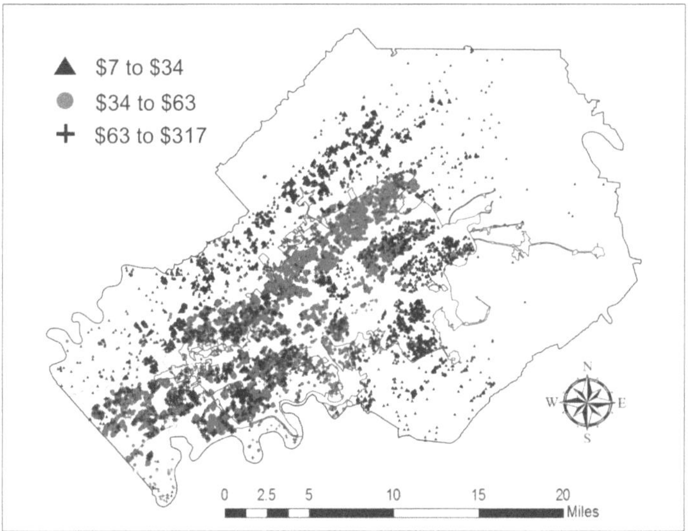

Fig. 3 Marginal implicit price of open space (10,000 square foot increase in open space)

Figure 4 shows that areas exist within Knox County where the demand for open space is upward sloping. This may be explained by speculative investing in open space in these regions. Generally, people invest in houses with rising values. These kinds of investments can result in an upward sloping demand curve for houses (Dusansky et al. 2004). Likewise, people may be inclined to invest in houses that are surrounded by open space of greater value for the same speculative purpose. Those

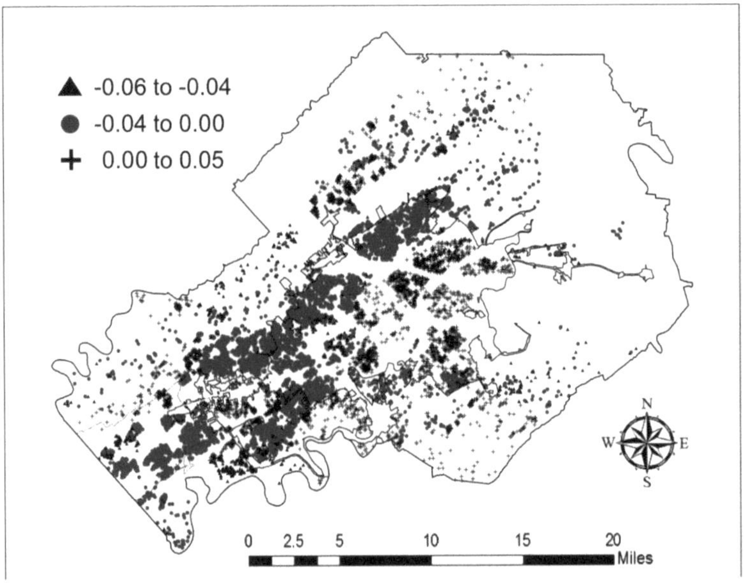

Fig. 4 Price elasticity of open-space demand

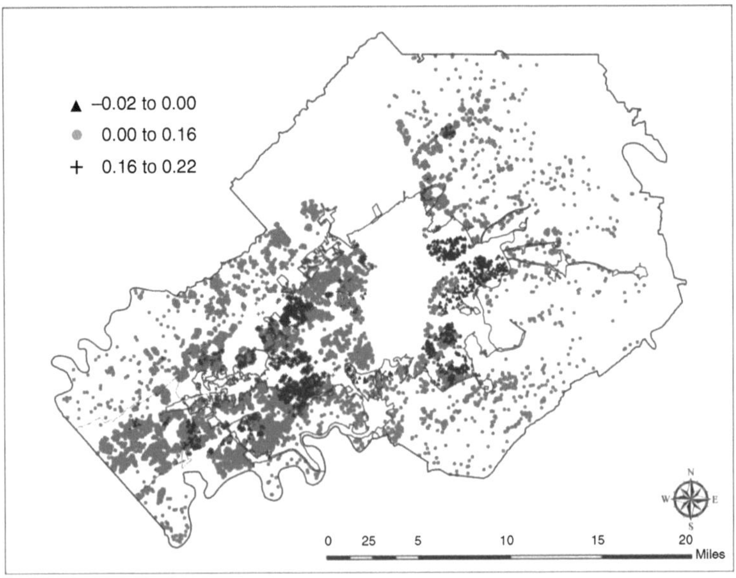

Fig. 5 Income elasticity of open-space demand

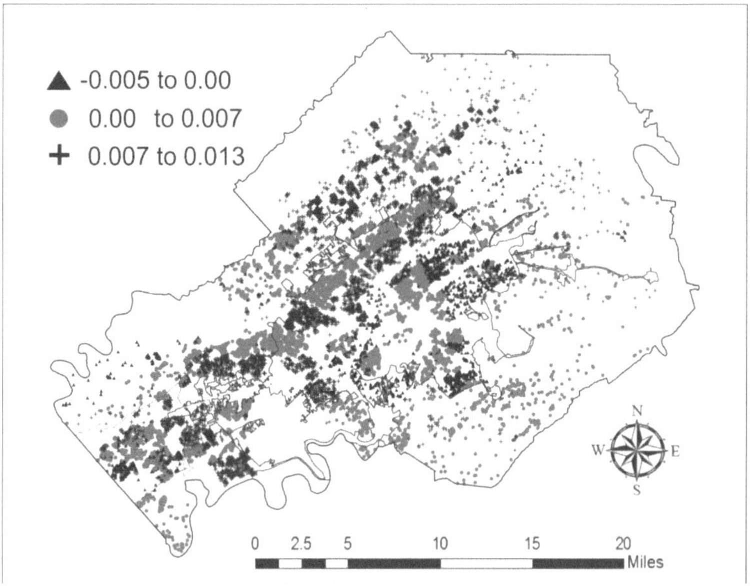

Fig. 6 Lot size elasticity of open-space demand

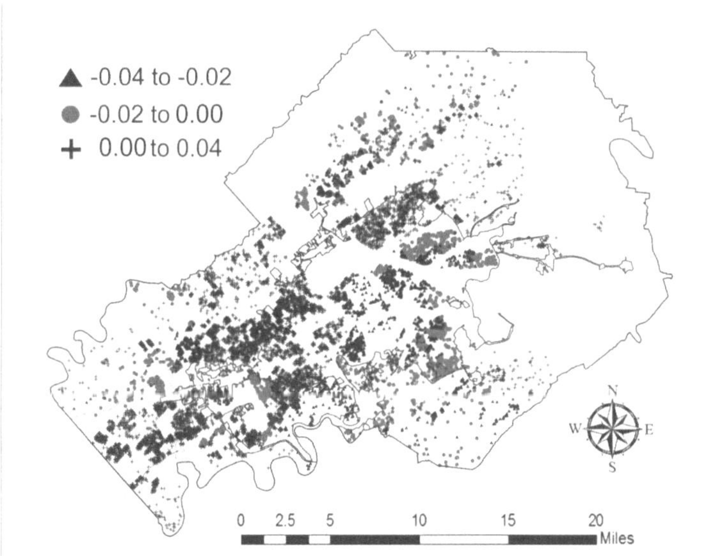

Fig. 7 Finished-area elasticity of open-space demand

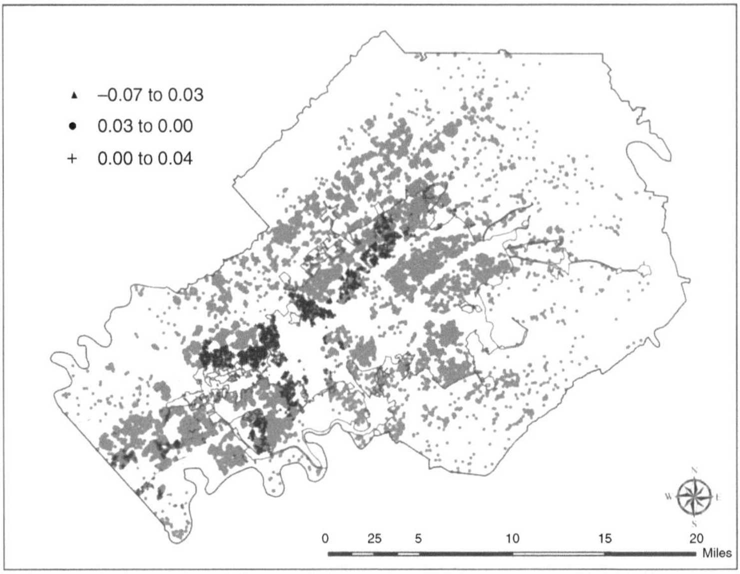

Fig. 8 Housing-density elasticity of open-space demand

areas are mostly inside Knoxville and Farragut, with some exceptions. Figure 5 shows that the demand for open space is more responsive to changes in income in the western end of Knoxville than in the rest of the County. The highly responsive demand for open space to changes in income in this area is consistent with the case of Connecticut communities (Bates and Santerre 2001).

The patterns in the southwest corner of Knox County and the town of Farragut probably result from this area being a bedroom community with affluent neighborhoods where many individuals work in private high-tech occupations; for example, scientists at the Oak Ridge National Laboratory or faculty at the University of Tennessee. This area has experienced rapid development of residential and commercial properties because of its location with respect to commuting. Demand for houses in this area is also driven by access to amenities such as shopping areas, parks, public infrastructure, and privacy on the urban fringe.

Open space area within a 1-mile buffer is positively associated with finished area and lot size at the 1% level. These results imply that properties with larger houses and lot sizes are likely to have greater open space within a 1-mile buffer. The finished-area (representing house size) elasticity of open-space demand is 0.03, and the lot-size elasticity of open-space demand is 0.03. Contrary to the findings of other studies where open space was a substitute for large residential lots (e.g., Thorsnes 2002), these results imply that house and lot sizes, and open space are complementary goods (on average) within the study area.

Figures 6 and 7 show regions with negative lot-size elasticities of open-space demand and negative finished-area elasticities of open-space demand mostly inside of Knoxville boundary, indicating substitutability between house and lot size, and open space. The results indicate that house and lot size, and open space can be both complimentary and substitute goods depending on the local area, with complementarities being the dominant relationship on average and substitutability existing inside the Knoxville boundary.

The open space area within a 1-mile buffer is negatively associated with housing density at the 1% level. The housing-density elasticity of open-space demand is −0.03. This result implies that houses located within areas of lower density housing are likely to have greater open space within a 1-mile buffer. Most areas, other than some area near the western end of Knoxville, have negative housing-density elasticities (Fig.8).

5 Conclusions

This case study examined the demand for open space in Knox County, Tennessee, United States. A GWR was modified to simultaneously model spatial heterogeneity and spatial error autocorrelation issues. The approach allows local elasticities of demand for open space to be measured and mapped. The empirical findings suggest that amenity values for open space are higher in west Knox County, and the demand for open space is more responsive to changes in income in the western end of Knoxville than the rest of the County. These patterns observed in the western end of Knoxville and in southwest Knox County coincide with the characteristics of preferences of persons employed by the Oak Ridge National Laboratory and the University of Tennessee. We also find that house and lot size, and open space can be complimentary or substitute goods, depending on the location, with complementarities being the dominant relationship on average while substitutability exists inside of city boundary.

Given the results, local officials may consider adopting location-specific policies. A smart growth policy encouraging higher-density housing with more surrounding open space might be fruitful in some parts of Knox County because the county has significant amenity values for open space as a whole. Local policymakers may encourage greater amounts of locally owned private and public open space in west Knox County, given the higher amenity values for open space. For example, some households may be more willing to pay into a fund designed to preserve neighborhood open space by purchasing development rights. Under this scenario, promoting compact development by stimulating demand for locally owned open space will likely be more successful in areas where demand for open space is higher because households would be more inclined to endorse and participate in programs or policies preserving open space, at least in the short term. However, with greater supply flexibility and occupier mobility, alongside growing open space demand, households can move to locations with more open space in the medium-to-long term. This mobility could give rise to sprawl over the longer term as households demand greater amounts of open space farther from the city center. Thus, the purchase of development rights may be a limited short-term solution unless sprawl is counteracted by regulatory policy.

A tool for achieving smart growth, such as conservation subdivisions that uses substitutability of public open space for larger residential lots is more likely to be successful inside the Knoxville boundary because house size, lot size, and open space are substitute goods; thus households may be more receptive to policies emphasizing substituting public open space for larger private lots.

Given the higher income elasiticies of demand for open space in the western end of Knoxville, increasing demand for locally owned private and public open space is expected to be higher in this area of growing economy activity. Because dynamic market forces in the western end of Knoxville are rapidly transforming open space into residential and commercial uses, open space may be in short supply compared with the growing demand for open space. Economic theory suggests that a market failure may arise when individuals do not consider the intangible benefits of open space, the social costs of excessive commuting, or the marginal social infrastructure costs of new development when engaging in private activities (Bates and Santerre 2001). Brueckner (2000) recommends greater imposition of regulatory types of conservation policies, i.e., development taxes, congestion tolls, and impact fees, to prevent any further urban sprawl and to preserve open space. These types of policies may be useful to planners in Knoxville for the medium to long term.

The empirical results suggest another possible interpretation of the hedonic regression. Households located in higher income areas place relatively higher value on open space. Because demand for open space is present, it may be that households in the higher income areas, that can afford the value of open space, end up being so-called "pioneers" living at, or beyond, the edge of the urban area in suburban and exurban lots surrounded by open space. This sort of market embodies the initial conditions for the provision of additional amenities and demand for them in the exurbs. This in turn may give rise to sprawl (e.g., residential, retail, services) over the medium to long term.[9]

One limitation of using the implicit price of open space as a proxy for the actual open-space price is that the implicit price of open space is a function of nearby house prices, ceteris paribus. The housing and open land prices were moderately correlated (0.48). As a result, the value of open space quality (i.e., composition, shape, and historic value) was not entirely represented in the implicit price of open space. In addition, the GWR-SEM econometric approach applied in this analysis addresses autocorrelation between disturbances. Future development of the general class of GWR models will focus on accommodating spatial lag processes (SAR) using an instrumental variables approach. Introduction of an autoregressive lag term in the GWR framework would allow for spatially heterogeneous clustering effects, suggesting that spatial dependencies in some regions are stronger than in others. We leave this model for further consideration, as well as development of the appropriate statistical tests to diagnose such processes in the GWR framework.

---

[9] We thank the editor for this insight.

References

Acharya G, Bennett LL (2001) Valuing open space and land-use patterns in urban watersheds. J R Estate Finance Econ 22:221–237
Anselin L (1988) Spatial econometrics: methods and models. Kluwer, Dordrecht
Bates, LJ, Santerre RE (2001) The public demand for open space: The case of Connecticut communities. J Urban Econ 50:97–111
Bin O, Polasky S (2004) Effects of flood hazards on property values: evidence before and after Hurricane Floyd. Land Econ 80:490–500
Blaine TW, Lichtkoppler FR, Stanbro R (2003) An assessment of residents' willingness to pay for green space and farmland preservation conservation easements using the contingent valuation method (CVM). J Ext 41 Available online: http://www.joe.org/joe/2003august/a3.shtml, 1 Oct. 2009
Blakely EJ (1994) Planning local economic development: Theory and practice, 2nd edn. SAGE, London
Breffle W, Morey E, Lodder T (1998) Using contingent valuation to estimate a neighborhood's willingness to pay to preserve undeveloped urban land. Urban Stud 35:715–727
Brueckner JK (2000) Urban sprawl: diagnosis and remedies. Int Region Sci Rev 23:160–171
Brunsdon C, Fotheringham A, Charlton M (1996) Geographically weighted regression: a method for exploring spatial nonstationarity. Geogr Anal 28:281–298
Carruthers JI, Ulfarsson GF (2002) Fragmentation and sprawl: evidence from interregional analysis. Growth Change 33:312–340
Cho S, Clark CD, Park WM (2006) Two dimensions of the spatial distribution of housing: dependency and heterogeneity across Tennessee's six metropolitan statistical areas. J Agr Appl Econ 38:299–316
Cho S, Roberts RK (2007) Cure for urban sprawl: measuring the ratio of marginal implicit prices of density-to-lot-size. Rev Agr Econ 29:572–579
Cummings RO, Taylor LO (1999) Unbiased value estimates for environmental goods: a cheap talk design for the contingent valuation method. Am Econ Rev 89:649–665
Daniels T (2001) Smart growth: a new American approach to regional planning. Plann Pract Res 16:271–279
Dusansky R, Koc C, Onur I (2004) Is the demand curve for housing upward sloping? Working paper. Department of Economics/University of Texas at Austin
ESRI (2004) ESRI Data & Maps 2004. http://www.esri.com/library/whitepapers/pdfs/datamaps2004.pdf, 1 Oct. 2009.
Ewing R, Pendall K, Chen D (2002) Measuring sprawl and its impact. Smart Growth America, Washington, DC
Flores NE, Carson RT (1997) The relationship between the income elasticities of demand and willingness to pay. J Environ Econ Manage 33:287–295
Fotheringham AS, Brunsdon C, Charlton M (2002) Geographically weighted regression: the analysis of spatially varying relationships. Wiley, New Jersey
Frumkin H (2002) Urban sprawl and public health. Publ Health Re 117:201–217
Geoghegan J, Lynch L, Bucholtz S (2003) Capitalization of open spaces into housing values and the residential property tax revenue impacts of agricultural easement programs. Agr Res Econ Rev 32:33–45
Gordon P, Richardson HW (1998) Prove it. Brookings Rev 16:23–26
Gordon P, Richardson HW (2000) Critiquing sprawl's critics. Policy Analysis No. 365, January 24
Gordon P, Richardson HW (2001a) Transportation and land use. Chapter 3 In: Holcombe R, Staley S (eds) Smarter growth: market-based strategies for land use planning in the 21st century. Greenwood Press, Westport, CT
Gordon P, Richardson HW (2001b) The sprawl debate: let markets plan. Publius J Federalism 31:131–149
Gujarati D (1995) Basic Econometrics, 3rd edn. McGraw-Hill, New York

Handy S (2005) Smart growth and the transportation-land use connection: what does the research tell us? Int Region Sci Rev 28:146–167
Hanham R, Spiker JS (2005) Urban sprawl detection using satellite imagery and geographically weighted regression. In: Geospatial technologies in urban economics. Springer, Berlinthebe, pp 137–151
International City /County Management Association (2008) http://icma.org/, 1 Oct. 2009
Irwin EG (2002) The effects of open space on residential property values. Land Econ 78:465–80
Irwin EG, Bockstael NE (2001) The problem of identifying land use spillovers: Measuring the effects of open space on residential property values. Am J Agr Econ 83:698–704
Iwata S, Murao H, Wang Q (2000) Nonparametric assessment of the effect of neighborhood land uses on residential house values. In: Thomas F, Hill RC (eds) Advances in econometrics: applying kernel and nonparametric estimation to economic topics. JAI Press, Stamford, CT, pp 229–257
Kelejian HH, Prucha IR (1998) A generalized spatial two-stage least squares procedure for estimating a spatial autoregressive model with autoregressive disturbances. J R Estate Finance Econ 17:99–121
KGIS (2009) Knox net where. Knoxville, Knox County, Knoxville Utilities Board Geographic Information System. http://www.kgis.org/KnoxNetWhere, 1 Oct. 2009
Krieger A (2005) The costs – And benefits? – of sprawl. In Saunders WS (ed) Sprawl and suburbia. University of Minnesota Press, Minneapolis, pp 44–56
Lichtenberg E, Tra C, Hardie I (2007) Land use regulation and the provision of open space in suburban residential subdivisions. J Environ Econ Manage 54:199–213
Maddala GS (1983) Limited dependent and qualitative variables in econometrics. Cambridge University Press, New York
Maddala GS (1992) Introduction to econometrics. Prentice Hall, Upper Saddle River, NJ
Mahan BL, Polasky S, Adams RM (2000) Valuing urban wetlands: A property price approach. Land Econ 76:100–113
McConnell V, Walls M (2005) The value of open space: Evidence from studies of non-market benefits. Working Paper, Resources for the Future, Washington DC
MPC (2001) Metropolitan Planning Commission. Tennessee public chapter 1101: growth plan for knoxville, Knox County, and Farragut, Tennessee. http://www.knoxmpc.org
Nechyba TJ, Walsh RP (2004) Urban sprawl. J Econ Perspect 18:177–200
Nelson N, Kramer E, Dorfman J, Bumback B (2004) Estimating the economic benefit of landscape pattern: an hedonic analysis of spatial landscape indices. Institute of Ecology, The University of Georgia, Athens, GA. Available online at http://www.rivercenter.uga.edu/publications.htm, 1 Oct. 2009
NLCD (2001) National Land Cover Database 2001. http://gisdata.usgs.net/website/MRLC/viewer.php, 1 Oct. 2009
OFHEO (2006) Office of federal housing enterprise oversight. http://www.ofheo.gov, 1 Oct. 2009
Open Space Inventory (1999) State of New York. www.nysaccny.org/content/cacinfo/article12f.pdf, 1 Oct. 2009
Rosenberger RS, Walsh RG (1997) Nonmarket value of western valley. Ranchland using contingent valuation. J Agr Res Eco 22:296–309
Skaburskis A (2000) Housing prices and housing density: Do higher prices make cities more compact? Can J Reg Sci 23:455–487
Sorg CF, Loomis JB, Donnelly DM, Peterson GL, Nelson LJ (1985) Net economic value of cold and warm water fishing in Idaho. USDA Forest Service, Resource Bulletin RM–11, p 26
Stevens T (1990) The economic value of bald eagles, wild turkeys, Atlantic salmon, and coyotes in New England. Resources and Environment: Management Choices. November report. Department of Resource Economics, University of Massachusetts, Amherst
Stone L, Gibbins R. (2002) Tightening our beltways: urban sprawl in western Canada. A Western Cities Project Discussion Paper, the Canada West foundation, October 2002
Thorsnes P (2002) The value of a suburban forest preserve: Estimates from sales of vacant residential building lots. Land Econ 78:426–441

Tracy S (2003) Smart Growth Zoning Codes: A Resource Guide. Local Government Commission
Tyrväinen L, Väänänen H (1998) The economic value of urban forest amenities: an application of the contingent valuation method. Landsc Urban Plann 43:105–118
US Census Bureau (2002) Census 2000 Summary File 1 (Sf 1) 100-percent data, Tables pctl2 and pctl2b. http://www.factfinder.census.gov, Oct 1, 2009.
Walsh R (2007) Endogenous open space amenities in a locational equilibrium. J Urban Econ 61:319–344
Weitz J (1999) From quiet revolution to smart growth: atate growth management programs, 1960 to 1999. J Plann Lit 14:266–337

Multilevel Models of Commute Times for Men and Women

Edmund J. Zolnik

1 Introduction

The commuting time discrepancy between men and women is known as the commuting time gender gap. Empirical evidence for the gender gap seems to be conclusive. However, recent research on commuting times in San Francisco (Gossen and Purvis 2005) and Philadelphia (Weinberger 2007) suggests that the gender gap is less ubiquitous than previously thought. To test whether or not the attenuation of the gender gap is idiosyncratic to single-city analyses of commuting times, national data is used to specify three statistical models of private-vehicle commuting times for men-only, women-only, and pooled men–women subsamples from the 2001 National Household Travel Survey (NHTS). The first goal of this chapter is to ascertain what personal characteristics of men and women and what locational characteristics of cities have the greatest affect on private-vehicle commuting times. The second goal of this chapter is to ascertain how much of the variation in commuting times for men and women originates within cities and how much originates between cities.

2 Review of the Literature

Empirical evidence on the shorter commute times of women in the US is extensive. Early in the twentieth century, Pratt (1911) found that women's commute times were shorter than men's in New York City even after controlling for hours worked. Later in the twentieth century, Ericksen (1977) found that residential location within a metropolitan area had a dramatic effect on commute times for women in the US. Women who resided in central cities took the longest to get to work, while women,

E.J. Zolnik
Department of Geography and Geoinformation Science, George Mason University, Fairfax, VA 22030, USA,
e-mail: ezolnik@gmu.edu

A. Páez et al. (eds.), *Progress in Spatial Analysis*, Advances in Spatial Science,
DOI 10.1007/978-3-642-03326-1_10,

who resided outside of central cities, but within a metropolitan area, took longer to get to work than women who resided outside of metropolitan areas. Race/ethnicity and residential location were also shown to have some association with differences in commute times. Black women took longer to get to work than white women regardless of residential location, and blacks living outside of metropolitan areas had longer commutes than whites living in central cities. However, results from a regression analysis with commute time as the dependent variable, and race, marital status, residence, age of youngest child, and type of transportation as independent variables, showed that after controlling for type of transportation, commute times for central-city women were not longer, on average, and that suburban women had the longest commutes (Ericksen 1977). In addition, married women had shorter work trips than unmarried women, and women with very young children were shown to have shorter commute times. Overall, Ericksen (1977) found that women with more household responsibilities commuted shorter distances and suburban women had the longest commute times. Given that women still have less access to private vehicles than men (Lansing and Hendricks 1967; Doyle and Taylor 2000), the suburbanization of employment confined women to smaller geographic areas in their job search. The net result was that women were unable to compete with men for jobs across the full extent of spatial labor markets. Subsequent research on spatial containment by McLafferty and Preston (1991), England (1993), Hanson and Pratt (1995), and Wyly (1998) also found that women's commute times were shorter than men's.

The unanimity in the empirical evidence on the shorter commutes of women is in sharp contrast to the ambiguity in economic theory on the origins of the commuting time gender gap (White 1977; White 1986). Economic theory divides commuting costs into monetary costs and time costs. Monetary costs refer to outlays to own and operate a private vehicle if the work trip is by automobile or fares if the work trip is by bus, train, or boat, for example. Time costs are dependent on the wages of the commuter. If the commuter earns a higher wage, they should value their commute time more highly and be less willing to commute. Because women typically earn lower wages than men (Rosenbloom 2006), economic theory suggests that women should value their commute time less and be more willing to take longer commutes. However, because women earn less in total than men, monetary costs consume a larger proportion of women's incomes which may compel them to commute less. Besides wages and income, housing prices and other socioeconomic variables are also predictors of commute times. Because commute times are manifestations of access to home and work, women's shorter commutes may also be attributable to differences in home and work locations between men and women. Assuming households are the same with regard to skills and preferences in a monocentric city, if work locations are fixed in the central city, workers will opt for longer commutes if compensated with lower housing prices. Relaxing the assumption of equal skills, higher-income workers will prefer to locate a greater distance from the central city than lower-income workers who will ultimately experience shorter commutes to jobs in the central city. If home locations are fixed, workers will opt for longer commutes if compensated with higher wages. If home and work locational decisions occur simultaneously in a monocentric city, then longer commutes will be synonymous with lower housing prices and higher wages.

Economic theory suggests that men and women with the same preferences and socioeconomic characteristics will exhibit the same commuting behavior. However, traditionally, men and women differed in their socioeconomic characteristics, e.g., educational attainment, incomes, and occupations (Johnston-Anumonwo 1992). Research by Hanson and Johnston (1985) suggested that female-dominated jobs were more uniformly distributed within Baltimore than were male-dominated jobs which may help explain why women's commutes were shorter than men's. Likewise, because women still shoulder more household commitments than men, wages associated with longer commutes may not be valued as highly as time away from home. Therefore, women may commute less because of the nonmonetary costs of commuting attributable to their household commitments (Madden 1981). A review of the available evidence by Turner and Niemeier (1997) found less than universal empirical support for the household responsibility hypothesis as in Gordon et al. (1989). However, their analysis of a subset of data on 13,074 commutes from the 1990 National Personal Transportation Survey (NPTS) tended to support the household responsibility hypothesis as an explanation for women's shorter commutes.

Nevertheless, theory on the spatial segmentation of labor markets (Hanson and Pratt 1988b; 1991) continues to show that commute times for women are shorter than commute times for men regardless of occupation. For example, Weinberger (2007) found that male-dominated work locations were less evenly distributed spatially than were female-dominated work locations in Philadelphia. Because home locations were more evenly distributed spatially, Weinberger expected commute times for women employed in male-dominated industries to rival commute times for men employed in male-dominated industries. Unexpectedly, commute times for women were shorter than commute times for men whether the woman was fully employed in a male-dominated, neutral, or female-dominated industry or not. Results from Weinberger suggest that sex is still a better determinant of commute times than the spatial distribution of male-dominated versus female-dominated jobs.

More recent research on the commuting gender gap seems to support the contention of Rosenbloom (1978) that as women work more, differences in travel behavior between men and women may disappear (Crane 2007). Research by Doyle and Taylor (2000) supports results from McLafferty and Preston (1991) which show that race or ethnicity interact with gender to affect commute times. Gossen and Purvis (2005) found that, except for 50–59 year olds, commute times for working men and women were approximately the same in San Francisco in 2000. Weinberger (2007) found that the commute time discrepancy between men and women who were fully employed decreased by thirty seconds between 1990 and 2000 in Philadelphia. Likewise, in 2000, the gender gap narrowed for women employed in male-dominated, neutral, or female-dominated jobs.

Results from single-city analyses of commuting seem to suggest that the commuting time gender gap is not as ubiquitous as previously thought. But what if the attenuation of the commuting time gender gap is idiosyncratic to San Francisco

(Gossen and Purvis 2005) or Philadelphia (Weinberger 2007)? As well, results from single-city analyses are only able to control for intra-city variation in commuting times between men and women. The emergence of polycentric labor markets and the continuous dispersion of jobs away from employment centers, e.g., may impact the commuting time gender gap differently in different cities. Indeed, the empirical evidence suggests that differences exist in the degree of job dispersion across US cities (Glaeser and Maré 2001). Research has yet to be conducted to gauge the influence of interurban differences in the degree of job dispersion and other urban spatial characteristics on the commuting time gender gap.

In order to test if the attenuation in the commute time gender gap is idiosyncratic to single-city analyses of commuting, equally sized subsamples of commuting data for men and women were extracted from a secondary data source on national travel behavior. Household data on commuters was then incorporated into statistical models which account for the spatial development patterns of cities throughout the United States. The specification of three models of private-vehicle commutes – one for men-only, one for women-only, and one for a pooled men–women subsample – provides an innovative approach for studying the variation in commute time differentials between men and women throughout the US urban system. If job sprawl, e.g., exacerbates the commuting time gender gap by lengthening commute times for men and women, then the chapter may provide empirical evidence to support urban planning which promotes the centralization of employment. The next two sections provide further details on the data and methodology, respectively.

3 Data

Data for private-vehicle commuters are from the 2001 NHTS. The 2001 NHTS was a cross-sectional survey of travel behavior in the United States. The 2001 NHTS used a non-clustered, probability sampling design to sample the travel behavior of the civilian, non-institutional population of the United States. To minimize sampling error, random-digit dialing was used to create a list of eligible telephone numbers for the 2001 NHTS. To protect the confidentiality of respondents, information to identify home and workplace locations for the 160,758 persons and 69,817 households in the 2001 NHTS was withheld.

To select subsamples of working households, only households whose respondent worked the week prior to the 2001 NHTS were eligible for selection. To focus on intrametropolitan commuters, only respondents whose home and work locations were in the same MSA were selected. To select private-vehicle commuters, only respondents who commuted to work by car, pickup truck, van, or sport utility vehicle, the week prior to the 2001 NHTS, were eligible for selection. To account for the nesting of households within MSAs, only households without missing data at the household- or MSA-level were included. Selection of households and MSAs from the 2001 NHTS, which met the above criteria, left a subsample of 4,011 male-respondent households nested within 43 MSAs, and a subsample of 4,793 female-respondent households nested within 43 MSAs. To ensure equal representation of male- and female-respondent households, 4,011 female-respondent households were randomly selected from the women-only subsample to match the sample size of the men-only subsample. Finally, the men-only ($n = 4{,}011$) and women-only ($n = 4{,}011$) subsamples were merged to create a pooled men-and-women ($n = 8{,}022$) subsample.

The dependent variable was the self-reported commute time in minutes it usually took respondents to get from home to work the week prior to the 2001 NHTS. Data for an equal number of male- and female-respondent households in the men-only ($n = 4{,}011$), women-only ($n = 4{,}011$), and pooled men-and-women ($n = 8{,}022$) subsamples includes information on the respondent and their household. Information on the respondent includes: age; ethnicity; occupation; and employment status (full-time or part-time). Information on the household includes: total income; life-cycle stage; and the ratio of vehicles to workers. Data for MSAs ($n = 43$) includes: congestion; land area in square kilometers; population size; census region; sprawl; and a dummy variable indicating the presence of commuter rail. The measure of congestion is the travel time index (TTI) from the Texas Transportation Institute (2008). The TTI is a unitless ratio of travel times during peak periods to travel times during free-flow periods (Schrank and Lomax 2007). The geography of the TTI is the urbanized area within MSAs. To account for the difference in geography between the TTI and the other MSA-level independent variables, the vast majority, i.e., over 80%, of male- and female-respondent households in the 2001 NHTS subsamples reside in the urbanized areas of their respective MSAs. The four measures of sprawl are from Ewing et al. (2002; 2003), and reflect multiple dimensions of the phenomena – density, land use, centering, and accessibility. Each dimension represents a characteristic of sprawl from the social science literature and includes: low residential density; segregated land uses; lack of significant centers; and poor street accessibility.

The mean score for the residential density-, land use mix-, degree of centering-, and street accessibility-scores was 100.00. A score below 100.00 is indicative of more sprawl, while a score above 100.00 indicates less sprawl. Of the 50 MSAs in the 2001 NHTS, data for congestion and land area were unavailable for Greensboro, Norfolk, and West Palm Beach, and sprawl data was unavailable for Charlotte, Las Vegas, Louisville, and Nashville. The exclusion of these MSAs left a final MSA level sample size of 43 (Fig. 1).

4 Methodology

Analyzing the commuting time gender gap with the latest release of national data on travel behavior in the United States won't make an especially noteworthy contribution to the commuting literature. After all, Gordon et al. (1989) used 1977 and 1983 national data from the NPTS to study the gender differences in metropolitan travel behavior. Likewise, Turner and Niemeier (1997) used national data from the 1990

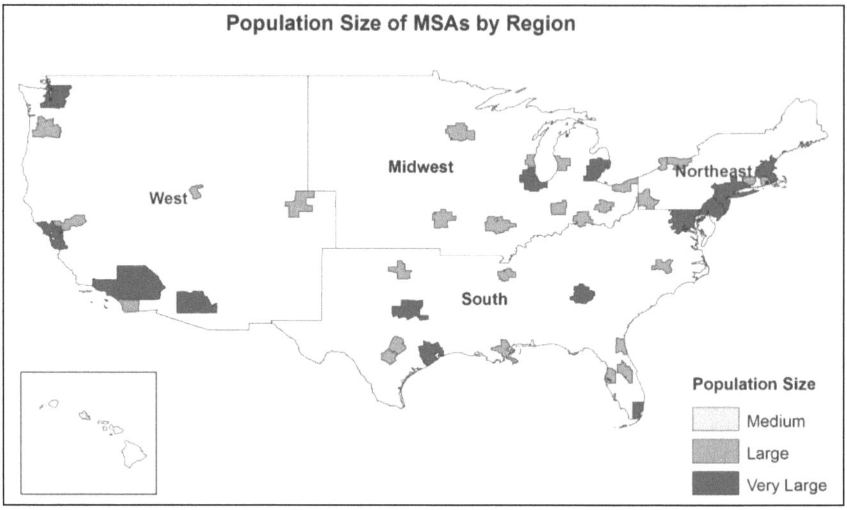

Fig. 1 Population size of MSAs (n = 43) by region

NPTS to study the household responsibility hypothesis. However, as Wyly (1998) notes:

> [q]uantitative analyses of national samples of commuting data often overlook interurban variations, focusing instead on aggregate behavioral regularities; and some datasets (e.g., the National Personal Transportation Survey) invite national-level analysis but provide insufficient samples for comparisons across different cities. (p. 398)

To overcome the intraurban sample size limitations of the NPTS, Wyly used national data from the 1990 Integrated Public Use Microdata Series to study the spatial containment of women in the US urban system. In this chapter, a multilevel approach is adopted to undertake a quantitative analysis of national travel behavior data that examines interurban variations in the commuting time gender gap. Examples of multilevel models of commuting in the Untied States (Bhat 2000; Weber and Kwan 2003) and the Netherlands (Smit 1997; Snellen et al. 2002; Schwanen et al. 2004) are evident in the literature; however examples of multilevel models of the commuting time gender gap are not.

The different multilevel model specifications in this chapter reflect divergent goals as well as an attempt to take advantage of applying a multilevel approach to studying the commuting time gender gap. The first advantage of a multilevel approach is its ability to account for nesting in the data structure of commuting events for men and women – commuters located within MSAs. Second, the interurban sample size limitations of national travel behavior data sources, such as the older NPTS and the newer NHTS, are not as problematic with a multilevel approach because the sample sizes required at higher levels of analysis are usually more restrictive than are sample sizes at lower levels of analysis (Snijders and Bosker 1999). The minimum sample size required at the higher level of analysis to avoid bias in the estimation of the higher level standard errors is not unequivocal. Snijders and Bosker (1999) suggest a minimum sample size at the higher level of analysis of ten, while Maas and Hox (2004) suggest a minimum sample size at the higher level of analysis of fifty. Therefore, the sample sizes at the higher levels of analysis ($n = 43$) in the multilevel models reported here are closer to the high end of the minimum sample sizes suggested in the literature. Third, multilevel models allow for the decomposition of variation in commuting times across household- and MSA-levels of analysis. Therefore, a multilevel model partitions variation in commute times into between-household, within-MSA and between-MSA components. The ability to partition variation between household (compositional determinants), and MSA covariates (contextual determinants) provides a methodology to estimate the proportion of variation in commuting times that is attributable to the characteristics of commuters, and the proportion attributable to the characteristics of the places where people commute. Fourth, multilevel models of commute times avoid violations of untenable homoscedasticity assumptions.

Each commute time multilevel model is a two-level model of households (h) at the micro-level nested within MSAs (m) at the macro-level (Raudenbush and Bryk 2002). Within each MSA, commute times are modeled as a function of household-level independent variables plus a household-level error term:

$$Y_{hm} = \beta_{0m} + \beta_{1m} X_{1hm} + \beta_{2m} X_{2hm} + \ldots + \beta_{Pm} X_{Phm} + r_{hm} \quad (1)$$

where:

- Y_{hm} is the commute time of household h in MSA m;
- β_{0m} is the y-intercept term in MSA m;
- X_{Phm} are $p = 1, \ldots, P$ household-level predictors of commute time;
- β_{Pm} are the level-1 coefficients that indicate the direction and strength of the association between each household characteristic, X_{Ph}, and the outcome in MSA m; and
- r_{hm} is a level-1 random effect term that represents the deviation of households hm's commute time from the predicted commute time based on the household-level model. These residual household effects are assumed to be normally distributed with a mean of zero and a variance of σ^2.

Each multilevel model is a random-intercepts model, i.e., the y-intercept is random and all of the regression coefficients at level-1 are fixed. The model for variation between MSAs is as follows. For the household effect β_{0m},

$$\beta_{0m} = \gamma_{00} + \gamma_{01} W_{1m} + \gamma_{02} W_{2m} + \ldots + \gamma_{0Q} W_{Qm} + u_{0m} \quad (2)$$

where:

- γ_{00} is the y-intercept term in the MSA-level model for β_{0m};
- W_{Qm} is an MSA-level predictor of the household effect β_{0m};
- γ_{0Q} is the corresponding level-2 coefficient that represents the direction and strength of the association between MSA characteristic W_{Qm} and β_{0m}; and

- u_{0m} is a level-2 random effect term that represents the deviation of MSA m's coefficient, β_{0m}, from its predicted value based on the MSA-level model.

Each commute time multilevel model was estimated using HLM 6.06 (Raudenbush et al. 2004). The estimation method was restricted maximum likelihood. The following sections discuss the analysis and conclusions as to the effect of interurban variations in MSA-level characteristics on the commuting time gender gap in the US urban system.

5 Findings

To control for variation in commute times for male and female private-vehicle commuters in the US urban system, separate multilevel models of private-vehicle commute times were specified and estimated for men-only, women-only, and pooled men–women subsamples from the 2001 NHTS. The major advantage of this multilevel approach to the analysis of the commuting time gender gap is that it accounts for the nested structure of national travel behavior survey data (Goldstein 1991). That is, male private-vehicle commuters (n = 4,011) nested within MSAs (n = 43), female private-vehicle commuters (n = 4,011) nested within MSAs (n = 43), and male and female private-vehicle commuters (n = 8,022) nested within MSAs (n = 43).

The next four subsections report descriptive statistics for the household- and MSA-level dependent and independent variables, as well as the model estimation results. In the first subsection, descriptive statistics for the household- and MSA-level dependent and independent variables are reported. In the second, third, and fourth subsections, results from men-only, women-only, and pooled men–women multilevel models are reported. The fifth and sixth subsections report on the analysis of the geographic variation in the commuting time gender gap and the share of commute time variation explained by the various levels of the specified models.

5.1 Descriptive Statistics for the Household- and MSA-Level Dependent and Independent Variables

Table 1 provides descriptive statistics for the household-level dependent variables in the men-only (n = 4,011), women-only (n = 4,011), and pooled men–women (n = 8,022) subsamples from the 2001 NHTS. As expected, commute times were 3:07 min longer for men than for women. On average, commute times for men were longest in the Northeast (27:15 min) and shortest in the Midwest (23:51 min), while commute times for women were longest in the South (24:43 min) and shortest in the Midwest (20:16 min). Commute times were longer for men than for women in all regions. The largest gender gap was in the Northeast (4:58 min) and the smallest gender gap was in the West (1:51 min). Regional comparison of commute times was examined by estimating the coefficient of variation, expressed as a percentage, nationally and for each regional subsample. Nationally, coefficients of variation in commute times for men (66.69%) and women (67.19%) were approximately the same. Regionally, however, coefficients of variation in commute times were lower for men than women in the Northeast (69.72% vs. 72.05%) and West (66.04% vs. 69.17%) and higher in the Midwest (65.28% vs. 60.78%) and South (63.73% vs. 63.27%). The detected regional differences in commuting behavior support the argument to control for regional variation in private-vehicle commute times in the multilevel models.

Table 1 also provides descriptive statistics for the household-level independent variables for all subsamples. The descriptive statistics for continuous independent variables are the mean and standard deviation, while the descriptive statistic for discrete independent variables is the percentage. Except for income, the categories with the highest percentages in the men-only, women-only, and pooled men–women subsamples were approximately the same. That is, the income category with the highest percentage of men was $50,000 to $74,999 (25.60%), while the income category with the highest percentage of women was $25,000 to $49,999 (30.27%).

Such a result reaffirms the notion that women earn less than men. On the one hand, the typical respondents in the men-only, women-only, and pooled men–women subsamples were:

- 43 years old;
- white;
- married/partnered with children; and
- employed full-time in professional occupations.

Consistent with empirical evidence on gender differences in access to private vehicles, the typical male respondent lived in a household with a 1.31 vehicle to worker ratio, while the typical female respondent lived in a household with a 1.14 vehicle to worker ratio. The other household characteristics that most distinguished men and women in the subsamples, besides income and the ratio of vehicles to workers were occupation and employment status (full-time or part-time). As expected, more women were employed in clerical occupations, while more men were employed in manufacturing occupations. Also as expected, more men worked full-time and more women worked part-time. Overall, however, these descriptive data lend support to the hypothesis that the socioeconomic characteristics of men and women are presently more alike than in the past.

Table 2 provides descriptive statistics for the MSA-level independent variables. The typical MSA:

- had a congestion measure of 1.22, 30:00-min trips during free-flow periods took 37:00 min during peak periods;
- had a land area of 1,746.26 km^2;
- had a population size greater than or equal to 3,000,000;

Table 1 Descriptive statistics for household-level dependent and independent variables for men-only, women-only, and pooled men–women subsamples

Variable Category	Men	Women	Men–Women
Commute time (min)	25.91(17.28)	22.80(15.32)	25.35(16.40)
Age	43.25(11.35)	42.99(11.11)	43.12(11.23)
Ethnicity			
White	78.16%	77.36%	77.76%
African American	5.76%	8.80%	7.28%
Asian	6.56%	4.34%	5.45%
AI/AN[a]	0.27%	0.60%	0.44%
NH/PI[b]	0.80%	0.82%	0.81%
Hispanic/Mexican	3.24%	2.99%	3.12%
Other	5.21%	5.06%	5.15%
Income			
<$25,000	6.13%	9.37%	7.75%
$25,000 to $49,999	25.18%	30.27%	27.72%
$50,000 to $74,999	25.60%	23.59%	24.59%
$75,000 to $99,999	20.17%	18.05%	19.11%
≥$100,000	22.91%	18.72%	20.82%
Life cycle			
1 Adult – no children	18.72%	18.85%	18.79%
1 Adult – children	2.79%	9.60%	6.20%
2+ Adults – no children	35.18%	31.26%	33.22%
2+ Adults – children	43.31%	40.29%	41.80%
Occupation			
Service	20.62%	20.67%	20.64%
Clerical	3.64%	24.41%	14.02%
Manufacturing	22.69%	4.14%	13.41%
Professional	53.05%	50.79%	51.92%
Work			
Full-time	94.56%	80.88%	87.72%
Part-time	4.94%	18.67%	11.81%
Multiple jobs	0.50%	0.45%	0.47%
Vehicles to workers	1.31(0.65)	1.14(0.45)	1.22(0.57)

Descriptive statistics reflect means and standard deviations for continuous variables and percentages for discrete variables. The standard deviations appear in parentheses after the means for the continuous variables and the percentages refer to the share of the subsample in each category of the discrete variables.

[a] American Indian/Alaskan Native.

[b] Native Hawaiian/Pacific Islander.

- was located in the Southern region of the coterminous US;
- had a residential density score of 104.37, a land use mix score of 99.97, a degree of centering score of 101.59, and a street accessibility score of 102.12; and
- had commuter rail.

Table 2 Descriptive statistics for MSA-level independent variables for men-only, women-only, and pooled men–women subsamples

Variable Category	Men	Women	Men–Women
Congestion			
Travel time index	1.22(0.10)	1.22(0.10)	1.22(0.10)
Land area (1,000 km^2)	1.75(1.35)	1.75(1.35)	1.75(1.35)
Population size			
Medium (0.5M to <1M)	6.68%	5.98%	6.33%
Large (≥1M to 3M)	31.71%	34.51%	33.11%
Very large (≥3M)	61.61%	59.51%	60.56%
Region			
Northeast	29.69%	29.97%	29.83%
Midwest	17.73%	19.20%	18.46%
South	30.62%	30.82%	30.72%
West	21.96%	20.02%	20.99%
Sprawl			
Residential density	104.37(27.47)	104.37(27.47)	104.37(27.47)
Land use mix	99.97(20.04)	99.97(20.04)	99.97(20.04)
Degree of centering	101.59(22.17)	101.59(22.17)	101.59(22.17)
Street accessibility	102.12(26.22)	102.12(26.22)	102.12(26.22)
Commuter rail (Yes)	50.91%	50.69%	50.80%
Commuter rail (No)	49.09%	49.31%	49.20%

Descriptive statistics reflect means and standard deviations for continuous variables and percentages for discrete variables. The standard deviations appear in parentheses after the means for the continuous variables and the percentages refer to the share of the subsample in each category of the discrete variables.

5.2 *Men-Only Multilevel Model*

Results from the household-level of the men-only multilevel model appear in the Men column of Table 3. Only statistically significant coefficients are reported. The referent category for each discrete independent variable represents the typical male respondent. At the household-level, men whose total income was less than \$25,000 and \$25,000 to \$49,999 commuted 4:02 and 3:35 min less, respectively, than men whose total income was \$50,000 to \$74,999. Single men with no children commuted 2:37 min less than married/partnered men with children. Men who worked part-time commuted 3:35 min less than men who worked full-time. Finally, a one unit increase in the vehicle to worker ratio increased commute times for men by 1:20 min.

Results from the MSA-level of the men-only multilevel model appear in the Men column of Table 4. Only statistically-significant coefficients are reported. The referent category for the discrete independent variables population size, region, and commuter rail represents the typical MSA. At the MSA-level, a one unit increase in the value of the congestion measure and land area is associated with an increase in the commute times of men of 7:07 and 0:01 min, respectively. Commute times

Table 3 Household-level coefficients and standard errors for men-only, women-only, and pooled men–women multilevel models

Variable Category	Men	Women	Men–Women
Y-Intercept	27.69 (0.46)***	26.96 (0.77)***	2.48 (0.01)***
Age			
Ethnicity			
White	Referent	Referent	Referent
African American			
Asian			
AI/AN[a]			
NH/PI[b]		6.40 (2.93)**	
Hispanic/Mexican			
Other			0.05 (0.02)**
Income			
<$25,000	−4.03(1.13)***	−2.19 (0.79)***	−0.05(0.03)*
$25,000 to $49,999	−3.59(0.99)***	Referent	Referent
$50,000 to $74,999	Referent		0.05 (0.02)***
$75,000 to $99,999		2.15 (0.70)***	0.11 (0.02)***
≥$100,000		3.11 (0.47)***	0.13 (0.02)***
Life cycle			
1 Adult – no children	−2.62(0.64)***		
1 Adult – children			
2+ Adults – no children			
2+ Adults – children	Referent	Referent	Referent
Occupation			
Service		−3.29(0.75)***	−0.07(0.02)***
Clerical		−1.34(0.67)**	−0.05(0.02)**
Manufacturing		1.94 (0.74)**	
Professional	Referent	Referent	Referent
Work			
Full-time	Referent	Referent	Referent
Part-time	−3.59(0.48)***	−3.01 (0.45)***	−0.15(0.01)***
Multiple jobs			
Vehicles to workers	1.33 (0.48)**	0.91 (0.50)*	0.04 (0.01)***

Referent category represents typical respondent. Standard errors appear in parentheses after coefficients. *, **, and *** indicate 90%, 95%, and 99% significance levels, respectively.
[a] American Indian/Alaskan Native.
[b] Native Hawaiian/Pacific Islander.

for men were, on average, 2:36 min shorter in large population-sized MSAs than in very large population-sized MSAs. Finally, a one unit increase in residential density score is associated with a decrease in commute times of men of 0:01 min. Overall, results from the men-only multilevel model suggest that:

- low-income men commute less than middle-income men;
- single men without children commute less than married/partnered men with children;

Table 4 MSA-level coefficients and standard errors for men-only, women-only, and pooled men–women multilevel models

Variable Category	Men	Women	Men–Women
Congestion			
Travel time index	7.11 (3.18)**	−11.18 (4.76)**	0.18 (0.07)**
Land area (km^2)	0.001 (2e-4)***	0.001 (3e-4)***	0.00003 (5e-6)***
Population size			
Medium (0.5M to <1M)			
Large (≥1M to <3M)	2.95 (0.67)***		−0.08 (0.02)***
Very large (≥3M)	Referent	Referent	Referent
Region			
Northeast		−4.79 (0.03)***	
Midwest		−3.66 (0.60)***	
South	Referent	Referent	Referent
West			
Sprawl			
Residential density	−0.02 (0.01)***	−0.04 (0.01)**	−0.001 (2e-4)***
Land use mix			−0.001 (4e-4)**
Degree of centering			
Street accessibility			
Commuter rail (Yes)	Referent	Referent	Referent
Commuter rail (No)		−2.85 (0.59)***	

Referent category represents typical respondent. Standard errors appear in parentheses after coefficients. *, **, and *** indicate 90%, 95%, and 99% significance levels, respectively.

- men who work part-time commute less than men who work full-time; and
- men with more access to private vehicles commute more than men with less access to private vehicles.

On average, commute times in the men-only subsample were shorter in less congested, large population-sized MSAs.

5.3 *Women-Only Multilevel Model*

Results from the household-level of the women-only multilevel model appear in the Women column of Table 3. Only statistically-significant coefficients are reported. The referent category for each discrete independent variable represents the typical female respondent. At the household-level, Native Hawaiian/Pacific Islander women commuted 6:24 min more than white women. Women whose total income was less than $25,000 commuted 2:11 min less than women whose total income was $25,000 to $49,999, while women whose total income was $75,000 to $99,999 and greater than or equal to $100,000 commuted 2:09 and 3:07 min more, respectively, than women whose total income was $25,000 to $49,999. Women employed in service and clerical occupations commuted 3:17 and 1:20 min less, respectively, than women employed in professional occupations, while women employed in manufacturing occupations commuted 1:56 min more than women employed in professional occupations. Women who worked part-time commuted 3:01 min less than women who worked full-time. Finally, a one unit increase in the vehicle to worker ratio increased commute times for women by 0:55 min.

Results from the MSA-level of the women-only multilevel model appear in the Women column of Table 4. Only statistically-significant coefficients are reported. The referent category for the discrete independent variables population size, region, and commuter rail represents the typical MSA. At the MSA-level, a one unit increase in the value of the congestion measure is associated with a decrease in the commute times of women of 11:11 min, while a one unit increase in land area is associated with an increase in the commute times of women of 0:01 min. Commute times for women were, on average, 4:47 and 3:40 min shorter in Northeastern and Midwestern MSAs, respectively, than in Southern MSAs. A one unit increase in residential density score is associated with a decrease in the commute times of women of 0:02 min. Finally, commute times for women were 2:51 min shorter in MSAs that had commuter rail than in MSAs that did not have commuter rail.

Overall, results from the women-only multilevel model suggest that:

- Native Hawaiian/Pacific Islander women commute more than white women;
- high-income women commute more than low-income women;
- women with manufacturing jobs commute more than women with service jobs;
- women who work part-time commute less than women who work full-time; and
- women with more access to private vehicles commute more than women with less access to private vehicles.

On average, commute times in the women-only subsample were shorter in more congested, Northeastern and Midwestern MSAs that had commuter rail.

5.4 Pooled Men–Women Multilevel Model

Results from the household-level of the pooled men–women multilevel model appear in the Men–Women column of Table 3. Only statistically significant coefficients are reported. The referent category for each discrete independent variable represents the typical male and female respondent. At the household-level, men and women in the "Other" ethnic category, which includes men and women who self identify with two or more ethnic categories, commuted 1:49 min more than white men and women. Men and women whose total income was less than $25,000 commuted 1:18 min less than men and women whose total income was $25,000 to $49,999, while men and women whose total income was $50,000 to $74,999, $75,000 to $99,999, and greater than or equal to $100,000 commuted 1:37, 3:31, and 3:58 min more, respectively, than men and women whose total income was $25,000 to $49,999. Men and women employed in service and clerical occupations commuted 1:55 and 1:38 min less, respectively, than men and women employed in professional occupations. Men and women who worked part-time commuted 3:55 min less than men and women who worked full-time. Finally, a one unit increase in the vehicle to worker ratio increased commute times for men and women by 1:19 min.

Results from the MSA-level of the pooled men–women multilevel model appear in the Men–Women column of Table 4. Only statistically significant coefficients are reported. The referent categories for the discrete independent variables population size, region, and commuter rail represent the typical MSA. At the MSA-level, a one unit increase in the value of the congestion measure and land area is associated with an increase in the commute times of men and women of 5:08 and 0:01 min, respectively. Commute times for men and women were, on average, 2:53 min shorter in large population-sized MSAs than in very large population-sized MSAs. Finally, a one unit increase in residential density and land use mix score is associated with a decrease in the commute times of men and women of 0:01 and 0:01 min, respectively. Overall, results from the pooled men–women multilevel model suggest that:

- men and women who self identify with two or more ethnic categories commute more than white men and women;
- high-income men and women commute more than low-income men and women;
- men and women with service jobs commute less than men and women with professional jobs;
- men and women who work part-time commute less than men and women who work full-time; and
- men and women with more access to private vehicles commute slightly more than men and women with less access to private vehicles.

On average, commute times for the typical male and female in the pooled men–women subsample were shorter in less congested, large population-sized MSAs.

5.5 Analysis of MSA-Level Residuals from Multilevel Models

Analysis of the MSA-level residuals from the men-only, women-only, and pooled men–women multilevel models offers valuable information on the geographic variation in the commuting time gender gap. On the one hand, in four MSAs – Austin, Buffalo, Minneapolis, and San Francisco – commute times for men were longer than predicted based on the men-only multilevel model *and* commute times for women were shorter than predicted based on the women-only multilevel model (Fig. 2). These four men-longer commute and women-shorter commute MSAs are located in all four census regions. In seven MSAs – Atlanta, Kansas City, Miami, New Orleans, Oklahoma City, Philadelphia, and San Antonio – commute times for men fell into the predicted category based on the men-only multilevel model *and* commute times for women were shorter than predicted based on the women-only multilevel model. These seven men-moderate commute and women-shorter commute MSAs are mostly located in the South. On the other hand, in one MSA – Los Angeles – commute times for men were shorter than predicted based on the men-only multilevel model *and* commute times for women were longer than predicted based on the women-only multilevel model. In two MSAs – Pittsburgh and Portland – commute times for men were shorter than predicted based on the men-only multilevel model *and* commute times for women fell into the predicted category based on the men-only multilevel model. Overall, two of the three men-shorter commute and women-longer or moderate commute MSAs are located in the West. Finally, in two MSAs – Saint Louis and Washington – commute times for both men and women were longer than predicted; in 11 MSAs – Columbus, Dallas, Hartford, Houston, Indianapolis, Jacksonville, New York, Orlando, Providence, Seattle, and Tampa – commute times for both men and women fell into the predicted category; and in five MSAs – Denver, Detroit, Grand Rapids, Milwaukee, and Phoenix – commute times for both men and women were shorter than predicted based on the men-only and women-only multilevel models, respectively. Analyzing MSA-specific residuals extends previous research on geographic variation in the commuting time gender gap (Wyly 1998) by showing that women's commute times are longer than predicted and men's commute times are shorter than predicted in the West; particularly in Los Angeles. Analysis of the MSA-level residuals also shows that men's commute times were longer than predicted and women's commute times were shorter than predicted in San Francisco, which runs counter to Gossen and Purvis' (2005) finding of an attenuation in the commuting time gender gap.

5.6 Proportion of Variance Between and Within MSAs

The intraclass correlation coefficient (ICC) is used here to measure the proportion of variance in private-vehicle commute times between MSAs (Raudenbush and Bryk 2002; Snijders and Bosker 1999). The ICC is applicable only to random-intercept models such as the men-only, women-only, and pooled men–women multilevel models reported in this chapter. To estimate the ICC, estimates of between-MSA and within-MSA variability are substituted for the parameters in the following equation:

$$\rho = \frac{\tau_{00}}{\tau_{00} + \sigma^2} \tag{3}$$

where:

ρ is the ICC;
τ_{00} captures between-MSA variability; and
σ^2 captures within-MSA variability.

Estimation of the ICCs for the men-only, women-only, and pooled men–women multilevel models reveal that 0.03%, 0.07%, and 0.05%, respectively, of the variance in private-vehicle commute times was between MSAs. Thus, just as Schwanen

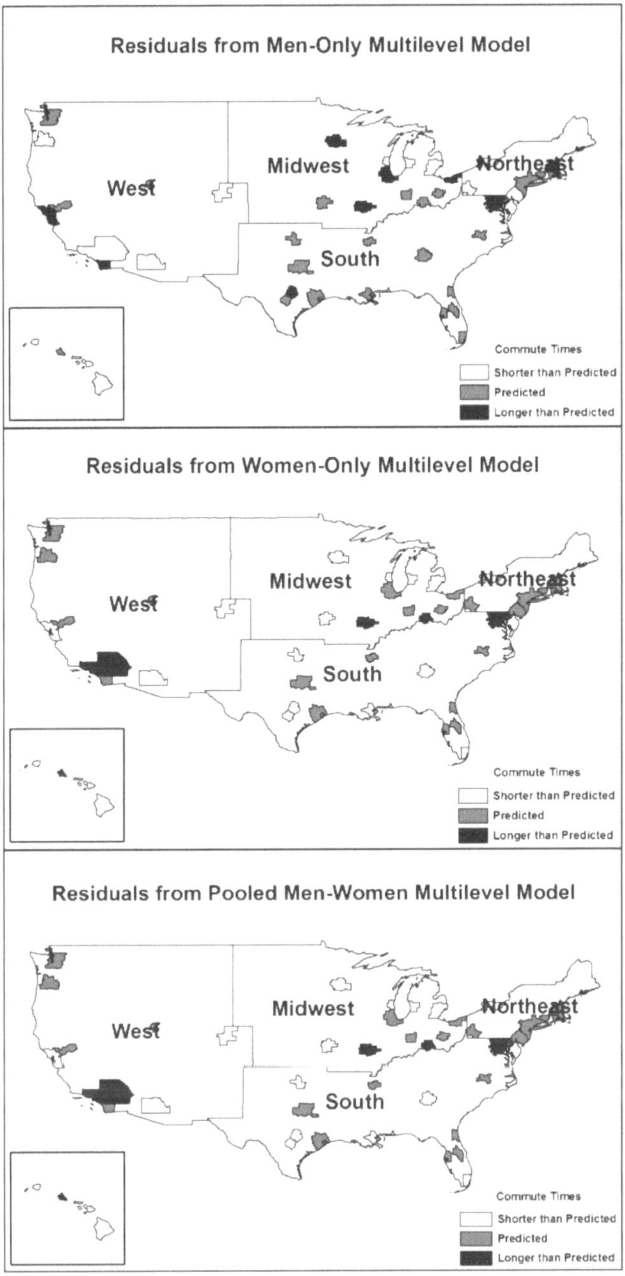

Fig. 2 Regional differences in commute times from men-only, women-only, and pooled men–women multilevel models

et al. (2004) found, place/spatial characteristics account for a small proportion of the total variation in private-vehicle commuting outcomes.

6 Discussion

Comparison of results at the household-level between the men-only, women-only, and pooled men–women multilevel models tend to support the contention of economic theory (White 1977; 1986) that commute time differences are at least partially attributable to income and occupational differences between men and women. Total incomes were, on average, lower in the women subsample than in the men subsample, and men and women in low-income categories commuted less than men and women in middle- and high-income categories. However, commute times for women in the highest-income categories were longer by 3:07 and 2:09 min, respectively, in comparison to commute times for women in the middle-income category. This finding suggests that higher incomes were more synonymous with longer commutes for women than for men. Occupation had different effects on commute times for men and women. None of the occupational categories had an effect on commute times for men. But, as expected, commute times for women employed in female-dominated industries such as service and clerical were shorter than for women employed in professional occupations. Such a result is consistent with other studies where shorter commute times for women employed in female-dominated industries have been reported (Wyly 1998). Interestingly, commute times for women employed in male-dominated, manufacturing occupations were longer than for women employed in professional occupations. Such a result is not consistent with results from a study conducted in Philadelphia, for example, where commute times for women were found to be shorter than commute times for men, regardless of the gender-industry pairing (Weinberger 2007). The results also suggest that access to private vehicles contributes to private-vehicle commute time differences between men and women. Access to private vehicles was lower for women in the women-only subsample than for men in the men-only subsample. However, the association between private-vehicle commute times and the ratio of vehicles to workers was positive for both men and women, and greater access to private vehicles appeared to lengthen commute times for men more than for women.

Comparison of household-level findings across all models offers no support for the household responsibility hypothesis (Turner and Niemeier 1997); none of the lifecycle-stage categories for women were statistically significant in the women-only multilevel model. Interestingly, even though the percentage of women in the women-only subsample who worked part-time was higher than the percentage of men in the men-only subsample who worked part-time, private-vehicle commute times were shorter for men (3:35 min) and women (3:01 min) who worked part-time by approximately the same amount. Such a result suggests that part-time work had the same effect on private-vehicle commute times for men and women.

Comparing model results at the MSA-level indicates that congestion had the largest differential effect on the commute times of men and women. Congestion, measured using the TTI, is associated with an increase of 7:07 min in private-vehicle commute times for men in the men-only subsample and a decrease of 11:11 min for women in the women-only subsample. In contrast to these results, commute times for women were, on average, shorter in Northeastern and Midwestern MSAs than in Western MSAs where congestion was highest. Taken together, the large differential effect of congestion on commute times for men and women appears to be a phenomenon specific to Los Angeles, where congestion was ranked first among the 43 MSAs, and where commute times were shorter than expected for men and longer than expected for women based on the men-only and women-only multilevel models, respectively. Land area was associated with an increase in commute times for men and women by the same amount. As expected, sprawl, measured here as a function of residential density and land use mix, had negative effects on commute times for men and women. That is, higher residential density and better land use mix appears to lower commute times.

The coefficient estimates for land area, residential density, and land use mix are statistically significant, but the strength of the associations between each of these variables and private-vehicle commute times, for men and women, is very small. Interestingly, the degree of centering, which predominantly reflects job sprawl, was not statistically significant. Such a result tends to contradict arguments that job sprawl in urban labor markets contributes to the commuting time gender gap (Wyly 1998). Finally, commute times for women in MSAs that did not have commuter rail were shorter than commute times for women in MSAs that did have commuter rail. Such a result suggests that the absence of public alternatives leads to more private-vehicle commuting which is more time efficient for longer distance commutes.

7 Conclusions

A reassessment of home-work linkages by Hanson and Pratt (1988a) underscores the need to consider how the, "home-work link functions for a variety of diverse household types in a variety of local contexts" and "to make explanations scale specific" (p. 318). In total, the results reported in this chapter suggest that compositional effects such as income and occupation have a greater impact on variation in commute times for men and women than contextual spatial effects such as sprawl. As such, results from the chapter are more supportive of economic rather than household responsibility explanations for the commuting time gender gap. Further, the results point away from polycentricity and job sprawl as major contributors to the commuting time gender gap (Rosenbloom 2006). Nonetheless, one contextual effect – congestion – has a large differential impact on commute times for men and women especially in Los Angeles. Likewise, regional variations in commute times for women are evident – commute times are shorter for women in the South and longer for women in the West. Taken together, a multilevel approach to the commuting time gender gap shows that a single-city analysis is suitable to study intraurban variations in commute times for men and women, but that a multi-city analysis is able to capture interurban and interregional variations in congestion that greatly impact the commuting time gender gap.

References

Bhat C (2000) A multi-level cross-classified model for discrete response variables. Transp Res B 34:567–582

Crane R (2007) Is there a quiet revolution in women's travel? Revisiting the gender gap in commuting. J Am Plann Assoc 73:298–316

Doyle D, Taylor B (2000) Variation in metropolitan travel behavior by sex and ethnicity. In: Final report: travel patterns of people of color. Federal Highway Administration, Washington, pp 181–244

England K (1993) Suburban pink collar ghettos: the spatial entrapment of women? Ann Assoc Am Geogr 83:225–242

Ericksen J (1977) An analysis of the journey to work for women. Soc Probl 24:428–435

Ewing R, Pendall R, Chen D (2002) Measuring sprawl and its impact. Smart Growth America, Washington

Ewing R, Pendall R, Chen D (2003) Measuring sprawl and its transportation impacts. Transp Res Rec 1831:175–183

Glaeser E, Kahn M, Chu C (2001) Job sprawl: employment location in U.S. metropolitan areas. The Brookings Institution, Washington

Goldstein H (1991) Multilevel modeling of survey data. Statistician 40:235–244

Gordon P, Kumar A, Richardson H (1989) Gender differences in metropolitan travel behaviour. Reg Stud 23:499–510

Gossen R, Purvis C (2005) Activities, time, and travel: changes in women's travel time expenditures, 1990–2000. In: Research on women's issues in transportation. Transportation Review Board, Washington, pp 21–29

Hanson S, Johnston I (1985) Gender differences in work-trip length: explanations and implications. Urban Geogr 6:193–219

Hanson S, Pratt G (1988a) Reconceptualizing the links between home and work in urban geography. Econ Geogr 64:299–321

Hanson S, Pratt G (1988b) Spatial dimensions of the gender division of labor in a local labor market. Urban Geogr 9:180–202

Hanson S, Pratt G (1991) Job search and occupational segregation of women. Ann Assoc Am Geogr 81:229–253

Hanson S, Pratt G (1995) Gender, work, and space. Routledge, New York

Johnston-Anumonwo I (1992) The influence of household type on gender differences in work trip distance. Prof Geogr 44:161–169

Lansing J, Hendricks G (1967) Automobile ownership and residential density. University of Michigan, Ann Arbor

Maas C, Hox J (2004) Robustness issues in multilevel regression analysis. Stat Neerl 58:127–137

Madden J (1981) Why women work close to home. Urban Stud 18:181–194

McLafferty S, Preston V (1991) Gender, race, and commuting among service sector workers. Prof Geogr 43:1–14

Pratt E (1911) Industrial causes of congestion and pollution in New York City. Columbia University Press, New York

Raudenbush S, Bryk A (2002) Hierarchical linear models: applications and data analysis methods. Sage, Thousand Oaks
Raudenbush S, Bryk A, Cheong Y, Congdon R, du Toit M (2004) HLM 6: hierarchical linear and nonlinear modeling. Scientific Software International, Lincolnwood
Rosenbloom S (1978) The need for study of women's travel issues. Transportation 7:347–350
Rosenbloom S (2006) Understanding women's and men's travel patterns: the research challenge. In: Research on women's issues in transportation. Transportation Review Board, Washington, pp 7–28
Schrank D, Lomax T (2007) The 2007 urban mobility report. Texas Transportation Institute, College Station
Schwanen T, Dieleman F, Dijst M (2004) The impact of metropolitan structure on commute behavior in the Netherlands: a multilevel approach. Growth Change 35:304–333
Smit L (1997) Changing commuter distances in the Netherlands: a macro-micro perspective. In: Westert G, Verhoeff R (eds) Places and people: multilevel modelling in geographical research. The Royal Dutch Geographical Society, Utrecht, 86–99
Snellen D, Borgers A, Timmermans H (2002) Urban form, road network type, and mode choice for frequently conducted activities: a multilevel analysis using quasi-experimental design data. Environ Plann A 34:1207–1220
Snijders T, Bosker R (1999) Multilevel analysis: an introduction to basic and advanced multilevel modeling. Sage, Thousand Oaks
Texas Transportation Institute (2008) Congestion data for your city, 2008. http://mobility.tamu.edu/ums/ums/congestion_data/. Accessed 14 July 2008
Turner T, Niemeier D (1997) Travel to work and household responsibility: new evidence. Transportation 24:397–419
Weber J, Kwan M (2003) Evaluating the effects of geographic contexts on individual accessibility: a multilevel approach. Urban Geogr 24:647–671
Weinberger R (2007) Men, women, job sprawl, and journey to work in the Philadelphia region. Publ Works Manag Pol 11:177–193
White M (1977) A model of residential location choice and commuting by men and women workers. J Reg Sci 17:41–52
White M (1986) Sex differences in urban commuting patterns. Am Econ Rev 76:368–372
Wyly E (1998) Containment and mismatch: gender differences in commuting in metropolitan labor markets. Urban Geogr 19:395–430

Walkability as a Summary Measure in a Spatially Autoregressive Mode Choice Model: An Instrumental Variable Approach

Frank Goetzke and Patrick M. Andrade

1 Introduction

In recent years it has become more common to include social interactions or neighborhood effects (also called social network effects) in transportation modeling. These models are typically in the tradition of Brock and Durlauf (2001, 2002) who were among the first to propose a discrete choice model that includes social interactions and neighborhood effects. However, their approach is inherently non-spatial, while the topology of social interactions and neighborhood effects can be best captured spatially (Leenders 2002; Páez et al. 2008a). Therefore, some of the latest articles in transportation modeling have moved towards explicitly incorporating the spatially autoregressive structure of social network effects into their models (e.g. Dugundi and Walker 2005; Páez and Scott 2007; Goetzke 2008). This new direction in transportation research is not all that surprising, given the success of spatial econometrics as an emerging modeling method across social science disciplines.

The econometric strategy to implement an independent variable representing social interactions and neighborhood effects, as proposed by Brock and Durlauf (2001, 2002), is to use the group mean of the observed dependent choice variable, as defined by social interactions and neighborhood effects. This approach can be spatially extended if the group mean is based on spatial relations, as in traffic analysis zones (Dugundi and Walker 2005), a spatial weight matrix (Goetzke 2008), or a matrix based on personal relations (Páez and Scott 2007). Therefore, choices could be modeled as a function of the typical choice determinants in travel behavior analysis (e.g. personal, household, trip and mode characteristics), or as choices of either a non-spatial group or spatial neighbors. Empirical work dealing with mode choice decisions by Dugundi and Walker (2005), and Goetzke (2008) give evidence that the mode choice decisions of spatial neighbors are indeed associated with the mode choice decision of the individual.

F. Goetzke (✉)
Department of Urban and Public Affairs, University of Louisville, 426 W. Bloom Street, Louisville, KY 40208, USA,
e-mail: f0goet01@louisville.edu

A. Páez et al. (eds.), *Progress in Spatial Analysis*, Advances in Spatial Science, 217
DOI 10.1007/978-3-642-03326-1_11,

However, do we really know if a positive regression coefficient for network effects is an indication of the existence of social interactions and neighborhood effects? Manski (2000), in his analysis of social interactions, differentiates between endogenous interactions and contextual interactions, as well as correlated effects. Only endogenous and contextual interactions are identified as true social phenomena, while correlated effects are described as nonsocial in nature.

The difference between endogenous and contextual interactions is that an agent's behavior varies, in the first case, with the behavior of the group, and in the second case, with the social characteristics of the group. For example, applying Manski's concept to walking mode choice decision making, endogenous interactions are at work if a person's probability to be a pedestrian increases with the number of walking neighbors, independent of the neighbor's social characteristics. Contextual interactions exist if a person's probability to be a pedestrian rather depends on the social characteristics of those neighbors. Manski shows that both endogenous and contextual interactions illustrate dissimilar ways of how a group influences the actions of a group member. At the same time, he also proves that it is impossible to empirically separate between the two kinds of interactions using econometric modeling, because the mean of that group's behavior is determined by the mean of the social characteristics of the group members.

The difficulty to distinguish between endogenous and contextual interactions is the first part of what Manski calls the reflection problem. The other part of the reflection problem, is the identification of social interaction in contrast to correlated effects. Correlated effects are at play if members of a group behave similarly because they face the same environment, such as increased walking mode share caused by better pedestrian transportation infrastructure. The isolation of social interactions is only possible by controlling for environmental effects in all completeness. However, in an econometric model this will only be successful if the environmental effects vary between groups and/or neighborhoods, and also between individuals within groups and/or neighborhoods. The latter condition it is not so difficult to fulfill, but meeting the first one may be almost impossible. An alternative condition would be a homogeneous environment.

In this chapter, we contend that it is important to include social interactions and correlated effects in the mode choice model as one combined spatial spillover variable. The reasons for this are twofold: The spatial spillover variable serves the purpose to avoid a possible omitted variable bias; and, in addition, the spatial spillover variable can be seen as a proxy for the mode-friendliness in the neighborhood (Goetzke 2003). Within the context of the choice to walk, the distinction between endogenous and contextual interactions may be important for designing certain policies to change pedestrian mode share: For example, a campaign that promotes walking would exhibit social spillovers only for endogenous interactions, but not for contextual interactions. In a different situation, however, when the objective is to identify the existence of social interactions in walking, the first part of the reflection problem does not play a role, as long as we are not trying to determine the nature of the interaction.

Following this premise, the first objective of this research is to investigate the spatially autoregressive structure in mode choice modeling. In technical terms, this necessitates an endogenous spatially lagged term which, in concert with the non-linearity of logit and probit models, complicates all efforts to set up an appropriate and also solvable maximum likelihood function. One way around the problem is what Anselin (2002) called the *conditional* spatially autoregressive discrete choice model, where the statistical inference is restricted because of the assumption that the spatial effect is determined prior to the dependent variable, and not simultaneously. While this assumption may not reflect the true model in mode choice decision making, and may generate inconsistent regression coefficient estimates, it is where the latest research in transportation modeling stands. Outside of the transportation field, Fleming (2004) is one of the few who addresses the problem of endogeneity in spatially autoregressive discrete choice models, and discusses complex techniques to account for the spatial lag term, but all of those suggested methods are highly computationally intensive and thus, difficult to implement.

This brings us to the second objective of this chapter, which is methodological in nature: We propose to circumvent the above described computational difficulties by using an instrumental variable approach for estimating the spatial lag term regression coefficient. We implement this instrumental variable approach both in conjunction with a linear probability mode choice model, and in conjunction with a logit mode choice model. We find in both cases, that the walkability variable improves the regression, and that the walkability regression coefficient is positive and significantly different from *zero*, indicating that walking exhibits social interactions and/or correlated effects. Furthermore, we find some evidence that the inclusion of the walkability variable avoided an omitted variable bias.

A case study using the 1997/1998 New York Metropolitan Transportation Council (NYMTC) household survey data allows us to develop a binomial mode choice model for home-based work trips less than 2 miles, where walking is one choice and all remaining modes the other choice. We are going to interpret the spatial spillover variable as a summary measure of the neighborhood's "walkability," essentially combining the social interactions with correlated effects. Using New York as a case study has the added advantage that transportation infrastructure for all modes is more or less homogeneous for the central boroughs of the city. This, in turn, contributes to isolated social interactions (social network effects) by providing better control for the correlated effects (transportation environment).

2 Econometric Model

The starting point of our spatially autoregressive mode choice model is research by Evans et al. (1992) and Brueckner and Largey (2006) that makes use of an instrumental variable to account for the endogeneity of a social network variable in a discrete choice model, the spatially autoregressive 2-stage least square method (2-SLS) with a continuous dependent variable of Anselin (1988) and Land and

Deane (1992), as well as Fleming's (2004) non-linear least square approach for spatially autoregressive discrete choice models. The basic idea behind our methodology is the following: First we estimate walkability as an instrumental variable by regressing the spatially weighted neighbor's pedestrian mode share on all social characteristics of both the observed household and the spatially lagged neighbors. Then, in a second step, we use the instrumental variable as the autoregressive term to estimate the mode choice model, both as a linear probability model as well as a logit model. The linear probability model (LPM) estimation with the instrumental variable follows essentially the 2-SLS approach; however, it also needs to be corrected for heteroskedasticity using the weighted least square method.

While the literature offers plenty of spatially autocorrelated 2-SLS models (Anselin 1988; Land and Deane 1992), as well as linear probability models that were corrected for heteroskedasticity using a weighted least square approach (Wooldridge 2005), to the best of our knowledge these ideas have not been synthesized into one; there is certainly no such model in the field of transportation mode choice modeling. Furthermore, while Evans et al. (1992) used an instrumental variable approach in combination with a logit model to account for endogeneity, it is new to do so for the spatially autoregressive term in a logit model. Both concepts are not only innovative in solving the endogeneity problem of the spatially autoregressive term in the context of discrete choice models, but are also simple to implement.

The mode choice model to be estimated is as follows:

$$\mathbf{y} = \alpha + \mathbf{Wy}\lambda + \mathbf{X}\beta + \mathbf{e} \tag{1}$$

Depending on whether the model is an LPM or a logit model with n observations, $\mathbf{y}$ is the $n\ x\ 1$ vector of either the observed, chosen mode (walking or not) or the latent variable representing the unobserved utility of the chosen mode, $\mathbf{Wy}$ is the spatial lag term with $\mathbf{W}$ being the spatial weight matrix, $\mathbf{X}$ is the $n\ x\ m$ matrix of m personal, household trip and mode characteristics, and $\mathbf{e}$ is the $n\ x\ 1$ vector of random error terms. The constant term is α, λ is the regression coefficient for the spatial lag term and the $m\ x\ 1$ vector of regression coefficient for $\mathbf{X}$ is β.

Before we can run the mode choice model in (1), we need to first estimate $\mathbf{Wy}$ by regressing it on feasible instruments in addition to the personal and household characteristics ($\mathbf{X}$), and to then use in (1) the fitted values $\mathbf{y}^*$ instead of $\mathbf{Wy}$. This is necessary in order to take care of the endogeneity of the spatial lag term, where $\mathbf{Wy}$ is determined simultaneously with $\mathbf{y}$.

In the instrumental variable regression, $\mathbf{Wy}$ becomes the dependent variable, which is derived by spatially weighting the three closest neighbors by distance. The instruments are the personal and household characteristics of these three neighbors. We decided to restrict the number of neighbors to three in order to avoid too much multicollinearity. Therefore, instrumental variable regression looks as follows:

$$\mathbf{Wy} = \gamma + \mathbf{X}\delta + \mathbf{X}_{(-1)}\eta + \mathbf{X}_{(-2)}\theta + \mathbf{X}_{(-3)}\xi + \mathbf{u} \tag{2}$$

where $\mathbf{X}$ is the *n x k* matrix of *k* personal and household characteristics, $\mathbf{X}_{(i)}$ is the *n x k* matrix of *k* personal and household characteristics for the *i*-th spatially lagged neighbor and $\mathbf{u}$ is the *n x 1* vector of random error terms. The constant term is γ, and δ, η, θ as well as ξ are the corresponding *k x 1 vectors of* regression coefficients. After running (2) and calculating the fitted values

$$\mathbf{y}^* = \gamma + \mathbf{X}\delta + \mathbf{X}_{(-1)}\eta + \mathbf{X}_{(-2)}\theta + \mathbf{X}_{(-3)}\xi \tag{3}$$

we can use $\mathbf{y}^*$ to finally estimate equation (1):

$$\mathbf{y} = \alpha + \mathbf{y}^*\lambda + \mathbf{X}\beta + \mathbf{e} \tag{4}$$

The regression coefficients should now be unbiased as long as (4) is estimated as a linear probability model or logit model. If (4) is a linear probability model, however, we need in a third step to further correct for heteroskedasticity, so that the estimation results are also efficient, meaning that we can trust standard errors and t-values for evaluating the significance of the regression coefficients. Since the LHS variable is binary, a binomial distribution can be assumed, which means that for observation *i* the conditional variance of y_i given the vector $\mathbf{x_i}$ is:

$$\mathrm{Var}(y_i|\mathbf{x_i}) = \mathrm{p}(\mathbf{x_i})[1 - \mathrm{p}(\mathbf{x_i})] \tag{5}$$

Since $\mathrm{p}(\mathbf{x_i}) = y_i{}^{**}$ can be derived by computing the fitted probability values in (4), (5) can be rewritten as

$$h_i = y_i{}^{**}(1 - y_i{}^{**}) \tag{6}$$

where $1/\sqrt{h_i}$ is then applied as the weight in a weighted least square estimation of (4). With this procedure, the estimation becomes both consistent and efficient.

3 Data and Instrumental Variables

The data employed for this model comes from the 1997/1998 comprehensive regional household travel diary conducted for the Best-Practice Travel Demand Forecasting Model by the New York Metropolitan Transportation Council (NYMTC 2004). The data was collected for the metropolitan areas of 28 counties, which include, besides the five central city boroughs, counties of upstate New York, Long Island, New Jersey and Connecticut. For our study, however, we only use data from four New York City boroughs: Manhattan, Queens Brooklyn, and Staten Island. The main reasons for choosing these four boroughs are:

- The New York City area has a relatively high density of surveyed households, allowing the calculated spatially weighted mean pedestrian share to be meaningful for the mode choice model. Outside of the New York City area, distances between neighbors included in the survey become rather large.

- Only in New York City is the number of pedestrians high enough to get the variability in the dataset desired for econometric analysis. Also, within New York City we do not have to be too concerned about differences in the walking infrastructure or the service quality of alternate modes.

For the mode choice model, only trips from home to work within a walking distance of 2 miles (40 min) were included. The literature in transportation modeling distinguishes between three typical trip purposes: Home-based work (HBW) trips, non-home based (NHB) trips, and home-based other (HBO) trips. Each trip purpose should be modeled separately, since mode choice determinants are expected to differ. Purely for practical reasons we decided to restrict the analysis only to HBW trips. After applying all these restrictions, we decided to exclude the Bronx because the number of observations became too small, so the final sample size was 541 observations. Figure 1 shows a map with the household locations of all the included trips.

If the binary choice variable was set to *one*, it indicates that the trip was done by walking. All remaining modes, such as automobile or public transit were set equal to *zero*. As shown in Table 1, a little more than half of the people included in the dataset walk to work. The following personal, household, trip and mode characteristics were included as explanatory variables: Race ("Black," "Asian" or "White"), gender ("Male"), "With disability," household "Income less than $50k," age ("Under

Fig. 1 Map with the household locations of all the included trips

Table 1 Descriptive statistics of all included variables

	Mean	Standard deviation
Walking	0.503	0.501
Black	0.076	0.265
Asian	0.059	0.236
White	0.697	0.460
Male	0.521	0.500
With disability	0.009	0.096
Income less than $50k	0.325	0.469
Under 30 years old	0.194	0.396
Over 55 years old	0.141	0.348
Distance to work	0.914	0.588
No car	0.434	0.496
Walkability	0.513	0.404

30 years old" or "Over 55 years old"), "Distance to work," as well as whether the person has no access to a car ("No car") and "Walkability."

Approximately 7.5% of those included are black, about 6% are Asian and 70% are white. More than half are male and a bit less than 1% have a disability. About one third of the households make less than $50,000 annually. Roughly 20% are younger than 30 years and almost 15% are older than 55 years. The mean travel distance to work is just below 1 mile.

The number of households with access to a car is more than 40%. However, like the walkability variable, automobile ownership and mode choice may be determined simultaneously. Therefore, we have modeled the "No car" variable as an instrumental variable as well. We regressed the "No car" variable on the personal and household characteristics (race, gender, disability status, income and age), using both, a heteroskedasticity-corrected linear probability model and logit model. The predicted group membership (whether or not the person has access to a car) were exactly the same for both models, therefore, the estimation method did not have an impact on the results. Since we used the predicted group membership, rather than the probabilities of the LPM, the instrumental variable will not be a linear combination of already included variables, despite the lack of true instruments.

The walkability variable was derived as the, by distance, spatially weighted mean walking mode share of the three closest neighbors. The mean value of the walkability variable is just over 50%. For the instrumental variable estimation we used as instruments the personal and household characteristics (race, gender, disability status, income and age) of the three closest neighbors. We restricted the number of spatial lags to three for two reasons: First, only the mode choice of the nearest neighbors gives information about walkability and a larger number of spatial lags would have increased the distance of the neighbors by quite a bit. Second, the larger the number of spatial lags, the more multicollinearity will be encountered in the estimation.

Table 1 exhibits the summary statistics (mean and standard deviation) of all the dependent, explanatory and instrumental variables included in the mode choice model.

4 Discussion

Model (1) and Model (2) are the heteroskedasticity-corrected weighted least square linear probability mode choice models; Model (1) is without the walkability variable and Model (2) is with the walkability variable. Both models have a reasonably good fit for an LPM. For Model (2) the adjusted-R^2 value of 0.501 is somewhat better than for Model (1) with an adjusted-R^2 value of 0.457. In both models, only the regression coefficients of the constant terms, as well as for the variables "Over 55 years old" and "Distance to work" are significantly different from *zero* at least at the 10% significance level. In addition the "Walkability" coefficient in Model (2) is significantly different from *zero* as well. They also have all the expected signs. So do all the remaining coefficient estimates. The regression results for the two linear probability models are summarized in Table 2.

Both models show that the longer the trip is the lower is the probability for a person to walk, as indicated by the negative sign of "Distance to work". These regression coefficients are significantly different from *zero* at the 1% level. Also, "Over 55 years old people" are less likely to walk compared to the reference

Table 2 Linear probability regression model results

	Model (1)		Model (2)	
Observations	541		541	
Adjusted R^2	0.457		0.501	
F-test	46.449***		50.347***	
Constant term	0.751***	(0.062)	0.596***	(0.076)
Black	−0.086	(0.062)	−0.077	(0.051)
Asian	0.009	(0.094)	−0.013	(0.087)
White	0.084	(0.055)	0.059	(0.053)
Male	−0.017	(0.033)	−0.006	(0.029)
With disability	−0.106	(0.180)	−0.069	(0.179)
Income less than $50k	0.034	(0.055)	0.022	(0.037)
Under 30 years old	0.040	(0.054)	0.038	(0.053)
Over 55 years old	−0.067*	(0.040)	−0.076**	(0.035)
Distance to work	−0.346***	(0.029)	−0.319***	(0.030)
No car (IV)	0.023	(0.069)	0.044	(0.063)
Walkability (IV)			0.293***	(0.081)
% correctly predicted: total	0.699		0.688	
Walk = 0 (n = 272)	0.676		0.691	
Walk = 1 (n = 269)	0.721		0.688	

* $p < 0.1$, ** $p < 0.05$, *** $p < 0.01$, standard errors are in parenthesis

category of 30 to 55 years old people. Again, the sign coefficient estimate is negative. While the significance level for this coefficient is only at the 10% level in Model (1), it is significant at the 5% level in Model (2).

The most surprising result may be that "No car" is not significantly different from *zero*. However, recalling that the variable is an instrumental variable, this means that by removing all the unexplained shocks in the error term, the variable is made exogenous. Therefore, once the simultaneity is accounted for, the interpretation is clearly that automobile access does not matter for the decision to walk, everything else being equal. The same is true for race, gender, disability status and income. Also, people "Under 30 years old" do not have a significantly different propensity to walk than the reference category of "30 to 55 years old."

In Model (2), the regression coefficient of the added "Walkability" variable turns out to be positive and significantly different from *zero* at the 1% level. It does not only improve the overall fit, but also the value of the F-test. Therefore, it can be said, that "Walkability" adds information to the model.

Comparing the two models, the main issue at hand is the constant term. They are both positive and significantly different from *zero* at the 1% level. However, while all the other coefficient estimates remain largely unchanged, this is not completely true for the constant terms. In Model (2) the constant term is lower than in Model (1) so that the effect coming from the walkability variable can be accounted for. For ease let's assume that all variables, except "Walkability" take on the value of *zero*, then we have in Model (1) a probability to walk of 0.751. In Model (2) however, the probability depends on the value of "Walkability". If "Walkability" is equal to *zero* then the probability for walking is 0.596, more than 15% points less than in Model (1), but if "Walkability" is equal to *one* then the probability for walking is 0.889, almost 14% points more than in Model (2). The probability for walking is the same in both models if the value for "Walkability" is 0.529.

Therefore, the model shows that walking does not only depend on personal, household and trip characteristics, but also on the walkability variable. The probability for walking increases with a higher walking mode share, because of social interactions between neighbors (Network effects) and/or correlated effects (common environment and shared infrastructure).

Finally, in a simulation we have calculated the predicted group membership for both models. We find that while the two models have almost the same predictive power overall (just under 70%), Model (1) is marginally better in predicting walking while Model (2) is better at predicting non-walking trips. This is not surprising, given the lack of the walkability variable in combination with a higher constant term makes it more difficult in Model (1) to reach low probability values.

Table 3 shows the results of the logit mode choice models. Model (3) does not include the walkability variable and Model (4) includes it. In terms of the signs and significance of the coefficient estimates, the results of the logit mode choice models are very similar to the results of the linear probability mode choice models. Again, they both have a reasonably good fit for a logit model, as measured by the Pseudo-R^2. The fit for the model with the walkability variable, Model (4), is again marginally better than for the one without, Model (3), 0.276 as

Table 3 Logit regression model results

	Model (3)		Model (4)	
Observations	541		541	
−2 Log likelihood	633.789		624.459	
Pseudo-R^2	0.258		0.276	
Constant term	1.261***	(0.316)	0.365	(0.431)
Black	−0.705	(0.448)	−0.608	(0.452)
Asian	0.042	(0.473)	−0.020	(0.478)
White	0.408	(0.289)	0.256	(0.296)
Male	−0.012	(0.193)	−0.024	(0.195)
With disability	−0.393	(1.171)	−0.179	(1.163)
Income less then $50k	0.195	(.291)	0.204	(.294)
Under 30 years old	0.282	(.284)	0.343	(.285)
Over 55 years old	−0.344	(.311)	−0.264	(.316)
Distance to work	−1.703***	(.185)	−1.570***	(.189)
No car (IV)	0.002	(.395)	0.054	(.398)
Walkability (IV)			1.648***	(.544)
% correctly predicted: total	0.693		0.697	
Walk = 0 (n = 272)	0.658		0.691	
Walk = 1 (n = 269)	0.729		0.702	

$^{*}p < 0.1$, $^{**}p < 0.05$, $^{***}p < 0.01$, standard errors are in parenthesis

opposed to 0.258. Only the "Distance to work" coefficient is significantly different from *zero* in both models. As expected, the sign is negative and the significance is at the 1-percent level. Therefore the propensity to walk again decreases with the distance to work. While in Model (3) the constant term is positive and significantly different from *zero* that is not true for Model (4), where the constant term is positive but not significantly different from *zero*. Instead, however, the "Walkability" coefficient in Model (4) is positive and significantly different from *zero*.

Models (3) and (4) tell exactly the same story as Models (1) and (2), but in an even more pronounced way: The signs of all logit mode choice model coefficient estimates are the same as for the linear probability mode choice model, except for "Asian," whose coefficient estimate is not significantly different from *zero*. Also, the regression coefficients between Models (3) and (4) change neither sign nor magnitude, except for the constant term. In a logit mode choice model, the constant term can be interpreted as the relative preference towards the alternative-specific mode. This means that while in Model (3) the reference group (values for personal and household characteristics are *zero*) prefers walking for HBW trips in New York City below 2 miles, this bias towards walking vanishes in Model (4) as soon as walkability is included as an explanatory variable.

The simulation results for Models (3) and (4) are also comparable to the ones for Models (1) and (2). This time Model (4) performs marginally better than Model (3),

Table 4 Observed and forecasted walking mode share for the whole dataset

	Overall	Walkability < 0.5	Walkability ≥ 0.5
Observed Probabilities:	0.503	0.332	0.663
Model (1)	0.491	0.462	0.519
Model (2)	**0.498**	**0.442**	**0.549**
Model (3)	**0.503**	0.469	0.534
Model (4)	**0.503**	**0.441**	**0.560**
Group membership:			
Model (1)	0.516	0.450	0.577
Model (2)	**0.597**	**0.385**	**0.601**
Model (3)	0.529	0.458	0.595
Model (4)	**0.506**	**0.400**	**0.605**

Predicted values closest to observed values are in bold

but it is still the case that Model (3) better predicts walking trips, while Model (4) better predicts non-walking trips.

Analyzing a final performance measure, the superiority of Models (2) and (4) is further consolidated. As shown in Table 4, the models with the spatially autoregressive term do not only have enhanced forecasts for overall walking mode share, as measured either in probabilities or predicted group membership, but this result is consistent if we split the dataset at a walkability value of 0.5. Models (2) and (4) forecasts improve over the models without the walkability variable, sometimes considerably. This means that the spatially autocorrelated mode choice models perform especially better if the area has varying walking mode shares. The models without walkability encounter a systematic bias: while all models overestimate areas with low walkability and underestimate areas with high walkability, the models without the spatially autocorrelated term are worse in their estimates than the spatially autoregressive models. This result is consistent with the omitted variable bias reflected in the constant term.

In summary, it can be said that both Models (2) and (4) are an improvement over Models (1) and (3), because they not only better forecast overall walking mode share, but also because they account for walkability and, therefore, avoid an omitted variable bias, especially in the constant term. The general results in the regression coefficient estimates do not differ, whether the model is a linear probability mode choice model or logit mode choice model.

5 Conclusion

In this chapter we have shown that the mode choice for walking in the city of New York is spatially autocorrelated, and that this autocorrelation can be modeled employing an instrumental variable approach. The instrumental variable is a composite measure for social interactions (interpreted as changes in individual behavior as a function of group behavior) as well as correlated effects (interpreted as changes in individual behavior as a function of the common environment and shared infrastructure). Overall, this effect is a summary of walkability. The approach presented here is without any major computational problems and, therefore, straightforward to implement. While the predictive power of the model with walkability is similar to the model without walkability, the model fit improves with the inclusion of the autoregressive term, and it avoid omitted variable bias, especially in the constant term.

We tested the instrumental variable approach for the autoregressive term using both a heteroskedasticity-corrected weighted least square linear probability mode choice model and a logit mode choice model, and found no major differences in their results. The instrumental variable approach for the linear probability model is essentially equivalent to the well-established 2-SLS estimation method.

The innovation is in the application to a discrete choice model with a spatially autoregressive term. Linear probability models, however, have one major downside, namely that they are restricted to just two choices. This, however, is not true for logit models. Therefore, by showing that the performance of both models is compatible, it would be easy to imagine that this approach can be successfully applied to McFadden-type conditional model choice models, as well as multinomial choice and nested mode choice models (McFadden 1974; Train 2003). Especially in the context of multinomial mode choice models, it may be possible to use the autoregressive mode-friendliness term as a mode-specific variable, which differs for each mode alternative (see Páez et al. 2008b).

Finally, the spatially autocorrelated mode choice model is not only technically superior, but also preferable for evaluating policies which do not refer to the walkability variable. This is especially important for transportation planning, where major investment decisions are based on a travel demand forecasting model. If the mode choice component of the travel demand forecasting model exhibits a systematic bias stemming from an omitted variable (spatial lag term), then derived investment decisions may be not economically efficient anymore (Goetzke 2003, 2008).

References

Anselin L (1988) Spatial econometrics: methods and models. Kluver, Dordrecht

Brock WA, Durlauf SN (2001) Discrete choice with social interactions. Rev Econ Stud 66:235–260

Brock WA, Durlauf SN (2002) A multinomial-choice model with neighborhood effects. Am Econ Rev 92:298–303

Brueckner JK, Largey AG (2006) Social interaction and urban sprawl. CESifo Working Paper Series No. 1843, Munich

Dugundi ER, Walker JL (2005) Discrete choice with social and spatial network interdependencies. Transp Res Rec 1921:70–78

Evans WN, Oates WE, Schwab RM (1992) Measuring peer group effects: a study of teenage behavior. J Polit Econ 100:966–991

Fleming MM (2004) Techniques for estimating spatially dependent discrete choice models. In: Anselin L et al. (eds) Advances in spatial econometrics: methodology, tools and applications. Springer, Berlin, pp 145–168

Goetzke F (2003) Are travel demand forecasting models biased because of uncorrected spatial autocorrelation? Regional Research Institute Research Paper 2003–10, West Virginia University, Morgantown

Goetzke F (2008) Network effects in public transit use: evidence from a spatially autoregressive mode choice model for New York. Urban Stud 45:407–417

Land KC, Deane G (1992) On the large-sample estimation of regression models with spatial or network effects terms: a two-stage least squares approach. Sociol Methodol 22:221–248

Leenders RTAJ (2002) Modeling social influence through network autocorrelation: constructing a weight matrix. Soc Networks 24:21–47

Manski CF (2000) Economic analysis of social interactions. J Eco Perspect 14:115–136

McFadden D (1974) Conditional logit analysis of qualitative choice behavior. In: Zarembka P (ed) Frontiers in econometrics. Academic, New York, pp 105–162

NYMTC (2004) New York best practice model (BPM) for regional travel demand forecasting. Technical report, New York

Páez A, Scott DM (2007) Social influence on travel behavior: a simulation example of the decision to telecommute. Environ Plann A 39:647–665

Páez A, Scott DM, Volz E (2008a) Weight matrices for social influence analysis: an investigation of measurement errors and their effect on model identification and estimation quality. Soc Networks 30:309–317

Páez A, Scott DM, Volz E (2008b) A discrete choice approach to modeling social influence on individual decision making. Environ Plann B 35:1055–1069

Train K (2003) Discrete choice methods with simulation. Cambridge University Press, Cambridge, UK

Wooldridge JM (2005) Introductory econometrics: a modern approach. South-Western College Publishing, Florence, KY

Part III
Economic and Political Geography

Employment Density in Ile-de-France: Evidence from Local Regressions

Rachel Guillain and Julie Le Gallo

1 Introduction

In recent decades, cities have experienced a particularly intense phase of urban sprawl. Urban growth has been characterized by the spatial concentration of population in urban areas and the concomitant extension of those urban areas (Nechyba and Walsh 2004). Urban sprawl has also been accompanied by major reorganizations of urban areas with regard to the location choices of households and firms. More specifically, most cities in developed countries have experienced several waves of suburbanization of economic activities: "an economic definition of suburbanization is a reduction in the fraction of a metropolitan area's population or employment that is located in the central city (corresponding to increased activity in surrounding suburbs)" (Mills 1999).

Suburbanization of economic activities has an impact on urban structure: cities are not exclusively organized with a *Central Business District* (CBD) around which land values, employment, and population densities decrease with distance. On the contrary, they are more and more characterized by a polycentric organization: employment is concentrated in several centers within urban areas. Strategic activities (headquarters and high-order producer services) play a major role in this process by locating themselves selectively in these various centers. The development of peripheral employment centers – where a significant proportion of these activities are located, reproducing the functions of the CBD – is accordingly viewed as the decline of the CBD (Stanback 1991).

However, some studies have challenged this idea that suburbanization of strategic activities implies the decline of the CBD. On the contrary, such reorganization may reinforce the supremacy of the CBD with more pronounced specialization in the high-order services in finance, insurance and legal services. This phenomenon has been observed in North-American cities (Coffey and Shearmur 2002) and in Europe,

R. Guillain (✉)
LEG-UMR 5118, Université de Bourgogne, Pôle d'Economie et de Gestion, BP 26611, 21066 Dijon Cedex, France,
e-mail: guillain@u-burgogne.fr

A. Páez et al. (eds.), *Progress in Spatial Analysis*, Advances in Spatial Science,
DOI 10.1007/978-3-642-03326-1_12, © Springer-Verlag Berlin Heidelberg 2010

and more particularly in the Paris urban area in France (Guillain et al. 2006). These results are found by the following method. First, employment centers are identified by measuring the spatial agglomeration of economic activities, with global and local spatial autocorrelation statistics. Second, a sectoral analysis of the centers is conducted, characterizing the specialization of these centers and their attractiveness for strategic activities.

The capacity of attraction of employment centers is not their only characteristic. They should also be able to structure their environment by shaping the economic organization of employment and population densities within the urban area. Another way to identify urban employment centers, then, is to measure this influence characteristic (McMillen and McDonald 1998; McMillen 2001).

In this context, we aim at determining whether the Paris CBD has any structuring power over the economic activities in Ile-de-France, which is the French capital region encompassing the city of Paris. Indeed, since Guillain et al. (2006) identified the CBD in Paris as an employment center using techniques measuring the agglomeration of economic activities, it remains to be seen whether it also shapes its environment. If so, the CBD will combine the two attributes of a center in a city, namely attracting activities and influencing the organization of economic activities around.

We set out, then, to answer two questions: Does the CBD still influence employment distribution in Ile-de-France? Does that influence differ by sector and by direction from the CBD?

Focusing on the influence of Paris CBD does not mean we claim that the region of Paris has to be perceived as monocentric. On the contrary, the identification of other employment subcenters that Guillain et al. (2006) have undertaken leads to perceive the region of Paris as a polycentric space. However, our point in this chapter is that a particular focus on the CBD is required because of the history of this center. Indeed, over the period 1965–2000, several regional plans have been applied to organize and support decentralization of economic activities because of the well-known hypertrophy of the Paris CBD. While planning policies did not aim at reducing the influence of Paris CBD, a possible unexpected impact could be that this CBD does not influence its environment anymore.

The two questions are addressed in two steps. The first step involves estimating the density gradient, which is the proportion rate at which density falls with distance, using global regressions. A significant positive density gradient would corroborate findings by Guillain et al. (2006) showing that the CBD is still powerful in Ile-de-France. We perform these analyses for total employment and for six sectors to determine whether the CBD's influence differs by sector. Spatial econometric specifications for the density functions are used (Anselin 1988, 2006). The second step is to perform local regressions, using Geographically Weighted Regression (GWR), where one density gradient is estimated for each observation. Indeed, global regressions imply a constant influence with the distance and direction to the CBD. More complex influences could be relevant: density gradient may vary with distance to the CBD and the density gradient distribution may be anisotropic. To perform these local regressions, we use the specifications suggested by Páez et al. (2002a,b), who place GWR within the context of a spatial model of error variance heterogeneity. In this framework, locational heterogeneity and the form of spatial autocorrelation can be tested for and the appropriate GWR model with spatial effects is estimated using the maximum likelihood method.

This chapter is organized as follows. The data and the spatial weights matrix are presented in the next section. Then, econometric results obtained with global and local regressions are presented. The last section concludes.

2 Data and Spatial Weights Matrix

With almost 11 million people and some five million jobs, Ile-de-France is the biggest region in France and is also the French capital region. It represents 18.8% of the national population and produces 29% of national GDP, so that GDP per inhabitant in this region exceeds the national average by 55%. By comparison, the GDP in Ile-de-France is the highest of the six main economic regions in Europe (Brussels in Belgium, London in United Kingdom, Ile-de-France, Randstadt, Rhin-Main, Rhin-Rhur in Germany) and the Ile-de-France region is similar to the regions of London and Rhin-Ruhr in terms of employment and population (IAURIF 1999).[1] With about 600,000 employees in the industrial sector, the Ile-de-France region is not only one of the most industrial region in France – even if a loss of about 555,000 employees has been observed during the 1978–1997 period – but also in Europe: the region is more industrialized than the Brussels or London region but less than the Rhin-Main and Rhin-Rhur. However, the Ile-de-France economy is largely oriented towards the service sector: 80% of the regional employment is in this sector, versus 72% at the national level (IAURIF 2001). Head offices are very present in Ile-de-France and reveal the economic power of the region: they represent about 40% of the regional establishments and one company with 100 employees or more in three has its head office in Ile-de-France and more precisely in the CBD of Paris (IAURIF (1999). Not only is the Ile-de-France the administrative French capital but it is also the core of the French and European economies.

The region covers 12,000 km^2, which is 2.2% of the land area of France. It consists of 1,280 communes (French municipalities) and the 20 districts (*arrondissements*) of the city of Paris. Since 1964, the metropolitan region has been divided into eight departments: Paris, Seine-et-Marne, Yvelines, Essonne, Hauts-de-Seine, Seine-Saint-Denis, Val-de-Marne and Val-d'Oise. Figure 1 shows the 1,300 geographic areas of our sample and the eight departments.

Historically, the CBD of the Ile-de-France is considered to be formed by the 1st, 2nd, 8th, 9th and 17th *arrondissements* of Paris because firms traditionally located mainly in this part of the city of Paris (IAURIF 1999). These *arrondissements* have

[1] The comparisons has been made by the Group for European Metropolitan Areas Comparative Analysis in 1996 by using data of 1994 for GDP, data of 1995 for population and data of 1996 for employment (IAURIF 1999).

Fig. 1 Departments and communes in Ile-de-France. Scale: 1:9,000

been identified as an employment center by Guillain et al. (2006). As a consequence, it is relevant to still consider these *arrondissements* as the CBD and to investigate the influence of the CBD on the surrounding areas. For this purpose, the centroid of the first *arrondissement* is considered as the representative point of the CBD but the results are quite similar if any other centroid of the CBD is considered. The suburban areas are the areas outside the city of Paris: the departments of Hauts-de-Seine, Val-de-Marne, and Seine-Saint-Denis form the inner ring (*Première Couronne*) and the departments of Essonne, Seine-et-Marne, Val-d'Oise, and Yvelines form the outer ring (*Seconde Couronne*).

The influence of CBD has to be viewed in the light of active planning policy carried in Ile-de-France since 1965. The spatial organization of Ile-de-France has long been considered as typically monocentric with the development of economic activities mainly focused in and immediately around Paris. To relieve congestion of Paris associated with the expected growth of population and employment, the authorities organized a polycentric expansion with the development of La Défense[2] and five new towns (*villes nouvelles*). In this context, a hypothesis has to be considered: the CBD could have lost its power to shape the organization of economic activities. Our study provides an answer to that question both for total employment and for the key economic sectors. Figure 2 shows the locations of the CBD, the new towns, and the two main airports.

[2] *La Défense* is an area located west of Paris and the intention was to create a second CBD for Ile-de-France because of the hypertrophy of the existing CBD (Piercy, 1999).

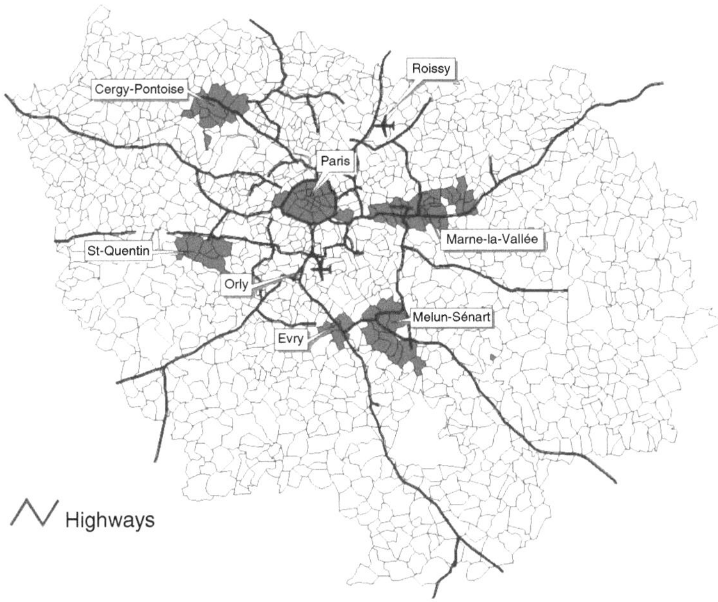

Fig. 2 CBD, new towns and highways. Scale: 1:9,000

In conducting our analysis, we use the Population Censuses compiled by the French National Institute of Statistics and Economic Studies (INSEE) for 1999. The employment data are measured at the commune level and classified by INSEE's industrial classification NAF 700 (*Nomenclature d'Activités Française*) for 1999. For reasons of reliability, data cannot be used at commune level for 700 sectors. Obviously, there is a trade-off between the level of spatial disaggregation and the level of sectoral disaggregation. Since our analysis is more meaningful with a fine spatial scale, we have used a sectoral disaggregation of total employment into six sectors, covering both manufacturing and service activities: industry, high-tech, finance-insurance, high-order services, consumer services, standard services. We thus obtain a general picture of the distribution of employment in Ile-de-France and a more detailed picture for certain key sectors.

We distinguish high-order services and standard services because of their structural differences. High-order services require high levels of information and qualification whereas standard services require fewer levels of qualification and are less information-dependent. For example, legal services, management consulting and advertising are high-order services while cleaning or security services are standard services. Moreover, we gain insight into potentially varied behavior for manufacturing versus services sectors and for business-oriented versus population-oriented services.

Finally, in order to implement the spatial statistical and econometric analysis, spatial interdependence between observations needs to be modeled by means of a row-standardized spatial weights matrix W. From an applied perspective, we have based our choice on the geographical characteristics of the spatial units, and more specifically on their heterogeneity in size. This leads us to choose a nearest-neighbors matrix. These matrices are computed from the distance between the units' centroïds and imply that each spatial unit, regardless of location, is connected to the same number k of neighbors. The general form of a k-nearest neighbors weights matrix $W(k)$ is defined as:

$$\begin{cases} w_{ij}^{*}(k) = 0 \text{ if } i = j, \ \forall k \\ w_{ij}^{*}(k) = 1 \text{ if } d_{ij} \leq d_i(k) \\ w_{ij}^{*}(k) = 0 \text{ if } d_{ij} > d_i(k) \end{cases} \quad \text{and } w_{ij}(k) = w_{ij}^{*}(k) / \sum_j w_{ij}^{*}(k) \tag{1}$$

where $w_{ij}^{*}(k)$ is an element of the unstandardized weights matrix; $w_{ij}(k)$ is an element of the standardized weights matrix and $d_i(k)$ is a critical cut-off distance defined for each unit i : it is the kth order smallest distance between unit i and all the other units such as each unit i has exactly k neighbors. Since, using a contiguity criteria, the average number of neighbors in our sample is 5.80, we present the results with k = 6.[3] However, we have evaluated the robustness of our results to the choice of the weights matrix. Therefore, as Guillain et al. (2006), we have also used a simple contiguity weights matrix and a distance-based matrix. All our results are robust to that choice.

3 Global Results

A first look of these data (cf. Table 1) indicates that total employment is rather evenly distributed between Paris, the inner ring and the outer ring. The situation is different, however, from one sector to another, and the weight of Paris in total employment of Ile-de-France appears to be quite variable. The sectors that are the most concentrated in Paris are finance-insurance (54.33% of total employment in finance-insurance) and to a certain extent consumer services (36.64% of total employment in consumer services). However, the high-order services and the standard services are mostly located in the inner ring with respectively 39.62% and 39.22% of total employment in this area. Finally, the outer ring concentrates a large proportion of industrial employment (40.07% of total industrial employment) and high-tech employment, this sector being located mainly outside Paris since only 9.20% of high-tech employment is located there.

The analysis of urban structures is usually conducted using employment density functions including the distance from the CBD as an explanatory factor. Although they are commonly used for population densities, density functions have been applied to employment by Erickson (1982), Waddell and Shukla (1993), McMillen and McDonald (1998) or Sridhar (2007), among others.

[3] The maximum distance to the sixth nearest neighbor represents 7% of the diameter or the region.

Table 1 Distribution of employment in Ile-de-France

	Total jobs in Ile-de-France	% in Paris	% in the inner ring	% in the outer ring
Total employment	3,314,495	30,87	36,73	32,39
Industry	371,262	26.44	33.48	40.07
High-Tech	229,279	9.20	41.65	49.15
Finance-Insurance	256,205	54.33	30.97	14.70
High-order services	637,510	35.97	39.62	24.41
Consumer services	547,927	36.64	29.78	33.59
Standard services	185,133	30.42	39.22	30.36

Many functional forms can be used to model urban densities.[4] In this chapter, we use the negative exponential function as a starting point. Indeed, while most complex models could be considered, we argue below that the local analysis that will be undertaken in the next section will better capture the irregularities of employment densities in Ile-de-France. The negative exponential function is written as follows:

$$D_i = D_{CBD} e^{-\gamma u_i + \varepsilon_i} \tag{2}$$

where D_i is the employment density of observation i, measured as the number of employees per square meter; D_{CBD} is the employment density at the CBD; γ is the density gradient and measures the proportional rate at which population density falls with distance, u_i is the distance of observation u from the CBD and ε_i is the error term with the usual properties, $i = 1, \ldots 1{,}300$. All distances are measured in straight-line kilometers from the centroïd of the first *arrondissement*. The function is then estimated by taking logs of (2) on both sides:

$$\ln D_i = \ln D_{CBD} - \gamma u_i + \varepsilon_i \tag{3}$$

As pointed out by Anselin and Can (1986), Griffith and Can (1995), Baumont et al. (2004) or Griffith and Wong (2007), the reliability of inference made using density functions may be affected by the presence of spatial autocorrelation. Therefore, in order to detect the appropriate form of spatial autocorrelation, we use the classical "specific to general" specification search approach outlined in Anselin (1995) using tests described in Anselin et al. (1996). When a formal theory is lacking, this strategy provides ways to discriminate between a spatial lag and a spatial error model using the Ordinary Least Squares (OLS) residuals. More specifically, Anselin et al. (1996) suggest Lagrange Multiplier (LM) tests (resp. LMERR and LMLAG) and their robust versions (resp. R-LMERR and R-LMLAG). The decision rule used to choose the most appropriate specification is as follows: if LMLAG (resp. LMERR) is more significant than LMERR (resp. LMLAG) and R-LMLAG (resp. R-LMERR) is significant whereas R-LMERR (resp. R-LMLAG) is not, then

[4] See for instance McDonald (1989) for a literature review on urban population density functions.

Table 2 Spatial autocorrelation LM tests for model (3), total employment

	LM-LAG	R-LMLAG	LMERR	R-LMERR
Total employment	769.466	0.189	776.352	7.075
	(0.000)	(0.663)	(0.000)	(0.008)

$n = 1,300$ observations. P-values are in brackets. LMLAG stands for the Lagrange Multiplier test for a spatial lag and R-LMLAG is its robust version. LMERR stands for the Lagrange Multiplier test for residual spatial autocorrelation and R-LMERR is its robust version (Anselin et al. 1996)

the most appropriate model is the spatial autoregressive model (resp. the spatial error model).

We have therefore estimated the model described in (3) by OLS and computed the LM tests. Following the decision rule described above, the results, displayed in Table 2 for total employment, show that the spatial error model is preferable to a spatial lag model. It is also the case for the different sectors. We therefore adopt the following error structure, in matrix form:

$$\varepsilon = \lambda W \varepsilon + \xi \tag{4}$$

where λ is the spatial error coefficient and $\xi \sim iid(0, \sigma^2 I_n)$ with $n = 1{,}300$. We have estimated (3) with spatial error autocorrelation as in (4) using maximum likelihood (ML). However, because non-normality of the error terms and heteroscedasticity may affect the results, we have also estimated all the models using Generalized Methods of Moments (Kelejian and Prucha 1999) or with a non-parametric heteroscedasticity and autocorrelation consistent estimator of the variance-covariance matrix in a spatial context (Kelejian and Prucha 2007). The results obtained are qualitatively and quantitatively similar to those presented here.[5]

The results obtained using ML estimation for total employment and for the six sectors are displayed in Tables 3 and 4. It appears that the spatial coefficient λ is always positive and significant. Spatial autocorrelation is more important for total employment ($\hat{\lambda} = 0.675$) than for the six sectors considered (with $\hat{\lambda}$ ranging from 0.259 to 0.465).

The density gradient for total employment is positive and significant, indicating that overall the CBD still influences employment distribution in Ile-de-France. More specifically, the 8.5% value of the estimated gradient indicates that the employment density decreases by 8.5% for each kilometer from the CBD. However, the situation is very different from one sector to another. Indeed, the value of the density gradient is significant for all sectors except for employment in high-tech and standard services. Three cases must be distinguished.

Firstly, for employment in industry, high-order services, and consumer services, the density gradient is positive and significant, although it is not as large as for total

[5] The results obtained in this section have been obtained using the spatial econometrics toolbox in Matlab (LeSage 1999).

Table 3 ML estimation results for global employment density functions (1)

	Total employment	Industrial employment	High-Tech employment	Finance-insurance employment
Constant	1.401	−1.503	−2.259	−2.633
	(0.000)	(0.000)	(0.000)	(0.000)
Distance from CBD ($-\gamma$)	−0.085	−0.028	−0.004	0.017
	(0.000)	(0.000)	(0.399)	(0.000)
λ	0.675	0.304	0.261	0.318
	(0.000)	(0.000)	(0.000)	(0.000)
σ^2	1.655	5.196	5.924	6.021
Sq. corr.	0.718	0.113	0.044	0.065

$n = 1{,}300$ observations. P-values are in brackets. Sq. corr. is the squared correlation between predicted values and actual values

Table 4 ML estimation results for global employment density functions (2)

	High-order services employment	Consumer services employment	Standard services employment
Constant	−1.496	−1.044	−2.404
	(0.000)	(0.129)	(0.000)
Distance from CBD ($-\gamma$)	−0.033	−0.044	−0.002
	(0.000)	(0.000)	(0.582)
λ	0.465	0.365	0.259
	(0.000)	(0.000)	(0.000)
σ^2	5.076	2.587	6.153
Sq. corr.	0.214	0.215	0.044

$n = 1{,}300$ observations. P-values in brackets. Sq. corr. is the squared correlation between predicted values and actual values

employment. It is larger for consumer services than for industry and high-order services. In other words, employment in consumer services decreases more quickly with distance from the CBD that does employment in industry or employment in high-order services.

Secondly, the gradient is not significant for high-tech employment and employment in standard services. This seems to indicate that the CBD does not influence the distribution of employment in Ile-de-France for these two sectors so location of employment in these sectors is not governed by distance from the CBD.

Thirdly, we do not obtain the expected positive sign for employment in finance-insurance: it is negative and significant. This seems to indicate the farther a commune is located from the CBD, the more employment there is in this commune for this sector. This would imply a repellent effect of the CBD. This counter-intuitive result requires closer scrutiny since the analysis by Guillain et al. (2006) points to a pronounced location of employment in finance-insurance in and immediately around the CBD.

A local analysis of density gradients is required to explain this counter-intuitive result for finance-insurance and the absence of influence of the CBD on high-tech employment and employment in standard services. Moreover, even when the sign of the global estimated density gradient is the expected one, a local analysis is relevant. Indeed, global gradients may mask large local disparities: different patterns may be observed for different distances and/or different directions from the CBD.

4 Local Results

Numerous studies have been undertaken to better capture the irregularities of population and employment densities in urban areas (McDonald 1989). For example, Anderson (1985) or Alperovich (1995) suggests using cubic spline specifications when population densities do not decrease homogeneously with distance from the CBD. Brueckner (1986) estimates distance-oriented density functions, with an unknown number of possible regimes, using switching regressions. Alperovich and Deutsch (2002) find evidence of two distinct regimes in the urban area of Tel-Aviv. Baumont et al. (2004) use a spline-exponential function. In fact, all these studies take account of spatial heterogeneity in different ways: the estimated coefficients differ depending on their distance from the CBD or on the spatial regime they belong to. However, all these different solutions suppose that the form of spatial heterogeneity is known a priori. On the other hand, misspecification of spatial heterogeneity may affect estimations.

Therefore, rather than imposing a structure on the form taken by spatial heterogeneity by extending the negative exponential function to more complex specifications, we use a generic and more flexible specification based on GWR (Fotheringham et al. 2004; Leung et al. 2000) yielding locally linear estimates. Indeed, as López et al. (2008) show using Monte-Carlo simulations, local approaches are useful when the heterogeneity in the data is high and when the appropriate functional form is not known. This is the case in our sample, as we consider a relatively large urban area and six different sectors, the amount of heterogeneity is therefore very important.

The local version of the negative exponential model can be written as:

$$\ln D_o = \ln D_{CBD,o} - \gamma_o u_o + \varepsilon_o \tag{5}$$

where o is a specific geographical location, which could be any point included among the sample observations. In addition, we follow Páez et al. (2002a) by considering GWR as a model of error variance heterogeneity, with heterogeneity having a precise geographical interpretation, which is labeled *locational heterogeneity*. More precisely, the variance-covariance matrix of the error terms of (5) is defined as a (n, n) matrix such as: $\Omega_o = E\left[\varepsilon_o \varepsilon_o'\right] = \sigma_o^2 G_o$ with the diagonal elements:

$$\omega_{oi} = \sigma_o^2 \exp(\gamma_o d_{oi}^2) \tag{6}$$

where d_{oi} is the distance between a focal point o and commune i for which the data are available, with $i = 1, \ldots n$ and $n = 1{,}300$. In this case, the variance is a

Table 5 LM tests (maximum) of spatial autocorrelation and locational heterogeneity

	LMLAG*	LMERR*	LM-LH
Total employment	852.594*	864.551*	36.671*
Industrial employment	150.654*	152.297*	146.290*
High-Tech employment	76.031*	77.164*	52.712*
Finance-insurance employment	67.831*	69.946*	10.178*
High-order services employment	364.404*	367.659*	168.072*
Consumer services employment	261.085*	263.780*	188.899*
Standard services employment	103.326*	106.226*	87.567*

$n = 1,300$ observations. * denotes significance at 5% using the adjustment by Páez et al. (2002a). LMLAG* is the LM test for an omitted spatial lag in the GWR model, LMERR* is the LM test for an omitted spatial error autocorrelation in the GWR model and LM-LH is the LM test for locational heterogeneity in the GWR-SEA model

function of two parameters, σ_o^2 and γ_o, it ensures the usual regularity conditions and has a geographical interpretation. The underlying model of homogeneity is provided by the case where $\gamma_o = 0$ since in this case the model reduces to the usual constant variance. Páez et al. (2002a) suggest an estimation method for models described in (5) and (6) based on maximum likelihood and a LM test for locational heterogeneity, which is in fact a test of heteroscedasticity, $H_0 : \gamma_o = 0$, one for each focal point o. In addition, this model can easily be extended to incorporate spatial autocorrelation, either in the form of a spatial lag or a spatial error term (Páez et al. 2002b), also estimated using maximum likelihood. This framework further allows testing for the presence of several forms of misspecification, i.e. locational heterogeneity in global spatial models and the presence of spatial autocorrelation in GWR models, using LM tests, one for each focal point o.

We have therefore estimated the model described in (5) with error variance as in (6) using maximum likelihood at every location, giving a total of 1,300 local models.[6] First, we computed the LM tests for an omitted spatial lag (LMLAG*) and an omitted spatial error (LMERR*) in the GWR model. Because of a problem of multiple comparisons, their level of significance was adjusted for simultaneous inference by a procedure described in Páez et al. (2002a). Table 5 shows the maximum values among the 1,300 local tests for LMLAG* and LMERR*. For total employment and all six sectors, all Lagrange multiplier tests are significant at the 5% level and the size of the tests suggests that an omitted spatial error is the dominant effect, similarly to what is found with global models.[7]

We have therefore estimated a GWR model incorporating a spatial error structure, labeled GWR-SEA:

[6] We thank A. Páez for sharing the Matlab programs used to estimate the models.

[7] Although the difference between the values of LM-LAG and LM-ERR is small, we select the spatial error model since it is consistent with the global analysis.

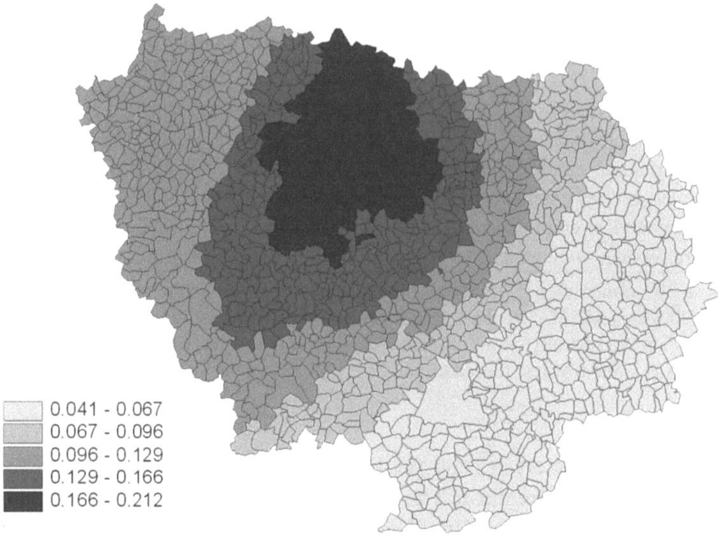

Fig. 3 Geographic distribution of the density gradient for total employment. Scale 1:9,000

$$\varepsilon_o = \lambda_o \sum_{j=1}^{n} w_{oj}\varepsilon_j + \mu_o \tag{7}$$

where λ_o is the spatial error coefficient for focal point o and commune i and the error term μ_o has a variance structure as in (6). The maximum value of the Lagrange multiplier test for locational heterogeneity ($\gamma_o = 0$) in a GWR-SEA model is reported in the third column of Table 5, where the significance level has again been adjusted. It is significant at 5%, indicating that variance, and consequently all the other parameters of the model, do indeed depend on location.

Our parameter of interest is the density gradient. We therefore display the local density gradients in Fig. 3 for total employment and in Figs. 4–9 for the six sectors, obtained using maximum likelihood to estimate the parameters of model described in (7) for a total of 1,300 local models. Only statistically significant local density gradients are represented, their values being color coded. Non-significant density gradients are associated with the communes left in blank.

The map for total employment (Fig. 3) shows that the CBD still influences the spatial organization of employment in Ile-de-France. Indeed, clusters of similar values of local density gradients that are all significant are observed around the CBD while gradient values decline progressively in Ile-de-France. In other words, the decrease in total employment is even less important with distance from the CBD. However, the geographic distribution of local density gradients is not concentric: the decline of the density gradient is not uniform in all directions; it is more pronounced along a south-north corridor, through the CBD. Moreover, density gradients decline more rapidly north of the CBD, possibly because this is a declining area, with

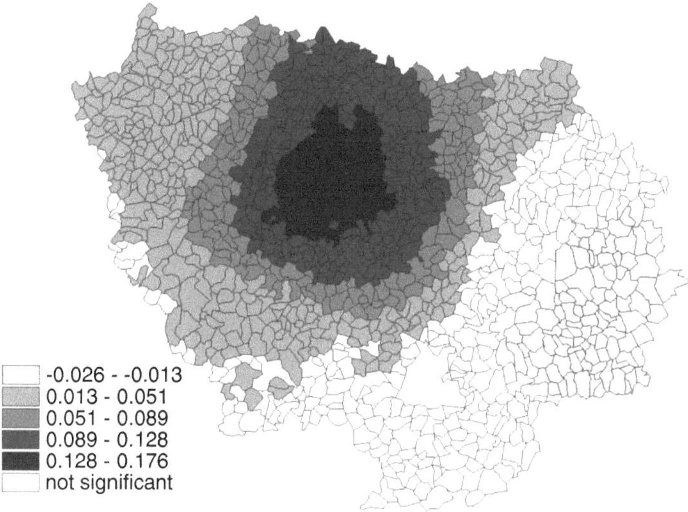

Fig. 4 Geographic distribution of the density gradient for industrial employment. Scale 1:9,000

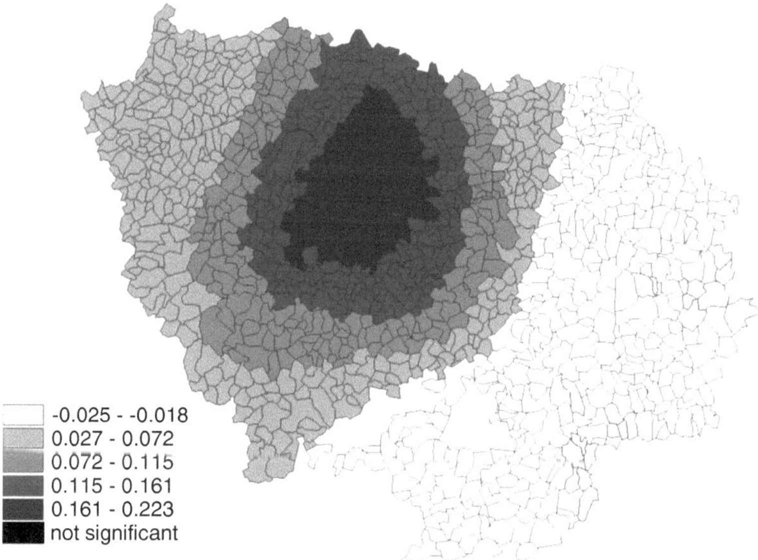

Fig. 5 Geographic distribution of the density gradient for high-order services employment. Scale 1:9,000

numerous firms closing and industrial wastelands that have not yet been redeveloped (IAURIF 2003).

Analyzing the distribution of the density gradients for each of the six sectors, we can distinguish several cases. In particular, the distribution of local density gradients

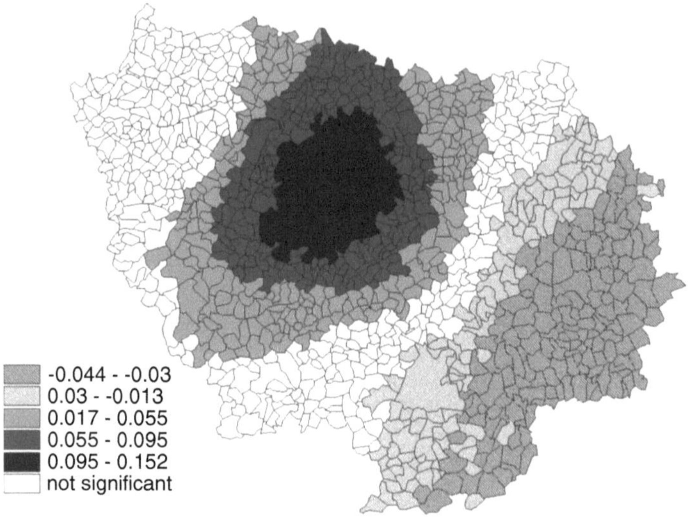

Fig. 6 Geographic distribution of the density gradient for high-tech employment. Scale 1:9,000

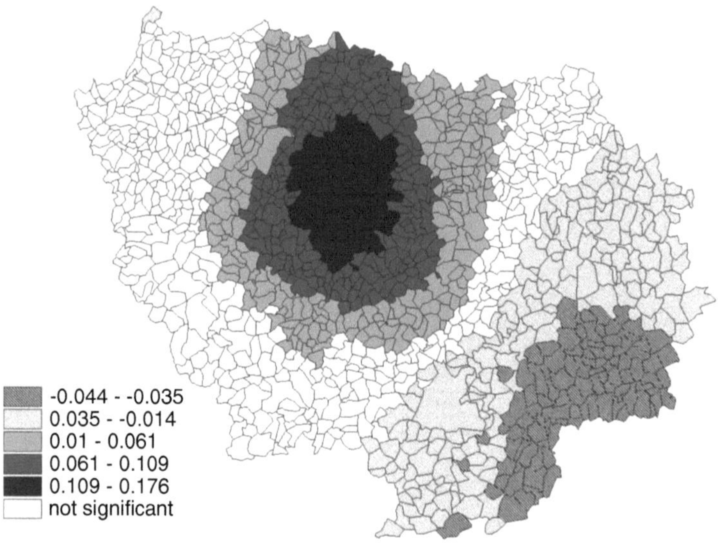

Fig. 7 Geographic distribution of the density gradient for standard services employment. Scale 1:9,000

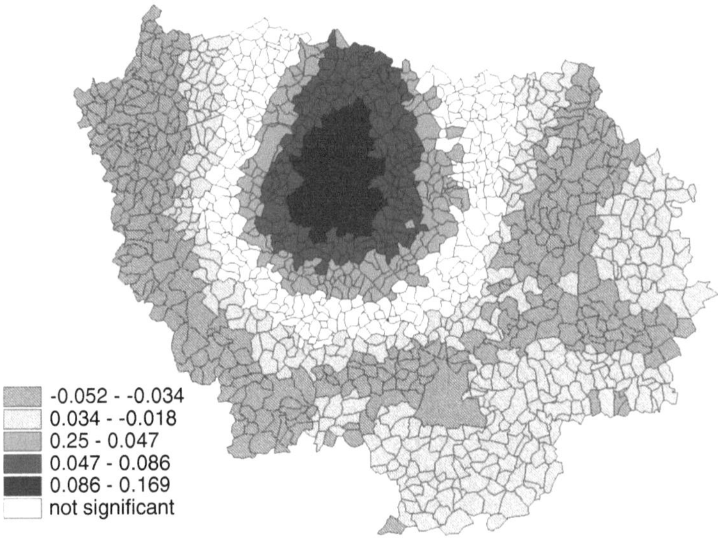

Fig. 8 Geographic distribution of the density gradient for finance-insurance employment. Scale 1:9,000

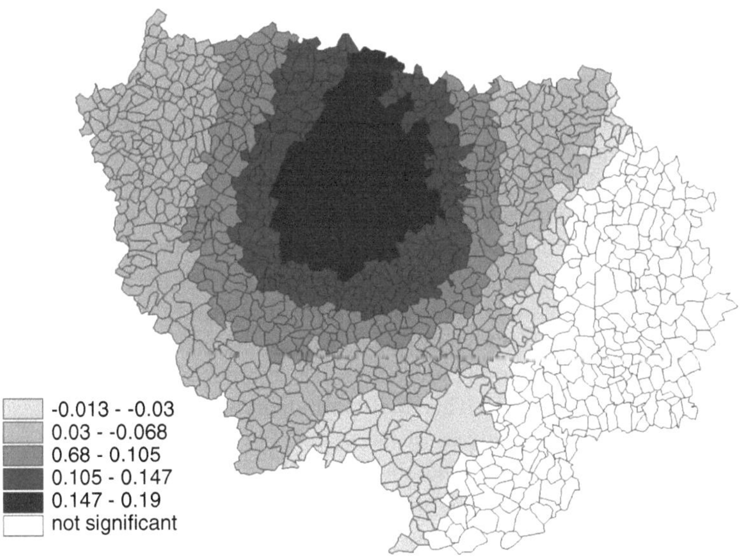

Fig. 9 Geographic distribution of the density gradient for consumer services employment. Scale 1:9,000

for industrial employment (Fig. 4) is similar to that of total employment. Again, we observe an organization of local gradients along a south–north corridor with higher density gradients values north of Paris, corroborating the idea of a zone in industrial decline. This decline has led to complete or partial closing-down of industrial sites, which were not entirely rehabilitated in 1999. The opening of "Stade de France" (sportive complex) in the North of Paris (Plaine Saint-Denis) in 1998 for the soccer world championship is part of the rehabilitation of this area. However, contrary to total employment, there is a zone of non-significant values east of Paris, in the outer ring, and then a zone in which the estimated density gradients are positive. This observation may be linked to a weak presence of industrial employment in the eastern part of the outer ring.

Similar comments apply for the geographic distribution of the density gradient for employment in high-order services (Fig. 5).

The geographic distribution of the density gradient for high-tech employment (Fig. 6) is also similar to that of total employment albeit with some important differences. Firstly, the zone of non-significant values is not limited to a strip east of Ile-de-France but surrounds the entire zone of significant values. Secondly, in the east of the outer ring is a large zone where local density gradient have unexpected negative values. Moreover, this zone is relatively large and the gradient increases more than for industrial employment. These two features combined explain why the global density gradient is not significant. However, the results are more intuitive when looking at the distribution of local values: the CBD remains a center around which the density gradients are organized decreasingly while the repellent effect is only effective on the outskirts of Ile-de-France. These characteristics are globally similar for employment in standard services (Fig. 7). As for high-tech employment, non-significant values of the density gradient and significant negative density gradients are located in the outer ring.

The geographic distribution of the density gradient for employment in finance-insurance (Fig. 8) is the most different from all the other distributions. The zone of positive and significant density gradients is located mainly in the inner ring and is made up of concentric circles. Then, we observe a circle of non-significant values of density gradients and circles of increasing negative values of the density gradient. This pattern of density gradient explains why, at global level, the sign is not the expected one. However, as in the preceding case, there is a decrease of local density gradients around the CBD. This density gradient pattern for finance-insurance must be linked to specific features of the sector. For one thing, the CBD is very attractive for finance-insurance (Guillain et al. 2006), due to the prestige implied by an address in this center, and more particularly because markets are internationally oriented (Coffey et al. 1996). This attraction to the CBD implies a sudden sharp decrease in density gradients. For another thing, one of the components of this sector is not linked to high-order services but to services to the population (Alvergne and Shearmur 2003). The available data do not allow these activities to be distinguished. Therefore, the geographical distribution of these gradients may reflect the dichotomy of the finance-insurance sector.

Finally, for consumer services (Fig. 9), a concentric distribution pattern of distribution of density gradient is observed, even though the decrease in density gradients is higher in the north of Ile-de-France. This distribution globally follows that of the population. There is here a feature typical of European cities: city centers not only attract employment, as do North-American cities, but they are also preferred areas for residential choice because of the diversity of their cultural heritage and amenities (Brueckner et al. 1999).

To sum up, the analysis of local density gradients made it possible to reveal contrasted situations in the value of these gradients, which are concealed when a total gradient alone is considered. Undertaking a study on various sectors is not sufficient to reveal the full complexity of the geographical distribution of the employment density gradients. On the contrary, an analysis of the local values allows highlighting situations that are differentiated not only in terms of the geographical distributions of the values from the local gradients but also in the spatial orientations considered around the traditional business district.

5 Conclusion

Urban sprawl, suburbanization of economic activities, and increased congestion costs and land values in the downtown areas are factors that have challenged the role and the importance of the traditional CBD in cities. In this chapter, we have analyzed whether and to what extent the CBD of Paris still shapes employment in Ile-de-France or whether, on the contrary, in the face of these various changes, its influence has become more limited. To that end, we chose to supplement the paper by Guillain et al. (2006), which identified employment centers using methods based on agglomeration of activities, with an econometric analysis. This analysis involves estimating employment density functions, linking employment density to distance from the first *arrondissement* of Paris. Analysis of the sign and the significance of the associated coefficient – the employment density gradient – provides a measure of the CBD's influence. We depart from earlier papers on this type of functions in three respects.

First, we consider the CBD's influence on total employment at the same time as its on six sectors, whose location choices are likely to be different. Second, to allow for the fact that density gradients may differ with distance (heterogeneous distribution) and direction (anisotropic distribution) from the CBD, we estimate local regressions, with one estimated density gradient for each commune. Third this analysis of heterogeneity is coupled with spatial autocorrelation, using the framework devised by Páez et al. (2002a,b).

The main results of our study indicate that the CBD still influences total employment in Ile-de-France but that its influence is indeed very different depending on sectors and depending on the distance and direction of the commune from the CBD. While this influence is overall manifest for total employment, industrial employment, employment in high-order services, and consumer employment, the CBD seems to have a repellent effect for finance-insurance employment and no effect for either high-tech employment or employment in standard services. However, these results hide wide geographic disparities in the distributions of employment density gradients. In particular, density gradients values decrease more or less quickly in Ile-de-France depending on sectors and along a south–north corridor. The repellent effect and the absence of effect globally observed for one of the six sectors is in fact implied by a major decrease of the gradient around the CBD, which is interpreted as a strong influence of this CBD on its immediate environment and a repellent effect on the fringes of Ile-de-France.

Finally, the CBD still influences the location of employment in Ile-de-France, but this influence is variable and complex. This study can be extended. In particular, the functional form used is a negative exponential function. Other specifications have been suggested for population density functions (McDonald 1989) and we will analyze the robustness of the results obtained with the selected specification. Moreover, the polycentric structure of Ile-de-France will have to be taken into account, by using polycentric density functions so that the influence of the secondary centers of employment in Ile-de-France can also be analyzed.

Estimations of global and local densities around the centers in a city can provide insights for planners about the location strategies of economic activities. Indeed, since urban sprawl of activities and population increases the commuting in the city, congestion and pollution are the counterparts of this movement. As a consequence, opportunities of densification are discussed now, which require to determine the distribution of densities.

References

Alperovich G (1995) The effectiveness of spline urban density functions: an empirical investigation. Urban Stud 32:1537–1548

Alperovich G, Deutsch J (2002) An application of a switching regimes regression to the study of urban structure. Pap Reg Sci 81:83–98

Alvergne C, Shearmur R (2003) Regional planning policy and the location of employment in the Ile-de-France: does policy matter? Urban Aff Rev 39:3–31

Anderson J (1985) The changing structure of a city: temporal changes in cubic spline urban density patterns. J Reg Sci 25:413–425

Anselin L (1988) Spatial econometrics: methods and models. Kluwer, Dordrecht

Anselin L (2006) Spatial econometrics. In: Mills TC, Patterson K (eds) Handbook of econometrics: volume 1, econometric theory. Palgrave MacMillan, Berlin, pp 901–969

Anselin L, Can A (1986) Model comparison and model validation issues in empirical work on urban density functions. Geogr Anal 18:179–197

Anselin L, Florax RJGM (1995) Small sample properties of tests for spatial dependence in regression models. In: Anselin L, Florax RJGM (eds) New directions in spatial econometrics. Springer, Berlin, pp 21–74

Anselin L, Bera A, Florax RJGM, Yoon M (1996) Simple diagnostic tests for spatial dependence. Reg Sci Urban Econ 26:77–104

Baumont C, Ertur C, Le Gallo J (2004) Spatial analysis of employment and population: the case of the agglomeration of Dijon, 1999. Geogr Anal 36:146–176

Brueckner JK (1986) A switching regression analysis of urban population densities. J Urban Econ 19:174–189

Brueckner JK, Thisse JF, Zenou Y (1999) Why is central Paris rich and downtown Detroit poor? An amenity-based theory. Eur Econ Rev 43:91–107
Coffey WJ, Shearmur R (2002) Agglomeration and dispersion of high-order service employment in the Montreal metropolitan region, 1981–1996. Urban Stud 39:359–378
Coffey WJ, Drolet R, Polese M (1996) The intrametropolitan location of high-order services: patterns, factors and mobility in Montreal. Pap Reg Sci 75:293–323
Erickson RA (1982) Employment density variation in the Baltimore metropolitan area. Environ Plann A 14:591–601
Fotheringham AS, Brundson C, Charlton M (2004) Geographically weighted regression: the analysis of spatially varying relationships. Wiley, Chichester
Griffith DA, Can A (1995) Spatial statistical-econometric versions of simple urban population density models. In: Griffith DA, Arlinghaus SL (eds) Handbook of spatial statistics. CRC Press, Boca Raton, pp 231–259
Griffith DA, Wong DW (2007) Modeling population density across major US cities: a polycentric spatial regression approach. J Geogr Syst 9:53–75
Guillain R, Le Gallo J, Boiteux-Orain (2006) Changes in spatial and sectoral patterns of employment in Ile-de-France, 1978–1997. Urban Stud 43:2075–2098
IAURIF (1999) Enjeux économiques pour l'Ile-de-France, du régional au local (Les Cahiers de l'IAURIF n° 124). IAURIF, Paris
IAURIF (2001) 40 ans en Ile-de-France. Rétrospective 1960–2000 (Etudes et Documents). IAURIF, Paris
IAURIF (2003) Franges des métropoles – Des territoires de projets. (Les Cahiers de l'IAURIF n° 136). IAURIF, Paris
Kelejian HH, Prucha IR (1999) A generalized moments estimator for the autoregressive parameter in a spatial model. Int Econ Rev 40:509–533
Kelejian HH, Prucha IR (2007) HAC estimation in a spatial framework. J Econom 140:131–154
LeSage JP (1999) The theory and practice of spatial econometrics. Mimeo, University of Toledo
Leung Y, Mei C, Zhang W (2000) Statistical tests for spatial non-stationarity based on the geographically weighted regression model. Environ Plann A 32:9–32
López F, Mur J, Angulo A (2008) Local estimation of spatial autocorrelation processes. This volume
McDonald JF (1989) Econometric studies of urban population density: a survey. J Urban Econ 26:361–385
McMillen DP (2001) Nonparametric employment subcenter identification. J Urban Econ 50: 448–473
McMillen DP, McDonald JF (1998) Suburban subcenters and employment density in metropolitan Chicago. J Urban Econ 43:157–180
Mills E (1999) Truly smart smart growth. Illinois Real Estate Letter. Office Real Estate Research, University of Illinois, Urbana-Champaign
Nechyba TJ, Walsh RP (2004) Urban sprawl. J Econ Perspect 18:177–200
Páez A, Uchida T, Miyamoto K (2002a) A general framework for estimation and inference of geographically weighted regression models: 1. Location-specific kernels bandwiths and a test for locational heterogeneity. Environ Plann A 34:733–754
Páez A, Uchida T, Miyamoto K (2002b) A general framework for estimation and inference of geographically weighted regression models: 2. Spatial association and model specification tests. Environ Plann A 34:883–904
Piercy P (1999) La Défense: 1958–1998, de la banlieue au pôle majeur de la région capitale. L'Information Géographique 1:33–36
Sridhar KS (2007) Density gradients and their determinants. Reg Sci Urban Econ 37:314–344
Stanback Jr, TM (1991) The new suburbanization. Challenge to the central city. Westview Press, Oxford
Waddell P, Shukla V (1993) Employment dynamics, spatial restructuring and the business cycle. Geogr Anal 25:35–52

The Geographic Dimensions of Electoral Polarization in the 2004 U.S. Presidential Vote

Ian Sue Wing and Joan L. Walker

1 Introduction

The 2004 U.S. presidential election was one of the most divisive in recent history (Pew Research Center 2004). The divisions in the electorate are popularly seen as the culmination of a process of political polarization underway since the 1970s (e.g., Frank 2004), and are epitomized by the now-ubiquitous map of the United States which shows swaths of red (i.e., majority Republican) states in the center of the country surrounded by blue (i.e., majority Democratic) states on the east and west coasts and in the north central region. In this chapter we investigate the geographic dimensions of political polarization in the United States through the lens of the 2004 election. We elucidate the principal contours of the divisions in the electorate, and characterize the manner in which the effects of the correlates of voting behavior cluster regionally. We take an ecological approach, using spatial econometrics to estimate the interregional divergence in the influences of the characteristics of populations and places on the odds of the Republican vote. To this end we employ aggregated data on 3,106 counties in the lower 48 states, which is the finest spatial scale at which both electoral returns and a variety of demographic and contextual variables are readily available.

Our goal is to push the limits of ecological analysis in electoral geography. We first develop a theoretical framework in which geography plays a central role in electoral polarization. Our central hypothesis, which draws on themes in the political science literature (Johnston et al. 2004; Cho and Rudolph 2008), is that a number of social processes that operate at fine spatial scales tend to push individuals voters' views into closer alignment with the ideological preferences of their geographically proximate majority – a phenomenon we call "localized entrenchment." Drawing on the sociological literature on polarization (DiMaggio et al. 1996; Evans 2003), we circumvent the well-documented handicap of weak

I. Sue Wing (✉)
Department of Geography & Environment, Boston University, 675 Commonwealth Avenue, Boston, MA 02215, USA,
e-mail: isw@bu.edu

A. Páez et al. (eds.), *Progress in Spatial Analysis*, Advances in Spatial Science,
DOI 10.1007/978-3-642-03326-1_13,

correlation between demographic attributes and ideology by employing a richer array of explanatory variables than prior spatial statistical analyses (e.g., O'Loughlin et al. 1994). We then apply spatial statistical techniques that exploit the spatial interrelationships among the electoral returns and our set of covariates, and find strong indications of entrenchment. Finally, we employ advanced methods to characterize the spatial heterogeneity in our estimated relationships – rather than re-estimate our aggregate statistical model on different regional sub-samples, we use geographically weighted regression (GWR). This technique enables us to exploit the spatial interdependencies among the entire universe of counties to estimate the fine-scale geographic variation in our covariates' influences on the 2004 presidential vote, while simultaneously controlling for the underlying spatial distributions of the characteristics of people and places. The patterns of agglomeration in the resulting influences on voting behavior are consistent with our explanation of how local entrenchment might induce polarization of the electorate.

We report three sets of results. First, we construct Local Moran's I statistics to analyze the spatial clustering of county election returns. We find substantial spatial autocorrelation in voting patterns, evidence that the returns for democrats and Republicans were significantly clustered in different regions of the United States, and indications of divergence among different sub-populations' vote distributions based on the spatial clustering of their demographic characteristics. Second, we perform spatial regressions at the aggregate level which identify the demographic and contextual factors that significantly impact the odds of voting Republican. We partition this propensity into direct influences associated with the attributes of counties and their populations, and indirect influences associated with the voting behavior and demographic characteristics of neighboring jurisdictions. The latter effects are particularly large, in many cases outweighing the former, and highlight the importance of the geographically varying contextual factors that are central to the predictions of our core hypothesis. Finally, our GWR results indicate considerable heterogeneity in the influence of several of our explanatory variables, and, most tellingly, regional agglomeration in their signs, which suggests that electoral polarization manifests itself as cross-cutting divisions in the U.S. electorate, not between population sub-groups but within sub-groups over space.

Given the nature of our analysis and results, we raise two caveats at the outset. First, when it comes to uncovering the mechanisms through which polarization occurs, we barely scratch the surface. Our more modest objectives are to clarify the irreducible spatial components of the divisions in the American electorate, and to outline their broad contours as the first phase of a program of more rigorous statistical testing. The second caution concerns the ecological fallacy. In particular, we take pains to distinguish what we do find: divergent patterns of spatial clustering in the impacts of the characteristics of counties and their populations on the vote, from what we do not: how the sign and magnitude of the effects of individuals' characteristics on their own voting behavior vary over space. The distinction between these inferences cannot be too sharply drawn (Goodman 1953; Hanushek et al. 1974).

The remainder of the chapter is organized into four sections. In Sect. 2 we set the stage by discussing our motivations and framing our inquiry. In Sect. 3 we describe the sources of data used in our analysis, and illustrate the spatial heterogeneity in key variables. We outline our methods of analysis in Sect. 4 and present and discuss the results in Sect. 5. We conclude in Sect. 6 with directions for future research.

2 Entrenchment: Geography's Role in Political Polarization

Political polarization is the segregation of the electorate along issue opinion and/or ideological lines, with concentration of voters about opposing extreme positions and concomitant erosion of moderate "centrist" preferences. The phenomenon is illustrated in Fig. 1, which plots the distribution of preferences in the electorate on a left-leaning (liberal) versus right-leaning (conservative) scale. In panel A, which draws on Fiorina and Abrams (2008), distribution A-I is not polarized, and exhibits the classic "single peaked" preferences of a centrist majority. By contrast, the bimodal distribution A-II, shown by the dashed line, illustrates the polarization of voters into equal opposing factions. The gray Distribution A-III, about which we say more below, is intermediate between A-I and A-II, with fatter tails and a less distinct peak indicating voters' movement away from the center toward the extremes.

The reality of the U.S. electoral landscape is far more complex that this picture suggests, however. Despite considerable heterogeneity in American voters' attitudes and beliefs, there is no evidence that the distribution of the electorate is either bimodal, or has recently become substantially more disperse – especially in light of the long view of history (Ansolabehere et al. 2006; Fiorina et al. 2006; Fiorina and Abrams 2008; Klinkner 2004; Klinkner and Hapanowicz 2005). There is, however, abundant evidence that the parties' candidates and activists alike have become increasingly partisan, and have staked out increasingly divergent positions on a

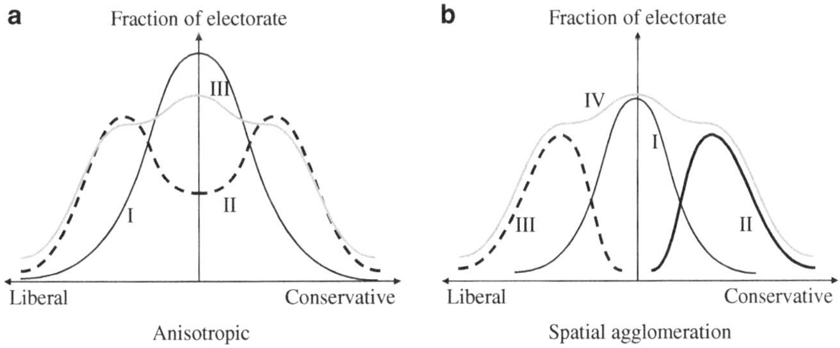

Fig. 1 Electoral polarization: a conceptual framework

range of issues (Bartels 2000; Fleisher and Bond 2004; Mellow and Trubowitz 2005; Poole and Rosenthal 2001; Stonecash et al. 2002).[1] Also, there are at best only weak indications of a rising intensity of opposing political views among the general electorate, and then only for a handful of "hot-button" issues such as abortion or homosexual persons' right to marry (DiMaggio et al. 1996; Evans 2003; Fiorina and Abrams 2008), while the distribution of liberal and conservative views on the broad spectrum of issues appears to have remained fairly even (Fiorina et al. 2006). What has occurred is a public redefinition of the labels "liberal" and "conservative," which, along with polarized choices offered in the political arena, has served to heighten issues' salience to voters, and has induced them to self-categorize and align more closely with one or the other party despite unchanged underlying preferences (Hetherington 2001; Baldassarri and Gelman 2008; Miller and Hoffmann 1999). These dynamics have led many analysts to conclude that Americans feel more polarized than they in fact are.

Our own view is that while the "red versus blue state" conception of polarization is undeniably simplistic, to claim that the electorate is not divided is to deny the essential geographic dimension of the phenomenon.[2] Klinkner (2004) cites a particularly apposite example which captures the essence of the phenomenon. In 1972, New Yorker magazine contributor Pauline Kael expressed surprise at Richard Nixon's re-election as president, saying "Nobody I know voted for him." The same could be said in 2004. Despite the fact that Republican incumbent George Bush was returned to office with 52% of the national electorate, few people in Washington, DC knew anyone who voted for him – he gained just 7% of the electorate there. Likewise, few people in Idaho's Madison County knew anyone who voted for Democratic challenger John Kerry, whose record there was similarly dismal. And although Bush won by a margin of 60% or greater in 54% of counties while Kerry enjoyed a similar margin in only 5% of counties, these "landslide" jurisdictions were home to 47% of the electorate. Moreover, 38 out of 50 states were carried by one or the other candidate with a margin of 5 percentage points or greater, with a stark

[1] Glaeser et al. (2005) develop a theory of strategic extremism which illustrates the incentives political parties have to divide on issues in order to increase their chances of winning at the polls. Partisanship turns on two key elements: among voters, the existence of an intensive margin where the level of support matters (e.g., turnout or donations, as opposed to the extensive margin of voting) and which parties can activate by taking extreme positions that appeal to their respective bases, and the ability of parties to target extreme statements to their own supporters while bypassing those of the opposition, thereby avoiding a backlash. Below, we note that this sort of targeting becomes easier the more the electorate is ideologically segregated along geographical lines.

[2] This is an example of the modifiable areal unit problem (Openshaw 1984). Differences between the number of Democratic and Republican votes were as large between red and blue counties within some states as they were between some red and blue states. Using counties as the unit of analysis is attractive precisely because, unlike states, congressional districts or electoral precincts, their geographic boundaries are independent of electoral processes relevant to the presidential vote. The consequent absence of selection bias makes us confident in exploiting county characteristics as strictly exogenous covariates in our subsequent analyses.

divergence in the attitudes and beliefs espoused by the voters in states with Republican and Democratic majorities (e.g., Abramowitz and Saunders 2008: Table 6; Glaeser and Ward 2006: Table 1).

The geographic evidence typically adduced in support of the no-polarization thesis is that county-level returns exhibit variances and indices of dissimilarity that are low and stable, as well as indices of isolation for each party's turnout that are similar in magnitude and fluctuate with no apparent long-run trend (Glaeser and Ward 2006; Klinkner and Hapanowicz 2005). But these same data indicate substantial geographic clustering of voting patterns in recent presidential elections (Kim et al. 2003), a phenomenon which persists into 2004. Democratic and Republican voters were more likely than not to be exposed only to individuals who voted in a similar way, with the result that one fifth of those supporting either party would have needed to relocate for the distribution of votes to be spatially uniform. The latter is the highest percentage since the 1940s (Glaeser and Ward 2006: Fig. 2).[3]

The statistical manifestation of this sort of division is shown in Fig. 1b, which illustrates a hypothetical situation in which the electorate is divided among two disjoint regions. Distribution B-I indicates the preferences of centrist voters, whose members are distributed among both regions. A conservative-leaning sub-population of voters with distribution B-II resides in one region, while a liberal-leaning sub-population with distribution B-III resides in the other. It is easy to see that in this society the aggregate preferences B-IV are the same as the intermediate distribution A-III, with zero mean and fair degree of central tendency, but with an electorate that feels – and is – polarized, but along geographic lines.

Our main contention is that this picture describes the 2004 presidential election, not in Kael's literal sense of the regional distributions of electoral returns, but rather in terms of the preferences that DiMaggio et al. (1996), Evans (2003) and others have sought to measure.[4] Because of the ecological nature of our data, our indicators of preference boil down to the influences of the characteristics of populations and places on the propensity to vote Republican or Democratic. Indeed, we demonstrate that along a number of key dimensions the influence of characteristics on the propensity to vote exhibit substantial spatial agglomeration, with geographic clustering of counties with divergent preferences, as in Fig. 1b.

In conducting our investigation we take up the gauntlet thrown down by Fiorina and Abrams (2008), demonstrating the strength and stability of the associations between the voting behavior on one hand and the characteristics of populations and

---

[3] These statistics were computed for our sample of 3,106 counties in the lower 48 states. Indices of isolation measure the likelihood of Republicans' and democrats' exposure to the opposing group at 55% and 51%, respectively, while non-uniformity in the pattern of votes is given by the index of dissimilarity at 22%.

[4] Note that B-IV's variance and excess kurtosis are larger than A-II's These authors test whether these two moments of the distributions of survey respondents' attitudes on a diverse array of social issues have increased over time.

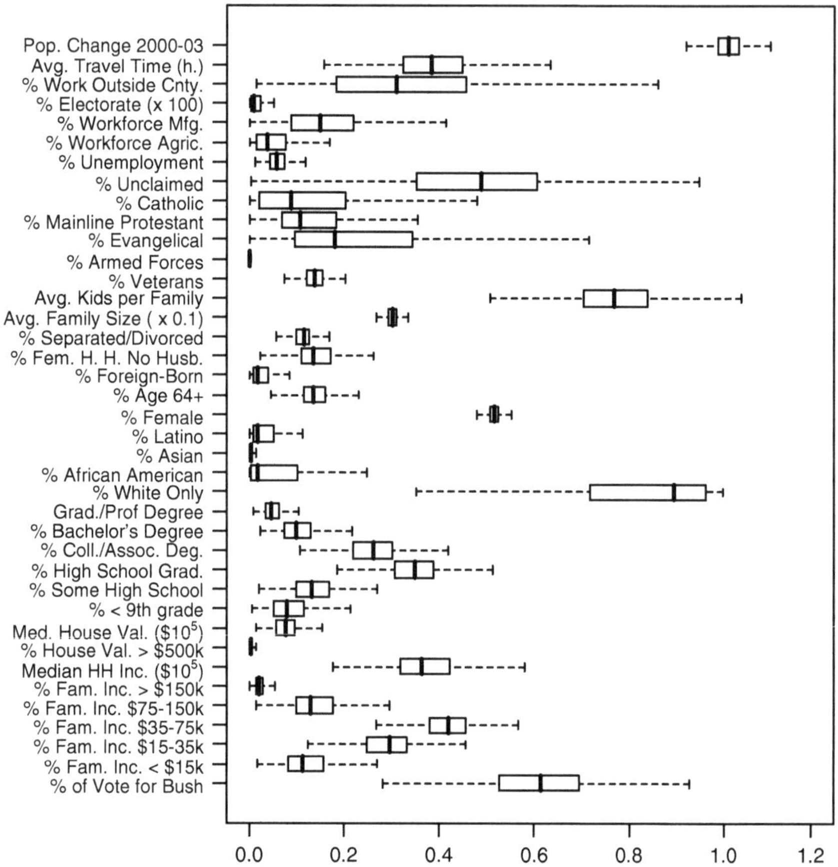

Fig. 2 Box plot of descriptive statistics of the dataset

places on the other.[5] Consistent with our interest in the segregation of the electorate over space, and cognizant of the strictures imposed by the cross-sectional data at our disposal, we reinterpret their (temporal) notion of stability to focus on how the correlates of voter behavior vary geographically.

Our first task is to articulate testable propositions about how regionally segregated voter distributions like B-II and B-III might arise. Shifts in the American electorate at broad geographic scales have been well documented, with sorting and clustering of individuals with similar ideological leanings arising as unintended consequence of interstate migration (Frey 2000; Gimpel and Schuknecht 2001), as well

[5] "contrasts in individual sociocultural characteristics are not direct indicators of political polarization. Hence, contrasts in such characteristics may or may not constitute evidence of polarization. Analysts must provide additional information about the strength of the links between social characteristics and relevant political variables, as well as information about the stability of such linkages." (p. 568)

as (intentional) ideological realignment in various regions of the United States.[6] But it seems unlikely that these forces by themselves are strong enough to generate either the regional homogeneity or intensity of preferences that underpin Fig. 1b.

In our view, the key element is how individuals' social and political values are shaped – and reinforced – by context and agglomeration at finer geographic scales.[7] Our thesis is that the same social forces that facilitate political participation contribute to ideological reinforcement at the local level. Cho and Rudolph (2008) identify four processes which are relevant in this regard:[8]

1. Residential self-selection, whereby citizens' characteristics jointly predict their residential and ideological preferences, and individuals choose to live near to others who are socially and demographically similar to them, leading to spatial clustering of voting tendencies.[9]
2. Voter mobilization, in which partisan elites selectively target segments of the electorate on the basis of demographic attributes which are spatially clustered, especially in closely contested states or electoral districts. Spatially homogenous preferences facilitate targeting of extreme political statements to demographic groups to which they may have particular salience, catalyzing party alignment and turnout (cf. fn. 5).[10]

---

[6] For example, southern conservative voters switching from Democratic to Republican (Schreckhise and Shields 2003; Bullock et al. 2005; Valentino and Sears 2005), northeastern voters becoming increasingly liberal (Speel 1998), and the rise of the mountain west as a conservative voting bloc (Marchant-Shapiro and Patterson 1995).

[7] For example, Glaeser and Ward (2006, p. 131A): "These differences in beliefs within the United States drive home a central point about how politically relevant beliefs are formed. People in different states have been exposed to quite similar evidence through national media outlets, but they have reached radically different conclusions, and continue to hold these conclusions despite being aware that others disagree. This disagreement requires either different prior beliefs or some other deviation from Bayesian reasoning. One natural alternative model is that people base opinions mostly on the views of those around them. As such, local interactions are critical, and these provide plenty of possibility for wide geographic variation..."

[8] See also Johnston et al. (2004), who develop a slightly different taxonomy.

[9] Despite anecdotal evidence (e.g., Bishop 2008; Bishop and Cushing 2004) and statistical indications particular kinds of neighborhood environments influence their residents' ideological leanings, irrespective of demographic composition (Williamson 2008), the political sources and consequences of self-selection have yet to be thoroughly investigated.

[10] For example, Mutz (2002, p. 852): "Homogeneous environments are ideal for purposes of encouraging political mobilization. Like-minded people can encourage one another in their viewpoints, promote recognition of common problems, and spur one another on to collective action. Heterogeneity makes these same activities much harder. Participation and involvement are best encouraged by social environments that offer reinforcement and encouragement, not ones that raise the social costs of political engagement." Also, Williamson (2008, pp. 20–21): "... the spatial sorting of residents by political ideology, once it reaches a sufficiently advanced stage, may help create what Lazarsfeld, Berelson, and Gaudet (1944) termed a 'reinforcement effect'; not only might residents of a very conservative suburb be less likely to hear a liberal viewpoint from their neighbors but such areas will likely be targeted and contacted frequently by conservative political activists while being relatively ignored by liberal political activists, further reinforcing the relationship between spatial context and individual political outlook." Homogeneity facilitates a political campaign's ability to mobilize voters by reducing the cost of what Lazarsfeld et al. (1944) refer to

3. Social interaction, the set of mutually-responsive behaviors adopted by individuals in social networks. Social interactions may amplify ideological divisions because organized networks such as civic associations are a particularly effective mechanism for the exchange of political information (McClurg 2003), but such information tends to be systematical biased due to homophily – the propensity of individuals to interact with others who are similar to them (e.g., McPherson et al. 2001). Social interactions also promote ideological homogeneity through the process of social learning, with views that are consonant with (dissonant from) those of the majority of network participants receiving positive (negative) reinforcement, leading to closer alignment of preferences within the network.[11] Finally, the fact that these effects transpire through direct interpersonal contact (and even non-verbal cues) suggests that the phenomenon of closer individual alignment with the local majority should only persist over a limited spatial domain.
4. So-called "casual observation," the indirect, often involuntary, social interaction induced by the characteristics of an individual's environment.[12] A key implication is that the physical attributes of citizens' action spaces are likely to significantly influence their ideological preferences, independent of neighborhood demographics (cf. Williamson 2008). Non-political, day-to-day social interactions remain a key source of political information for Americans (Klofstad et al. 2006), with the workplace being the principal forum in which they are exposed to dissonant political views (Mutz and Mondak 2006). This suggests that the spatial domain of political influence is not limited to the neighborhood in which an individual resides, and may extend well beyond her commuting distance.

At a minimum, these processes imply that voters' preferences should be influenced by those of the citizens around them. But, in view of the reinforcing character of the first three processes, we further claim that the likely outcome will be a phenomenon which we term "localized entrenchment": in the absence of exogenous shocks, communities remain locked in a cycle of reinforcement of the values held by their ideological majorities, with corresponding suppression of the inward diffusion of countervailing viewpoints and ideas, leading to entrenchment of attitudes, beliefs and, ultimately, voting behavior. Our view of entrenchment as closer alignment

as activation ("not to form new opinions but raise old opinions over the thresholds of awareness and decision," p. 74), and reinforcement ("to secure and stabilize and solidify [. . .] vote intention and finally to translate it into an actual vote," p. 88).

[11] For example, Huckfeldt and Sprague (1995). For formal models of this process see Baldassarri and Bearman (2007), Dixit and Weibull (2007) and Glaeser and Sunstein (2009).

[12] For example, Cho and Rudolph (2008, p. 277): "Casual observation exposes citizens to meaningful information through low-intensity neighbourhood cues such as the display of yard signs, bumper stickers, or simple observations and biases created by how neighbors dress and behave, what types of cars they drive, or how well their garden is groomed. Such low-intensity cues may influence behavior by subtly communicating information about the prevailing norms and sentiments within a community. In particular, they may provide signals about a local community's political culture and ethic or the nature and distribution of political preferences within that community."

between the individual vote and the local majority vote is consistent with evidence of increased party identification by voters (Miller and Hoffmann 1999), ideological cleavages along geographic lines (Abramowitz and Saunders 2008; fn. 6) and regional concentration of the Democratic and Republican parties' representation in the U.S. Congress (Mellow and Trubowitz 2005).

The econometric consequences of local entrenchment are spatial correlation and endogeneity. These are anticipated by the economic literature on social interactions (Manski 1993, 2002; Glaeser et al. 2003), which suggests that the group of citizens in the zone of political influence around a particular individual will impact her vote decision in three ways. The first is endogenous effects, where the group's average voting behavior affects the individual's vote, which could potentially reflect the influences of any or all of the four processes above. The second is contextual or exogenous effects, where the group's average (exogenous) characteristics affect the individual's vote. This might reflect processes 3 and/or 4, as well as the spatial clustering of citizens with similar characteristics for reasons other than self-selection. The third is correlated effects, where the individual's error term is correlated with the error terms of members of the group because of similar characteristics not observed by the econometrician (process 4), sorting or selection of individuals based on who they are (process 1), or exposure to common shocks (process 2, and the polarized character of choices in the political arena more generally).

An additional consideration is that our ecological data on citizens' characteristics and votes at the level of the county (not the individual) forces us to reinterpret these effects in terms of areal units and their neighboring jurisdictions. We argue that even though this invariably introduces aggregation bias of unknown magnitude and sign, such a reinterpretation is still valid because of exogeneity in the boundaries of our areal units (see fn. 2) and the potentially long spatial reach of processes of casual observation. Moreover, the fact that we know the location of each observation means that the three effects above neatly correspond to the components of different spatial econometric models. Endogenous effects are captured by the coefficient on the spatially lagged county vote in a spatial autoregressive model; contextual effects are captured by the coefficients on spatial lags of the covariates; and correlated effects are indicated by the coefficient on the spatially lagged error term in a spatial error model. Quite likely, all three effects are simultaneously at work in our dataset, which presents a challenge for estimation.

3 Data

Our dependent variable is the vector of votes cast for Bush as a share of total votes at the county level. We estimate the size of total electorate as the sum of ballots cast for Bush, Kerry and independent candidate Ralph Nader, data for which were downloaded from the CBS News election 2004 website.[13]

[13] These data are of necessity approximate, not being adjusted for the results of recounts in Ohio and New Mexico. There were additional independent candidates on the ballot in each state, but the

For explanatory variables we selected a broad spectrum of demographic characteristics that are likely to have influenced individuals' voting decisions, which we organized into categories similar to those used in prior analyses of political polarization (DiMaggio et al. 1996; Evans 2003).

We employ four sets of demographic variables at the county level. These are the distribution of income, measured over four income categories; housing costs; the distribution of educational attainment, measured over five grades; racial and ethnic composition; and age, sex and national origin. These data were obtained the 2000 U.S. Census and Current Population Estimates data files, and are coded as percentages of either the total or the voting-age population within each county.

We also employ two categories of variables on economic characteristics of places: median household income, unemployment and the composition of employment; and local geographic characteristics such as population growth, the size of the local electorate, whether the county belonged to the core (urban) or outlying (suburban) region of a metropolitan statistical area (MSA), travel time to work and prevalence of commuting outside one's county of residence. Unemployment and wage data were compiled from U.S. Bureau of Labor Statistics Local Area Unemployment Statistics and U.S. Bureau of Economic Analysis Regional Economic Information System data files, respectively, while local geographic variables were collected from the 2000 Census and Current Population Estimates.

We include three additional sets of variables in an attempt to proxy for issues which exit polls indicate played an important role in the election: the war on terror and U.S. military intervention in Iraq, and "moral" or "family" values. Based on social interaction theory, we hypothesize that attitudes toward the former issue among the general population will be most strongly shaped by personal knowledge of – and face-to-face interaction with – individuals who have served or are currently serving in the armed forces, and that the diffusion of attitudes will increase with geographic proximity to clusters of this sub-group (e.g., counties which host or immediately surround active military bases). Accordingly, we code for attitudes toward the war on terror using Census data on the fractions of veterans and active military personnel in counties' population.

Like no other set of issues, moral values are the bellwether of electoral polarization as the reflection a so-called "culture war" (Hunter 1992; Miller and Hoffmann 1999; Evans and Nunn 2005). Pew Research Center (2004) notes that moral values are not precisely defined, but encompass conservative views on subjects as diverse the appropriate role of religious expression and proselytization in public life, marriage and divorce, women's fertility and right of access to abortion, and child-rearing in a traditional nuclear family setting. The multivariate and ambiguous character of values, coupled with the fact that they are not directly observable even at the individual level, means that our ability to precisely identify their effects using aggregate data is weak at best.

numbers of votes cast for them were small. Neither of these factors seems likely to significantly change our main results.

With regard to religion, a useful indicator is the distribution of adherents to different faiths – particularly evangelical Christians – among the population. We use data from Glenmary Research Center (2004) to construct the distribution of individuals with different religious affiliations by county, which we code as shares of the population.[14] We proxy for attitudes to marriage using data on the fraction of population separated or divorced tabulated by the Census. Although some data are available on rates of teen pregnancy, out-of-wedlock births and abortion rates, they are not disaggregated to the county level, and were not used in our analyses.[15] To proxy for attitudes to fertility and child-rearing practices, we employ Census data on average family size, the average number of children per household, and the percentage of households headed by a female with no husband present.[16]

Our final set of covariates captures an important aspect of the debate over values which played out in the 2004 election, namely, the polarization of attitudes toward homosexuals, especially the legalization of same-sex marriage, civil unions or domestic partnership benefits. There is a dearth of data on either the geographic distribution of either gay persons or general attitudes toward them. However, during 2004 eleven states enacted ballot initiatives to ban same-sex marriages.[17] In an attempt to capture the effect of related attitudes on the vote, we treat these initiatives as an exogenous shock, and construct a dummy variable for gay marriage bans (GMB), coding the counties in these states as ones and the remaining counties in our sample as zeros. Following Campbell and Monson (2008), we include the interaction between this dummy and the percentage of Evangelical Christians in the population as a proxy for the potentially galvanizing influence of the ballot initiatives on turnout by evangelical voters for the Republican party.

We restrict our analysis to the contiguous counties of the lower 48 states, dropping counties in Alaska (for which disaggregate election returns are not tabulated) and Hawaii. Remaining counties for which one or more variables were missing were also dropped. Our final sample consists of 3,106 observations (denoted below by N), which we geo-coded using the county centroids from the Census 2000 gazetteer files.

Figure 2 presents the distributions of the variables as box plots to facilitate comparison. A few covariates, such as the percentages of Asian Americans or

[14] Campbell and Monson (2008) note that this database suffers from a number of problems, principally non-response bias in survey questionnaires, omission of non-denominational churches – which account for an increasing share of religious participation, and an inability to track the number of residents of one county who attend church in another.

[15] These data are available online from the Alan Guttmacher Institute. Preliminary regressions indicated that the state-level incidence of abortion and teen pregnancy were not significant predictors of the odds of voting Republican, in part because of their collinearity with state fixed effects.

[16] Our use of the proportions of divorced persons and households headed by single females is admittedly crude. In particular, it is hard to know whether the statistical effect of these variables on electoral outcomes is driven by the voting behavior of people in these groups or by morally conservative voters' negative reactions to the former.

[17] The states are: Arkansas, Georgia, Kentucky, Michigan, Mississippi, Montana, North Dakota, Ohio, Oklahoma, Oregon and Utah.

persons on active duty in the armed forces, have small magnitudes and exhibit very little variation. Conversely, other variables, such as the percentage of evangelicals and non-adherents, persons of solely Caucasian background, the average number of children per family, and the share of counties' electorates voting for Bush, all vary substantially across counties. In the working paper version of the article (Sue Wing and Walker 2005) we provide additional descriptive statistics that show that these variables exhibit significant interregional heterogeneity.

Our aim is to identify the association between the dependent variable and independent variables above, and then characterize the spatial variations in these relationships. To do this we turn to our spatial econometric toolkit.

4 Methods

Our analysis proceeds in three phases, the algebraic details of which can be found in an appendix to the preliminary version of this chapter (Sue Wing and Walker 2005). Our first task is to characterize the degree of spatial polarization in the vote by examining the intensity and geographic scope of spatial clustering in county-level returns. Following Kim et al. (2003), we compute the vector of Local Moran's I statistics for Bush's share of the electorate in each county. Rather than use their method of employing county-to-county commuting flows as a spatial weighting variable, we construct a symmetric spatial weighting matrix (W) based on a simpler distance-based scheme.[18] We use the results of this calculation to generate maps the regions of statistically significant spatial clustering of votes.

Our second task is to test the predictions of our local entrenchment hypothesis by investigating the effects of the explanatory variables in Sect. 3 on the odds of voting Republican at the national level. We estimate the following linear-in-logarithms logistic model:

$$Y = X\beta + u, \tag{1}$$

in which the dependent variable is an $N \times 1$ vector of the log-odds ratios of each county voting Republican,

$$y_c = \log\left(p_c - (1 - p_c)\right), \quad c \in N \tag{2}$$

where the subscript c indicates counties and p_c is the probability of c's Republican vote, estimated by Bush's share of the total votes cast, X is an $N \times k$ matrix of covariates given by the logarithms of the continuous independent variables in Fig. 1,

[18] Consistent with our discussion of the prominent role of social interactions, we defined the neighborhood of each county as a radius of 200 km, which is approximately twice the distance traveled at the highest state-mandated speed limit (75 mph) for the maximum average commute time in Fig. 2. The advantage of this scheme is that every row in W has at least one non-zero off-diagonal element, which allowed us to row-standardize the resulting matrix of distances without having to worry about divide-by-zero errors.

as well as the dummy variables and interaction terms described in the previous section. We interpret the coefficient β as the vector of elasticities of the odds of voting Republican with respect to the k covariates at the county level.

The overall explanatory power of the basic model in (1) was good, but (not surprisingly) tests of the residuals indicated that the disturbance vector u exhibits significant spatial autocorrelation (without state dummies, the Moran's I standard deviate = 57.55, $p < 0.01$). The likely culprits are spatial sorting and selection of voters on the basis of demographics, as well the omission of contextual variables, common shocks – especially congressional and gubernatorial elections which were simultaneously being held in each state, and the endogenous effects of surrounding counties' votes.

To test how much of the spatial dependency in the errors could be explained by omitted contextual factors, we included fixed effects for each state. This dramatically improved the fit and mitigated the degree of spatial autocorrelation, but tests of the residuals still indicated problems (Moran's I standard deviate = 33.44, $p < 0.01$). Moreover, Lagrange multiplier tests of the residuals for an omitted spatially lagged dependent variable (LM_ρ) and spatially autocorrelated errors (LM_λ) led us to reject the linear model, with or without state dummies.[19]

Encouraged by these results, we turned to more sophisticated estimators capable of capturing the effects of interest: the spatial lag model

$$Y = \rho W Y + X\beta + \varepsilon, \tag{3}$$

and the spatial error model, which augments (1) with:

$$u = \lambda W u + \varepsilon. \tag{4}$$

The variable ε is an $N \times 1$ vector of i.i.d. errors, ρ is a spatial lag correlation parameter, λ is a spatial error correlation parameter, and W is the $N \times N$ matrix of spatial weights described above. The models corresponding to (3) and (4) were estimated by maximum likelihood.

There is little a priori guidance as to which of these models is more appropriate. Spatial lag models are more common in the political science literature, and assume that the effects of a county's attributes on the odds of voting Republican are influenced by neighboring counties votes (i.e., endogenous effects), via the parameter ρ. On the other hand, the spatial error model assumes that spatial autocorrelation can be explained by aggregation bias, sorting and selection, or spatially varying omitted variables (i.e., correlated effects), captured by the parameter λ.

Our estimates of the two parameters indicate a high degree of spatial dependency in the data ($\rho = 0.40$ and $\lambda = 0.96$, both $p < 0.01$), and the challenge we faced was to apportion this dependency among the three effects above. To this end, we

[19] For the basic model, $LM_\rho = 2695.41$ ($p < 0.01$) and $LM_\lambda = 1044.49$ ($p < 0.01$), while for the fixed effects model, $LM_\rho = 523.08$ ($p < 0.01$) and $LM_\lambda = 278.27$ ($p < 0.01$).

employed Anselin et al.'s (1996) Lagrange multiplier tests of the spatial lag and spatial error specifications being mutually contaminated by each other, but both the test for error dependence in the possible presence of a missing lagged dependent variable (LM_λ*), and the test for a missing lagged dependent variable in the possible presence of spatially correlated disturbances (LM_ρ*), had power against each other. In both tests the null was rejected for the basic as well as the fixed-effects models,[20] but the test of robustness of the spatial error model against contamination by a spatially lagged dependent variable saw rejection of the null at the higher level of significance, apparently favoring the spatial error model. The log-likelihood and AIC statistics supported this conclusion, indicating that the spatial error model has the better fit to the data, however the extent of autocorrelation in the spatial error model remained a concern, especially since λ subsumed both endogenous and contextual effects.

In light of McMillen's (2003) critique that spatial dependence in the error term might simply indicate misspecification – especially given our potential omission of spatially correlated right hand side variables, we decided to pursue a third alternative: the unconstrained spatial Durbin model:

$$Y = \rho WY + X\beta + WX\gamma + \varepsilon \tag{5}$$

This model nests both our lag and error specifications through the restrictions $\gamma = \mathbf{0}$ and $\gamma = -\rho\beta$, respectively (Anselin 2002). The latter "common factor hypothesis" (Burridge 1981) is decisively rejected by a likelihood ratio test (LR $= 336.3$, $p < 0.01$), suggesting that residual spatial autocorrelation in the error term of (3) arises as a consequence of omitted spatial lags of the covariates (i.e., contextual effects). Accordingly, we relied on the results of (5) for our insights regarding the aggregate-level correlates of voting patterns in 2004, subject to the caveat that our results likely overstate endogenous and contextual effects while giving short shrift to correlated effects.

Our third task is to bring the results of the previous phases together to elucidate the implications of local entrenchment for the polarization of the U.S. electorate. In the preliminary phase of our analysis we re-estimated eq. (5) on contiguous subsamples of counties defined by the nine U.S. census divisions. The parameter estimates varied markedly among regions in magnitude, sign and significance, indicating that the national-level estimates mask substantial spatial heterogeneity. But given the hypothesized importance of local environmental influences for counties' voting behavior, we sought a way to systematically characterize how the parameters of (5) vary over fine geographic scales.

Accordingly, to capture the full extent of spatial non-stationarity in our data we re-estimated our model as a GWR, a nonparametric technique that generates intercept and slope parameters for every county by running a sequence of locally linear

---

[20] For the basic model $LM_\rho^* = 172.99$ ($p < 0.01$) and $LM_\lambda^* = 1823.91$ ($p < 0.01$), while for the fixed-effects model $LM_\rho^* = 73.96$ ($p < 0.01$) and $LM_\lambda^* = 318.77$ ($p < 0.01$).

regressions on a sub-sample of data from nearby counties (Brunsdon et al. 1996; Fotheringham et al. 1997, 2002). The GWR model can be written:

$$Z_c Y = Z_c X \theta_c + v_c \tag{6}$$

in which Z_c is the matrix of local spatial weights centered around the cth county,[21] and θ_c is a spatially-varying $k \times 1$ vector of parameters associated with observation c. The latter allows us to map and analyze the spatial variation and clustering in our aggregate results. The fact that the GWR method estimates an intercept for each county drastically diminishes the ability of state dummies and spatial lags to capture unobserved contextual effects, and in any case, the computational exigencies of estimating many additional parameters overwhelmed our computing resources.[22] We therefore used GWR to estimate only our basic linear model. Our final step was to test for polarization by examining whether the resulting vector of local odds elasticities θ_c exhibited significant spatial clustering along each of its dimensions (indicating entrenchment), and whether the clusters gave rise to distributions of effects similar to Fig. 1b.

5 Results

5.1 *The Spatial Clustering of Votes and Covariates*

Applying Moran's test to county vote returns reveals significant global spatial autocorrelation in the election results (Moran's I standard deviate = 109.68, $p < 0.01$). We compute local Moran's I statistics for the Republican share of the vote and key independent variables, and plot the results as a series of significance maps, shown in Fig. 3. A two-tailed test of significance ($p < 0.05$) allowed us to classify each observation as one which exhibited significant spatial clustering of voting returns for Bush above (dark gray) or below (light gray) the sample means.

Significant clustering in the share of the electorate voting Republican, shown in panel A, is comparable to that found by Kim et al. (2003, p. 749, Fig 2b), with clustering above the national average in large swaths of the Midwest, West Central and upper Mountain regions, as well as pockets in Appalachia, and clustering below the average in the Northeast and North Central regions, as well as in pockets along the Pacific coast and in southern Texas and Florida. Such agglomeration is precisely

[21] Specifically, $Zc = diag[Z_{1c}, \ldots, Z_{Nc}]$ is an $N \times N$ diagonal matrix of c's distance-based weights expressed as a local kernel, $Z_{jc} = \exp\left(-0.5\left(d_{jc}/\boldsymbol{h}\right)^2\right)$, in which d_{jc} is the distance between c and other counties (j), and the spatial interaction radius is given by a fixed bandwidth parameter, $\boldsymbol{h}$, that we estimate using a crossvalidation procedure.

[22] All our analyses were performed using the spatial packages for the R statistical language (Bivand 2006; Bivand and Brunstad 2006).

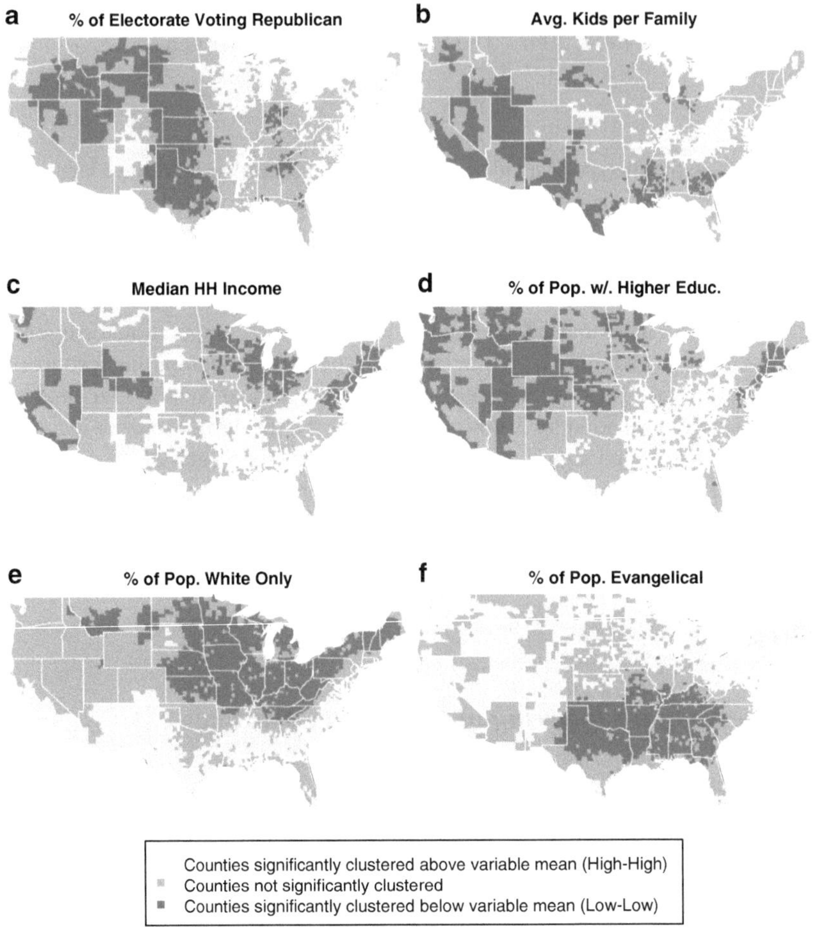

Fig. 3 Local Moran's I significance maps of votes and key covariates

what one would expect to be associated with an electorate that is polarized over space. It is natural to inquire into the factors on which such clustering might depend. For example, the ideological sorting of populations could be based on any number of factors such as income, race or religion. Preliminary insight into this question can be gained by visually inspecting the patterns of spatial clustering of the explanatory variables.

Panels B–F show the results of computing Local Moran's I statistic for a subset of the covariates in Fig. 2. The average number of children per family is clustered above the mean in pockets around the Great Lakes and across the South and Southwest, and clusters below the mean in the Ohio River valley and pockets in Florida and the Midwest. Median household in-come exhibits significant positive clustering in the northeast and upper Midwest, as well as in pockets in the mountain west and along the Pacific coast, with significant negative clustering in pockets throughout the Mississippi river valley, the west and the southeast of the country. The proportion of the population with post-secondary education is clustered above the mean in the northeast and in swaths across the upper Mountain West and coastal California, and clusters below the mean across the South. The share of Caucasians in the population is positively clustered in the North and East and negatively clustered across the South, particularly in the Southwest, the share of African Americans is positively clustered in the Southeast and negatively clustered in the North Central region, while the share of Evangelicals is positively clustered in the South and negatively clustered in the Northeast, Mountain and Western regions.

The relationships between the clustering of voting returns and the covariates are not obvious. To shed light on these associations, in Fig. 4 we follow Ansolabehere et al. (2006) and plot the distributions of the log odds ratio at the county level (see (1)) for different subsets of the data based on the clustering of the variables in Fig. 3. We examine how the propensity to vote Republican is distributed across counties which exhibit significant spatial clustering above or below the means of our subset of variables by constructing separate kernel density estimates, weighted according to the distribution of the electorate across the counties in each sample. A rightward (leftward) shift of the distributions thus indicates citizens' propensity to vote Republican (Democratic). To facilitate comparison we superimpose the plots of the densities on the reference vote distribution of the national electorate, shown in gray, whose unimodality at zero is often taken as prima facie evidence against polarization (e.g., Ansolabehere et al. 2006).

In panel A, the positive clustering of the vote in the center of the country indicates strongly Republican preferences in the less populous counties there, with counties that do not exhibit significant clustering leaning slightly Republican, and the more populous counties that cluster negatively on the east and west coasts with exhibiting preferences that are moderately Democratic but with a negatively skewed distribution. A very different picture emerges if we segment the electorate according to the spatial clustering of fertility, however. In panel B, the large mass of non-clustered counties mirrors the shape of the aggregate vote distribution, while regions with families that have less than the average number of children tend to lean slightly Republican. Surprisingly, clusters of counties with greater-than-average numbers of kids per family have a bimodal distribution, with similar numbers of voters leaning Republican and Democratic. A similar pattern is exhibited by the influence of Caucasian populations (panel E), with a higher propensity to vote Republican in clusters of less racially diverse counties, centrist preferences is counties that are not clustered, and a bimodal distribution in clusters of counties with smaller-than-average white populations.

Segmenting the electorate based on the clustering of household income and educational attainment (panels C and D) yields similar results. Counties belonging to low-income and low-education clusters seem to have fairly strong Republican leanings, non-clustered counties show a slight propensity in this direction, and the preferences of counties in high-income and high-education clusters are largely centrist, with slight Democratic leanings. In panel F, clustering of counties with

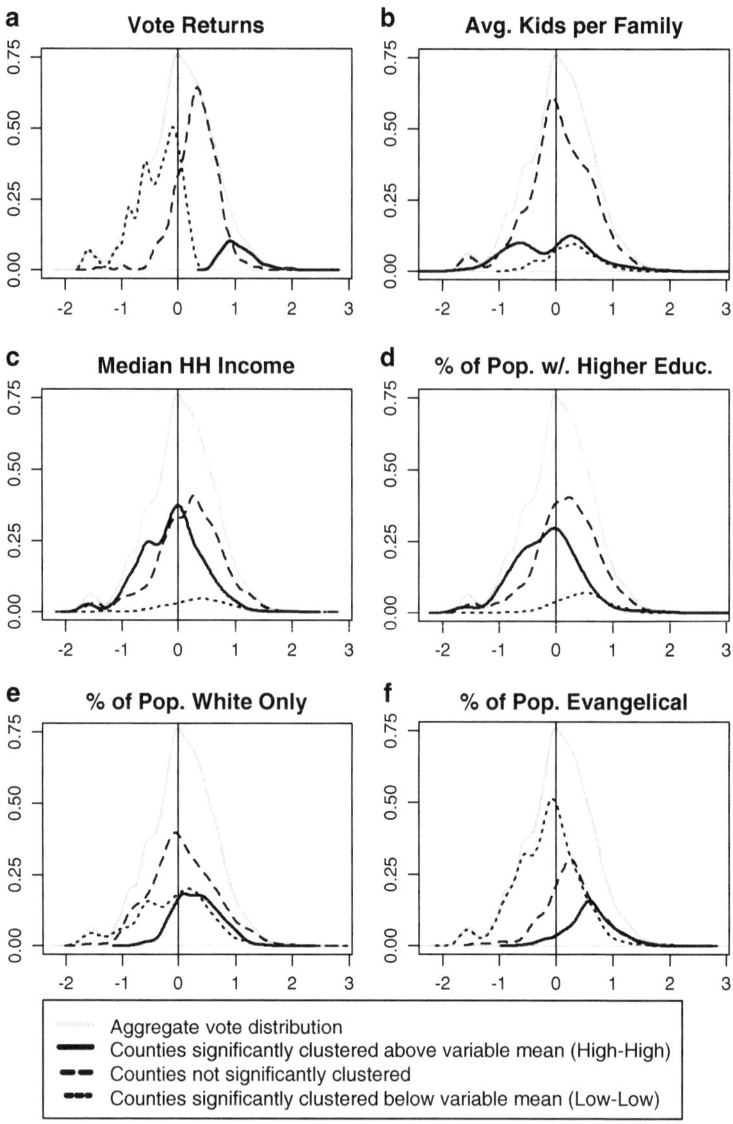

Fig. 4 Log-odds of voting republican by county clusters

larger-than-average proportions of evangelical adherents is strongly associated with voting Republican. This influence is less strong but still substantial for counties where clustering is not significant, are the large number of voters in clusters of counties with lower-than-average populations of evangelical Christians seem to have centrist or slightly Democratic preferences. Qualitatively similar patterns emerge on the basis of indicators local geographic context (not shown). The vote distributions for urban and non-urban counties are markedly different, with rural, and especially suburban, contexts exhibiting a strong propensity to vote Republican (cf. McKee 2007, 2008; Williamson 2008).

Overall, the spatial agglomeration in both voting returns and selected covariates is broadly consistent with the predictions of our local entrenchment hypothesis. But even though agglomeration appears to be somewhat related to divisions in voting behavior, the precise association is not readily discernable. With the exception of panel F, Fig. 4 clearly indicates the dominance of the non-clustered subsample's influence on the aggregate vote distribution. Thus, although entrenchment might well be occurring, there is no polarization of electoral returns across easily observable demographic segments of counties' populations, as anticipated by Fiorina and Abrams (2008).[23] But the key issue is whether, and if so, how, entrenchment might be affecting the propensity to vote, controlling for demographic characteristics. Our ability to draw inferences in this regard is limited by the univariate character of Fig. 4's distributions, which fails to capture the simultaneous influences of multiple spatially clustered variables on the vote distribution. To address this issue we turn to our regression model, which rigorously establishes the statistical associations between the vote distribution and all of our covariates, controlling for the myriad patterns of spatial clustering in the data.

5.2 Aggregate-Level Regression Results

Our aggregate-level estimation results are summarized in Table 1. For our preferred specification, a Lagrange multiplier test of the spatial Durbin model's errors did not indicate significant residual spatial autocorrelation ($LM = 2.6$, $p > 0.10$), which in our opinion vindicates our statistical approach. The spatial autoregressive parameter is positive and significant ($\rho = 0.3$, $p < 0.01$), and its magnitude suggests that the spatial clustering of voting behavior is accompanied by substantial endogenous effects, even after demographic and contextual influences are controlled for.[24] Moreover, with few exceptions the contextual influences associated with spatial lags of the covariates share the same sign as their direct counterparts. Thus, although

[23] For example, as in racially polarized voting, where whites and non-whites have divergent ideological preferences which push their vote distributions in opposite directions away from the mean, like B-II and B-III in Fig 1**b**.

[24] Our results suggest that the "social multiplier" associated with voting in the 2004 U.S. presidential election is around 1.4. This is substantially smaller than the values found by Glaeser et al. (2003) for the peer effects of college roommates, criminal behavior in cities, or the human capital spillovers in urban labor markets. This outcome is not surprising given that ballots are secret, and that even with early voting, individuals are only exposed to the influence of neighbors' self-announced behavior for at most three weeks. (Although more prolonged exposure might result from proximity to intensely partisan voters.) It therefore seems more plausible that ρ is picking up the influence of correlated effects associated with counties' common exposure to political campaigns, and the reflection of that stimulus in their residents' everyday social interactions.

Table 1 Spatial Durbin model results

	Direct effects (β)	Spatial lag effects (γ)
% Fam. inc. <$15k	−0.010 (0.034)	−0.543 (0.233)**
% Fam. inc. $15–35k	0.193 (0.058)***	0.592(0.371)
% Fam. inc. $35–75k	−0.363 (0.063)***	−1.300 (0.413)***
% Fam. inc. $75–150k	−0.095 (0.033)***	0.184(0.241)
%<9th grade	0.064 (0.023)***	0.198(0.126)
% Some high school	0.255 (0.035)***	−0.211(0.212)
% High school grad.	0.512 (0.059)***	0.404(0.352)
% Some college	0.473 (0.052)***	0.565 (0.324)*
% Bachelor's degree	0.091 (0.031)***	0.332 (0.192)*
% White only	0.953 (0.037)***	−0.322 (0.153)**
% Latino	0.009 (0.010)	0.086 (0.045)*
% Foreign-born	−0.006 (0.009)	−0.159 (0.049)***
% Fem. H. H. No husb.	−0.646 (0.037)***	−0.040(0.217)
Avg. family size	1.478 (0.380)***	−0.989(2.326)
Avg. kids per family	0.426 (0.122)***	1.364 (0.781)*
% Veterans	−0.068 (0.039)*	0.217(0.214)
% Evangelical	0.003 (0.005)	0.080 (0.029)***
% Mainline protestant	0.006 (0.006)	−0.153 (0.041)***
% Catholic	−0.004 (0.002)*	0.016(0.022)
% Unclaimed	0.027 (0.012)**	0.232 (0.063)***
Median HH inc.	0.383 (0.092)***	0.314(0.498)
% Unemployment	−0.079 (0.018)***	0.012(0.085)
% Workforce agric.	0.026 (0.008)***	−0.106 (0.052)**
% Workforce mfg.	0.038 (0.010)***	0.071(0.053)
% of nat'l electorate	−0.005(0.009)	−0.227 (0.056)***
% Work outside cnty.	−0.025 (0.012)**	−0.293 (0.081)***
Avg. travel time	−0.276 (0.040)***	−0.474 (0.258)*
Pop. change 2000–03	0.646 (0.166)***	1.314(0.907)
Suburban county	−0.070 (0.017)***	−0.236(0.183)
Rural county	−0.052 (0.015)***	−0.099(0.141)
GMB	−0.063(0.182)	−1.735 (0.396)***
GMB × % Evangelical	0.473 (0.079)***	0.797 (0.438)*
ρ	0.297 (0.063)***	
LR test	18.759 ***	
Log likelihood	−43.62	
ML residual variance	0.060	
AIC	405.23	
LM resid. autocorrelation	2.622	

The dependent variable is the log odds of voting Republican (eq. 2); asymptotic standard errors in parentheses; * $p < 0.1$, ** $p < 0.05$, *** $p < 0.01$

the various characteristics of populations and places influence county-level voting returns in different directions, these effects are almost uniformly amplified by their geographic context. These, we argue, are powerful pieces of evidence in support of the local entrenchment thesis.

Income, Income Distribution and Housing: Counties with higher median household incomes are significantly more likely to vote Republican, consistent with Kim et al. (2003). Simultaneously, however, having high proportions of families with moderately high and particularly middle incomes (\$35–\$150,000) significantly lowers the odds of voting Republican, as does proximity to larger populations of the poorest families (<\$15,000). Larger shares of low-income households (\$15–\$35,000) have the reverse effect. The fact that we drop the proportion of families with the highest income (>\$150,000) from the regression to avoid collinearity then suggests that the propensity to vote Republican varies with income according to a U-shaped distribution. Once the spatial dependence in the data is accounted for, housing values do not have a significant effect.

Education: Surprisingly, larger proportions of persons at all levels of educational attainment are significantly associated with higher odds of voting Republican. As before, we interpret this result in light of the fact that we drop the proportion of highest-attaining persons (those with postgraduate training) to avoid collinearity. The suggestion is that the propensity to vote Republican varies with education according to an inverted U-shaped distribution, with larger proportions of very low and very high attaining individuals substantially reducing the propensity to vote Republican. Interestingly, contextual influences amplify these forces in both directions, with the odds of voting for Bush reinforced by clustering of individuals with some post-secondary education and attenuated by clustering of college graduates.

Race and Ethnicity: The influences of Asian- and African-American populations were tiny and not statistically significant, which led us to drop these variables from the model. Proximity to higher proportions of persons of Latin American origin is associated with increased odds of voting Republican. The proportion of persons reporting purely Caucasian origins has a similar influence, but its magnitude is an order of magnitude larger. Interestingly, the coefficient on the spatial lag of this sub-population has the opposite sign. The likely reason is that core metro counties, which on average had larger minority populations, tended to vote for Kerry in significantly higher numbers relative to their surrounding suburban counties, which had a significantly higher proportion of white residents.[25]

Age, Sex and National Origin: The percentages of the elderly and voting age females in the population did not appear to significantly influence the 2004 vote. However, proximity to clusters of persons born outside the United States had a small negative impact on the odds of voting Republican.

Moral/Family Values: The effect of the share of divorced or separated individuals in the population was not significant, in line with findings of the broad acceptance of this social phenomenon (e.g., Thornton and Young-DeMarco 2001). The proportion

[25] Kruskal-Wallis rank-sum tests indicated significant differences between suburban and core metro counties' distributions of the vote and the proportion of the population self-identifying as white only.

of households headed by a single female was strongly associated with lower odds of voting for Bush, while the corresponding spatial lagged variable is not significant. This result appears less consistent with the "values-voter" hypothesis than with interest group behavior by poor single mothers, who, as the principal beneficiaries of the American welfare system (Gensler 1996) were directly impacted by Republican-initiated conservative social policies such as accelerated welfare-to-work transitions (see, e.g., Allard 2007). The direct effects of higher average fertility and, especially, larger family sizes were significant and positive. The odds elasticities for family size and proximity to large populations of children are both particularly large, and the similarity in their magnitudes is not surprising given these variables' high correlation.

Iraq/War on Terror: Our proxies for the spatial distribution of attitudes to U.S. foreign policy perform poorly. The proportion of active duty personnel in the population is not significant, while the effect of the proportion of veterans is positive and significant, but small. Thus, bearing in mind the significant limitations of our data, we find little evidence that in 2004 security concerns trumped values in influencing voter behavior (cf. Hillygus and Shields 2005).

Religious Affiliation: We do not find that the shares of adherents to various religious denominations and substantially increase the odds of voting Republican. The share of Catholics and the spatially lagged percentage of mainline Protestants in the population both have small, negative and significant effects, while the spatial lag of the proportion of Evangelicals is significant, positive and not as large, and the fraction of persons with no religious affiliation is significantly positive in the spatial lag and spatial error models. These results are consistent with previous evidence of low turnout among conservative protestants (Manza and Brooks 1997; Woodberry and Smith 1998), as well as the conclusion that religious issues on their own made little difference to the outcome of the election (Hillygus and Shields 2005; Campbell and Monson 2008).

Employment: High unemployment rates are associated with significant reductions in the odds of voting Republican (consistent with Kim et al. 2003), while the fractions of the workforce in agriculture and manufacturing both have the opposite effect. As well, the coefficient on the spatial lag of agricultural employment is negative, which appears to reflect the fact that suburban counties have significantly lower agricultural employment than their surrounding rural counties without significant differences in their vote distributions.[26]

Local Geographic Factors: The rate of population increase has a large, positive and significant effect of the odds of voting for Bush, whereas travel time to work, the fraction of population commuting outside the county, being a suburban or rural county, or neighboring a more populous county all have smaller negative impacts. The association between a higher propensity to vote Republican voting behavior and rapid growth of the population reflects the effect of migration on spatial sorting along ideological lines, and is consistent with Gimpel and Schuknecht's (2001)

[26] Kruskal-Wallis tests indicated significant differences between suburban and rural counties' distributions of the proportion of jobs in agriculture, but not their voting patterns.

finding that interstate in-migration has aided Republicans whereas out-migration has aided democrats. The influence of commute time may simply reflect the fact that voters in the highly urbanized and strongly Democratic areas of the northeast and the west coast live closer to where they work and thus enjoy shorter commutes. But it also suggests a higher probability of counties' residents being exposed to social contexts that are potentially differ from their own neighborhoods, with consequent inward diffusion of a diversity of political ideas and beliefs (cf. our discussion of process 4 in the previous section), which would tend to mitigate local entrenchment. The sign of the effects of suburban and rural dummies is at odds with previous findings (McKee 2007, 2008; Williamson 2008), indicating that once contextual and endogenous factors are controlled for, the environment in these types of locales does not appear to strongly increase – or, for that matter, reduce – the propensity to vote Republican.

Gay Marriage Bans: Surprisingly, we find that at the national level, the GMB dummy has a consistently negative, but indirect, impact on the odds of voting Republican.[27] Nevertheless, the interaction between the GMB dummy and the evangelical population has strongly positive direct and indirect effects. These two results, taken together with our prior finding that the concentration of evangelical adherents appears to have little effect on its own, are consistent with Campbell and Monson's (2008) conclusion that the state ballot initiatives to ban same-sex marriage served to increase turnout among moral conservatives. If one compares the magnitudes of the two sets of coefficients, the most interesting feature is the suggestion that at the aggregate level the ballot initiatives may have provoked a backlash which was big enough to compensate for its positive (direct and indirect) influences on evangelical turnout.

These results confirm both the existence and importance of effects that are consistent with local entrenchment. Furthermore, the fact that the prevalence of minority populations and our proxies for voters' economic status and orientation on moral values end up having the strongest influence suggests that these are key dimensions along which there was significant segmentation of the American electorate in the 2004 election. Nevertheless, our conclusions are tempered by the fact that the elasticities in Table 1 are national averages that do not indicate how these divisions might have played out spatially. To shed light on this question we turn to the results of our GWR analysis.

6 Geographically Weighted Regression

Our GWR results are summarized in Fig. 5. Estimation of the model was plagued by multicollinearity between the average number of kids and the average family size by county, which led us to drop the former variable from our specification. The residual

[27] Given our coding of GMB as a state dummy variable, the significance of the indirect spatial lag (as opposed to the direct) coefficient is to be expected, as it is by definition a wide-area effect.

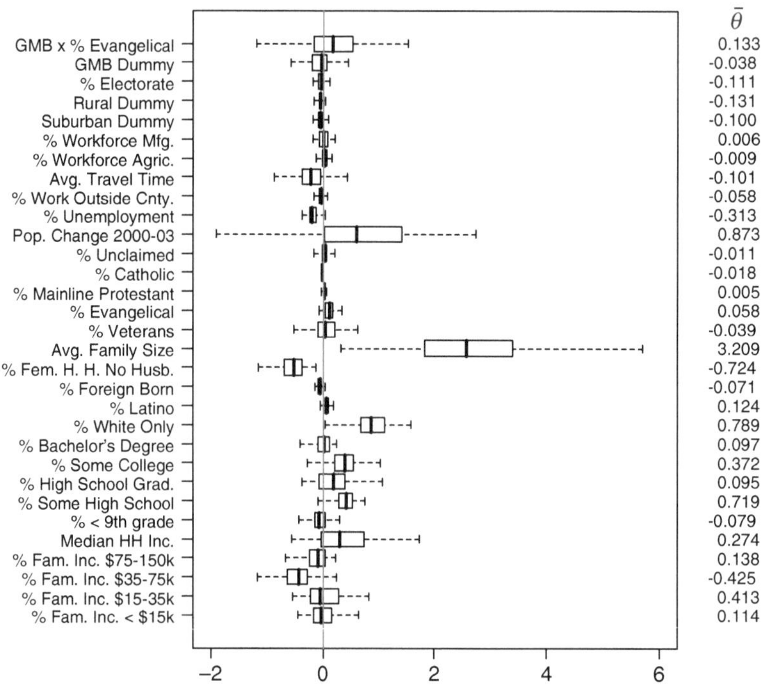

Obs.	3106	σ^2(ML)	0.049
Bandwidth	281.27	AICc	481.461
Effective no. of parameters	604.83	AIC	-120.04
Effective degrees of freedom	2501.17	Resid. sum of squares	151.48

Notes: the dependent variable is the log odds of voting Republican (eq. 2).

Fig. 5 Geographically weighted regression results

variance and the AIC statistic indicate an improvement in the fit over the spatial Durbin model above, and the distribution of local R^2 values, which ranges from 0.47 to 0.97 with a mean of 0.73, suggests that the GWR model's overall ability to account for the local spatial variation in the dependent variable is quite good. Our optimal bandwidth estimated through crossvalidation is substantially larger than the one used to compute the weights in the spatial Durbin model, with the result that the global values of the GWR parameters $(\bar{\theta})$ differ slightly from the odds elasticities of the previous section.[28] Even so, the overall results are basically the same. With the exception of family size, all of the parameters are less than one in

[28] Globally, the signs and relative magnitudes of the estimates are similar. However, the magnitudes of almost half of the estimates shrink while the rest increase. The median values of the parameter distributions are in closer agreement with the signs of our spatial Durbin estimates, though slight differences in their magnitudes persist.

absolute magnitude, which suggests that on average most covariates do not exert overwhelmingly large effects on the electoral returns.

Looking beyond averages, we see that the most important features of the county-level results are the variance of the parameter distributions, and the considerable spatial heterogeneity in the magnitude and sign of the influences of our covariates on the election returns. Our estimates can be classified into three types: (i) variables whose overall impact is large enough to be definitively signed, especially population change, household income and family size, (ii) those whose global impact is negligible but whose spatial variation is large, such as the proportions of veterans and very poor households, and GMB and its interaction term, and (iii) those whose average impact and its cross-county variance are both small – the category into which the majority of variables falls.

None of the covariates whose average estimates are significantly positive or negative exhibits tight clustering of their effects on individual counties' propensity to vote. Furthermore, once we control for the influences of other attributes, GWR does not produce not a simple relationship between the cross-county variation in a particular factor (i.e., column of X) and the variation exhibited by that factor's odds elasticities (i.e., the corresponding column of θ_c in (6)). These results are consistent with our previous findings, and reinforce the point that the electorate is not polarized along easily observable demographic lines. While characteristics such as median household income, average commute time, and the proportions of Caucasians, persons in middle income families, and households headed by single females come closest in this regard, with average effects that are large and either significantly positive or negative in at least three-quarters of our sample, even they exhibit substantial non-stationarity.

Category-(i) covariates (above) which seem most likely to shift the vote in a particular direction exert the opposite impact in a substantial minority of counties, and category-(ii) covariates exhibit effects of similar intensity but opposite sign in equally large numbers of counties. Due to the local character of the regression, these attributes are the ones for which the dependence of counties' odds of voting Republican on similar propensities among their neighbors will be the most obvious.

To test this proposition we examine the patterns of agglomeration in the odds elasticities with large spatial variation. Our strategy is to once again compute local Moran's I statistics, this time for the vector of parameters associated with each covariate (θ_c), and, as in Fig. 3, display the results as a series of significance maps.[29] The results appear in Fig. 6, which illustrates the striking spatial trends of agglomeration in our estimates. The effect of each of the six characteristics on the propensity of a particular county to vote Republican or Democratic depends on how that attribute influences the ideological leaning of its neighbors, which exert an amplifying effect within the clusters. While our GWR model is not able to identify either the precise channels through which these feedbacks operate, or how their

[29] In conducting these analyses we employ our original 200-km bandwidth kernel.

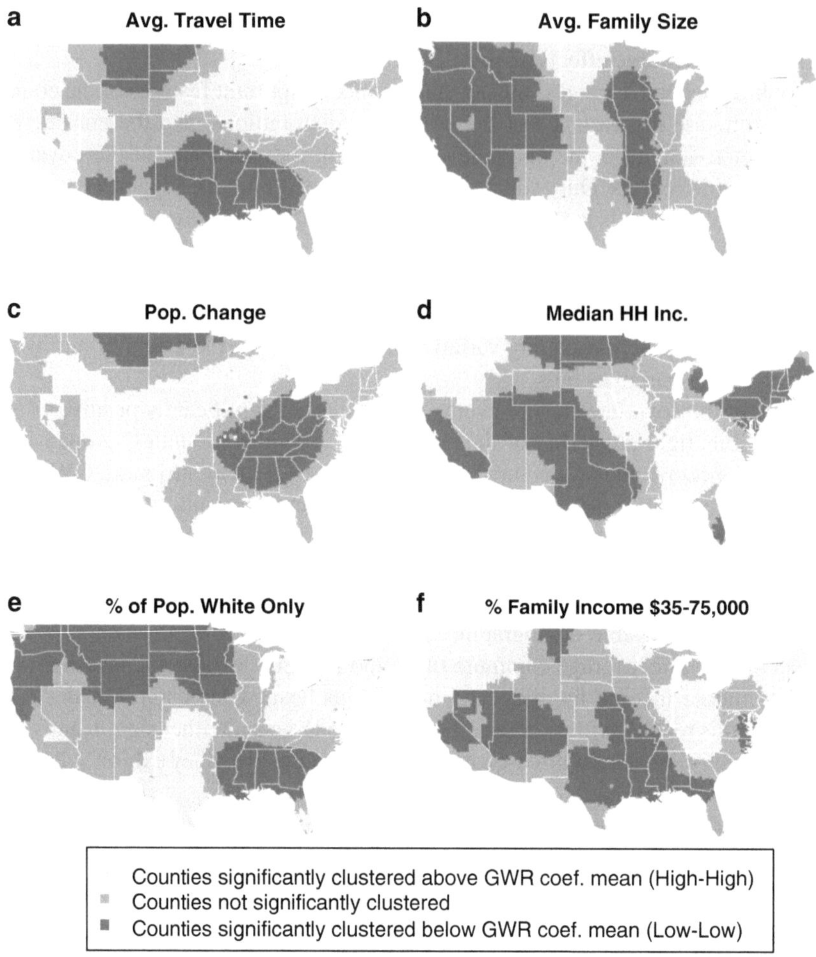

Fig. 6 Local Moran's I significance maps of GWR odds elasticities

signs and intensities vary, it is nonetheless clear that the clustering in Fig. 6 reflects the influences of the spatially lagged variables seen in the previous section.

It is a challenge to even describe – not to mention intuitively account for the origins of – these patterns (Sue Wing and Walker 2005). Accordingly, we move directly to comparing the distributions of the elasticity values for counties within and outside the various clusters in Fig. 6. The latter results are summarized in Fig. 7 as kernel density estimates of the distributions of odds elasticities that are weighted by counties' shares of the national electorate and segmented according to their propensity to cluster significantly above or below their respective means. The distributions bear a striking resemblance to Fig. 1b, especially the influence of family size (panel B),

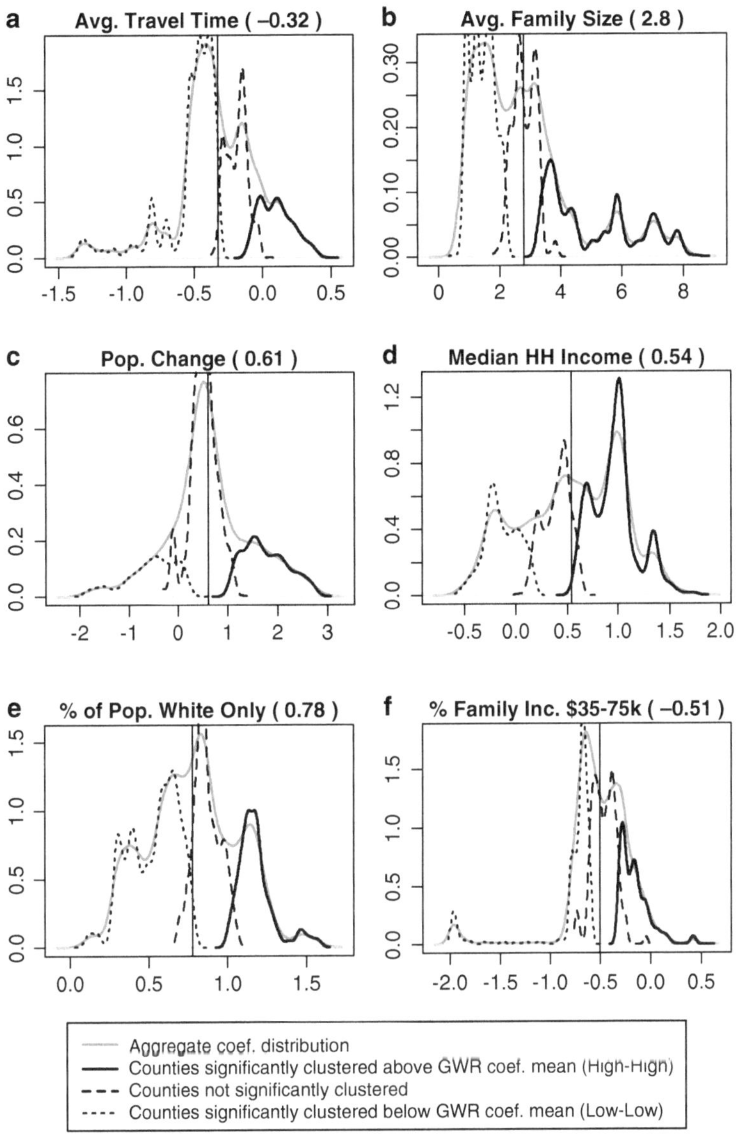

Fig. 7 GWR odds elasticities of voting republican by county clusters

for which 54% of the electorate tended to cluster below the mean and 17% clustered above, and median household income (panel D), for which the corresponding proportions are 24 and 48%. We note that these variables have multimodal aggregate distributions with fairly large variances, especially the effect of family size, with its long upper tail. The distributions of spatially-clustered elasticities for commute time and the proportion of Caucasians (panels A and E) are also bimodal, with

18% (21%) of the probability mass in the former (latter) case clustered above the mean and 55% (46%) clustered below. By contrast, the distributions for population change and the proportion of middle income families (panels C and F) exhibit a greater degree of central tendency, with the majority of voters in residing counties that are not significantly clustered in one way or the other.

These results support our hypothesis that local entrenchment is associated with polarization of the electorate. We uncover similar evidence in the distributions of spatial clustering of other, less wide-ranging odds elasticities, but we leave the elaboration of these details to future work. Our findings underscore the subtle point that polarization is an inherently multidimensional phenomenon. Stepping back, it is clear that the overall picture is not as simple as the one articulated in popular discourse – we show that despite the fact that votes for different parties cluster regionally, they are not concentrated in disjoint subsets of the electorate. Rather, along dimensions such as race, income, and indicia of family values the United States appears to be divided into disjoint swaths of geographically contiguous counties, with the same attribute amplifying the propensity to vote Republican in one set of regions while simultaneously exerting the opposite influence in another.

Given this, it is easy to see why a simple pattern of red and blue states does not arise: it is not the case that the same counties cluster above or below the mean along all, or even most, dimensions of the space of characteristics. A particular county's odds elasticities might be in a "high-high" cluster in some dimensions, while in other dimensions they might be in a "low-low" cluster – or not belong to a cluster at all. The county's ultimate propensity to vote one way or another is the scalar product of these varying local odds elasticities and its actual characteristics (i.e., $X_c\theta_c$), which generally differs from the influence that an individual characteristic (given by the relevant element of θ_c) might have. The implication is that in the U.S. context, electoral polarization should be thought of as a series of cross-cutting divisions that manifest themselves not between population sub-groups but within individual sub-groups over space.

We close by qualifying this conclusion with an important caveat. Wheeler and Tiefelsdorf (2005) find that the GWR algorithm can potentially induce spatial bias in the local parameter estimate that is sufficiently large to invalidate their meaningful interpretation.[30] The considerable spatial dependence in our GWR results might then lead one to question the extent to which we are able to trust the local values of the odds elasticities, and in particular their patterns of clustering which are the basis of our inferences about polarization. To address this issue, we follow Wheeler and Tiefelsdorf's (2005) recommendations and examine the extent of local and global

[30] These authors find that a simple model with two independent variables, the coefficients associated with each covariate may exhibit collinearity even if the underlying exogenous variables in the data generating process are uncorrelated, and a high degree of spatial correlation between two covariates increases the potential for the two sets of coefficients to exhibit interdependent, spatially opposing patterns of effects. In both cases the upshot is spurious spatial trends in the GWR estimates.

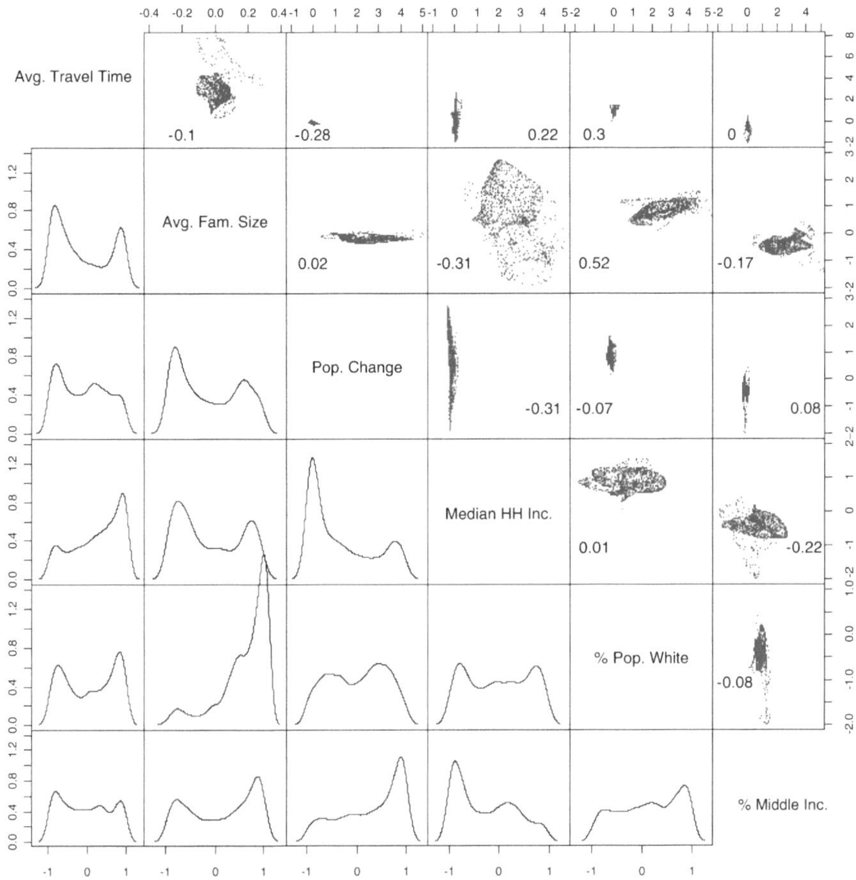

Fig. 8 GWR odds elasticities: global and local correlations

correlation among the six sets of odds elasticities in Figs. 6 and 7. The results of our robustness tests are summarized in Fig. 8 as scatterplots of the coefficient estimates (upper panels) and distributions of their local correlations (lower panels). The odds elasticities are not globally correlated, but there are indications of correlation at the local level, particularly between the effects of average family size and the proportion of Caucasians in the population, and population change and median household income. These tests are not conclusive. But in the absence of strong prima facie evidence of bias we are confident that our results stand. In any event, there is no easy way to remedy the effects of spatial multicollinearity within the analytic framework developed here. Quite likely, efforts in this regard will require an entirely separate program of analysis and testing (e.g., along the lines of Wheeler 2007). The best we can do given the constraints of available space is to flag this issue as a priority for future research.

7 Conclusion

This paper sheds new light on the fundamental role of geography in determining both the outcome of the 2004 U.S. presidential election, and the polarized character of the American electorate more generally. Our guiding hypothesis is that polarization of the U.S. electorate has occurred over space and is attributable to a process of local entrenchment, whereby a variety of social forces amplify county populations' propensity to vote Republican or Democratic.

Analyzing data from a large sample of counties in the lower 48 states, we find the influences on voting behavior associated with contextual and endogenous factors to be broadly consistent with the predictions of our thesis, with considerable spatial clustering in both electoral returns and the characteristics of populations and places clearly pointing to the amplification of the effects on the vote of the attributes of populations and places. A much richer picture emerges when we explicitly account for the geographic variations in these estimates. At the global level, our GWR odds-elasticities basically agree with the results of our aggregate-level analyses, while at the local level exhibiting substantial heterogeneity in both magnitude and sign, and strong spatial trends. The latter imply that in the U.S. context electoral polarization is not synonymous with segmentation across population sub-groups following observable demographic characteristics. Rather, polarization appears to be a phenomenon which occurs within individual sub-groups across space. Furthermore, geography matters in ways that are crucial, but not easily explained using aggregate data analysis. It is not simply the case that the spatial distribution of population characteristics drives the interregional differences of voting patterns observed in the 2004 presidential elections. Rather, the latter emerge from the reinforcing influence of the local social context on the effects of the racial composition, income, and, less tangibly, social values of counties' populations.

Our hope is that this study will motivate geographers and political scientists alike to employ disaggregate individual data to account for the detailed social mechanisms that give rise to these broad spatial trends.

References

Abramowitz A, Saunders K (2008) Is polarization a myth? J Polit 70:542–555

Allard, SW (2007) The changing face of welfare during the Bush administration. Publius 37: 304–332

Ansolabehere S, Rodden J, Snyder JM (2006) Purple America. J Econ Perspect 20:97–118

Anselin L (2002) Under the hood: issues in the specification and interpretation of spatial regression models. Agric Econ 27:247–267

Anselin L, Bera AK, Florax RJGM, Yoon MJ (1996) Simple diagnostic tests for spatial dependence. Reg Sci Urban Econ 26:77–104

Baldassarri D, Bearman P (2007) Dynamics of political polarization. Am Sociol Rev 72:784–811

Baldassarri D, Gelman A (2008) Partisans without constraint: political polarization and trends in American public opinion. Am J Sociol 114:408–446

Bartels LM (2000) Partisanship and voting behavior, 1952–1996. Am J Polit Sci 44:35–50

Bivand R (2006) Implementing spatial data analysis software tools in R. Geogr Anal 38:23–40

Bivand R, Brunstad R (2006) Regional growth in Western Europe: detecting spatial misspecification using the R environment. Pap Reg Sci 85:277–297
Bishop W (2008) The big sort: why the clustering of like-minded America is tearing us apart. Houghton-Mifflin, New York
Bishop W, Cushing RG (2004) Response to Philip A. Klinkner's "red and blue scare: the continuing diversity of the American electoral landscape." Forum 2(2):Art 8
Brunsdon C, Fotheringham AS, Charlton ME (1996) Geographically weighted regression: a method for exploring spatial non-stationarity. Geogr Anal 28:281–298
Bullock CS, Hoffman DR, Gaddie RK (2005) The consolidation of the white southern congressional vote. Polit Res Q 58:231–243
Burridge P (1981) Testing for a common factor in a spatial autoregression model. Environ Plann A 13:795–800
Campbell DE, Monson JQ (2008) The religion card: gay marriage and the 2004 presidential election. Public Opin Q 72:399–419
Cho WKT, Rudolph TJ (2008) Emanating political participation: untangling the spatial structure behind participation. Br J Polit Sci 38:273–289
DiMaggio P, Evans J, Bryson B (1996) Have Americans' social attitudes become more polarized? Am J Sociol 102:690–755
Dixit AK, Weibull JW (2007) Political polarization. Proc Natl Acad Sci 104:7351–7356
Evans JH (2003) Have Americans' attitudes become more polarized? – an update. Soc Sci Q 84: 71–90
Evans JH, Nunn LM (2005) The deeper "culture wars" questions. Forum 3(2):Art 3
Fiorina MP, Abrams SJ, Pope JC (2006) Culture war? the myth of a polarized America, 2nd edn. Longman, New York
Fiorina MP, Abrams SJ (2008) Political polarisation in the American Public. Ann Rev Polit Sci 11:563–588
Fleisher R, Bond J (2004) The shrinking middle in the U.S. congress. Br J Polit Sci 34:429–451
Fotheringham AS, Brunsdon C, Charlton M (2002) Geographically weighted regression: the analysis of spatially varying relationships. Wiley, Chichester
Fotheringham AS, Charlton ME Brunsdon C (1997) Measuring spatial variations in relationships with geographically weighted regression. In: Fischer MM, Getis A (eds) Recent developments in spatial analysis. Springer, London, pp 60–82
Frank T (2004) What's the matter with Kansas? How conservatives won the heart of America. Metropolitan Books, New York
Frey WH (2000) Regional shifts in America's voting-aged population: what do they mean for national politics? Population Studies Center Report No. 00–459, Institute of Social Research, University of Michigan
Gensler H (ed) (1996) The American welfare system: origins, structure, and effects. Praeger, Westport
Gimpel JG, Schuknecht JE (2001) Interstate migration and electoral politics. J Polit 63:207–231
Glaeser EL, Ward BA (2006) Myths and realities of American political geography. J Econ Perspect 20:119–144
Glaeser EL, Sacerdote BI, Scheinkman JA (2003) The social multiplier. J Eur Econ Assoc 1: 345–353
Glaeser EL, Sunstein CR (2009) Extremism and Social Learning. J Legal Anal 1:263–324
Glenmary Research Center (2004) Religious congregations and membership in the United States: 2000. Nashville
Goodman L (1953) Ecological regression and the behavior of individuals. Am Sociol Rev 18: 663–664
Hanushek EA, Jackson JE, Kain JF (1974) Model specification, use of aggregate data, and the ecological correlation fallacy. Polit Method 1:87–107
Hetherington MJ (2001) Resurgent mass partisanship: the role of elite polarization. Am Polit Sci Rev 95:619–631

Hillygus DS, Shields TG (2005) Moral issues and voter decision making in the 2004 presidential election. PS: Polit Sci Pol 38:201–210
Huckfeldt R, Sprague J (1995) Citizens, politics, and social communication: information and influence in an election campaign. Cambridge University Press, New York
Hunter JD (1992) Culture wars: the struggle to define America. Basic Books, New York
Johnston R, Jones K, Sarker R, Propper C, Burgess C, Bolster A (2004) Party support and the neighbourhood effect: spatial polarisation of the British electorate, 1991–2001. Polit Geogr 23:367–402
Kim J, Elliott E, Wang DM (2003) A spatial analysis of county-level outcomes in U.S. Presidential elections: 1988–2000. Elect Stud 22:741–761
Klinkner PA (2004) Red and blue scare: the continuing diversity of the American electoral landscape. Forum 2(2):Art 2
Klinkner PA, Hapanowicz A (2005) Red and blue Déjà Vu: measuring political polarization in the 2004 election. Forum 3(2):Art 2
Klofstad C, McClurg SD, Rolfe M (2006) Family members, friends, and neighbors: differences in personal and political networks. Paper presented at the 2006 annual meeting of the The Midwest Political Science Association, Palmer House Hilton, Chicago, IL
Lazarsfeld P, Berelson B, Gaudet H (1944) The people's choice: how the voter makes up his mind in a political campaign. Columbia University Press, New York
Manski CF (1993) Identification of endogenous social effects: the reflection problem. Rev Econ Stud 60:531–542
Manski CF (2002) Economic analysis of social interactions. J Econ Perspect 14:115–136
Manza J, Brooks C (1997) The religious factor in US presidential elections, 1960–1992. Am J Sociol 103:38–81
Marchant-Shapiro T, Patterson KD (1995) Partisan change in the mountain west. Polit Behav 17:359–378
McClurg SD (2003) Social networks and political participation: the role of social interaction in explaining political participation. Polit Res Q 56:449–464
McKee SC (2007) Rural voters in presidential elections, 1992–2004. Forum 5(2):Art 2
McKee SC (2008) Rural voters and the polarization of American presidential elections. Polit Sci Pol 41: 101–108
McMillen DP (2003) Spatial autocorrelation or model misspecification? Int Reg Sci Rev 26: 208–217
McPherson M, Smith-Lovin L, Cook JM (2001) Birds of a feather: homophily in social networks. Annu Rev Sociol 27:415–444
Mellow N, Trubowitz P (2005) Red versus blue: American electoral geography and congressional bipartisanship: 1898–2002. Polit Geogr 24:659–677
Miller AS, Hoffmann JP (1999) The growing divisiveness: culture war or a war of words? Soc Forces 78:721–745
Mutz DC (2002) The consequences of cross-cutting networks for political participation. Am J Polit Sci 46:838–855
Mutz DC, Mondak JJ (2006) The workplace as a context for cross-cutting political discourse. J Polit 68:140–155
O'Loughlin J, Flint D, Anselin L (1994) The geography of the nazi vote: context, confession and class in the Reichstag election of 1930. Ann Assoc Am Geogr 84:351–380
Openshaw S (1984) The modifiable areal unit problem. Geo Books, Norwich
Pew Research Center (2004) Moral values: how important? voters liked campaign 2004, but too much "mud-slinging." Pew Research Center for the People and the Press Report, 11 November
Poole KT, Rosenthal H (2001) D-NOMINATE after 10 years: a comparative update to congress: a political-economic history of roll-call voting. Legis Stud Q 26:5–29
Schreckhise WD, Shields TG (2003) Ideological realignment in the contemporary U.S. electorate revisited. Soc Sci Q 84:596–612
Speel RW (1998) Changing patterns of voting in the northern United States: electoral realignment, 1952–1996. Pennsylvania State University Press, Philadelphia, PA

Stonecash JM, Brewer MD, Mariani MD (2002) Diverging parties: social change, realignment, and party polarization. Westview Press, Boulder

Sue Wing I, Walker JL (2005) The 2004 presidential election from a spatial perspective. Mimeo, Boston University

Thornton A, Young-DeMarco L (2001) Four decades of trends in attitudes toward family issues in the United States: the 1960s through the 1990s. J Marriage Fam 63:1009–1037

Valentino NA, Sears DO (2005) Old times there are not forgotten: race and partisan realignment in the contemporary south. Am J Polit Sci 49:672–688

Wheeler DC (2007) Diagnostic tools and a remedial method for collinearity in geographically weighted regression. Environ Plann A 39:2464–2481

Wheeler DC, Tiefelsdorf M (2005) Multicollinearity and correlation among local regression coefficients in geographically weighted regression. J Geogr Syst 7:161–187

Williamson T (2008) Sprawl, spatial location, and politics: how ideological identification tracks the built environment. Am Polit Res 36:903–933

Woodberry RD, Smith CS (1998) Fundamentalism et al.: conservative protestants in America. Annu Rev Sociol 24:25–56

Gender Wage Differentials and the Spatial Concentration of High-Technology Industries

Elsie Echeverri-Carroll and Sofía G. Ayala

1 Introduction

Moretti (2004) finds that the distribution of human capital across cities in the United States became more unequal during the 1990s. He believes that one reason for the increased concentration of human capital in some metropolitan areas was the high-tech boom of that decade, since it benefited a handful of already highly skilled cities. This trend reflects the decisions of skilled workers and the skill-intensive industries that employed them to colocate in the same cities or regions (high-tech clusters). Zucker et al. (1998), for instance, find that the entry decisions of new biotechnology firms in cities depends on the stock of human capital in outstanding scientists there, as measured by the number of relevant academic publications. Colocation benefits workers (who enjoy the productivity-enhancing effects associated with local learning processes) as well as high-tech firms (which profit from highly productive and creative workers who enhance the firms' innovation processes).

The primary cooperative linkages in high-technology clusters are those related to knowledge exchange. As Fingleton (2004) note, sharing knowledge is the key to the generation and maintenance of innovation flows that are particularly relevant in these clusters. A strong evidence of the learning networks-innovation relationship comes from studies showing that patents (a proxy for innovations) are more likely to emerge from the same states or metropolitan areas as the cited patents than one would expect based in the preexisting concentration of related research activity (Jaffe et al. 1993).

Several previous empirical studies show that workers are more productive and make higher wages in cities with a large concentration of human capital (Rauch 1993; Echeverri-Carroll and Ayala 2004, 2006; Glaeser and Maré 2001; Acemoglu and Angrist 2000; Ciccone and Peri 2006; Moretti 2004). These studies attribute the higher productivity of workers in cities rich in human capital to knowledge externalities that arise when the presence of educated workers makes other workers more

E. Echeverri-Carroll (✉)
IC2 Institute, University of Texas at Austin, 2815 San Gabriel, Austin, TX 78705, USA,
e-mail: e.carroll@mail.utexas.edu

A. Páez et al. (eds.), *Progress in Spatial Analysis*, Advances in Spatial Science,
DOI 10.1007/978-3-642-03326-1_14,

productive. Marshall (1890) was among the first to recognize that social interactions among workers create learning opportunities that enhance productivity.

As far as we are aware, all the work to date has considered the effect of a city's human-capital externalities on male wages only (Yankow 2006; Glaeser and Maré 2001; Acemoglu and Angrist 2000; Echeverri-Carroll and Ayala 2009), or on wages of a sample of workers of both genders combined (Rauch 1993; Glaeser et al. 1992; Black and Henderson 1999; Ciccone and Peri 2006; Moretti 2004). In the present study, we examine the effect of human-capital externalities separately for women and men in the United States on the basis of data from the 2000 Census of Population.

This chapter presents empirical evidence of the effects on wages of living in a city with a large endowment of human capital separately for college-educated men and women. A key difference of our study from previous ones is that we focus on human-capital externalities in high-tech relative to low-tech cities. Our interest is on technology-oriented learning processes that affect technical innovations. On the contrary, previous studies proxy human-capital externalities by variables such as the average level of education in the city, capturing the knowledge externalities occurring in any field – including those that are not technology oriented (such as music, theater, English, or cooking).

Three questions interest us. First, do wage differences between high-tech and low-tech cities exist for female workers as they do for male workers? Second, are the wage differences between high-tech and low-tech cities across genders statistically significant? Third, of the overall wage gap between male and female workers, what proportion is due to: (a) differences by gender in the patterns of high/low-tech city wage differentials, (b) differences in the distribution of male and female workers among high-tech cities, and (c) differences by gender in productivity-related factors?

Our research is close to the analysis by McCall (1998) on the effects of clustering high-technology manufacturing and service industries on wage premiums for men and women. She finds that regions specializing in high-tech manufacturing and services are associated with a higher absolute level of gender wage inequality among the college educated. In her view, existing transformations of the economy based on technology advances in manufacturing and services are biased toward well-educated male workers, even though women as well as men receive wage premiums relative to the average labor market. This finding is consistent with temporal trends that show smaller declines over time in the gender wage gap for college-educated workers (Blau and Kahn 1994).

McCall (1998) does not correct for the potential endogeneity of the high-tech employment concentration variable. Her results may reflect the selection effect of relatively more-educated or higher-ability male and female workers choosing to locate in cities with a large proportion of local employment in high-tech manufacturing and services. Moreover, she measures spatial clustering as the proportion of employment in high-tech manufacturing and services. This measure does not account for industry-city scale effects (explained later). Our empirical strategy deals with both of these issues. First, we manage the possible endogeneity of the high-tech-city variable by using an instrumental variable (IV) approach. Second, we use the horizontal clustering measure to estimate high-tech-employment concentration with respect to the national average.

For both sexes, we find that the high-tech-city coefficients are significant at the 1% level. In addition, high-tech-city effects appear approximately 30% greater for men than for women. Indeed, the high-tech-city wage elasticity is 0.13 for women and 0.17 for men. Using a standard t-test, however, we find no significant differences between the regression coefficients of both sexes. Hence, there do not seem to be gender disparities in the effect of high-tech city on wages, although there are statistically significant male and female wage premiums associated with living in a high-tech city.

Another indicator of gender differences comes from a decomposition of the overall gender wage gap using the Oaxaca (1973)–Blinder (1973) decomposition. We find that the overall gender wage gap, measured as the difference between mean log wages of male and female workers, stands at 0.25. This outcome indicates that the average female college-educated worker earns 75% of the mean male wage. Moreover, depending on which gender wage structure is used, the results illustrate that a very small percentage, between 0.02% and 0.03%, of the overall gender wage gap can be explained by the fact that (on average) women live in cities where the proportion of high-tech employment is lower. Our findings suggest that between 0.235% and 0.239% of the overall gender wage gap derives from differences between high-tech-city wage elasticities for men and women. The latter result should be interpreted with caution, however, because the wage elasticity in a high-tech city is not significantly different for both sexes.

The remainder of this chapter is organized as follows. Section 2 reviews the literature of human-capital externalities, interindustry gender wage differentials, and case studies of high-tech regions. In this context, it analyzes why we would expect gender skill-based wage differences across high-tech and low-tech cities. Section 3 describes our data and the variables used in the statistical analysis, including our definition of high-tech cities. Section 4 explains our econometric approach. Section 5 presents the results from our OLS and IV models of the effects of high-tech cities on urban wages by gender, and discusses the validity of our instruments. Section 6 introduces the Oaxaca (1973)–Blinder (1973) decomposition. Section 7 contains our conclusions.

2 High-Tech Cities and the Gender Gap

Starting in the 1980s, the economy bifurcated into two interrelated worlds of industries ruled by different economics. As Arthur (1996, p. 100) explains, "... diminishing returns hold sway in the traditional part of the economy – the processing industries. Increasing returns reign in the newer part – the knowledge-based industries."[1] In this regard, Teece (2002) observes that the economy has undergone a transformation largely from the processing of raw materials and the manufacturing of products to the processing of information along with the development, application, and transfer of new knowledge. As a consequence, diminishing-returns activities have been replaced by those involving increasing benefits in knowledge-based industries.

Teece (2002) notes that increasing returns relate to mechanisms of positive feedback that reinforce the winners and challenge the losers. In our view, these feedback processes are not only related to industry-specific externalities[2] but also to location-specific externalities (human-capital spillovers) that tend to perpetuate the agglomeration of knowledge-based firms in the same few core regions (Malecki 1981). Saxenian (1994) describes how knowledge externalities in Silicon Valley are associated with a region where relationships are easily formed and maintained, technical and market information is exchanged, new enterprises are conceived, and networks are developed. Storper and Venables (2003) point out that knowledge "rubs off" on people in places such as Silicon Valley or London. The result is that people in these "buzz cities" should be highly productive because they interact and cooperate with other high-ability people, are well placed to communicate complex ideas with them, and are highly motivated.[3] In their view, to be able to reap these benefits in full almost invariably requires colocation rather than occasional interludes of face-to-face contacts.

This chapter deals with gender wage differences in high-tech cities, defined as those cities with higher-than-expected employment in high-tech industries. In this regard, it is important first to review what we know about gender wage differentials across industries, which empirical analyses have shown to be large and persistent even after controlling for a broad set of worker characteristics. These studies report that workers of the same quality may receive different wages depending on the industry in which they work (Edin and Zetterberg 1992; Krueger and Summers 1988). Yet, as noted by Gannon et al. (2007), it is surprising to observe that the evidence regarding the interplay between gender wage gaps and interindustry wage differentials is limited. To our knowledge, only two previous studies have focused on this interplay – an earlier paper by Fields and Wolff (1995) and the more recent one by Gannon et al. (2007).

Using the 1988 U.S. Current Population Survey, Fields and Wolff (1995) find significant industry wage differentials for women and men after controlling

---

[1] *Returns to scale* refers to a technical property of production that examines changes in output subsequent to a proportional change in all inputs. There are constant returns to scale if output increases by that same proportional change, and increasing (decreasing) returns to scale if it increases by more (less) than that proportional change.

[2] Teece (2002) explains that positive industry-specific feedbacks are associated with standards and network consumption externalities. For instance, if standards are proprietary, ownership of a dominant standard can yield significant rents.

[3] The authors introduce the concept of "buzz cities" resulting from the increasing importance of colocation of economic activities that involve the exchange of tacit knowledge or complex ideas.

for productivity-related individual characteristics. Gannon et al. (2007) study interindustry wage differentials in six European countries: Belgium, Denmark, Ireland, Italy, Spain, and the United Kingdom. Their results show that even when controlling for working conditions, as well as for individual and firm characteristics, gender wage differentials exist between workers employed in different industries. They find that (on average) women have an interindustry wage differential of between 11% and 18% below that of men, and that industry effects explain between 0% and 29% of the overall wage gap.

Fields and Wolff (1995) review the literature on interindustry wage differentials and conclude that four models explain them. High wages will be paid in industries in which monitoring is difficult and the failure of workers to perform up to standards is costly (shirking model). High wages are also a firm's strategy to reduce labor turnover (turnover model), to attract a better-quality workforce (selection model), or to improve morale among workers (sociological model). Perhaps the most appropriate model to explain the well-documented fact that high-tech firms (and industries) pay higher wages than their low-tech counterparts is that they need to attract the best engineers and scientists to develop new products and processes continuously (selection model). These models do not, however, explain *gender* wage differences across industries, nor how the geography of knowledge spillovers impacts gender wage differences.

At least one reason why workers who live in a high-tech city might be observed to have higher wages than those who do not is that knowledge is a partially nonexcludable good. Thus, it generates externalities or spillovers. Therefore, creators or owners of knowledge cannot always exclude others from making unauthorized use of it (Grossman and Helpman 1991).[4] Living in a high-tech city facilitates access to tacit information (knowledge). Therefore, a tech-city wage premium exists for workers who might have above-average market productivity. Access to knowledge, to the extent that it increases labor productivity, translates into higher wages. Indeed, Echeverri-Carroll and Ayala (2009) present evidence of a tech-city wage premium of approximately 4.6% for male workers. Here we question: Is there a smaller/higher high-tech-city wage premium for female workers?

McCall (2001) finds that the average gender wage gap in the United States is significantly greater in high-tech services and high-tech manufacturing regions than in low-tech ones. She explains that if high-tech industries adopt technologies that are biased toward well-educated male workers, even though women as well as men received wage premiums relative to the average labor market, regions specializing in high-tech manufacturing and services would be associated with higher gender wage inequalities, especially among the college educated.

The literature reviewed in this section shows significant industry wage differentials for women and men. Thus, female and male workers of the same quality receive different wages in the same industry. These gender wage inequalities result not only from gender differences in skills but also from industry-specific gender-bias

[4] Excludability of a good reflects both legal and technological considerations.

strategies such as the adoption of technologies that favor men. Our focus in this chapter is on the effect on gender wage inequality of knowledge externalities associated with the "informal" networks prevalent in high-tech cities (Saxenian 1994). Our hypothesis is that the possible exclusion of women, fully or partially, from city-based "informal" networks could intensify the gender wage gap across high-tech and low-tech areas associated with industry-specific gender-bias policies. Empirically, it means that even after controlling for whether a worker works in a high-tech industry or not, we still find gender wage differences associated with living in a high-tech city.

3 Data and Variables

This two-part section describes the data used in our econometric analysis. In the first part, we explain variables at both the individual and city levels (non-high-tech related) and their data sources. In the second part, we define and provide the data sources for two high-tech variables: high-tech city and high-tech industry.

3.1 Non-High-Tech Variables

We use a sample obtained from the 5% Public Use Microdata Sample (PUMS) of the 2000 Census of Population. Our sample is comprised of male and female workers with both a college education and a strong attachment to the labor market (full-time workers): those aged 18–65 working full time (at least 35 hours per week), neither self-employed nor in the military, and who worked at least fourteen weeks in the year preceding the census. Using those parameters, we obtained samples of 484,899 and 396,143 college-educated male and female full-time workers respectively residing in the hundred most populous cities (MSAs/PMSAs). We chose only the top hundred metros because 76% of the total urban population in the United States lived in those metro areas in 2000.

Our dependent variable is the logarithm of hourly wages (annual wage and salary earnings divided by the product of weeks worked and usual weekly hours). Our independent variables (from human-capital theory) identify observed individual characteristics that affect wages, such as years of college education and its square,[5] potential level of experience (age minus years of schooling minus six), and potential experience squared. Other independent variables included in our analysis are five general occupational categories, with services being the omitted occupational

[5] Although our sample includes only college-educated workers, we thought that it was important to include the variable *years of education* to capture differential effects beyond undergraduate college education.

category.[6] *Marital status* is 1 if the individual is married, *race* is 1 if the individual is nonwhite, and *disability* is 1 if the individual has a personal health limitation. These variables are from the 5% PUMS of the 2000 Census of Population.

We also control for city-level variables shown by previous research to affect variation of wages among individuals (Rauch 1993; Echeverri-Carroll and Ayala 2009). In particular, we control for a city's climate as well its proximity to a coast or the Great Lakes. Climate is a composite score that ranges from 0 (poor weather) to 100 (mild weather). A dummy variable for coastal location equals 1 if an MSA/PMSA borders an ocean or any of the Great Lakes, and 0 if not. We also control for the arts endowment in a metro area and for the census region where the city is located. Art is a composite score that ranges from 0 (the lowest arts endowment) to 100 (the highest).[7] Data for climate, coast, and art comes from the *2000 Places Rated Almanac*.

In a recent paper, Henderson (2007) notes that most empirical work on urban knowledge spillovers does not distinguish between localized knowledge effects and agglomeration economies. In his view, failure to do so could confound and overestimate the actual effects on productivity of the selected proxy for knowledge externalities with other effects resulting from the agglomeration of people and economic activity. Indeed, he points out that a proxy for agglomeration economies is an essential regressor in econometric models (like ours) that try to measure the effects of localized knowledge externalities on wages. We control for agglomeration effects using MSA/PMSA population from the 2000 Census of Population.[8]

3.2 High-Tech Variables

For high-tech indicators, we employ two variables in our econometric analysis. The first is a control for whether the individual works for a high-tech industry – a dummy variable equal to 1 if this is the case (and 0 otherwise). The second, a variable that conceptualizes the purpose of our research, is a dummy equal to 1 if the individual lives in a high-tech city (and 0 otherwise). To understand more clearly the concept of high-tech cities, we first need to define *high-tech industries* because *high-tech cities* are simply cities with employment higher than expected in high-tech industries.

[6] Services includes the following occupations: health care support, protective services, food preparation and serving related occupations, building and grounds cleaning and maintenance, and personal care and services.

[7] The highest score is given to metropolitan areas with the larger number of art museums, annual museum attendance, per capita museum attendance, annual ballet performances, touring artist bookings, opera performances, professional theater performances, and symphony performances.

[8] Some evidence suggests that the productivity of all types of labor (and therefore wages) rises with the size of a city (Segal 1976; Shefer 1973; Sveikauskas 1975). Others, like Garofalo and Fogarty (1979), believe that only the productivity of skilled labor rises with city size. Thus, there is consensus that city size increases the productivity of skilled workers, but there is disagreement about its effects on unskilled labor.

Although there is no single authoritative definition of high-technology industries (or firms), there is wide agreement on their general characteristics (Hecker 1999). The Office of Technology Assessment (1982) describes high-technology firms as those engaged in the design, development, and introduction of new products and/or innovative manufacturing processes through the systematic application of scientific and technical knowledge. To classify firms or industries by their relative innovativeness, studies use a large variety of proxies for innovations (Chapple et al. 2004). In most current studies, though, the main proxy used is the employment of scientific and technical workers (Hecker 1999; Chapple et al. 2004; Yu 2004).[9] High-tech industries are those with a large proportion of workers in scientific and technical occupations or technology-oriented occupations (Richie et al. 1983).

Many studies from the U.S. Department of Labor define four Standard Occupational Categories as technology-oriented occupations: engineers, life and physical scientists, computer professionals and mathematicians (except actuaries), and engineering, computer, and scientific managers (Hadlock et al. 1991; Hecker 1999, 2005; Luker and Lyons 1997). Workers in these occupations need in-depth knowledge of theories and principles of science, engineering, and mathematics. Such knowledge is generally acquired through specialized post-high-school education – ranging from an associate degree to a doctorate – in some field of technology. Recent studies exclude from these four broad occupational categories those at the assistant and technician levels (Chapple et al. 2004; Yu 2004). We adopt the more recent definition under the premise that high-tech industries are defined mainly by their innovativeness, a variable more easily captured by technology-oriented workers with at least a college degree.

Following Hecker (1999) and Chapple et al. (2004), we calculate the number of technology-oriented workers (TOW) in each four-digit NAICs industry using the 2002 Occupational Employment Statistics (OES) from the Bureau of Labor Statistics. A limitation of the survey is that OES censors employment data for certain occupations. We find that the underestimation of TOW that arises from this censoring, however, is relatively small.[10]

Hecker (1999) classifies an industry as high tech if its percentage of TOW is at least *twice* the national average, but Chapple et al. (2004) require at least *three times* the national average. OES data show that the average of TOW, for all 294 industries (four-digit NAICS) in the United States, was 3.15% in 2002. Thus, we classify industries as high tech if the proportion of TOW is at least 6.3% (twice the national average). We find 33 high-tech sectors among the 154 manufacturing and service NAICS. To verify the robustness of our estimates, we also define high-tech

[9] This definition has gained wide acceptance partly because it closely matches a growing body of research suggesting that human capital (i.e., skilled labor) may be a better gauge and more important driver of economic development than other indicators (Yu 2004).

[10] Of the 154 manufacturing and services industries in the OES database (by four-digit 2002 NAICS), only 12 industries have more than 30% unreported employment in the four technology-oriented occupations. In contrast, 74 industries have less than 10% of unreported employment in these occupations. The rest (68 industries) have between 10% and 30% of unreported employment.

industries as those with at least three times the national average of TOW, or at least 9.45%. This stricter definition gives us 25 high-tech manufacturing and services NAICS.

We are now ready to build a measure that allows us to classify cities as high tech or low tech. We are not interested in finding overall measures of industry concentration such as via Gini coefficients (Krugman 1993), the dartboard approach (Ellison and Glaeser 1997), or indices of specialization (Midelfart-Knarvik et al. 2000). Instead, we need a measure that captures the absolute concentration of high-tech activity, such as the location quotient (LQ). As Ratanawaraha and Polenske (2007) note, however, the LQ indicates whether an area has a higher or lower share of a particular industry's employment (e.g., high-tech employment) than the national share, but it does not provide information regarding the absolute size of the industry in that area (industry-city scale effect).

The lack of scale sensitivity of the LQ led Fingleton (2004, 2006, 2007) to develop the Horizontal Cluster Location Quotient (HCLQ). They define a *horizontal cluster* simply as a spatial agglomeration of firms in a particular industry and the HCLQ as the number of jobs in the local industry that exceeds its expected number. The *expected number* is the number of jobs in the industry that would correspond to the area having the national share of the industry, and therefore producing an LQ equal to 1. The horizontal cluster measure is the difference between the actual and expected numbers of high-tech jobs, hence $HCLQ_g = E_g - \hat{E}_g$. In our case, E_g is the 2000 high-tech employment in city "g," whereas $\hat{E}_g$ is the expected high-tech employment in that city. Positive $HCLQ_g$ values indicate that the larger the actual high-tech employment (E_g) in city g is from the expected high-tech employment ($\hat{E}_g$), the more spatially concentrated the high-tech industry in that city will be. Negative values indicate that the high-tech industry employment in the city is less than the expected high-tech employment, indicating that the high-tech industry is not very concentrated in the city.

Using the Horizontal Cluster measure $\left(HCLQ_g\right)$, we develop a dummy variable equal to 1 if the worker lives in a city with a positive $HCLQ_g$ value, and 0 if the worker lives in a city with a negative $HCLQ_g$ value. High-tech cities are then those with higher-than-expected employment in high-tech industries. Low-tech cities are those that show lower-than-expected concentration of employment in high-tech industries.

4 Empirical Framework

In this section, we first describe how we obtain the estimates of our wage equations, accounting for potential endogenous explanatory variables. Then we explain our measure of the gender wage gap based on the Oaxaca (1973)–Blinder (1973) decomposition of differences in average wages.

4.1 Estimating the Wage Equation with Endogenous Regressors

Our log wage equation that accounts for the value of living in a high-tech city for each gender group g (male or female) is the traditional Mincer (1974) equation:

$$Log\ W_i^g = \mathrm{X}_i^{g\,\prime}\beta + \mathrm{Z}_{(i)}^{g\,\prime}\theta + \delta \mathrm{H}_{(i)}^g + \varepsilon_i^g \quad (1)$$

$Log\ W_i^g$ is the log hourly wage of individual i in gender group g. X_i^g is a vector of observed individual characteristics that affect wages in both gender groups. $Z_{(i)}^g$ is the vector that identifies general characteristics of the city in which individual i in gender group g resides that affect his or her productivity. $H_{(i)}^g$ is a dummy variable indicating whether individual i in gender group g lives in a high-tech city. And ε_i^g denotes the error term for individual i in gender group g.

The high-tech-city effect on wages could perhaps be a sign of a selection effect where relatively more-educated or higher-ability male and female workers choose to locate in these cities. In this case, there is a clear potential for correlation between factors that influence the decision to live in a high-tech city and the error term in the wage equation. Such factors remain unobserved to the analyst and hence become incorporated into the error term of the wage equation. Specifically, if the decision to live in a high-tech city is correlated with ability, and ability only enters the wage equation through an additive error term, then OLS produces estimates that are biased and inconsistent. This correlation gives rise to the need for estimators, such as the instrumental variable model, that account for the possible endogeneity of the high-tech-city variable.[11]

We use two city-level characteristics as excluded instruments that affect the decision of a college-educated worker to locate in a high-tech city: logarithm of venture-capital investment and logarithm of defense expenditures. Case studies of high-tech regions document high levels of defense expenditure (Newman 1998) and venture-capital investment (Belke et al. 2003; Florida and Smith 1993) in these cities. Venture-capital-investment data by metropolitan area for 2000 come from Thomson Venture Economics/NVCA. Data on total prime contract awards in dollars come from the Department of Defense.[12]

Our choice of excluded instruments complies with assumptions that are the basis for selecting good ones. The first assumption is that our excluded instruments will affect the decision of a worker to live in a high-tech city versus a low-tech city. Specifically, we assume that skilled male and female workers who choose to live in high-tech cities tend to do so because these cities increase their probability to work for a diversity of knowledge-intensive organizations, including venture-capital firms

[11] Our IV estimation is conducted using the *ivreg2* module programmed for Stata by Baum et al. (2003, 2007, 2007a).

[12] Data on prime contract awards state/county summary are available online at http://siadapp.dmdc.osd.mil/procurement/historical_reports/geographic/geostat.html. Data are for 2001 since county data were not published before this period.

(e.g., as management consultants) or companies receiving Department of Defense prime contract awards (e.g., as research scientists).[13]

The second assumption made in choosing our instruments is that they do not directly affect individual wages. In their study of the impact of venture-capital investment in employment at the country level, Belke et al. (2003, p. 26) note that dynamic models (rather than static ones) better support significant impact of venture-capital investment on the growth of employment. The long-term effect of venture capital on employment tends to be associated with the fact that venture capital finances mainly innovative new firms, which often have significant failure rates (Gorman and Sahlman 1989; Manigart et al. 2002). This empirical evidence supports the view that venture-capital investment in a city at time t_0 affects local wages at a future time t_n. Similarly, given the nature of multi-year Department of Defense contracts, employment effects tend to be long term as well. Thus, defense expenditure at time t_0 will also mainly affect employment and wages at a future time t_n.

Studies on the effects of high technology on wages usually need to control not only for the endogeneity of the high-tech-city indicators but also for sample selection bias. Self-selection into work may introduce bias in the estimation of wage equations, especially for female workers. In this case, bivariate selectivity models are normally used, in which the researcher models two simultaneous decisions – whether or not to work and whether or not to work in a high-tech city (see, for instance, Cutillo and Di Pietro 2006). The self-selection issue, however, is important only in the case where the female sample has very little attachment to the labor market, which is not our case. Our samples, from the 2000 Census of Population, of male and female workers with college education have a strong attachment to the labor market. These data show that 82% of college-educated women and 94% of male employees with similar education work full time.

4.2 Estimating the Gender Wage Gap

To complete our analysis, we decomposed the overall gender wage gap to assess what proportion is due to: (a) differences between male and female high-tech-city elasticities, (b) differences in the proportion of workers living in a high-tech city by gender, and (c) differences by gender in all the other factors. To do so, we used the decomposition procedure developed by Oaxaca (1973) and Blinder (1973), who show that the difference between the average hourly wage (in logarithms) of men and women can be decomposed as follows:

$$\bar{W}_m - \bar{W}_f = \left(\bar{H}_m - \bar{H}_f\right)\hat{\delta}_m + \bar{H}_f\left(\hat{\delta}_m - \hat{\delta}_f\right) + \left(\bar{X}_m - \bar{X}_f\right)\hat{\beta}_m + \bar{X}_f\left(\hat{\beta}_m - \dot{\beta}_f\right) \quad (2)$$

[13] Venture firms often provide capital and management expertise.

where the male wage structure is used to assess the gender wage gap, or alternatively:

$$\bar{W}_m - \bar{W}_f = \left(\bar{H}_m - \bar{H}_f\right)\hat{\delta}_f + \bar{H}_m\left(\hat{\delta}_m - \hat{\delta}_f\right) + \left(\bar{X}_m - \bar{X}_f\right)\hat{\beta}_f + \bar{X}_m\left(\hat{\beta}_m - \dot{\beta}_f\right) \quad (3)$$

where the female wage structure is used. The indices m and f refer respectively to male and female workers. $\bar{W}$ represents the average (Naperian logarithm) of the hourly wage, $\bar{H}$ is the average value of the high-tech city variable (proportion of gender-specific workers in high-tech cities), and $\bar{X}$ is a vector containing an intercept and the average value of the individual characteristics of the workers, the general characteristics of the city where they live, and the region where the city is located. The IV regression coefficients pertaining to H and X respectively are $\hat{\delta}$ and $\hat{\beta}$, as reported in Table 1 for male workers and Table 2 for female workers.

Table 1 Determinants of (log of) individual hourly wages for male workers

	Model 1 OLS	Model 2 OLS	Model 3 OLS	Model 4 IV-Fuller
Intercept	0.667**	1.168***	1.018***	1.190***
	(0.264)	(0.221)	(0.236)	(0.240)
Yrs of college education	0.145***	0.086***	0.075***	0.072***
	(0.028)	(0.023)	(0.022)	(0.022)
Yrs of college education sq	−0.003***	−0.001*	−0.001	−0.001
	(0.0007)	(0.0006)	(0.0006)	(0.0006)
Potential experience	0.043***	0.044***	0.044***	0.044***
	(0.001)	(0.001)	(0.0006)	(0.0007)
Potential experience squared	−0.001***	−0.001***	−0.001***	−0.001***
	(0.00001)	(0.00001)	(0.00001)	(0.00001)
Race (non-white = 1)	−0.122***	−0.128***	−0.148***	−0.150***
	(0.016)	(0.011)	(0.013)	(0.012)
Disability (limitation = 1)	−0.109***	−0.101***	−0.104***	−0.102***
	(0.004)	(0.004)	(0.004)	(0.004)
Marital status (married = 1)	0.190***	0.185***	0.195***	0.196***
	(0.010)	(0.010)	(0.007)	(0.007)
Professional or managerial	0.362***	0.285***	0.287***	0.287***
	(0.013)	(0.011)	(0.010)	(0.010)
Technologist	0.126***	0.081***	0.085***	0.085***
	(0.011)	(0.012)	(0.011)	(0.012)
Sales, administrative support	0.208***	0.150***	0.151***	0.151***
	(0.011)	(0.010)	(0.010)	(0.010)
Manual (craft/operator/laborer)	−0.049***	−0.080***	−0.075***	−0.076***
	(0.014)	(0.014)	(0.013)	(0.013)
High-tech industry		0.216***	0.211***	0.199***
		(0.011)	(0.010)	(0.010)

(continued)

Table 1 (continued)

	Model 1 OLS	Model 2 OLS	Model 3 OLS	Model 4 IV-Fuller
High-tech city		0.067***	0.057***	0.170***
		(0.025)	(0.017)	(0.043)
Climate score			0.0007**	0.0007
			(0.0003)	(0.0004)
Coast			0.062***	0.057**
			(0.018)	(0.023)
Arts score			0.002***	−0.0006
			(0.0006)	(0.001)
Northeast			0.034	0.029
			(0.025)	(0.031)
Midwest			0.010	−0.008
			(0.022)	(0.028)
West			0.017	−0.036
			(0.033)	(0.047)
Population			7.13e−9**	1.4e−8***
			(3.51e−9)	(4.82e−9)
Orthogonality of instruments:				
Hansen's J statistic				0.209
[p-value]				[0.648]
Relevance of Instruments:				
Partial R^2				0.202
F-stat (First-stage regression)				24.71
[p-value]				[0.000]
Kleibergen–Paap rk LM statistic				80,031
[p-value]				[0.000]
Kleibergen–Paap Wald rk F stat				79,504
[p-value]				[0.000]
Anderson–Rubin Wald test				11.96
[p-value]				[0.000]
Endogeneity – high-tech city:				
Hausman test				8.102
[p-value]				[0.004]
Observations[a]	484,899	484,899	484,899	481,222

Levels of statistical significance are represented as follows: ***$p \leq 0.01$, **$p \leq 0.05$, and *$p \leq 0.10$. Standard errors in parenthesis are robust to arbitrary heteroskedasticity and intra-group (MSA) correlation

Excluded instruments: log of venture capital investments and log of defense expenditures in the city

[a]Four MSAs (El Paso, TX; Gary, IN; McAllen-Edinburg-Mission, TX; Wichita, KS) were dropped from the sample employed in the IV models because of unreported venture capital investments

Table 2 Determinants of (log of) individual hourly wages for female workers

	Model 1 OLS	Model 2 OLS	Model 3 OLS	Model 4 IV-Fuller
Intercept	−0.796***	−1.114***	−1.325***	−1.267***
	(0.232)	(0.261)	(0.230)	(0.243)
Yrs of college education	0.280***	0.306***	0.301***	0.303***
	(0.024)	(0.027)	(0.025)	(0.026)
Yrs of college education sq	−0.006***	−0.007***	−0.007***	−0.007***
	(0.0006)	(0.0007)	(0.0006)	(0.0007)
Potential experience	0.037***	0.037***	0.038***	0.038***
	(0.0007)	(0.0008)	(0.0007)	(0.0007)
Potential experience squared	−0.001***	−0.001***	−0.001***	−0.001***
	(0.00001)	(0.00002)	(0.00001)	(0.00001)
Race (non-white = 1)	−0.028***	−0.031***	−0.054***	−0.054***
	(0.008)	(0.006)	(0.008)	(0.008)
Disability (limitation = 1)	−0.075***	−0.068***	−0.072***	−0.070***
	(0.004)	(0.003)	(0.004)	(0.004)
Marital status (married = 1)	0.031***	0.033***	0.042***	0.042***
	(0.006)	(0.006)	(0.004)	(0.004)
Professional or managerial	0.417***	0.376***	0.379***	0.380***
	(0.011)	(0.011)	(0.012)	(0.012)
Technologist	0.408***	0.397***	0.402***	0.404***
	(0.015)	(0.014)	(0.016)	(0.016)
Sales, administrative support	0.220***	0.174***	0.175***	0.175***
	(0.012)	(0.011)	(0.012)	(0.012)
Manual (craft/operator/laborer)	0.041**	−0.009	−0.004	−0.004
	(0.016)	(0.015)	(0.016)	(0.016)
High-tech industry		0.212***	0.204***	0.196***
		(0.010)	(0.008)	(0.008)
High-tech city		0.055*	0.052***	0.127***
		(0.030)	(0.017)	(0.043)
Climate score			0.001**	0.001**
			(0.0003)	(0.0004)
Coast			0.053**	0.050**
			(0.020)	(0.022)
Arts score			0.001**	0.0001
			(0.0006)	(0.001)
Northeast			0.042*	0.042
			(0.024)	(0.029)
Midwest			0.013	0.0006
			(0.022)	(0.023)
West			0.007	−0.026
			(0.033)	(0.047)
Population			1.2e−8***	1.6e−8***
			(3.72e−9)	(5.06*e*−9)
Orthogonality of Instruments:				
Hansen's J statistic				0.757
[p-value]				[0.384]

(continued)

Table 2 (continued)

	Model 1 OLS	Model 2 OLS	Model 3 OLS	Model 4 IV-Fuller
Relevance of Instruments:				
Partial R^2				0.204
F-stat (First-stage regression)				25.18
[p-value]				[0.000]
Kleibergen–Paap rk LM statistic				64,203
[p-value]				[0.000]
Kleibergen–Paap Wald rk F stat				61,583
[p-value]				[0.000]
Anderson–Rubin Wald test				6.70
[p-value]				[0.002]
Endogeneity – high-tech city:				
Hausman test				4.134
[p-value]				[0.042]
Observations[a]	396,143	396,143	396,143	392,860

See notes in Table 1

5 Empirical Results

This section is divided in two parts. Part one presents the results of our OLS and IV models for the male (Table 1) and female (Table 2) samples. All the models account for clustering (by metro area) and heteroskedasticity in the error terms.[14] Part two shows the analysis of the validity of the instrumental variables.

5.1 *Benchmark and IV Models*

Model 1 presents results of the log wage model controlling for human-capital characteristics and occupation of the male and female workers. The signs and magnitude of the coefficients on the observed individual characteristics for both male and female workers are as expected (see Tables 1 and 2). All the coefficients are significant at the 1% or 5% levels for both samples.

Model 2 expands the log wage model, introducing two new variables of particular interest in this study: whether the worker works in a high-tech industry and/or lives in a high-tech city. The variable that identifies whether the worker works for a high-tech industry is positive and statistically significant at the 1% level for both genders. College-educated, full-time male and female workers who

[14] Clustering arises in our case since it may be reasonable to assume that observations of individuals drawn from the same city (cluster) are correlated with each other, but individuals from different cities are not. The intraclass correlation may vary from cluster to cluster (the cluster analog to heteroskedasticity).

work in a high-tech industry make wages that are 21% higher than their respective counterparts in low-tech industries. Moreover, as predicted by the theory, male workers make 6.7% more when working in a high-tech city than their counterparts in low-tech cities (Table 1), while women in high-tech cities make 5.5% more than their counterparts in low-tech cities (Table 2).

Model 3 controls for other city-level variables that previous studies find to be important determinants of individual wages. Similar to a study by Rauch (1993), we find that the variable *coast/Great Lakes* is positive and significant at the 1% level for male and female workers. Rauch (1993) argues that wages should be higher in port cities due to their privileged access to the gains from international trade. The coefficients on climate and arts are also positive and statistically significant for both groups. Rauch (1993) finds similar productivity effects associated with mild climates and cities with large arts endowments. Population is positive and significant, indicating that large cities tend to pay higher wages for college-educated workers.

The coefficients for the dummies that identify the census region where the worker's city is located were not significant at conventional levels in explaining variations in individual wages. Controlling for city-level variables reduces the value of the tech-city wage premium from 6.7% to 5.7% for males (Table 1) and from 5.5% to 5.2% for females (Table 2). Thus, some of the tech-city effects on wages identified in Model 2 are really the effects of other observable city-level variables on wages. The high-tech-city coefficients for both samples are still positive and highly significant, however, supporting the hypothesis that there *is* a tech-city wage premium for both male and female workers.

Model 4 shows the estimated coefficients and standard errors for the IV model using the Fuller-modified LIML estimator for the male (Table 1) and female (Table 2) cases. The high-tech-city variable is significant for both groups at the 1% level. These effects are considerably larger than those obtained using OLS regressions. The high-tech-city coefficient increased for the male sample, from 0.057 in the OLS model to 0.17 in the IV model. A similar trend is observed in the female sample, where this coefficient also increased from 0.052 in the OLS model to 0.13. Moreover, the difference between OLS and IV high-tech-city coefficients for both genders is significant at conventional levels.[15]

High-tech-city IV coefficients that are substantially higher than the corresponding OLS estimates for both male and female workers may be explained by the existence of gender-specific heterogeneity in individual returns, as well as by the fact that our study is based on instruments influencing only the location decision

[15] Under the assumption that family ties could hinder women mobility, we run the female IV regression adding the number of children (NOC) as an excluded instrument. However, an LR redundancy test shows that the NOC is a redundant instrument indicating that the asymptotic efficiency of the estimation is not improved by using it. This result is in line with recent evidence showing that college-educated women are in fact more interregionally mobile than men (Faggian et al. 2007). As noted by Faggian et al. (2007), the overall evidence on the migration of women is, however, very limited and, more importantly, it largely ignores the interaction between human capital and migration.

of individuals with high marginal returns. This conclusion is also consistent with the Local Average Treatment Effect (LATE) interpretation of instrumental variables (Imbens and Angrist 1994), according to which IV identifies the average marginal return of those who comply with the assignment-to-treatment mechanism implied in the instruments. Thus, the estimate recovered by IV does not necessarily coincide with the average marginal return in the population but rather the average marginal return for the population subgroup affected by exogenous variation in the high-tech-city outcome.[16] These differences between OLS and IV estimates might also reflect measurement errors in our Horizontal Cluster indicator related to the aggregation of industries in two broad categories: high tech and low tech.

Our IV results for both samples are robust when using a horizontal cluster measure built on the assumption that high-tech industries are those with at least three times the national average of TOW. The results are also robust to alternative estimators – in particular, the limited information maximum likelihood (LIML), the two-stage efficient generalized method of moments (GMM2S), and the traditional IV two-stage least squares (2SLS).

5.2 Examining Instrument Validity

The Hansen J statistics (adjusted for the clustered-error structure) in Table 1 (for male workers) and Table 2 (for female workers) indicate that the instruments satisfy the orthogonality condition. In both samples, they are appropriately uncorrelated with the disturbance processes. The test of instrument relevance demonstrates that the set of instruments has acceptable strength for the log wage equation for males (Partial $R^2 = 0.202$) and females (Partial $R^2 = 0.204$) and F-statistics larger than 20 for both groups.[17] As noted by Baum (2006), however, the distribution of this F statistic is nonstandard. Other statistics are, therefore, required to test for the correlation between the instruments and the endogenous variable.

A more general approach for testing the relevance of instrumental variables is the *underidentification test,* or test for the rank of a matrix. Recently, Kleibergen and Paap (2006) have proposed two more general versions of these statistics that are robust to heteroskedasticity and intra-class correlation: the rk Wald statistic

[16] As Morgan and Winship (2008) explain, the new IV literature suggests that IV techniques are more effective for estimating narrowly defined causal effects than for estimating average causal effects. For instance, suppose that we could brainwash workers, erasing their location decisions at will and assign them to a high-tech or low-tech city using a lottery. The IV estimator identifies the average causal effect for the subset of workers that would chose to live in a high-tech city if wining the lottery (compliers) and the group of workers that would live in a low-tech city if not wining the lottery (defiers). It would *not* identify the average causal effect for workers that would live in a high-tech city even if not winning the lottery (always takers) or the group that would live in a low-tech city even if winning the lottery (never takers).

[17] Bound et al. (1995) suggest that the first-stage F-statistic must be larger than 10 for IV inference to be reliable.

and the rk LM statistic.[18] Rejection of the null implies full rank and identification, while failure to reject the null implies the matrix is rank deficient and the equation is underidentified (Baum et al. 2007). As Tables 1 and 2 show, the rk LM statistic ($p_m = 0.000$; $p_f = 0.000$) strongly reject the null, in both the male and female samples, implying full rank and that our models are identified.

As Baum et al. (2007) note, the weak instrument problem can arise even when the correlation between the endogenous variable and the instruments is significant at conventional levels, at the 1% or 5% levels and the researcher is using a large sample. Testing for weak identification is therefore necessary. In the presence of heteroskedasticity or clustering, Baum et al. (2007) propose to use the Kleibergen–Paap Wald rk F statistic. We rely on the critical values tabulated in Stock and Yogo (2005).[19] For our Fuller-LIML estimation with two instruments and one endogenous variable, the critical value for having at most 5% of the OLS bias left in the IV estimation is 13.46. The values for this statistic presented in Tables 1 and 2 are clearly larger than this value for both the male and female models, indicating that we do not seem to have a problem with weak instruments. This critical value nevertheless depends on an assumption of uncorrelated errors within cities that our data may violate.

We estimate an alternative statistic that is robust to the presence of weak instruments: the Anderson–Rubin (A–R) statistic.[20] The null hypothesis tested is that the coefficients of the endogenous regressors in the structural equation are jointly equal to zero.[21] As Tables 1 and 2 show, the A–R statistics ($p_m = 0.000$; $p_f = 0.002$) comfortably reject the null that all endogenous regressors are jointly equal to zero for the male and female samples at the 1% levels. This statistic signals that we do not have a weak instruments problem for any of the samples.

Finally, the asymptotic variance of the IV estimator is always larger, and sometimes much larger, than the asymptotic variance of the OLS estimator (Wooldridge 2006). This loss of efficiency is a price worth paying only if the OLS estimator is biased and inconsistent. The Hausman statistic tests the null hypothesis that the OLS estimator is consistent and fully efficient.[22] We find that endogeneity

[18] The LM version of the Kleibergen–Paap rk statistic can be considered a generalization of the Anderson canonical correlation rank statistic to the non-i.i.d. case. Similarly, the Wald version of the rk statistic reduces to the Cragg-Donald statistic when the errors are i.i.d.

[19] Critical values from Stock and Yogo (2005) are only available for i.i.d. errors. Baum et al. (2007) suggest that when using the *rk* statistic to test for weak identification, users should either apply with caution the critical values compiled by Stock and Yogo (2005) for the i.i.d. case, or refer to the older "rule of thumb" of Staiger and Stock (1997), that the F-statistic should be at least 10 before dismissing weak identification as a problem.

[20] The A–R statistic provides a Wald test, whereas the closely related Stock and Wright (LM) S statistic provides an LM or GMM distance test of the same hypothesis (Baum et al. 2007).

[21] Because our models are estimated with a robust covariance matrix estimator, both the A–R statistics (F and χ^2 versions) and the S statistic are correspondingly robust.

[22] The Hausman test is sensitive to several types of misspecification. In particular, the Hausman test performs poorly if the correlation between potentially endogenous variables and instruments is low. That is, the performance of the Hausman test in the presence of weak instruments is very poor (Chmelarova and Hill 2004).

does indeed exist among workers living in high-tech cities. The Hausman test statistic is positive and significantly different from zero ($\chi^2 = 8.102$; $p = 0.004$ for males; $\chi^2 = 4.134$; $p = 0.04$ for females). As already reported, once this endogeneity is accounted for, the effect of high-tech city on wages increases significantly for both male and female workers, giving support to the view that OLS estimates for high-tech-city wage premiums are biased in both samples.

6 Decomposition of the Gender Wage Gap

In the last section of our analysis, we consider the factors that account for the overall gender wage differences. We are particularly interested in the contribution played by dissimilarities by gender in both high-tech-city wage effects and the distribution of employment across high-tech cities. As already indicated, the overall gender wage gap can be decomposed into three elements: (a) the variation due to differences in the male and female estimated high-tech-city coefficients, (b) the variation due to the different employment distributions of men and women across high-tech cities, and (c) the variation explained by all the other factors (the difference in the male and female intercepts and the effects of all other variables besides high-tech city in the wage equation).

Table 3 shows the results. We find that the overall gender wage gap, measured as the difference between mean log wages of male and female workers, stands at 0.25. This figure means that the average female worker earns 75% of the mean male wage. Moreover, depending on which gender wage structure is used, results indicate that a mild proportion, between 0.02% and 0.03%, of the overall gender wage gap can be explained by the fact that (on average) women live in cities where the proportions of high-tech employment are lower.[23] Our results suggest that between 0.235% and 0.239% of the overall gender wage gap derives from differences between high-tech-city wage elasticities for men and women. The latter result, however, should be interpreted with caution because the high-tech-city wage elasticity is not significantly different for both sexes.

Table 3 Decomposition of the gender wage gap

Wage structure:	Overall gender wage gap: $\overline{W}_m - \overline{W}_f$	Percentage of overall wage gap due to differences in: Proportion of employment in high-tech cities: $(\overline{H}_m - \overline{H}_f)\,\hat{\delta}_{m(f)}$	High-tech-city wage elasticities: $\overline{H}_{f(m)}\left(\hat{\delta}_m - \hat{\delta}_f\right)$
Male wage structure	0.256	0.03%	0.235%
Female wage structure	0.256	0.02%	0.239%

[23] We find that at least 2% more male workers live in high-tech cities than female workers. Moreover, we find that this difference is statistically significant.

7 Conclusions

The Graduate School of Management of the University of California, Davis, recently released a report titled "2007 UC Davis Study of California Women Business Leaders: A Census of Women Directors and Executive Officers." The report details the presence of women at the very top of the 400 largest public companies in California. It states that Silicon Valley companies based in Santa Clara County (where San Jose is located) ranked last in the state, elevating fewer women to executive ranks and corporate boards than any other county in California. Only 9% of the companies in the county have promoted women to top posts, and a mere 7% of corporate boards include even one woman.

The report does not speculate on the reasons why women are not found in top executive positions. An obvious one is that fewer women choose to study engineering and science. After interviewing some experts in the community when reporting on the UC Davis study, however, the *San Jose Mercury News* suggested another reason: Networks favor men. "Silicon Valley is as much who you know as what you know. Men have broader networks because they have been in the field longer. And when they reach for their Rolodexes, they are more likely to find other men because the tech industry is dominated by men" (Schwanhausser 2007).

In a previous study, Echeverri-Carroll and Ayala (2009) found that, on average, college-educated male workers have a high-tech-city wage premium of approximately 6.2% (regardless of the industry). This evidence is consistent with the hypothesis that highly skilled workers do best in high-tech cities because they benefit from being around other highly skilled workers. Using a different measure of high-tech-employment clustering, we find (in this chapter) a much larger high-tech-city wage premium for college-educated male workers – 17.7% over their counterparts living in low-tech cities.

Although many studies present evidence of city-based human-capital externalities, they measure this effect for male workers only (or for a joint sample of male and female workers). To our knowledge, this paper is the first to address the issue of the effect of knowledge externalities on female workers. We find that indeed there is a female-specific high-tech-city wage premium of 14.6%. Thus, college-educated female workers who live in a high-tech city have wages that are on average 14.6% higher than their counterparts in low-tech cities. We find, however (using a standard t-test), that the difference between the high-tech-city coefficients for male and female workers (17.7 and 14.6, respectively) is not statistically significant.

Although results from a Oaxaca (1973)–Blinder (1973) decomposition present evidence that some of the average gender wage differences in our sample are explained by the fact that more women live in cities with a relatively smaller proportion of high-tech employment, this contribution is mild (0.02–0.03%). Most of the gender wage differences seem to come from other variables in our model, or perhaps from industry-specific strategies (e.g., glass-ceiling policies) that are unobserved in our sample, rather than from the exclusion of women from knowledge networks in high-tech cities, as the informal evidence suggests.

Acknowledgements This study is based on work supported by the National Science Foundation under Grant No. 0318174 and by the Bureau of Business Research, The University of Texas at Austin. Opinions, findings, and conclusions or recommendations are those of the authors and do not necessarily reflect the view of any of these organizations. The authors thank two anonymous reviewers and Robert A. Peterson for constructive comments.

References

Acemoglu D, Angrist J (2000) How large are human-capital externalities? Evidence from compulsory schooling laws. NBER Macroecon Annu 15:9–59

Arthur WB (1996) Increasing returns and the new world of business. Harv Bus Rev 74:99–109

Baum CF (2006) An introduction to modern econometrics using Stata. Stata Press, Texas

Baum CF, Schaffer ME, Stillman S (2007) Enhanced routines for instrumental variables/GMM estimation and testing, CERT Discussion Paper 0706, Centre for Economic Reform and Transformation, Heriot-Watt University

Baum CF, Schaffer ME, Stillman S (2007) IVREG2: Stata module for extended instrumental variables/ 2SLS, GMM and AC/HAC, LIML and k-class regression. Available online: http://ideas.repec.org/c/boc/bocode/s425401.html, 27 Feb. 2009

Baum CF, Schaffer ME, Stillman S (2003) Instrumental variables and GMM: estimation and testing. Stata J 3:1–31

Belke AH, Fehn R, Foster N (2003) Does venture capital investment spur employment growth? CESifo Working Paper Series No. 930. National Bureau of Economic Research, Cambridge, MA

Black D, Henderson V (1999) A theory of urban growth. J Polit Econ 107:252–284

Blau F, Kahn L (1994) Rising wage inequality and the U.S. gender gap. Am Econ Rev 84:23–28

Blinder AS (1973) Wage discrimination: reduced form and structural estimates. J Hum Resour 8:436–455

Bound J, Jaeger DA, Baker R (1995) Problems with instrumental variables estimation when the correlation between instruments and the endogenous explanatory variable is weak. J Am Stat Assoc 90:443–450

Chapple K, Markusen A, Schrock G, Yamamoto D, Yu P (2004) Gauging metropolitan "high-tech and i-tech" activity. Econ Dev Q 18:10–29

Chmelarova V, Hill RC (2004) Finite sample properties of the Hausman test. Southern Economic Association Meeting, New Orleans, LA. Available online: http://www.shsu.edu/~eco_www/resources/documents/Hausman10.pdf, 10 Aug. 2008

Ciccone A, Peri G (2006) Identifying human capital externalities: theory with applications. Rev Econ Stud 73:381–412

Cutillo A, Di Pietro G (2006) The effects of overeducation on wages in Italy: a bivariate selectivity approach. Int J Manpow 27:143–168

Echeverri-Carroll EL, Ayala SG (2004) Economic growth and linkages with Silicon Valley: the cases of Austin and Boston. Texas Business Review (December), Bureau of Business Research, Red McCombs School of Business, The University of Texas at Austin

Echeverri-Carroll EL, Ayala SG (2006) High-technology agglomeration and gender inequalities. Paper presented at the American Economic Association Meetings, Boston, MA

Echeverri-Carroll EL, Ayala SG (2009) Wage differentials and the spatial concentration of high-technology industries. Pap Reg Sci 88:623–641

Edin P-A, Zetterberg J (1992) Interindustry wage differentials: evidence from Sweden and comparison with the United States. Am Econ Rev 82:1341–1349

Ellison G, Glaeser EL (1997) Geographic concentration in U.S. manufacturing industries: a dartboard approach. J Polit Econ 105:889–927

Faggian A, McCann P, Sheppard S (2007) Some evidence that women are more mobile than men: gender differences in U.K. graduate migration behavior. J Reg Sci 47:517–539
Fingleton B, Igliori DC, Moore B (2006) Cluster dynamics: new evidence and projections for computing services in Great Britain. J Reg Sci 45:283–311
Fingleton B, Igliori DC, Moore B (2004) Employment growth of small high-technology firms and the role of horizontal clustering: evidence from computing services and R&D in Great Britain, 1991–2000. Urban Stud 41:773–799
Fingleton B, Igliori DC, Moore B, Odedra R (2007) Employment growth and clusters dynamics of creative industries in Great Britain. In: Polenske KR (ed) The economic geography of innovation. Cambridge University Press, Cambridge, pp 60–86
Fields J, Wolff EN (1995) Interindustry wage differentials and the gender wage gap. Ind Labor Relat Rev 49:105–120
Florida R, Smith Jr. DF (1993) Venture capital formation, investment and regional industrialization. Ann Assoc Am Geogr 83:434–451
Gannon B, Plasman R, Rycx F, Tojerow I (2007) Inter-industry wage differentials and the gender wage gap: evidence from European countries. Econ Soc Rev 38:135–155
Garofalo G, Fogarty MS (1979) Urban income distribution and the urban hierarchy-equality hypothesis. Rev Econ Stat 61:381–388
Glaeser EL, Maré DC (2001) Cities and skills. J Labor Econ 19:316–342
Glaeser EL, Kallal HD, Scheinkman JA, Shleifer A (1992) Growth in cities. J Polit Econ 105:889–927
Gorman M, Sahlman WA (1989) What do venture capitalists do? J Bus Venturing 4:231–248
Grossman GM, Helpman E (1991) Innovation and growth in the global economy. MIT, Cambridge, MA
Hadlock P, Hecker D, Gannon J (1991) High-technology employment: another view. Mon Labor Rev 114:26–30
Hecker DE (2005) High-technology employment: a NAICS-based update. Mon Labor Rev 128:57–72
Hecker DE (1999) High-tech employment: a broader view. Mon Labor Rev 122:18–28
Henderson VJ (2007) Understanding knowledge spillovers. Reg Sci Urban Econ 37:497–508
Imbens GW, Angrist JD (1994) Identification and estimation of local average treatment effects. Econometrica 62:467–75
Jaffe AB, Trajtenberg M, Henderson R (1993) Geographic localization of knowledge spillovers as evidenced from patent citations. Q J Econ 108:577–598
Kleibergen F, Paap R (2006) Generalized reduced rank tests using the singular value decomposition. J Econom 127:97–126
Krueger AB, Summers LH (1988) Efficiency wages and the inter-industry wage structure. Econometrica 56:259–293
Krugman PR (1993) Geography and trade. MIT, Cambridge, MA
Luker W, Lyons D (1997) Employment shifts in high-technology industries, 1988–1996. Mon Labor Rev 120:12–25
Malecki E (1981) Public and private sector interrelationships, technological change, and regional development. Pap Reg Sci Assoc 47:121–137
Manigart S, DeWaele K, Wright M, Robbie K, Desbričres P, Sapienza HJ, Beekman A (2002) Determinants of required return in venture capital investments: a five-country study. J Bus Venturing 17:291–312
Marshall A (1890) Principles of economics: an introductory text. McMillan, London
McCall L (2001) Complex inequality gender, class, and race in the new economy. Routledge, New York
McCall L (1998) Spatial routes to gender wage (in)equality: regional restructuring and wage differentials by gender and education. Econ Geogr 74:379–404
Midelfart-Knarvik K, Overman H, Redding S, Venables A (2000) The location of European industry. Economic Papers No. 142, European Commission, D-G for Economic and Financial Affairs, Brussels

Mincer J (1974) Schooling, experience and earnings. National Bureau of Economic Research, New York
Moretti E (2004) Human capital externalities in cities. In: Henderson V, Thisse JF (eds) Handbook of urban and regional economics, vol 4. North-Holland (Elsevier), Amsterdam, pp 2243–2292
Newman NS (1998) Net loss: government, technology and the political economy of community in the age of the internet. Ph.D. Dissertation, Department of Sociology, University of California, Berkeley
Oaxaca R (1973) Male-female wage differentials in urban labor markets. Int Econ Rev 14:693–709
Office of Technology Assessment (1982) Technology, innovation, and regional economic development. U.S. Congress, Washington, DC
Ratanawaraha A, Polenske KR (2007) Measuring the geography of innovation: literature review. In: Polenske KR (ed) The economic geography of innovation. Cambridge University Press, pp 30–59
Rauch JE (1993) Productivity gains from geographic concentration of human capital: evidence from the cities. J Urban Econ 34:380–400
Richie RW, Hecker DE, Burgan JU (1983) High-technology, today and tomorrow: a small slice of the employment pie. Mon Labor Rev 106:50–59
Saxenian A (1994) Regional advantage: culture and competition in Silicon Valley and Route 128. Harvard University Press, Cambridge, MA
Schwanhausser M (2007) In Silicon Valley, few women reach top jobs. San Jose Mercury News. Available online: http://www.mercurynews.com, 13 Aug. 2008
Segal D (1976) Are there returns to scale in city size? Rev Econ Stat 58:339–350
Shefer D (1973) Localization economies in SMSAs: a production function approach. J Reg Sci 13:55–64
Staiger D, Stock JH (1997) Instrumental variables regression with weak instruments. Econometrica 65:557–586
Stock JH, Yogo M (2005) Testing for weak instruments in linear IV regression. In: Andrews DW, Stock JH (eds) Identification and inference for econometric models: essays in honor of Thomas Rothenberg. Cambridge University Press, pp 80–108
Storper M, Venables AJ (2003) Buzz: face-to-face contact and the urban economy. J Econ Geogr 4:351–370
Sveikauskas L (1975) The productivity of cities. Q J Econ 89:393–413
Teece D (2002) Knowledge and competitiveness as strategic assets. In: Holsapple CW (ed) Handbook on knowledge management: knowledge matters, vol 1. Birkhäuser, Cambridge, MA, pp 129–152
Yu PD (2004) Focus on high-tech: what's in a name? Gauging high-tech activity. Reg Rev 14:6–9 (Federal Reserve Bank of Boston)
Wooldridge JM (2006) Introductory econometrics: a modern approach, 3rd edn. Thomson South-Western, Massachusetts
Yankow JJ (2006) Why do cities pay more? An empirical examination of some competing theories of the urban wage premium. J Urban Econ 60:139–161
Zucker LG, Darby MR, Brewer MB (1998) Intellectual human capital and the birth of the U.S. biotechnology enterprises. Am Econ Rev 88:290–306

Fiscal Policy and Interest Rates: The Role of Financial and Economic Integration

Peter Claeys, Rosina Moreno, and Jordi Suriñach

1 Introduction

A government running a deficit needs to turn to financial markets to place this additional public debt. Newly issued public bonds compete for financing with bonds issued by private agents. The additional demand created by the fiscal expansion pushes up interest rates, and eventually crowds out private investment. Not all economists agree that consolidating public finances would immediately reduce pressure on interest rates, however. Despite a vast literature testing crowding out, there is actually surprisingly little robust empirical support for this hypothesis.[1]

Interest rates are insulated from fiscal policy under two alternative conditions. The first explanation for a zero impact of deficits on aggregate macroeconomic variables is that economic agents anticipate paying down currently high deficits with higher taxes in the future. Under Ricardian Equivalence, private saving fully offsets the effect of higher public consumption (for a given level of taxation). Few economists would consider the assumptions underlying the Ricardian Equivalence null as realistic, however. More elaborate macroeconomic models that depart from the baseline Ricardian assumption easily find real economic effects of fiscal policy. There is by now also a large body of empirical evidence that clearly refutes the Ricardian hypothesis (Blanchard and Perotti 2002).

A second explanation for the lacking crowding out effect is capital mobility. Fiscal deficits need not be financed by domestic financial resources only. Capital flows between economically integrated economies, offsetting any interest rate differentials that follow upon an increase in the domestic supply of government bonds. Under full capital mobility, domestic and foreign agents alike diversify their asset

[1] See the contrasting arguments of the European Commission (2004) and the Bush Administration (Gale and Orszag 2003).

P. Claeys (✉)
AQR Research Group-IREA, University of Barcelona, Avinguda Diagonal 690, 08034 Barcelona, Spain,
e-mail: peter.claeys@ub.edu

A. Páez et al. (eds.), *Progress in Spatial Analysis*, Advances in Spatial Science,
DOI 10.1007/978-3-642-03326-1_15,

portfolio across borders. As a consequence, the budget decision of one government affects the financing conditions of all other governments on international capital markets. Domestic interest rates rise in proportion to the amount of bonds issued worldwide. For a small open economy, the crowding out effect would tend to zero. In practice, capital mobility is far from complete, as imperfect information, risk aversion and imperfect substitutability of domestic and foreign bonds introduce a home bias in portfolio decisions. As a consequence, the spillover is likely to be less than complete.

The empirical models that are used to assess the crowding out effect of fiscal policy fail to account for this spillover. A baseline test for crowding out typically regresses a domestic interest rate on some domestic fiscal indicator. Even simple extensions of this model to include the effects of fiscal policy in open economies require quite restrictive assumptions on parameterization. Often, one simply controls for a set of additional (foreign) explanatory variables. Usually, only a particular subset of countries is examined, or identical restrictions are imposed on the transmission of fiscal policy across all countries. Due to the dimension of open economy models, even simple extensions quickly exhaust the available degrees of freedom. In practice, the interactions are much more complex. Spillover works out on global financial markets, and affects a large group of countries contemporaneously.

In this chapter, we use spatial techniques to account for this spillover. We deliberately keep the baseline model as simple as possible to make the strongest possible case for spillover. We test a panel model that explains interest rates by fiscal variables to analyze crowding out of fiscal policy. The spatial model simply extends this baseline model for the spillover effect in all nearby foreign economies. In particular, we test the spillover of interest rates on financial markets in a spatial lag model. We then control also for spatially distributed economic linkages in a spatial error model. We test this model on a large cross section of OECD and emerging market economies over the period 1990–2005. Our main finding is that the domestic crowding effect of fiscal policy is sizeable. But the spillover on financial markets offsets the significant effects of larger deficits on interest rates. We cannot identify whether this spillover is the direct effect of financial market integration, or the by-product of the economic integration of countries. Spillover is much stronger in the mid-1990s when there were major crises, or policy actions were being coordinated between governments. Various measures of cross-country linkages give broadly similar results. The main findings are also robust to alternative specifications and data definitions. Finally, we find the spillover to be quite strong among EU countries.

The chapter is structured as follows. First, we provide a simple theoretical model for testing crowding out, and the effects of financial and economic integration. We consequently specify the spatial model for including this spillover. We then discuss the results of the baseline model of spillover of fiscal policies, and provide several robustness checks. The final section summarizes the main results, and discusses some policy implications.

2 Crowding Out and Spillover

Crowding out of interest rates is typically analyzed in a partial equilibrium "loanable funds" model (Barro 1992). This model determines the interest rate on assets from the equilibrium between the demand and supply of bonds. Both the private sector and the government turn to financial markets to look for finance. Firms invest the capital raised on stock or bond markets (b) to invest in new capital stock. Additional government financing is necessary when the government runs a deficit (d). Financial intermediaries channel the demand for bonds of the private sector, both at home (f) and abroad (f*), to match the total supply of domestic bonds (b + d):

$$b + d = f + f^* \tag{1}$$

We can illustrate all the major points of the analysis with this simple partial equilibrium model.[2] Consider the case in which the government runs a higher deficit d. For a given demand for bonds f + f*, the increased supply of debt will put downward pressure on the price of government bonds. Ceteris paribus, this rise in bond yields is making it more difficult for the private sector to seek finance on capital markets: government finance crowds out private bonds b on capital markets (Cebula 1998). However, the demand for additional bonds is likely affected by the government's decision to lend. Under two alternative theoretical conditions, private sector savings fully offset the additional supply of bonds.

First, if economic agents anticipate the pay down of higher deficits they set aside savings for the higher tax burden in future periods. In the limit, domestic private saving fully offsets the effect of the higher public dissaving d. Under this Ricardian Equivalence hypothesis, a higher deficit d does not have an impact on aggregate macroeconomic variables at all. Many economists consider Ricardian Equivalence as a reasonable starting point for the analysis of fiscal policy. Few would endorse it as a realistic description, however. The view that private savings offset the change in public savings is not based on a firm empirical rejection since Ricardian Equivalence is not directly testable. But plenty of empirical studies have examined the alternative hypothesis that fiscal policy has any real economic effects. Recent evidence seems to converge on at least some expansionary effects on major economic variables (Blanchard and Perotti 2002). More elaborate macroeconomic models that depart from the baseline Ricardian assumption easily find support for these real economic effects of fiscal policy. It therefore seems a safe assumption to reject the Ricardian Equivalence hypothesis.

Second, when financial markets are integrated across borders, the foreign demand for bonds f* can shift as well. In open economies that are economically integrated and do not impede trade or financial flows, capital flows move massively so as to

[2] These partial equilibrium models have been extended for intertemporal saving behaviour (Laubach 2003; Engen and Hubbard 2004). Dynamic macroeconomic models that include both debt non-neutrality and long term interest rates have not been developed yet, due to their complexity.

offset any interest rate differential. Under these circumstances, the supply of savings is very interest rate elastic: even a small rise in d is likely to trigger a large increase in f*. The higher is capital mobility, the weaker will be the reaction of domestic interest rates to a change in the supply of domestic bonds. Under the null of full capital mobility, the rise in interest rates is simply proportional to each country's total indebtedness on the global bond market. Consequently, due to the spillover on international bond markets, the crowding out effect of deficits is only a fraction of the total rise under autarchy. In reality, this dilution is likely to be less than complete as capital mobility is partial: domestic private agents prefer to invest in domestic financial assets. As a consequence, domestic savings and investment are highly correlated (the "Feldstein-Horioka" puzzle). This "home bias" depends on information imperfections on foreign financial markets. Neither are financial assets in different countries perfect substitutes, due to exchange rate, inflation and default risk. Due to differences in regulation across countries, risk averse agents may prefer to invest in domestic assets only. Moreover, private agents are likely to hold a larger portion of domestic public debt. Governments often prefer to place debt only domestically in order to avoid having to pay exchange rate premia, and as a commitment not to default on debt put with its own citizens. As a consequence, complete spillover is unlikely, yet it is hard to put a precise size on the spillover effect. The spillover is likely to be stronger between economies that are more closely integrated.

3 A Spatial Test for Crowding Out

The most common test for crowding out based on the loanable funds model takes a very simple form: it basically explains domestic interest rates by domestic fiscal balances (in this case, the surplus s_t).

$$i_t = \alpha + \beta s_t + \varepsilon_t \tag{2a}$$

We measure by the coefficient β the degree of crowding out. The large number of studies that have employed various definitions of the government surplus, interest rates, econometric approaches and data sets to test (2a) can basically give support for any view.[3] There are two cases in which we would not reject $\beta = 0$. First, we would not find a significant crowding out effect under Ricardian Equivalence. Second, we may not be able to reject $\beta = 0$ if there is capital mobility between open economies.

However, a specification like (2a) does not account for these spillover effects of financial markets. Basically, the test for crowding out only considers domestic variables, but does not account for the integration of the domestic economy with

[3] See the references in Barth et al. (1991) and the overview article by the European Commission (2004).

foreign economies. Existing empirical evidence on the effect of integration is little, and uses alternative ways for netting out the international linkages from the domestic crowding out effect. Due to the omission of these foreign variables that explain the spillover effect, β would be biased downward. The reasoning is that the domestic effect of fiscal expansions will likely be larger once a proxy X_t for the foreign demand for bonds f* is introduced.

$$i_t = \alpha + \beta s_t + \phi X_t + \varepsilon_t \tag{2b}$$

One alternative is to condition the relation between interest rates and deficits on foreign capital inflows (Cebula and Koch 1994). Another is to assess directly the effects of deficits on interest rates abroad. In (2c), we explain interest rates in country A by the crowding out by deficits in country A, and in addition a direct crowding out effect of deficits in country B.

$$i_{A,t} = \alpha + \beta s_{A,t} + \phi s_{B,t} + \varepsilon_t \tag{2c}$$

The coefficient ϕ measures the size of the spillover effect. In order to estimate a specification similar to (2c), plenty of identifying restrictions are necessary that severely reduce the dimension of the open economy model. Usually, only a particular subset of countries can be examined, or identical restrictions are imposed on the transmission across all countries in a panel. Cohen and Garnier (1991) find a positive effect of US deficits on interest rates in several G7 countries. Ardagna et al. (2007) find significant crowding out effects from both domestic and foreign fiscal expansions in a panel of OECD countries. Marcellino (2002) or Giuliodori and Beetsma (2005) consider the impact of shocks to German fiscal policy on the French and Italian economy. Paesani et al. (2006) take a somewhat different approach by identifying spillover from shocks to bond markets on internationally linked capital markets. This allows them also to consider the direction of the spillover.

An alternative control that models the linkages between domestic and foreign bond markets is to include the level of foreign interest rates in (2b).

$$i_{A,t} = \alpha + \beta s_{A,t} + \phi i_{B,t} + \varepsilon_t \tag{2d}$$

Quite a few papers include foreign interest rates in the analysis of crowding out. Chinn and Frankel (2007) take the German long-term rate as the benchmark in their study of US fiscal policy. Caporale and Williams (2002) or Paesani et al. (2006) reduce their sample to a few G7 economies and use in turn the interest rate from the other country as a benchmark. The assumption that domestic interest rates directly depends on a single foreign benchmark rate, is rather strong. Ideally, one would like to control for the level of interest rates in various countries. Most papers construct as the benchmark interest rate an aggregate "world" interest rate. Tanzi and Lutz (1993) argue that at the world level all spillover effects should cancel out and a significant crowding out effect of fiscal policy restored. They aggregate all domestic deficits and examine the effect on global interest rates. Ford and Laxton (1999) and

De Haan and Knot (1995) do the same for OECD and EU countries respectively. Faini (2006) calculates an average euro area interest rate, and considers the fiscal effect on interest rates at home and at EMU level contemporaneously in a panel framework. In empirical applications, modeling the transmission across financial markets requires quite restrictive assumptions. Assumptions like these likely bias the direction and the strength of the spillover effect. We would expect that on financial markets, spillover is at work between different markets contemporaneously. Spillover should also be stronger between markets that are more closely connected.

A convenient way to think of these complex linkages is with an exogenously specified matrix W that specifies the structure and intensity of the "closeness" of different observations. The element w_{ij} of W represents the proximity between two observations i and j. A common specification for this weight matrix W is physical contiguity. Bordering regions are believed to have closer links. It is straightforward to find other W's that reflect either economic distance between countries, such as geographic distance, trade, level of economic development, cultural or institutional differences. A pattern of spatial interaction in a variable implies that the distribution of this variable across observations is not random, and therefore the co-movement of interest rates on integrated financial markets will bias the OLS estimates of (2a). With spatial spillover, parameter estimates are biased, inefficient and inconsistent (Anselin 1988). This bias may explain the mixed findings in the empirical literature testing crowding out and spillover.

By introducing spatial lags (i.e. interest rates in neighboring countries) we directly control for the interaction with the level of interest rates in close by units. We can rewrite (2d) more generally as a spatial autoregressive model:

$$i_{n,t} = \alpha + \beta s_{n,t} + \rho W i_{n,t} + \varepsilon_{n,t} \tag{3a}$$

In specification (3a) we control the crowding out effect in country n for the interaction with interest rates in all neighboring countries (the term Wi_n). This is a weighted measure of interest rates in the countries with which a country has "economic links." This spatial lag term has to be treated as an endogenous variable; (3a) can be estimated with ML-techniques. A positive (and significant) coefficient ρ indicates spillover. We can also recast the test of significance of ρ as a test for the degree of financial market integration: the larger ρ, the more integrated are financial markets. A significant spatial lag also reduces the bias in the estimate of the direct crowding out effect β.

Financial integration is not the only factor behind the co-movement of bond markets. Real economic integration affects macroeconomic conditions globally, and there is quite some evidence for increased synchronization of business cycles (Doyle and Faust 2002). Instead of a direct spillover on financial markets, the spillover could alternatively be due to a co-movement of some omitted economic variables, unrelated to financial market integration, that vary across space. There are two possible ways to account for these effects of real economic integration. First, and as in (2b), we proxy the economic co-movement across countries by introducing other (domestic) macroeconomic variables. Although an OLS regression of (2b) still yields unbiased estimates, inference may be misleading since the precision of the estimates is affected. Alternatively, we can introduce such spatial links in the residuals of the empirical model. Hence, we qualify model (2a) to include a spatial correlation structure of the error term ε_n. We then rewrite (2a) as follows:

$$\begin{aligned} i_{n,t} &= \alpha + \beta s_{n,t} + \varepsilon_{n,t} \\ \varepsilon_{n,t} &= \lambda W \varepsilon_{n,t} + \nu_{n,t} \end{aligned} \tag{3b}$$

In this spatial error model, $\nu_{n,t}$ is a white noise error term. The parameter λ in (3b) is the spatial autoregressive parameter, and reflects the spillover across countries due to economic integration.

Financial and real economic integration probably work in the same direction. Co-movements in macroeconomic factors also drive the synchronization of financial markets. It is likely that economies with close economic (trade) links also have more tightly linked financial markets. As a consequence, the spillover effects of financial markets and economic integration may be similar, and hard to distinguish. This observational equivalence may cause the spatial lag and spatial error model to give very similar results (Kaminsky and Reinhart 2000).[4]

3.1 Specification

We argue that we can recover significant crowding out effects of fiscal policy on interest rates if we use a spatial extension of the simple baseline model (2a) to analyze the spillover effect of fiscal policy. Hence, we depart from a model as (2a) and use spatial techniques to test and model the crowding out and spillover effect of fiscal policy.

We are interested in two effects in the spatial model. In first instance, we test the crowding out effect of domestic deficits on domestic interest rates (β). We expect that a higher surplus (lower deficit) has a significantly negative effect on interest rates. Secondly, the spatial model allows testing for the effect of interest rates abroad. The more interest rates rise in other countries, the higher will be the domestic interest rate as well. If interest rates are very close, this suggests that global credit markets are fairly integrated. The pool of loanable funds any government draws from, exceeds the available funds in the domestic credit market only. We expect that this spillover effect will be significant, and positive.

The interpretation of the spillover effect is different in the spatial lag or error model. First, in the spatial lag model, we look at the aggregate effect of all other countries' interest rates on the home country's interest rate. We interpret a significant spatial lag as evidence of a direct spillover of fiscal policy on financial markets. The

[4] Models (3a) and (3b) can be combined in a general specification that encompasses both effects. We do not test a general spatial model in this paper.

spatial lag parameter ρ is the slope of this reaction function, and measures the degree of financial integration. Positive spatial correlation in interest rates exists if $\rho > 0$, whereas there is evidence of negative spatial correlation if $\rho < 0$.[5]

The economic interpretation of the model (3b) with spatial links in the errors is slightly different. A shock to the domestic interest rate, which is not explained by fiscal policy, spills over to all other observations that are "close." Spatial correlation in the error term reflects a similar reaction of countries' interest rates to shocks, because of omitted variables that are spatially correlated. This indicates important economic channels of spillover, but is not related to fiscal policy per se. A positive λ indicates positive spatial correlation of the shocks; negative λ shows that shocks are of opposite sign.

3.2 Data

We estimate these two spatial models on a panel of 101 countries, for which we have annual data on interest rates and fiscal policy covering the period 1990–2005. A panel model allows combining the typical analysis of domestic crowding out in a time series model, with the spillover of interest rates in the cross-section dimension. We tie both dimensions and the structure of linkages across countries, with a weight matrix W that reflects geographical distance.[6] Countries that are more distant are assumed to have weaker economic links. Geographic distance usually is a good proxy for economic linkages (as in the gravity model, for example). We show that our results do not depend on this specific assumption, and we check our results for different definitions of W.

Spatial panel data models have only recently been developed, and not all their properties have been examined. Our starting point is the fixed effect panel model in which subsequently spatial dependence is accounted for by including a spatially lagged term of the dependent variable. This is the expression given in (3a), including a country specific fixed effect. The standard estimation method for the fixed effect model is to eliminate the intercept term of the regression by expressing all variables as a deviation from their time average, and then using standard OLS estimators. In presence of spatial autocorrelation it is common practice to use maximum likelihood methods to estimate the demeaned equation. These spatial autoregressive models

[5] A spatial test for financial market integration could be equally applied to other financial assets. We look at the spillover effects of government bonds on interest rates for two reasons. First, spillover on government bond markets is policy relevant. In contrast to private bond issues, fiscal policy could introduce distortions on government bond markets. That is, there are not only pecuniary implications for other domestic or foreign issuers. Second, we can include many countries in our sample as government bonds are the most comparable asset across countries. They are usually the least risky asset and are traded on the most liquid market.

[6] We assign a centre to each country, and use its coordinates to calculate the distance between these centroids. We use a GIS software for these calculations.

Table 1 Data sources

Series	Definition	Source
Interest rate	1- to 5-year government bond yield or corresponding (%)	IMF/IFS Central Banks
Surplus	Surplus/GDP ratio (%)	IMF General government Statistics
Public debt	Debt/GDP ratio (%)	IMF General government Statistics
Short-term interest rate	Central Bank, T-bill 3 months or corresponding (%)	IMF
Long-term interest rate	10-year government bond yield or corresponding (%)	IMF
GDP per capita	US $ PPP	Penn World Tables
Exports	Export CIF	IMF Dots
Imports	Import FOB	IMF Dots
Distance	Great circle distance	Matlab
Common border		Authors
Dummies	Trade agreement	WTO
Country characteristics	Latitude/longitude	Rose (2000)

are estimated through the maximum likelihood method of estimation developed by Elhorst (2003).[7]

We use a nominal long-term interest rate (5–10 years) as our dependent variable. Deficits are usually argued to affect long-term interest rates. Not that many countries outside the OECD have been able to issue long-term bonds, however. Most emerging market economies have financed deficits with short-term bonds at a 5 years horizon at most. Similarly, fiscal data for many countries are available only over recent years. The surplus to GDP ratios all come from the IMF Government Statistics Database. As we prefer working with balanced panels over the full sample period, we consequently had to eliminate a large number of countries from the study. Due to variable data quality, we also decided to remove some outlier observations. We first run a simple pooled estimate and quitted the observation if the standard error exceeds three times the residual variance. Eventually, we have 496 observations on 31 countries over 16 years. This keeps in the sample mostly OECD countries and a few emerging market economies.[8] The sources of data are shown in Table 1.

[7] Consistent estimation of the individual fixed effects is not possible as n grows large, due to the incidental parameter problem. Anselin and Le Gallo (2008) argue that "since spatial models rely on the asymptotics in the cross-sectional dimension to obtain consistency and asymptotic normality of estimators, this would preclude the fixed effects model from being extended with a spatial lag." However, Anselin and Le Gallo (2008) show that for consistent estimates of β, the demeaned spatial regressions from ML estimation like in Elhorst (2003) are appropriate. One complication with this is that the variance covariance matrix of the demeaned error term is different from the usual one. Alternative approaches to the Elhorst estimation are still a topic of ongoing research.

[8] The sample includes the following EU countries (Austria, Belgium, Denmark, Finland, France, Germany, Hungary, Ireland, Italy, Luxembourg, Netherlands, Portugal, Spain, the UK), some other OECD countries (Australia, Canada, Japan, Korea, New Zealand, Switzerland, Turkey, the

4 Results for the Baseline Model

4.1 Spillover on Financial Markets: The Spatial Lag Model

We first estimate a simple pooled OLS regression of (2a), without any further restrictions on the spatial effects, to replicate the crowding out effect of similar studies. Table 1 reveals a very significant and quite strong crowding out effect: a rise in the surplus of 1% of GDP leads to a 109 basis points fall in the interest rate. This result continues to hold even if we impose more structure on the pooled model. A priori, we would prefer to use a fixed effects estimator. First, we include a specific group of countries. Even if we draw a number of countries from the global sample of economies, this draw is not random (Baltagi 2001). Second, the specification (2a) is rather basic, and we do not control for other relevant determinants of interest rates. The setting of fiscal policy is rather heterogeneous across countries. As a consequence, the country-specific effect is likely correlated with the explanatory variable (the surplus). The Hausmann test indicates that a fixed effects estimator is indeed preferable. It has also been more common in this literature to use simple pooled estimates or panel fixed effects estimators (Frankel and Chinn 2005; Kinoshita 2006). We report both, and find that the panel estimates of β in the fixed or random effects model are very similar (Table 2).

Our estimate is on the high end of the range of estimates found in the literature. In the overview study of the European Commission (2004), the crowding out effect varies between about 20 and 100 basis points.[9] For the United States, the crowding out effect is usually estimated to be around 40 basis points (Canzoneri et al. 2002; Laubach 2003; Engen and Hubbard 2004). This result is also confirmed by VAR studies on US data, like Dai and Phillipon (2005). For EU countries, the crowding out effect is mostly smaller in magnitude (Faini 2006). What explains the strong

Table 2 Baseline model, pooled and panel estimates; and spatial panel lag model (W-matrix = distance)

		β	t-stat	ρ	t-stat
	Pooled	−1.09	−5.22	–	–
Panel	Panel, fixed effects	−1.16	−5.30	–	–
	Panel, random effects	−1.06	−5.20	–	–
	Hausmann test statistic	15.05 (0.00)			
	Panel, fixed effects	−0.45	−4.54	0.55	8.09
Spatial lag	Panel, random effects	−0.46	2.02	0.51	0.00
	Panel, spatial + time period fixed effects	−0.43	−4.09	−0.05	−0.37

United States) and emerging markets (Colombia, Lebanon, Mexico, Pakistan, Peru, Philippines, Singapore, South Africa, Thailand).

[9] See in particular the overview table in European Commission (2004) on pp. 153–55.

crowding out effect in our estimates then? Most other papers examine the crowding out effect of deficits in a single country. In contrast, panel studies, like ours, have found much stronger effects. Chinn and Frankel (2007) estimate a crowding out of interest rates between 150 and 200 basis points in a panel of the United States and the largest EU countries. Similarly, Ardagna et al. (2007) use panel VAR techniques to look at the impact of deficits on interest rates in a panel of OECD countries and find a rise of 150 basis points after 10 years. De Haan and Knot (1995) reach similar conclusions for the large EU countries. Hence, the inclusion of more countries in a cross-section analysis of deficits and interest rates typically delivers stronger crowding out effects. We actually observe a similar effect in single country studies in which control variables for international capital flows are included. Cebula and Koch (1994) find that interest rates rise by more than 60 basis points after a 1% increase in the deficit ratio, whereas capital flows reduce the effect by about 24 points. Chinn and Frankel (2007) find a stronger impact on rates, once foreign interest rates are controlled for. Tanzi and Lutz (1993) aggregate data for the G7 and find a rise in long-term rates of about 150 basis points. These results suggest that a control for the spillover effect from other countries is important. Omission of linkages on international financial markets biases the findings of crowding out.

As the estimate of β is quite likely biased, inefficient and inconsistent, we now introduce the spatial extension. In the second panel of Table 2, we present the estimates of different versions of the spatial lag model. The baseline estimate is the spatial lag model with fixed effects. We find that the crowding out effect halves in case a spatial lag is included: a deficit of 1% of GDP pushes up interest rates by 45 basis points. The spillover effect is very significant and quite large: a 1% rise in interest rates abroad also raises domestic rates by about 0.55%. The consequence is that an increase in the deficit of 1% will cause domestic interest rates to rise by 45 basis points. Consequently, the second round effect of the deficit is to push up interest rates abroad by a further 25 ($\approx$0.55%*45 pp) basis points. A government creating a deficit still faces a quite steep increase in domestic rates, but part of this increase spills over abroad.

The crowding out effect in the spatial lag model is more in line with the results of the empirical studies of single countries. This suggests that the control for the spatial links indeed corrects the initial panel estimates. As regards our findings on spillover, it is slightly harder to compare its size. Most studies simply report that the domestic crowding out effect is larger than the foreign spillover effect. Caporale and Williams (2002) find this result for the United States; and Faini (2006) reports similar results for the EU countries. Ardagna et al. (2007) report that the aggregate (world) deficit affects domestic interest rates, but its impact is less relevant than that of domestic fiscal policy. In different settings, other studies have found close connections between interest rates across borders (Minford and Peel 2007). Nonetheless, country-specific factors still play a role in explaining the deviation of domestic interest rates from the evolution in worldwide interest rates (Breedon et al.

Table 3 Baseline model, spatial panel error model (W-matrix = distance)

	β	t-stat	λ	t-stat
Panel, fixed effects	−0.44	−4.17	0.56	8.42
Panel, random effects	−0.48	−4.26	0.52	5.28
Panel, spatial + time period fixed effects	−0.43	−4.10	−0.06	−0.43

1999). One of the main reasons is a change in the fiscal policy stance.[10] Few studies report the impact that international capital flows or foreign interest rates have on domestic interest rates. Cebula and Koch (1994) find a similarly strong reduction in interest rates (24 pp) as we do.

The co-movement of interest rates may not just reflect the integration of financial markets. Economic integration makes countries susceptible to global economic developments. Trade, financial integration and similar economic structures raise the co-movement of business cycles internationally (Imbs 2004). Economics shocks that are common to a group of countries would display a close synchronization of economic variables. This might show up in a significant spillover effect. We introduce a time period fixed effect in the spatial panel to absorb these common shocks. We indeed find that the spillover effect is much smaller in this case, whereas the crowding out effect remains as strong (Table 2).

4.2 *Financial and Real Economic Integration*

An alternative possibility is that the co-movements of economic variables are also spatially distributed. Another way to model these economic links is to incorporate a spatial structure in the residuals of the baseline model. The assumption is that these economic factors, except interest rates, are spatially distributed across economies. We estimate this spatial error panel model (3b).

By controlling for these spatial linkages, we pick up a significant crowding out effect. Table 3 shows that the results are very similar to those of the spatial lag model. Moreover, the spillover effect causes a 1% rise in foreign rates to raise domestic rates by 56 basis points. We can not identify whether spillover is due to either financial market integration, or the co-movement of macroeconomic variables (Kaminsky and Reinhart 2000).

4.3 *Some Control Variables*

Alternatively, one may consider a correction of the baseline model (2a) with a spatial structure for the errors too naive. The factors that determine interest rates are plenty

[10] Note that for other assets than government bonds, most empirical papers find similar results on the importance of spillover. Ehrmann et al. (2005) find that asset prices react more strongly to domestic shocks, but still allows for a strong spillover between the US and EU markets.

Table 4 Augmented model, spatial panel lag model, spatial fixed effects, specifications (W-matrix = distance). See (4)

		β	t-stat	ρ	t-stat	θ	t-stat
	Baseline	−0.45	−4.54	0.55	8.09	–	–
$X_t =$	Debt	−0.16	−2.94	0.73	17.28	−0.00	2.07
	Short-term interest rate	0.00	0.00	−0.53	−2.98	0.04	2.78
	Inflation	0.01	0.29	0.34	3.37	0.01	0.28

of course, and the surplus is certainly not the only determinant of (long-term) interest rates. One often stated reason for the ambiguous findings regarding crowding out is the contemporaneous influence of monetary policy, automatic fiscal stabilizers, interest payments on outstanding debt and any economic effects of fiscal policy itself.[11] We test extensions of the spatial lag model that control for these additional regressors $X_{n,t}$, as in (4):[12]

$$i_{n,t} = \alpha + \beta s_{n,t} + \rho W i_{n,t} + \theta X_{n,t} + \varepsilon_{n,t} \tag{4}$$

It is quite common in the empirical literature on crowding out to directly test the effect of public debt on (long-term) interest rates, instead of using deficits. The argument is that public debt substitutes private capital, and hence it the stock of debt that has an impact on the level of interest rates (Engen and Hubbard 2004). Moreover, the initial fiscal position of countries matters for crowding out. Fiscal policy has non-linear effects. At higher levels of debt, interest rates typically react more strongly to higher deficits (Ardagna et al. 2007). In particular, emerging market economies start paying a higher risk premium for fiscal indiscipline (Zoli 2004).

Table 4 reports the estimates of the spatial panel lag model with fixed effects for a model augmented with public debt. Controlling for public debt gives an interesting result. The crowding out effect of the surplus becomes less strong: interest rates rise by a mere 16 basis points after a 1% rise in the deficit. Ardagna et al. (2007) find a short run effect of deficits of about 10 basis points, after controlling for debt. This effect accumulates over time to about 100 basis points, especially as the debt ratio rises. The impact of debt – albeit significant – is very small.

These results fall in a similar range as in the other studies. Single country studies find rather modest crowding out effect of higher public debt. The consensus estimate ranges between 2 and 7 basis points for the United States with a variety of methodologies (Ford and Laxton 1999; Canzoneri et al. 2002; Laubach 2003; Engen and

[11] These effects could cause some problems of endogeneity in (2a), but these feedback effects are likely small. IV estimates are not considered in most of the literature, however. Spatial panel models that control for endogeneity of the regressors have not been developed yet.

[12] In the spatial econometrics literature, the bottom-up approach for searching an adequate specification prevails. The so-called Hendry approach is not common. Florax et al. (2003) demonstrate that the specific-to-general approach slightly outperforms the Hendry approach in the case of the estimation of linear spatial models.

Hubbard 2004).[13] As for the impact of the surplus on interest rates, studies that control for links between different countries, or use a cross-section approach, find slightly stronger effects of debt. For example, the impact of debt is slightly stronger for the EU countries than in the United States (Faini 2006). Pooled estimates for a group of OECD countries show a rise of about 25 basis points after a rise in domestic debt (Ford and Laxton 1999; Orr et al. 1995). A similar effect is found by Tanzi and Lutz (1993).

Interestingly, the spillover effect is much stronger and is estimated very precisely. Three quarters of a 1% rise in the interest rates spills over to close by countries. After all crowding out and spillover effects have worked out, a 1% rise in the deficit will push up interest rates by 16 basis points at home, and by 12 basis points abroad.

Long-term interest rates are very much influenced by monetary policy in the short term.[14] We control in two different ways for its effect. First, we include a short-term interest rate in the specification. At short horizons, monetary policy sets interest rates to stabilize inflation and output. Central bank decisions directly influence the financing conditions of the government (and its interest payments on outstanding debt). The insignificance of the crowding out effect confirms that the short run impact of a higher deficit may be significant in raising interest rates, but it is not very important and it is blurred by the impact of monetary policy. But once a control for short-term rates is included, the spatial lag coefficient ρ is negative. Such a negative spillover effect can only be explained by a substantial spatial transmission of changes in short-term interest rates, which offset the co-movement of long-term interest rates between neighbouring countries. Other studies also illustrate this co-movement of short-term rates across borders (Minford and Peel 2007; Ehrmann et al. 2005). Second, we include also the inflation rate.[15] Higher inflation eases pressures on deficits as it erodes the real value of outstanding debt. We find that the spillover is not really affected by the spatial variations in inflation.

4.4 Time Variation in the Crowding Out Effect

Financial and economic integration can explain why changes in asset markets have large effects on other financial markets. Globalization is often argued to have

[13] The only exception is Friedman (2005), who finds that a 1% rise in the debt ratio increases interest rates by 90 basis points.

[14] Crowding out is obscured by static specification of the relation between deficits and interest rates in (2a). The reason is that government bonds are actually traded on financial markets. As financial markets are forward looking, it is the anticipation of upcoming deficits, rather than the current fiscal balance, that results in higher long term rates instantly. A few studies include expectations about the deficit or ratings, and directly analyse the effect of these budget projections on expected interest rates (Laubach 2003). These data are available for a limited sample only. Papers that look into the effect of deficit announcements by the government, or analyse the effect of deficits on risk premia usually ignore the spillover effects of fiscal policy, with the exception of Kitchen (1996).

[15] Data on inflation expectations are not available for all countries in the sample.

strengthened the spillover between economies in two different ways. On the one hand, as integration is a gradual process, we are likely to observe a change over time in the strength of spillover. On the other hand, there could be turbulence in the spillover channel due to financial or economic crises. Tranquil periods in which there is a normal degree of real and financial interdependence suddenly switch to an environment with wild co-movements during currency and financial crises (Claessens et al. 2001). Some authors argue this distinction is only apparent, and interdependence is determined by real factors that change only gradually over time (Boyer et al. 1999; Forbes and Rigobon 2002). The results in Table 2 showed that common shocks might be more important in explaining interdependencies across countries than a genuine spillover from other economies. If interest rates are indeed driven by some common factors in any given year, then we would not expect to see a spillover effect in a year-by-year estimation of the spatial lag model. All interdependencies would be absorbed by the constant term in this cross-section model. Note that if spatial links are predominantly determined by contagious crises across emerging economies, the annual frequency of fiscal data may not pick up the high frequency movements on financial markets due to sudden crises.

We turn again to the standard spatial lag model for explaining the variation in interest rates by fiscal variables but estimate it at a cross-sectional level for each year. Note that the efficiency of these cross-section estimates is smaller than in the panel case. Figure 1 plots the coefficients of an ML estimation of the baseline regression over the sample 1990–2005.

We have three major results. First, there is a crowding out effect of fiscal policy on interest rates: a fall in the surplus (higher deficit) raises interest rates. Second, the spillover effect is not particularly stable. The spatial lag coefficient varies in a rather large band between 0 and 40 basis points since the mid-1990s, but there are some strong drops in 1994 and 2004. Finally, both effects vary over time. We can distinguish three different episodes. In the first half of the 1990s, fiscal policy has hardly any crowding out effects. Foreign interest rates tend to go in the opposite direction of domestic rates. In a second period, which goes from the mid-1990s to the year 2000, crowding out is significant and large. At the same time, spillover becomes stronger as well. Starting in 1999, crowding out and spillover both become less pronounced. There is a gradual decline in the estimated coefficients β and ρ. These results are also corroborated by the findings of a cross-section estimation of the spatial error model.

These results teach us some important lessons. First, if we compare these findings with our panel estimates with time period effects, we cannot clearly attribute the smaller spillover effect to common shocks only. There is an important change in the crowding out effect as well.

It is not surprising that spatial links are increasingly important in explaining the transmission of interest rates across borders. Increasing globalization is believed to have spurred capital mobility and increased trade flows. Linkages on international markets have certainly become much stronger than they were a decade ago. Moreover, the 1990s have seen several large crises that have spread to other countries. The 1994 Tequila crisis in Mexico was the first big “fiscal crash.” The Asian Flu

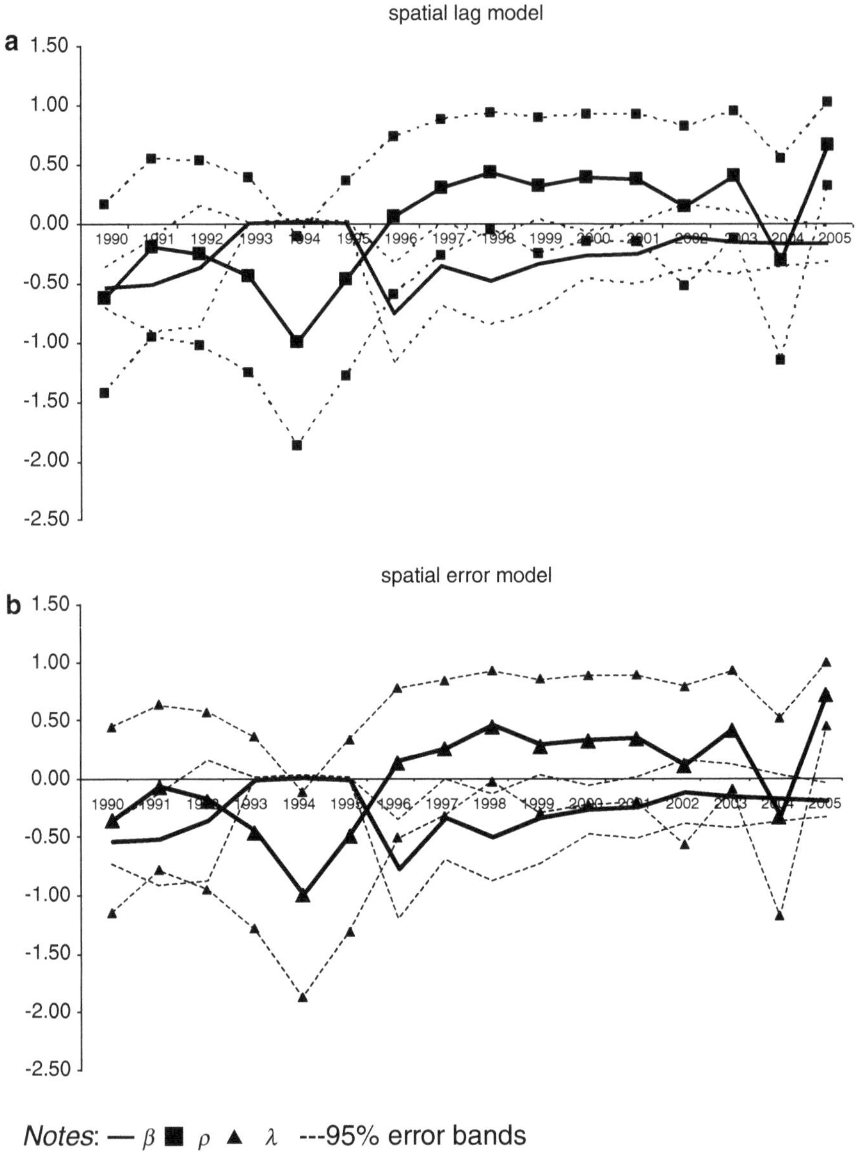

Fig. 1 Baseline model, spatial model estimates (W = distance matrix)

that started in 1997 in Thailand set off a series of problems in the Asian Tigers, but spread globally. Russia defaulted in 1998 after Brazil had devalued the real a few months before. Argentina defaulted in 2001 and Turkey experienced fiscal and monetary trouble in the same year. Since then, no major "emerging market" crisis has occurred. We find a break in spillover: there are no significant spatial links since

2001. In contrast to the 1990s, domestic crises have had much less impact abroad in recent years. There could be two reasons for this. First, domestic crises are less serious now than they were in the 1990s. Second, even though financial and economic integration are progressing, contagion is now much weaker. Other studies have also found that spillover has become much weaker in recent years. Forbes and Chinn (2004) also find evidence of stronger links over the period 1996–2000. Didier et al. (2006) show that the co-movement of emerging market bond spreads and returns have been declining since 2000. Mauro et al. (2006) present similar results.

The reasons for the changes in the crowding out effect are unclear. Large international crises can explain the large crowding out effect in the mid-1990s. In fact, some – but not all – of the emerging market crises started with domestic fiscal problems. High and rapidly growing public debt cast current monetary policy strategies into doubt, and meant high interest rates to prevent capital from fleeing the country. This is only a partial explanation, however. For lack of data, we have not been able to include many emerging markets in the sample. And even if these economies in crisis had a global impact, the mere size of their budget problems is probably too small to affect interest rates in industrialized economies. Instead, fiscal policy in both the United States and the EU was much more focused on debt consolidation in the 1990s. The Clinton Administration governed a 10% reduction in public debt in the span of 5 years, in part helped by the strongly booming economy. In addition, EU countries decided on a common fiscal retrenchment and a strict monetary policy stance to prepare for EMU. EU countries had to abide by the rules of the Stability and Growth Pact in order to qualify for the eurozone. After this joint consolidation effort, budget discipline has become less tight. It should not come as a surprise that after the entry in the EMU in 1999, crowding out is much smaller.

5 Some Robustness Checks

5.1 Global or Local Linkages

We immediately pick up on the previous explanation for the change in the crowding out effect. Fiscal consolidation in the EU countries might indeed be responsible for the large crowding out effect in the mid-1990s. There are additional reasons to expect a stronger spillover effect between EU member states over time. Strong interlinkages are the consequence of ongoing economic and financial integration, and this must have strengthened the spillover of economic policies between these countries. In addition, for those EU countries participating in monetary union, spillover may even be stronger. A common monetary policy has spurred financial integration and probably also trade links. If different governments borrow in the same currency, as in a monetary union, free riding makes each government disregard its own intertemporal budget constraint (Chari and Kehoe 2007). A variety of reasons may be invoked for the lack of credibility of the no bailout clause that prevents other governments (or the central bank) from rescuing the insolvent government. The offsetting interest rate effects do not need to materialize then, as default premia are spread out over all members of the union.[16] In the absence of agreements specifying the fiscal relations between governments, the crowding out effect depends – ceteris paribus – on the aggregate fiscal policy stance of all member states.

Could spatial links be stronger between particular groups of countries, or are capital markets truly global? We run the same baseline model for some subsamples of countries. We are particularly interested in the subgroup of EU countries. The results of the spatial tests must be taken with some caution since we include only 13 EU countries. The properties of spatial panel tests are instead asymptotically valid.

Table 5 summarizes the results of the different specifications of the spatial panel. Crowding out is much less significant for an EU country. In contrast to the "global" sample, a 1% deficit raises interest rates by only 10 basis points. Notwithstanding, the total effect on interest rates is much larger due to the spillover effect. Nearly 90% of an interest rate rise is transmitted to other EU countries. A 1% deficit raising domestic rates by 10 points will – in a second step – cause a rise in foreign rates of about 9 basis points. Hence, deficits will raise interest rates by nearly the same amount at home as in another EU country. As before, the panel with fixed or random effects gives very similar results. There is again some evidence that common shocks are driving interest rates. The results of the spatial panel model are somehow altered when accounting for time period fixed effects. The spillover effect

Table 5 Baseline model, spatial panel lag, country groups (W-matrix = distance)

EU15 (number of observations = 208)				
Spatial lag	β	t-stat	ρ	t-stat
Panel, fixed effects	−0.10	−2.54	0.86	36.46
Panel, random effects	−0.11	−2.02	0.86	33.37
Panel, spatial + time period fixed effects	−0.01	−0.09	0.29	2.56
Spatial error	β	t-stat	λ	t-stat
Panel, fixed effects	−0.03	−0.52	0.88	43.35
Panel, random effects	−0.21	−3.77	0.81	22.13
Panel, spatial + time period fixed effects	−0.00	−0.09	0.26	2.31
OECD (number of observations = 352)				
Panel, fixed effects	−0.45	−4.54	0.55	8.09
Panel, random effects	−0.46	−1.54	0.53	6.69
Panel, spatial + time period fixed effects	−0.43	−4.09	−0.05	−0.37
Spatial error	β	t-stat	λ	t-stat
Panel, fixed effects	−0.44	−4.17	0.56	8.42
Panel, random effects	−0.48	−4.26	0.52	5.28
Panel, spatial + time period fixed effects	−0.43	−4.10	−0.06	−0.43

[16] Yardstick comparisons across governments may partially undo this spillover, if the accumulation of debt by one government increases the relative creditworthiness of comparable governments.

is much weaker, and the crowding out effect is completely absent. As regards the source of the spillover, there is not much evidence to identify the role of financial or economic integration in the transmission of interest rates across EU countries. The estimates of the spatial error model show an important spillover effect, but no crowding out.

Most studies have examined spillover between OECD countries. We look at the industrialized economies, but exclude the most recently acceded member states. Our baseline results for the global sample are not much affected: crowding out effects are significant and spatial links are rather large. The estimates are of the same order as for the global sample.

5.2 *Different Weight Matrices*

For all previous results, we have used a measure of geographical distance as a proxy for cross country economic linkages. We check if the results are robust to other definitions of the weight matrix W, and focus on the spatial panel lag model with fixed effects.[17] We first try out some different measures of distance. We alternatively measure the (inverted) distance between countries as the distance between capital cities, or the great circle distance between country centroids.[18] Table 6 shows that the point estimates are very similar, and so is the significance of both effects. A more common choice of the weighting matrix in spatial studies is physical contiguity. Countries that share a common border are believed to transmit effects to their direct neighbors, but no effect at all to far-away countries. Under this type of transmission mechanism, crowding out is only marginally stronger, but the spatial effect is negative. The reason is that border links are an awkward choice, as there are plenty of missing observations in our sample. Only a few European countries share a common border, but most other economies are isolated from each other (i.e. there are many zeros in the weighting matrix). This downplays the importance of spatial transmission.

Physical distance is at best a proxy for the integration of countries' financial markets, but still gives little economic content to "being close." Our estimates of the spillover effect could be quite conservative as a consequence. We experiment with some more "economic" weight matrices. It is often argued that trade is a major channel for economic transmission across countries. We therefore use different possible weight matrices incorporating bilateral exports and imports. We scale total exports from country i to country j by total exports of country i.[19] Similarly, for

[17] This result is robust for the other panel models.

[18] A great circle is the shortest path between two points along the surface of a sphere.

[19] All data are in USD, trade data are FOB or CIF. Spatial panel models cannot handle time varying weight matrices. We arbitrarily fix exports and imports at a base year in the mid-of-sample (1998). Two countries are "close" if they have strong bilateral trade (relative to the other trading partners). In contrast to the literature on contagion, we do not use the competition for export shares on third markets (Forbes and Chinn 2004).

Table 6 Baseline model, spatial panel lag model, various weight matrices

	β	t-stat	ρ	t-stat
	Panel, fixed effects			
Inverted distance[a]	−0.45	−4.54	0.55	8.09
Inverted distance[b]	−0.48	−4.81	0.40	5.72
Circle distance	−0.45	−4.57	0.53	7.59
Border	−0.64	−5.62	−0.24	−3.86
Country size* distance	−0.45	−4.54	0.55	8.09
Exports	−0.45	−4.55	0.55	9.37
Imports	−0.45	−4.57	0.55	9.09
Trade	−0.45	−4.55	0.54	9.11
Free trade area	−0.54	−5.16	0.20	1.96
GDP per capita	−0.48	−4.76	0.41	5.83
	Panel, random effects			
Inverted distance[a]	−0.46	−2.02	0.51	0.00
Inverted distance[b]	−0.49	−2.09	0.37	5.15
Circle distance	−0.47	−1.63	0.50	6.62
Border	−0.53	−2.75	0.19	0.00
Country size* distance	−0.46	−2.05	0.53	0.00
Exports	−0.49	−2.30	0.40	5.15
Imports	−0.48	−2.22	0.43	0.00
Trade	−0.48	−2.24	0.44	0.00
Free trade area	−0.55	−2.30	0.18	3.52
GDP per capita	−0.49	−1.96	0.37	0.00
	Panel, spatial and time period fixed effects			
Inverted distance[a]	−0.43	−4.09	−0.05	−0.37
Inverted distance[b]	−0.43	−4.11	0.04	0.48
Circle distance	−0.43	−4.09	−0.05	−0.39
Border	−0.44	−4.09	−0.24	−3.82
Country size* distance	−0.43	−4.09	−0.05	−0.37
Exports	−0.43	−4.06	0.10	1.12
Imports	−0.43	−4.07	0.06	0.58
Trade	−0.43	−4.06	0.08	0.83
Free trade area	−0.41	−3.94	−0.11	−0.88
GDP per capita	−0.43	−4.10	0.01	0.14

[a]Distance between centroids of the country coordinates
[b]Distance between capital cities

imports of country j we scale by total imports of country j.[20] We also weigh by total trade, summing bilateral exports and imports, and dividing by total trade of the country. As a consequence, these weight matrices are asymmetric: the strength of the transmission depends on the size and importance of each country. For example, the United States may trade a lot with Colombia, yet the importance of this trade

[20] The two numbers do not match for statistical reasons. This is known as the "missing trade" problem.

volume for the US economy is tiny. In contrast, for Colombia, US trade is much more important. We would expect the spillover from the United States to Colombia to be much stronger than vice versa. This weight matrix better reflects the strength of transmission from large to small economies. Surprisingly, none of the results of the baseline model is altered very much. The crowding out effect is as large as before, and so is the spatial effect. Kelejian et al. (2006) similarly find little differences between the use of trade or distance matrices in their analysis of financial market spillover. This result confirms that distance is a good proxy for trade and economic relations in a gravity model.

One might argue that trade is endogenous to the strength of the economic links. We choose an alternative weight matrix that has a dummy if two countries are in a free trade agreement. This results in a slightly stronger crowding out effect, and weaker spatial links. However, the results could be biased. There are a few countries only that do not have some kind of bilateral trade agreement in our sample. As a consequence, the importance of spatial links is probably understated.

The panel model provides an average effect of fiscal policy on interest rates, while arguably these crowding out and spillover effects may differ across countries. Changes in fiscal policy in the large industrialized economies are likely to have a larger effect on smaller economies. The transmission of economic events is likely to run in one direction. For example, measured by great circle distance, Germany is equally distant from France and Hungary. The impact of changes in the German economy is likely to be large for both countries. Yet, the inverse impact of the French economy on Germany is almost certainly much larger than that of the Hungarian economy. We control for the direction of spillover and the importance of transmission between economies by multiplying country size (GDP in USD PPP terms) by physical distance. Nonetheless, the results do not change much if we use this asymmetric weight matrix.

Both industrialized and emerging market economies are increasingly open to financial markets. Financial integration between industrialized economies is gradually proceeding with economic integration. Instead, emerging market economies could be subject to contagious crises that spread from a crisis in another emerging market, but are unrelated to the economic fundamentals (and in particular the fiscal position) of the country itself. Economic crises may spread faster between emerging markets that are more exposed on financial markets, have similar macroeconomic characteristics or are prone to information asymmetries that trigger sunspot crises. As a final robustness check, we try to model these various channels of contagion. We capture the heterogeneity between industrialized and emerging economies by the difference in economic development. We use a weight matrix in which spatial links are stronger if the difference of (log) per capita income (in PPP USD) is smaller. We do not find significant differences in the crowding out effect, and the spatial effects remain as strong as with the other weight matrices.

Our weight matrix is a rather rough attempt to distinguish links between industrialized and emerging market economies. We have not attempted to model these channels of contagion with alternative definitions of the weight matrix.[21] It has been argued that these alternative channels may be less important than the classical transmission channels, such as trade (see the findings of Gerlach and Smets 1995). Eichengreen and Rose (2001) and Forbes and Chinn (2004) also find that real and financial linkages are predominantly determined by "real" trade integration. Kaminsky and Reinhart (2000) argue that trade or financial links are hard to distinguish, and hence often both specifications give very similar results.

5.3 Alternative Data Definitions

Fiscal policy is argued to affect interest rates at long horizons. This can be tested in most industrialized economies that are able to issue long-term bonds at a horizon of 10 or more years. Not that many countries outside the OECD have been able to issue long-term bonds, however. Most developing economies can only get finance on capital markets at short horizons, and have financed deficits with short-term bonds at a 5-year horizon at most. We add to the sample those countries that issue government bonds at a horizon shorter than 5 years, and also use a short- to medium-term interest rate (of 1–5 years) as our dependent variable for the countries in the initial sample. Table 7 gives the results of the spatial panel lag model with fixed effects. For the baseline model, the results barely change. If we augment the model with some additional control variables, the results still do not change very much. This is surprising, given that with a few exceptions (Barth et al. 1984), the literature usually does not find effects of deficits on short-term rates (Cebula 2000). We also add several control variables to this specification, and confirm the previous results. The crowding out effect is weaker when we add public debt, but the spillover effect becomes much stronger. If we add inflation, only the spillover effect is important. When data are available, we also add as a control the long-term rate. In this case, both the crowding out and spillover effect disappear.

The level of interest rates is determined by many other factors than fiscal policy. As argued before, it is the level of public debt that determines crowding out. The addition of new stock of public debt should instead put additional pressure on the change in interest rates. Hence, instead of using the level of interest rates as the dependent variable, one should use the first difference instead. Table 7 shows that this does not affect the estimates of the crowding out effect, nor of the spillover. The addition of control variables does not change our conclusions.

Finally, some other studies have used the yield to filter out any of the short-term effects of fiscal or monetary policy, and analyze the impact of expected deficits on interest rates. The argument is that higher future deficits translate into higher interest rates in the future, and hence, a steepening of the yield curve. By using the yield as

[21] Alternative definitions could be: interest rate spread, stock market index, exchange rate regime, real exchange rate, competition for bank lending, or international reserves. Data on these other channels are unfortunately not completely available for all countries in our sample.

Table 7 Augmented model, spatial panel lag model, spatial fixed effects, specifications (W-matrix = distance). See (4)

Dependent	Short-term interest rate	β	t-stat	ρ	t-stat
	Baseline	−0.52	−2.66	0.48	6.39
$X_t =$	Debt	−0.09	−2.03	0.77	21.39
	Long-term interest rate	0.03	0.16	0.14	1.23
	Inflation	0.01	0.24	0.45	5.65
Dependent	Δ long-term interest rate	β	t-stat	ρ	t-stat
	Baseline	−0.61	−2.72	0.44	5.44
$X_t =$	Debt	−0.09	−1.91	0.76	19.54
	Short-term interest rate	0.00	0.00	0.00	−0.01
	Inflation	0.01	0.24	0.42	5.09
Dependent	Δ interest rate	β	t-stat	ρ	t-stat
	Baseline	−0.56	−2.03	0.29	3.05
$X_t =$	Debt	−0.09	−2.03	0.77	21.39
	Long-term interest rate	0.14	0.65	0.21	1.94
	Inflation	0.05	1.62	0.44	5.77
Dependent	Yield	β	t-stat	ρ	t-stat
	Baseline	−0.52	−2.66	0.48	6.39
$X_t =$	Debt	−0.12	−2.94	0.73	17.95
	Inflation	0.03	1.03	0.35	3.84

the dependent variable, there is no need to model the factors that drive the level of interest rates. We subtract from the long-term interest rate the short-term rate, and use this yield as a dependent variable. The estimation of the baseline model does not show very different results, even if we add the usual control variables.

6 Conclusions

There is much discussion about the effect of fiscal expansions on interest rates. This variety in our opinion is due to the little robust empirical endorsement for crowding out. A lack of response of interest rates can be justified under two different theoretical conditions. First, under Ricardian Equivalence, deficits do not affect macroeconomic variables as economic agents anticipate the paydown of higher deficits with future taxes. Second, capital flows between economically integrated economies offset any interest rate differentials that follow upon an increase in the supply of government bonds. Fiscal deficits are not necessarily financed by domestic financial resources only. Financing conditions of governments depend on international capital markets.

Theoretical models of open economies displaying non-neutrality of debt and including long-term interest rates have not been developed. With little theoretical guidance, the robustness of the empirical tests is important. In this chapter, we concentrate on modeling the crowding out effect of deficits on interest rates in open economies. We extend a simple empirical model for testing crowding out and test it with spatial techniques. Spatial models impose few restrictions on the spillover, as the contemporaneous interactions on many capital markets is taken into account.

Our main finding is that there are significant and robust effects of larger deficits on interest rates. Spillover mitigates this effect. It is not obvious that this spillover is related to fiscal policy directly, as both financial and economic integration may drive the spillover. Spillover is much stronger in the mid-1990s when there are major crises, or policy actions have been coordinated between governments. We find the spillover effect to be particularly strong among EU countries.

Our results have some implications for fiscal policy. The argument for coordination of fiscal policy is perhaps not convincing in case the spillover occurs on capital markets. After all, the mitigating effect of financial markets is a purely pecuniary externality and does not really require international coordination. The allocation of savings to the public or private sector, whether at home or abroad, is efficient. Nonetheless, in case spillover is related to contagion on financial markets (in the case of emerging economies) or to monetary union (in the case of EMU) this distorts capital markets. Some institutional correction mechanisms might then be necessary. Given that crowding out, and spillover, is quite strong in the EMU, this justifies fiscal rules as in the Stability and Growth Pact.

Acknowledgements We would like to thank Anna Giribet for excellent research assistance. Peter Claeys acknowledges support by a Marie Curie Intra-European Fellowship within the 6th European Community Framework Programme.

References

Anselin L (1988) Lagrange Multiplier test diagnostics for spatial dependence and spatial heterogeneity. Geogr Anal 20:1–17

Anselin L, Le Gallo J, Jayet H (2008) Spatial panel econometrics. In: The econometrics of panel data, fundamentals and recent developments in theory and practice (3rd Edition), eds. L. Matyas and P. Sevestre, Springer-Verlag, 627–662

Ardagna S, Caselli F, Lane T (2007) Fiscal discipline and the cost of public debt service: some estimates for OECD countries. J Macroecon 7:15–41

Baltagi B (2001) Econometric Analysis of Panel Data. John Wiley, Chichester

Barro R (1992) World interest rates and investment. Scand J Econ 94:323–342

Barth J, Iden G, Russek F (1984) Do federal deficits really matter? Contemp Policy Issues 3:79–95

Barth J, Iden G, Russek F, Wohar M (1991) The effects of federal budget deficits on interest rates and the composition of domestic output. In: Penner R. (ed) The great fiscal experiment, Urban Institute Press, Washington, pp 69–141

Blanchard O, Perotti R (2002) An empirical characterization of the dynamic effects of changes in government spending and taxes on output. Q J Econ 117:1329–1368

Boyer B, Gibson M, Loretan M (1999) Pitfalls in tests for changes in correlation. Federal Reserve Board, International Finance Division, working paper 597

Breedon F, Henry B, Williams G (1999) Long-term real interest rates: evidence on the global capital market. Oxf Rev Econ Pol 15:128–142

Canzoneri M, Cumby R, Diba B (2002) Should the ECB and the Federal Reserve be concerned about fiscal policy? Federal Reserve Bank of Kansas City, Kansas City, pp 333–389

Caporale G, Williams X (2002) Long-term nominal interest rates and domestic fundamentals. Rev Financ Econ 11:119–130

Cebula R (1998) An empirical analysis of the impact of federal budget deficits on long-term nominal interest rate yields: using alternative expected inflation measures. Rev Financ Econ 7:55–64

Cebula R (2000) Impact of budget deficits on ex post real long-term interest rates. Appl Econ Lett 7:177–179

Cebula R, Koch J (1994) Federal budget deficits, interest rates, and international capital flows: a further note. Q Rev Econ Finance 34:117–120

Chari V, Kehoe P (2007) On the need for fiscal constraints in a monetary union. J Monetary Econ 54:2399–2408

Chinn M, Frankel J (2007) Debt and interest rates: the US and the Euro Area. E-conomics Discussion Paper 11

Claessens S, Dornbusch R, Park Y (2001) International financial contagion: how it spreads and how it can be stopped. In: Claessens S, Forbes K (eds) International financial contagion. Kluwer, Boston, pp 3–18

Cohen D, Garnier O (1991) The impact of forecasts of budget deficits on interest rates in the US and other G7 Countries. Mimeo, Board of Governors of the Federal Reserve System

Dai Q, Phillipon T (2005) Fiscal policy and the term structure of interest rates. NBER working paper 11574

De Haan J, Knot K (1995) Fiscal policy and interest rates in the European Community. Eur J Polit Econ 11:171–187

Didier T, Mauro P, Schmukler S (2006) Vanishing financial contagion. IMF Policy Discussion Paper 1

Doyle B, Faust J (2002) An investigation of co-movement among the growth rates of the G7 countries. Fed Reserv Bull 88:427–437

EC (2004) Public Finances in EMU, European Economy. DG for Economic and Financial Affairs, 3

Ehrmann M, Fratzscher M, Rigobon R (2005) Stocks, bonds, money markets and exchange rates: measuring international financial transmission. NBER working paper 11166

Eichengreen B, Rose A (2001) Staying afloat when the wind shifts: external factors and emerging-market banking crises. In: Calvo G (ed) Money, capital mobility, and trade. MIT, Cambridge, pp 171–205

Elhorst J (2003) Specification and estimation of spatial panel data models. Int Reg Sci Rev 26:244–268

Engen E, Hubbard R (2004) Federal government debt and interest rates. NBER Macroecon Annu 19: 83–138

Faini R (2006) Fiscal policy and interest rates in Europe. Econ Policy 21:443–489

Florax R, Homer H, Rey S (2003) Specification searches in spatial econometrics: The relevance of Hendry's methodology. Reg Sci Urban Econ 33:557–579

Forbes K, Rigobon R (2002) No contagion, only interdependence: measuring stock market co-movements. J Finance 57:2223–2261

Forbes K, Chinn M (2004) A decomposition of global linkages in financial markets over time. Rev Econ Stat 86(3):705–722

Ford R, Laxton D (1999) World public debt and real interest rates. Oxf Rev Econ Pol 15:77–94

Frankel J, Chinn M (2007) Debt and interest rates: The U.S. and the Euro Area," Economics Discussion Papers 2007-11, Kiel Institute for the World Economy

Friedman B (2005) Deficit and debt in the short and long run. NBER working paper 11630

Gale W, Orszag P (2003) The economic effects of long-term fiscal discipline. Urban-Brookings Tax Policy Center Discussion Paper 60

Gerlach S, Smets F (1995) Contagious speculative attacks. Eur J Polit Econ 11:45–63

Giuliodori M, Beetsma R (2005) What are the spillover from fiscal shocks in Europe? An empirical analysis. Economist 153:167–197
Imbs J (2004) Trade, finance, specialization, and synchronization. Rev Econ Stat 86:723–734
Kaminsky G, Reinhart C (2000) On crises, contagion and confusion. J Int Econ 51:145–168
Kelejian H, Tavlas G, Hondroyiannis G (2006) A spatial modeling approach to contagion among emerging economies. Open Econ Rev 17:423–441
Kinoshita N (2006) Government debt and long-term interest rates. IMF working paper 63
Kitchen J (1996) Domestic and international financial market responses to federal deficit announcements. J Int Money Finance 15:239–254
Laubach T (2003) New evidence on the interest rate effects of budget deficits and debt. Board of Governors of the Federal Reserve System, working paper 12
Marcellino M (2002) Some stylized facts on non-systematic fiscal policy in the Euro-area. CEPR discussion paper 3635
Mauro P, Nathan S, Yishay Y (2006) Emerging markets and financial globalization: sovereign bond spreads in 1870–1913 and today. Oxford University Press, Oxford
Minford P, Peel D (2007) On the equality of real interest rates across borders in integrated capital markets. Open Econ Rev 18:119–125
Orr A, Edey M, Kennedy M (1995) Real long-term interest rates: the evidence form pooled-time-series. OECD Econ Stud 25:75–107
Paesani P, Strauch R, Kremer M (2006) Public debt and long-term interest rates: the case of Germany, Italy and the USA. ECB working paper 656
Rose A (2000) One Money, One Market: Estimating the effect of common currencies on trade. Econ Policy 16(33):449–461
Tanzi V, Lutz M (1993) Interest rates and government debt: are the linkages global rather than national? In: Verbon H, Van Winden F (eds) The political economy of government debt, Elsevier, Amsterdam, pp 233–253
Zoli E (2004) How does fiscal policy affect monetary policy in emerging market countries? BIS working paper 174

Part IV
Spatial Analysis of Population and Health Issues

Spatial Models of Health Outcomes and Health Behaviors: The Role of Health Care Accessibility and Availability

Brigitte S. Waldorf and Susan E. Chen

1 Introduction

It is still open to debate whether increased availability and accessibility of physicians and health care services has a significant beneficial impact on the health status of populations in the United States. While there is convincing evidence that increased availability and accessibility has a significant beneficial impact on the health status of populations in developing countries (see, e.g., Lavy et al. 1996; Frankenberger 1995; Perry and Gesler 2000), a large body of literature suggests that additional resources spent on health do not significantly reduce mortality in the United States (Thornton 2002; Hadley 1982; Auster et al. 1969). A recent review of the literature on primary care and health in developed countries, however, suggests that the supply of primary care physicians is positively related to population health (Starfield et al. 2005). Moreover, medical care may not influence gross mortality but it may affect mortality rates of particular subgroups, the morbidity of the population, and preventative health behaviors (Anderson and Morrison 1989). In addition, spatial variations in the use and quality of medical care (Skinner 2006; Chan et al. 2006) may confound a simple link between access to health care and health care outcomes. The mixed evidence on the link between population health and health service provision and accessibility challenges policymakers who have to determine how to equitably allocate medical resources to improve public health, particularly in medically underserved areas.

Accessibility is a multidimensional concept and can be broadly defined as the ability of a population to obtain health care services. It varies across space because neither health professionals nor residents are uniformly distributed (Luo and Wang 2003). Previous research has addressed the connection between space and health but has generally ignored conceptual and methodological advancements in the spatial sciences. In particular, two issues have been treated in a rather rudimentary fashion.

B.S. Waldorf (✉)
Department of Agricultural Economics, Purdue University, 403 W. State Street, West Lafayette, IN 47907-2056, USA,
e-mail: bwaldorf@purdue.edu

A. Páez et al. (eds.), *Progress in Spatial Analysis*, Advances in Spatial Science,
DOI 10.1007/978-3-642-03326-1_16, © Springer-Verlag Berlin Heidelberg 2010

First, although there is a large literature on measuring the spatial accessibility of medical care,[1] studies that addressed the link between access and health status have used crude measure, such as a simple enumeration of physicians within an area. Thus, they de facto measure availability (the number of locally available service points from which a patient can choose) rather than accessibility which accounts for the distances people need to travel to take advantage of health care services. Second, previous research has used spatially referenced data but has not yet fully explored spatial dependencies. The Dartmouth Health Atlas, for example, is an effort to track and document geographic differences in health outcomes, health care utilization, and medical expenditure. The results of a number of the studies using the Dartmouth data have shown that there are large geographic differences in the three main outcomes (see, e.g., Wennberg et al. 2002). Moreover, the studies suggest that these differences may be generated by some unobserved process – perhaps of a spatial nature.

This study aims to rectify these oversights by choosing a traditional availability measure, plus a more sophisticated measure of spatial accessibility as the key explanatory variables of health outcomes and health behaviors. Moreover, we directly address spatial dependencies. Spatial dependence may arise when a county's health outcomes and behaviors are correlated with those in neighboring counties. It can be a statistical artifact, but it can also be grounded in behavioral processes such as imitation behavior and the spatial diffusion of cultural norms influencing health care utilization. They could also be a result of underlying factors such as poor labor market conditions which affect people's access to health insurance and thus ultimately people's health.

A traditional health production function is transferred to a spatial setting, amenable to spatially aggregated data where spatial accessibility and availability of health care serve as the key explanatory variables. To account for spatial processes and to avoid potential misspecifications, the model allows for spatial dependence. The empirical analysis focuses on vulnerable groups – the elderly, children, and pregnant women – and utilizes county-level data from the Indiana State Department of Health and the U.S. Census Bureau. The health outcome variables include mortality measures for infants and the elderly; the health behavior variables focus on pregnancy related behaviors.

The remainder of this chapter is organized as follows. The second section summarizes the literature and provides a conceptualization of the link between accessibility and health status, with special attention to the spatial manifestations of the linkages. The third section presents the empirical analysis. It begins with a profile of the study area, followed by a description and exploratory analysis of the variables used in the study, the model specifications, and the presentation of empirical results. We conclude with a discussion and implication of our findings.

---

[1] For an overview, see Guagliardo (2004).

2 Background

2.1 Literature Review

In 2005, the U.S. federal government spent over 2.25 billion dollars on programs designed to increase access to physicians and health care services in the United States. Despite the large amount of spending on these programs, there are still 3,032 Health Professionals Shortage Areas in the U.S. (GAO 2006). While it is clear that there is a spatially unequal distribution of both primary and specialized health care professionals in the United States, more evidence is needed to understand the association between access to care and health status.

Access to medical care includes affordability, accommodation, acceptability, availability, and accessibility (Penchansky and Thomas 1981). The first three dimensions, extensively reviewed in Wyszewianski (2002), are considered non-spatial and have been given considerable attention in the literature. They reflect health care financing arrangements and the access barriers of an economic, social and cultural nature. The final two dimensions are spatial in nature and reflect the adequacy of the supply of health care providers inside a region while taking into account such factors as distance, travel time, and the demand for services. In this study, we focus mainly on the spatial dimensions of accessibility.

Spatial accessibility is a necessary, albeit not sufficient pre-requisite for equitable high-quality health care services for all population segments of society, whether they reside in urban agglomerations or in peripheral rural areas. However, spatial barriers – most notably long travel distances to health care facilities – are significant factors contributing to the exclusion from high-quality medical care.[2] Indeed, McDonald and Coburn (1988) showed that prenatal care utilization decreases with increasing travel times to providers.

A review of the literature shows no consensus on the relationship between spatial accessibility and health status. Using data from the 1960s on individuals living in 39 city size strata, Newhouse and Friedlander (1980) related physiological measures of individual health status, such as diastolic blood pressure and cholesterol to an area's medical resources measured by various indicators, such as the number of physicians per residents and number of hospital beds. They did not find a relationship and concluded "that in the United States what an individual does for himself is probably more important to his health than the quantity of medical-care resources in his area of residence" (p. 214). Similarly, Krakauer et al. (1996) – using data aggregated to the Health Care Service Area for the Medicare population – could not find evidence

[2] Though extremely important on the supply side, lack of data does not allow us to refine supply to include physicians' willingness to accept different types of patients (e.g., Medicare or Medicaid). The focus is thus solely on physical proximity to health care providers.

that physician supply reduces mortality or morbidity, measured by ambulatory care sensitive conditions.[3]

Shi and Starfield (2001) focused on the effects of primary care physician supply, income inequality, and a battery of socio-economic variables on mortality among Blacks and Whites across U.S. metropolitan areas in 1990. They found that white, but not black mortality is significantly associated with primary care physician supply and income inequality. Shi et al. (2003) used data for the 50 U.S. states to investigate that same relationship during four time periods. Their results suggested that increased supply of primary care physicians significantly lowers mortality, in contrast to specialty care physician supply which has the opposite effect.

Mansfield et al. (1999) used county level data for the United States to investigate the factors influencing premature mortality, defined as the number of life-years lost before age 75. They found that the influence of socio-economic and demographic variables on premature mortality exceeds the influence of the supply of medical care. Moreover, the effect of primary physician supply on premature mortality depends on the geographic context, lowering premature mortality in metropolitan counties but increasing it in rural counties.

While the study by Mansfield et al. (1999) emphasizes that the link between medical resources and health may very well be context-specific, there is also evidence that improved resources may be beneficial for vulnerable population groups. That is, for particular subgroups such as infants and the elderly it does appear that health care resources matter. Hadley (1982), in a study of the Medicare population, found that a one percent increase in Medicare expenditure would lower mortality among black males by 16%. Similarly, Corman et al. (1987) found that approximately 56.5% of the decline in black neonatal mortality in the United States between 1964 and 1982 can be attributed to health care programs and medical innovations. Allen and Kamradt (1991) report decreasing infant mortality rates with increasing physician availability in Indiana counties.

The link between access to medical care and health has also been the focus of many studies conducted in other countries such as England and Canada. The results from these studies are important because the socialized healthcare systems in these countries remove the financial barriers to seeking health care. Interestingly, these studies also paint an inconclusive picture of the link between health care and population health status. Especially in Canada, where large areas are sparsely populated and very remote, accessibility takes on added significance. In these sparsely populated areas, there may also be rationing of health care resources within a health care facility as well.

In England, Goyder et al. (2000) used individual data on patients who had been diagnosed with diabetes, to investigate the factors that influence whether patients attended a hospital diabetes clinic or had a diabetes review subsequent to the diagnosis. They found that living in a deprived area negatively affected the chance of a

---

[3] Ambulatory care sensitive conditions are health care conditions evident when a patient is admitted for ambulatory care and presumed to be sensitive to the adequacy (availability) of ambulatory care.

diabetes review. Contrary to this finding, Gulliford et al. (2004) looked at the relationship between general practitioner supply and mortality in 99 health authorities in 1999 and did not find evidence to support the hypothesis that increased supply lowers mortality. In Canada, Veugelers et al. (2004) investigated a specific health condition – hypertension. They used individual data in combination with contextual information about the individual's neighborhood to investigate variations in diagnosis and treatment. Using the number of physician visits per person as the measure for health care delivery, they cannot establish a relationship between health care delivery and either diagnosis or treatment of hypertension.

This brief review of the literature shows a lack of consensus on the effect of medical resources on mortality. There does seem to be agreement that socioeconomic and behavioral factors are more strongly associated with mortality than access (Fuchs 1974; Joyce et al. 1992; Wolfe 1986). Mokdad et al. (2004) find that almost half of all deaths in the United States can be attributed to preventable factors that range from poor diet and physical inactivity to firearms and toxic agents. They also find that the estimated contribution of health care to a population's health status is small, only about 15%. This estimate is for the entire population and therefore does not accurately depict the effect that health care may have on more vulnerable groups in the population. Our study contributes to the literature by focusing explicitly on the health outcomes of vulnerable groups, infants and the elderly.

The literature review also reveals that existing studies differ with respect to the chosen health outcome/behavior/condition variables and the spatial scale. They are similar though in that almost all of them used cross-sectional data[4] to measure the association between medical resources and health outcomes, thus exploiting variation across space to examine the link between access and health. Another common thread is the reliance on availability measures rather than accessibility measures. Our study contributes to this literature by using both types of measures, and by explicitly accounting for spatial linkages. The following section outlines the spatial modeling approach.

2.2 *Modeling the Link Between Health Status and Accessibility*

Generally, the health production function hypothesizes health status to be dependent on a variety of inputs that refer to lifestyle, the environment, genetic endowments, and medical resources (Folland et al. 1997). Health status is usually measured by mortality or morbidity, and the inputs into the health production function include socioeconomic variables, behavioral factors, and factors measuring access to medical care.

---

[4] One notable exception is Shi et al. (2003) who used panel data to study the effect of primary care physician access on mortality.

In this study, we estimate an aggregate health production function to test the hypothesis that poor spatial accessibility to health care services leads to poor health status. Thus, the health status, **Y**, of region *i*'s population, is expressed as a linear function of regional characteristics, **X**. The linear predictor includes the spatial accessibility of health care services as the key explanatory variable and other regional characteristics consistent with the health production function, such as the population's socio-economic status and the area's contextual setting along the rural-urban continuum. The implied model, $\mathbf{Y} = f(\mathbf{X})$, suggests that health outcomes in a region are a function of local characteristics. However, in a multi-regional setting health outcomes may also be affected by the diffusion of norms and values that influence health behaviors and health outcomes (Rice and Smith 2001). Prime examples are cultural norms impacting health care usage or dietary and exercise habits. In particular for adolescents, peer influences can strongly affect their alcohol and tobacco consumption as well as their sexual behavior.

Spatially, the sphere of influence for such norms may not coincide with administrative boundaries. Moreover, the influence of norms and values is not confined within any fixed boundaries but may diffuse through space. The most obvious behavioral manifestation of such diffusion is imitation behavior: people mimic the activities and habits of those who live in close proximity. Ultimately, such diffusion and imitation will result in spatial spillovers (see Fig. 1) and the clustering of similar health behaviors and health outcomes across space.

The spillovers have implications for the specification of a spatial health production function. Omitting spillovers from the model implies that their effects will be erroneously attributed to the impact of the structural factors included in the health production function. Furthermore, the parameter estimates will be biased and the errors of the model will be spatially autocorrelated.

We thus adopt a two-stage strategy to model the spatial health production function. In the first step, we model the health status variable, **Y**, as a linear function of the predictor variables, **X**. If the residuals are spatially autocorrelated, we will correct for the spatial autocorrelation in the second step, by using maximum likelihood techniques to estimate either a spatial error (2a) or spatial lag (2b) model:

$$\text{Non-spatial Model}: \mathbf{Y} = \mathbf{X}\beta + \varepsilon, \varepsilon \sim N(0, \sigma^2) \quad (1)$$

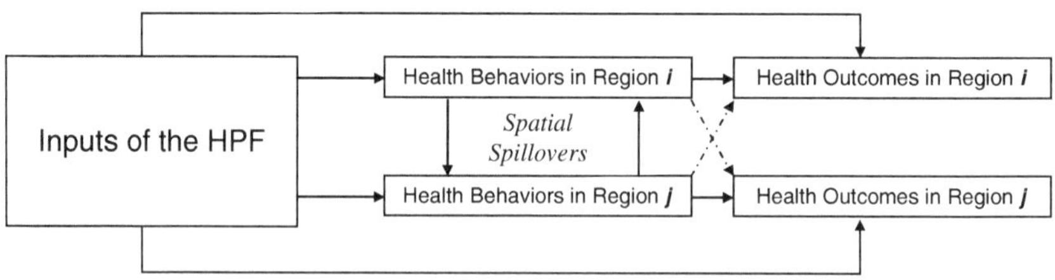

Fig. 1 Spatial linkages of a health production function (HPF)

$$\text{Spatial error model}: \mathbf{Y} = \mathbf{X}\beta + \varepsilon,\ \text{and}\ \varepsilon = \lambda\mathbf{W}\varepsilon + \mu, \mu \sim N(0, \sigma^2) \quad (2a)$$

$$\text{Spatial lag model}: \mathbf{Y} = \rho\mathbf{W}\mathbf{Y} + \mathbf{X}\beta + \varepsilon\ \text{or}\ \mathbf{Y} = (I - \rho\mathbf{W})^{-1}\{\mathbf{X}\beta + \varepsilon\} \quad (2b)$$

$\mathbf{Y}$ is the $n \times 1$ vector describing the health status in regions $i = 1, \ldots, n$; $\mathbf{X}$ is the $n \times k$ matrix of predictor variables, β is a $k \times 1$ vector of parameters and ε is the $n \times 1$ error vector. $\mathbf{W}$ is an exogenously specified $n \times n$ weight matrix that summarizes the spatial linkages relevant for spillovers. If the spillovers operate as contagious processes, then a first-order contiguity matrix will be a good approximation of the spatial linkages.

3 Empirical Analysis

3.1 Study Area

Indiana is a Midwestern state with a long tradition in both agriculture and manufacturing. Not surprisingly, thus, almost a third (29.2%) of Indiana's 6.3 million residents live in rural areas, compared to only 21% in the nation. Indiana's urban system follows a typical hierarchy with its centrally located capital city, Indianapolis. Indianapolis accounts for more than 10% of the total population and is complemented by a series of smaller regional centers that are almost uniformly distributed across the state. An anomaly to this almost perfect Christallerian city system is the northwestern region of the state. This region is part of the tri-state Chicago Consolidated Metropolitan Area, houses a very urban population and its economic base is comprised of the typical rustbelt industries.

Indiana's health care service provision ranks below the national average (Table 1). Using the number of physicians per capita as an indicator, Indiana ranks 39th among the 50 states. Indiana has only 213 physicians per 100,000 residents and is ranked far below the national average of 266 physicians per 100,000 residents. Massachusetts takes the lead with 450 physicians per 100,000 residents, followed by Maryland and New York State with 411 and 389, respectively. Even when comparing Indiana to other Midwestern states, it still ranks quite low. The deficit of physicians is slightly compensated by an above average number of nurses per capita. With 877 nurses per 100,000 residents, Indiana exceeds the national average of 824 nurses per 100,000 residents and ranks 25th among U.S. states.

Taking a closer look inside Indiana reveals stark disparities across the 92 counties. On average, there are 99 physicians per 100,000 county residents. With 306 physicians per 100,000 residents, Marion County (Indianapolis) has the highest number of physicians per capita, followed by Vanderburgh County (Evansville) with 289 physicians per 100,000. At the other end of the spectrum are small, predominantly rural counties with as few as only seven physicians per 100,000 residents. It is thus not surprising that the vast majority of a recent survey in rural Indiana rated the

Table 1 Physicians and nurses per 100,000 residents in 2004

Rank	State	Physicians per 100,000 residents	Rank	State	Nurses per 100,000 residents
Top Three					
1	Massachusetts	450	1	South Dakota	1,207
2	Maryland	411	2	North Dakota	1,179
3	New York	389	3	Massachusetts	1,177
Midwestern States					
10	Minnesota	281	5	Iowa	1,107
11	Illinois	272	13	Minnesota	1,018
18	Ohio	261	14	Missouri	997
22	Wisconsin	254	15	Ohio	985
27	Michigan	240	16	Wisconsin	939
29	Missouri	239	23	Illinois	895
39	Indiana	213	25	Indiana	877
46	Iowa	187	29	Michigan	841
United States		266	United States		824

Source: http://www.census.gov/compendia/statab/health_nutrition/health_care_resources/

lack of rural health care and health services as a top priority for State Government (PCRD 2006).

Within Indiana, the number of physicians per county resident is positively correlated with population size ($r = 0.654$). The bias towards the most urban centers is visualized in Fig. 2. It portrays the cumulative distribution of physicians relative to the cumulative population distribution after sorting the counties in ascending order by population size. The curves for each specialty area are below the 45° line, suggesting that the least populated counties house a disproportionately small share of physicians. The disparities are weakest for primary care and emergency medicine. They are, however, quite drastic for internal medicine: almost 39% of all internists are located in the largest county, Marion County. These patterns suggest that Indiana's mixture of urban and rural counties sets the stage for an inequitable provision of health care services with health care professionals being disproportionately located in urban centers.

4 Data and Measurements

The empirical analysis uses four types of variables: health outcome variables, health behavioral variables, measures of spatial access to health care services, and control variables measuring important socio-economic differences across Indiana counties. All variables are measured at the county level. Table 2 shows data sources and descriptive statistics.

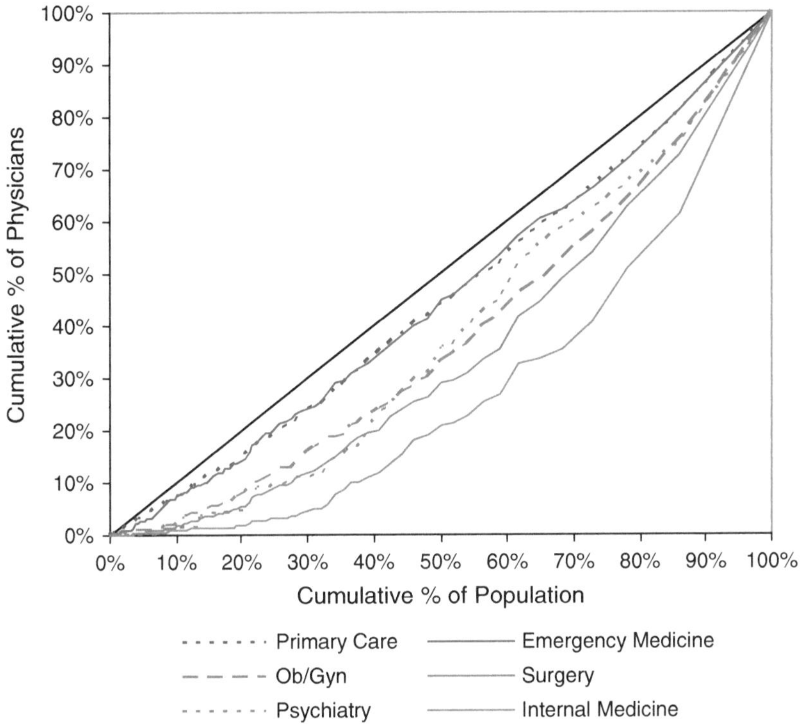

Fig. 2 Cumulative distribution of physicians relative to the cumulative population distribution across Indiana counties, 2003

4.1 Health Outcomes

Finding adequate measures of a population's health status is a pervasive problem in empirical research. Morbidity measures are certainly good proxies but they are difficult to obtain, especially when using spatially aggregated data. Most studies thus turn to mortality measures because they are reliable and easily obtainable from vital statistics.

We chose outcome measures for two vulnerable groups expected to be most sensitive to medical resources, namely infants and the elderly. For infants we chose the percentage of babies with low birth weight (less than 2,500 g) and infant mortality. Infant mortality is a direct reflection of geographic access to hospitals. Proximity to a neonatal care unit can vastly improve the outcomes of premature and low birth weight infants (Cifuentes et al. 2002). Arguably, advances in neonatal care (in particular the introduction of NICU in hospitals) accounted for much improvement in infant mortality over the last decade.

Typically, the infant mortality rate is defined as the number of babies dying prior to the first birthday per 1,000 live births. However, the United States is a low mortality country and infant mortality has become a rare event, at least from a statistical

Table 2 Variable definitions and descriptive statistics

Variable	Definition	Mean	Std. Dev.	Min	Max
IMR	Infant mortality rate, cumulative 1990–2003[a]	7.46	1.74	3.40	11.72
%LBW	% Low birth weight 2003[b]	7.55	1.48	4.10	10.60
MORT	Age-adjusted elderly(55+) mortality rate, 2003[b]	35.83	3.70	24.30	45.80
CVD	Cardiovascular disease deaths per 100,000 elderly (55+), 2003[b]	1512.8	266.6	792.3	2395.3
Cancer	Cancer deaths per 100,000 elderly (55+), 2003[b]	948.06	135.3	608.7	1294.1
Smoking	% mothers smoking during pregnancy 2003[a]	22.28	5.87	5.00	35.00
Prenatal	% mothers receiving prenatal care during 1st trimester, 2003[a]	82.81	6.63	46.50	93.50
Teenpreg	Teenage pregnancy rate, 2003[a]	42.08	12.06	14.80	70.90
Healthy	% Kids with a healthy beginning 2002[a]	21.52	3.56	9.50	29.80
Nurse	Nurses per 10,000 residents[c]	9.93	2.72	4.30	17.64
Hospital	Accessibility to hospital care, A_i[d]	0.25	0.21	0.00	1.00
Income	Medium household income [$1,000], 2003[c]	42.37	6.99	32.7	80.88
Education	% population (25+) with college degree, 2003[c]	14.56	6.65	7.60	48.90
Uninsured	% uninsured, 2003[c]	10.76	2.10	6.19	16.49
Rurality	Index of relative rurality, 2000[e]	0.40	0.10	0.12	0.58

[a]Kids Count Indiana
[b]Indiana Health Department
[c]Stats Indiana
[d]Unal et al. (2007)
[e]Waldorf (2007)

point of view. Thus, using *annual* data at a *small* spatial scale, such as at the county level, implies that many counties did not record any infant deaths. To circumvent the small-number problem, we analyzed the *cumulative* infant mortality rate for 1990 to 2003.

The key outcome variable for the older population is the age-adjusted elderly mortality rate. It is the sum of county-specific mortality rates for persons of age 55 or older, applied to a standard population.[5] In addition to overall elderly mortality,

[5] The standardization removes the effects of variations in mortality due to differences in the age composition.

we also focus on cause-specific mortality rates for the elderly. These include the two main causes of death, cancer and cardiovascular disease.[6] Both cancer and cardiovascular diseases are sensitive to early diagnosis and treatment and the associated mortality is expected to be influenced by access to health care.

4.2 Health Behavior Variables

Using cross-sectional data is useful in understanding geographic differences in health and can provide valuable information to local health policy decision makers on how to target scarce resources (e.g., Wennberg et al. 2002). However, using mortality as a measure of health status is not without problems in cross-sectional analysis. Mortality is responsive to the cumulative exposure to medical resources over an entire lifetime not just to the contemporaneous exposure. We thus extend our analysis to a battery of health behavior variables that are better measures of the effect of contemporaneous medical resources on health.

The health behavior variables include the percentage of mothers receiving prenatal care during the first trimester, the percentage of mothers who smoked during pregnancy, and the teenage pregnancy rate. In addition, we also use a composite measure that combines information on maternal health behavioral indicators. It is referred to as the percentage of kids with a healthy start and defined as the percent of total births with no birth characteristics that research has shown to negatively impact children's later school success: prenatal care beginning after the first trimester, maternal weight gain of less than 20 pounds, mother smoked during pregnancy, mother drank alcohol during pregnancy, three or more older siblings, and mother's last birth less than 19 months prior.

4.3 Measures of Spatial Accessibility

In the empirical analysis, we utilize two access variables. First, as an indicator of primary care, we use a nurse-to-population ratio, defined as the number of nurses per 10,000 county residents (NURSE). We chose nurses rather than primary care physicians because a good deal of primary care in rural areas is provided by nurses who are contracted with physicians who may have their main practice in a different

[6] In this study the cause-specific mortality rates for the elderly are defined as all deaths due to a disease (independent of age) per 100,000 residents of age 55 or older. Thus, the events in the numerator do not perfectly match the population-at-risk in the denominator. However, for cancer and CVD it is quite rare that the events in the numerator involve people under the age of 55. In 2000, 93% of all deaths due to cardiovascular disease and 87% of all cancer deaths were among persons age 55 or older (U.S. National Center for Health Statistics, Vital Statistics of the United States, annual and National Vital Statistics Report, NSVR).

county. Since primary care is typically consumed locally we specified it as an availability measure rather than an accessibility measure. Thus, the measure does not account for distance-discounted primary care services in other counties.

Second, following Luo and Wang (2003), we use a gravity-based measure of spatial accessibility to health care. It is based on the idea that any resident can utilize all health care services in the state, not just those in the county of residence. However, services in close proximity are more valuable to the user than those further away. In our study, services are operationalized as the number of hospital beds.[7] In total, 214 hospitals throughout Indiana were included in the study.

The specification of the gravity-based accessibility measure for Indiana counties is taken from Unal et al. (2007). For residents in county i, health services in county j are discounted by distance, d_{ij}. The accessibility measure also accounts for the demand from other users, discounted by distance. More formally, demand-adjusted accessibility, A_i,, for the population in county i, $i = 1, \ldots, m$ is defined as:

$$A_i = \sum_{j=1}^{n} \frac{S_j d_{ij}^{-1}}{V_j} \tag{3}$$

where S_j is the service capacity at provider location j, $j = 1, \ldots, n$, d_{ij} is the distance between the population in county i and provider location j, and the denominator, V_j, represents demand for the care facility at location j:

$$V_j = \sum_{k=1}^{m} P_k d_{kj}^{-1} \tag{4}$$

A critical component of the measure is the distance between the county population and the service provider. Instead of simply assuming that, on average, county residents are located at the county midpoint, we more accurately measure the distances by taking the spatial arrangement of the population within a county into account. The county population is assigned to three possible locations, $p = 1, 2, 3$. The three locations are the midpoint of the largest city within the county ($p = 1$), the midpoint of the second largest city within the county ($p = 2$), and the county midpoint ($p = 3$). The distance from county i to a service provider j, d_{ij}, is thus defined as the weighted average:

$$d_{ij} = \frac{1}{P_i} \sum_{p=1}^{3} P_{ip} d_{pj} \tag{5}$$

[7] In a preliminary analysis we also used an accessibility measure in which the service capacity is defined as a county's total number of physicians, allocated to the hospital locations in proportion to the size of the hospital. While this is an approximation of the internal (within-county) spatial distribution of physicians, it does account for the tendency of physicians to locate close to hospitals so as to take advantage of agglomeration economies. Using access to physicians yields similar results as the results reported in this chapter.

where P_{ip} is the population at location p of county i, and P_i is the total county population. Distances between population points and service providers, d_{pj}, were obtained in ArcGIS using hospitals' exact addresses.

4.4 Control Variables

The control variables address variation in health outcomes and health behaviors due to income, education, health insurance coverage, and degree of rurality.[8] Income is measured as median household income, and education is measured as the percentage of the adult county population with at least a bachelor's degree. Health insurance coverage, an important indicator for financial barriers to health care provision, is measured as the percentage of county residents who do not have health insurance. High income levels, a greater percentage of well educated county residents, and a low percentage of uninsured residents are hypothesized to have a beneficial effect on health outcomes and behaviors. Finally, the spatial setting is further characterized by the index of relative rurality (Waldorf 2007). The index provides a more nuanced differentiation of rural settings than the frequently used metro/non-metro dichotomy. It is a composite measure that combines population size, population density, and distance to the closest metropolitan area. The index indicates a county's position on a continuous scale bounded by zero (least rural) to one (most rural) and is calibrated using all U.S. counties (except those in Hawaii and Alaska).

Race and ethnic compositions are not included as controls. Indiana has a mostly (89%) white population. African Americans are the largest minority group (8.8%) and are highly segregated with more than 63% of African Americans living in only 2 of the 92 counties (Lake and Marion counties). Preliminary analysis showed that race and ethnicity do not have significant impacts on health outcomes and behaviors in a multivariate context. That is, after controlling for income, education, insurance coverage, and rurality, the racial and ethnic composition has no effect on health status at the county level.

4.5 Exploratory Spatial Data Analysis

In the spatial data analysis, performed using the GeoDa software, we defined the weight matrix, **W**, as a 92 × 92 spatial contiguity matrix for Indiana's counties. After row-standardization, **W** takes on the form:

$$w_{ij} = \begin{cases} 1/k_i \ if \ i \neq j \ share \ a \ common \ border \\ 0 \quad otherwise \end{cases} \tag{6}$$

[8] Note that, for the cumulative infant mortality model, the control variables are averages of the 1990 value and the value of the year specified in Table 2.

Table 3 Spatial autocorrelation (Moran's I) of variables across Indiana counties

	Variable	Moran's I	p-value
Health outcomes	IMR	0.068	0.108
	%LBW	−0.024	0.439
	MORT	0.073	0.102
	CVD	0.227	0.001
	Cancer	0.089	0.063
Health behaviors	Smoking	0.242	0.001
	Prenatal	0.177	0.003
	Teenpreg	0.007	0.369
	Healthy	0.049	0.178
Access	Nurse	0.300	0.001
	Hospital	0.068	0.116
Controls	Income	0.253	0.002
	Education	0.102	0.047
	Uninsured	0.139	0.010
	Rurality	0.140	0.011

where a "common border" is defined as sharing at least one point, and k_i is the number of counties bordering county i.

Table 3 reports Moran's I and the associated p-values[9] for all variables included in the analysis. Overall, the health-related variables show surprisingly little spatial autocorrelation, whereas the control variables are highly spatially correlated. Among the health outcome variables, cardiovascular disease mortality is the only variable that is highly spatially clustered whereas cancer mortality is weakly clustered. For all other health outcome variables, we cannot find sufficient evidence for a non-random pattern. Among the health behavior variables only variables related to the behavior of pregnant women are spatially clustered suggesting that the spatial diffusion of norms regarding healthy behaviors may well play a role for this sub-population.

Interestingly, for the availability and accessibility variables only the distribution of nurses per capita (availability) is spatially correlated whereas access to hospital care is randomly distributed across space. Finally, all of the control variables are spatially clustered. The clustering is most pronounced for the income variable. Moreover, much of the spatial sorting that created the distribution of the population by income and educational attainment level seems to be related to the degree of rurality. The highly educated tend to be clustered in the most urban settings (low IRR values), in a corridor that stretches from Tippecanoe County to Monroe County and includes Indianapolis and the surrounding counties.

The series of maps shown in Figs. 3–5 allow us to localize the spatial clusters in space. Figure 3 shows the spatial distributions of elderly cardiovascular disease mortality and cancer mortality. High cardiovascular disease mortality is clustered

[9] The p-value refers to the test of $H_0 : E(I) = -1/(n-1)$ versus $H_1 : E(I) \neq -1/(n-1)$ and is based on 999 permutations under the randomization assumption.

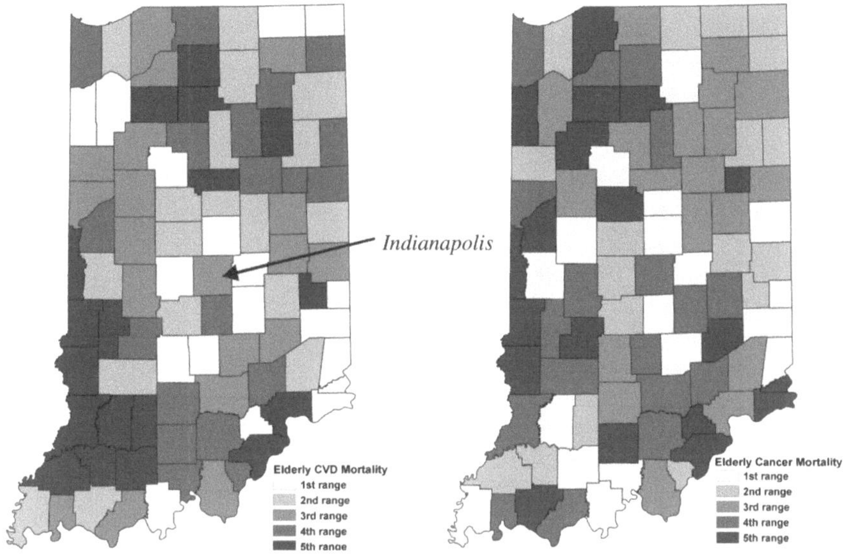

Fig. 3 Spatial distribution of elderly CVD mortality (*left*) and elderly cancer mortality (*right*)

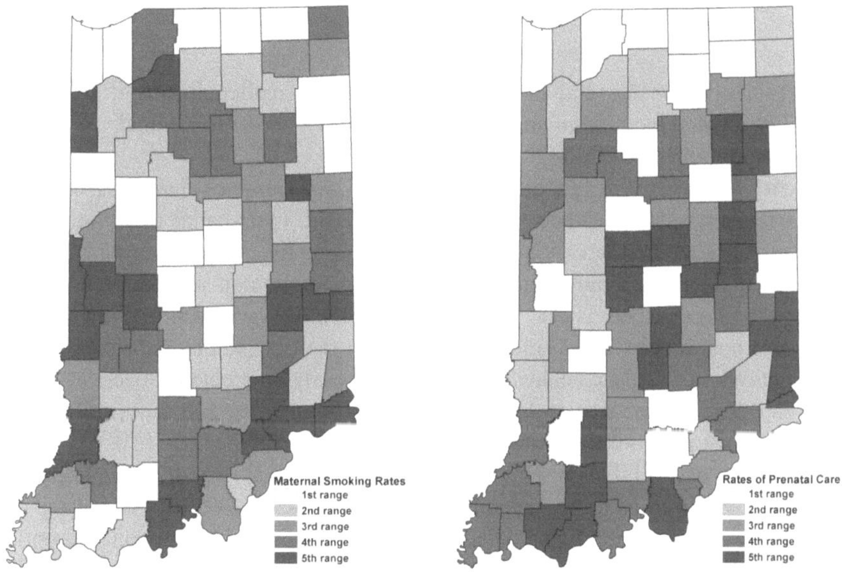

Fig. 4 Spatial distribution of maternal smoking rates (*left*) and rates of prenatal care (*right*)

in southwestern Indiana, as well as in a smaller group of counties in the north-central portion of Indiana. Counties with low cardiovascular disease mortality are concentrated in the southeast, the northeast, and in counties around Indianapolis. High cancer mortality is concentrated along the western border to Illinois and some

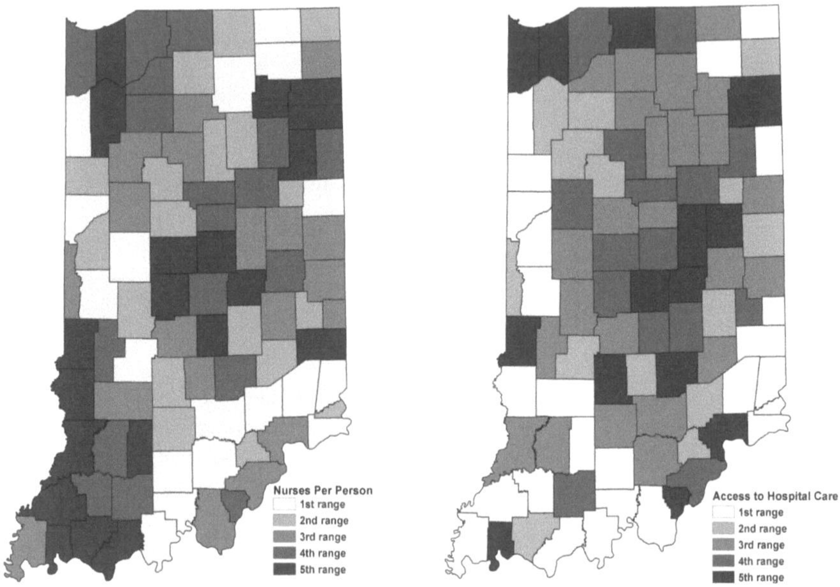

Fig. 5 Spatial distribution of nurses per person (*left*) and access to hospital care (*right*)

counties in northwestern Indiana. There are also three smaller clusters of low cancer mortality, located around Indianapolis, at the southern border to Kentucky, and at the southeastern border to Ohio.

Figure 4 shows the spatial distributions of the two variables characterizing pregnant women's health behaviors. In some areas, the two spatial distributions are almost contradictory. For example, the counties along the northern border have very low maternal smoking rates, but the rate of pregnant women seeking prenatal care during the first trimester is also very low. Overall, the bivariate correlation between these two health behaviors is close to zero ($r = 0.142$) and the maps suggest that different types of healthy behaviors do not necessarily co-locate in space.

Figure 5 shows the spatial distributions of the availability and accessibility variables. Statistically, only the availability variable (nurses per residents) is highly clustered in space (Table 3). However, Fig. 5 shows that, overall, availability and accessibility are co-locating, meaning that counties with good access to hospitals tend to have a high nurse per person ratio ($r = 0.373$). An obvious exception is the southwest corner of the state where several counties have poor access to hospital care but a high nurses-to-residents ratio. Counties with the best access to hospital care are located in and around Indianapolis as well as around the northern state border. Good access to hospital care is also prevalent in the regional centers, namely Allen County (Fort Wayne), Tippecanoe County (Lafayette), Monroe County (Bloomington) and Vanderburgh County (Evansville), Vigo County (Terre Haute), Jefferson County (New Albany), Madison County (Anderson), The most underserved counties with poor access to hospital care are concentrated in the southern rural portion of Indiana, especially along the Ohio River.

4.6 Estimation Results

Tables 4–7 show the estimation results for health outcomes and health behaviors, using the availability of primary care (NURSE) and accessibility to hospital care (HOSPITAL), respectively, as the key explanatory variable. In both tables, the upper panels show the results for the baseline models that do not take into account spatial dependencies. The lower panels provide the results for spatial lag models.

Before discussing the linkage between health care access, spatial spillovers, and health status, several observations on the systematic variation of health outcome and health behaviors can be made. Overall, the estimated health production functions do a much better job accounting for variations in health behaviors than for variations in health outcomes. For elderly mortality this is not surprising given that the model only accounts for contemporaneous conditions. Interestingly, the model performance is particularly poor for the broad measure of elderly mortality (MORT) but performs better for the cause-specific mortality measures, CVD and CANCER.

Table 4 Outcomes as a function of primary care availability (NURSE)[a]

	IMR	%LBW	MORT	CVD	Cancer
	b	b	b	b	b
Intercept	12.433(***)	12.554(***)	42.373(***)	2188.112(***)	1269.616(***)
Income	−0.079(**)	−0.041	−0.097	−17.560(***)	−7.567(***)
Education	−0.058	−0.082(**)	−0.125	−11.590(**)	−2.064
Uninsured	0.088	−0.062	−0.164	0.857	1.330
Rurality	−7.133(***)	−5.110(**)	−1.953	−431.092	−32.890
Nurse	0.077	0.063	0.192	40.267(***)	2.811
Diagnostics					
R^2	0.341(***)	0.144(**)	0.073	0.306(***)	0.188(***)
Moran's I	0.311	0.090	1.510	1.960(**)	1.990(**)
LM lag	0.098	0.278	0.956	4.227(**)	1.274
Robust LM lag	0.433	1.191	0.333	2.236	1.157
LM error	0.003	0.967	1.252	2.394	2.482
Robust LM error	0.532	0.967	0.630	0.403	2.365
				b	
Intercept				1771.614(***)	
Income				−14.675(***)	
Education				−12.195(**)	
Uninsured				0.300	
Rurality				−530.397(*)	
Nurse				33.900(***)	
ρ				0.271(**)	
R^2				0.346	

The number of observations is $n = 92$

[a]The asterisks identify significance at the 0.01, 0.05, and 0.10 level using ***, ** and *, respectively

Table 5 Behaviors as a function of primary care availability (NURSE)[a]

	Smoking	Prenatal	Teenpreg	Healthy
	b	b	b	b
Intercept	59.449(***)	106.195(***)	86.678(***)	20.594(***)
Income	−0.534(***)	−0.337(***)	−0.074	0.131(**)
Education	−0.212(**)	0.288(**)	−1.163(***)	0.116(*)
Uninsured	−1.008(***)	−2.338(***)	0.703	−0.629(***)
Rurality	0.765	15.283(**)	−59.72(***)	−0.011
Nurse	−0.091	0.581(***)	−0.839(*)	0.046
Diagnostics				
R^2	0.545(***)	0.563(***)	0.420(***)	0.464(***)
Moran's I	2.584(***)	1.866(*)	1.370	0.130
LM lag	6.133(**)	3.565(*)	0.063	0.305
Robust LM lag	1.665	1.457	1.283	0.464
LM error	4.581(**)	2.125	0.971	0.037
Robust LM error	0.112	0.016	2.191	0.197
	b	b		
Intercept	50.130(***)	87.820(*)		
Income	−0.476(***)	−0.309(***)		
Education	−0.216(***)	0.279		
Uninsured	−0.948(***)	−2.238(***)		
Rurality	−0.475	14.424(**)		
Nurse	−0.103	0.556(***)		
ρ	0.310(***)	0.204(*)		
R^2	0.584	0.582		

The number of observations is $n = 92$
[a]The asterisks identify significance at the 0.01, 0.05 and 0.10 level using ***, ** and *, respectively

Turning to the inputs of the health production function, the results suggest that income has, by and large, the expected beneficial effect on health outcomes and health behaviors. Infant mortality, cause-specific elderly mortality, and smoking prevalence among pregnant women decrease with increasing median household income. In addition, the percentage of kids with a healthy start is estimated to grow as a county's median income increases. Contrary to our expectation, median household income has a negative effect on the percentage of pregnant women seeking prenatal care during the first trimester. Interestingly, the income variable has no impact on teenage pregnancy rates. This may be because the variable only accounts for pregnancies that resulted in a live birth. Other pregnancy outcomes, i.e., miscarriages, induced abortions and still births, are not included in the variable.

The education variable does not affect health outcomes but has a beneficial effect on health behaviors. The higher the percentage of the college educated population, the lower the prevalence of smoking during pregnancy, the higher the percentage of pregnant women seeking prenatal care during the first trimester, and the lower the teenage pregnancy rate. The estimations suggest quite disturbing impacts of financial barriers on health behaviors. As the share of the uninsured

Table 6 Outcome as a function of accessibility of hospital care (HOSPITAL)[a]

	IMR	%LBW	MORT	CVD	Cancer
	b	b	b	b	b
Intercept	13.688(***)	13.954(***)	40.313(***)	2448.863(***)	1228.031(***)
Income	−0.068(**)	−0.038	−0.060	−12.929(**)	−6.986(**)
Education	−0.038	−0.073(**)	−0.137	−9.812	−2.316
Uninsured	0.072	−0.071	−0.199	−5.628	0.806
Rurality	−9.993(***)	−6.895(**)	2.935	−537.732	57.337
Hospital	−0.571	−0.895	4.163	117.476	73.432
Diagnostics					
R^2	0.333(***)	0.140(**)	0.076	0.198(***)	0.189(***)
Moran's I	0.897	0.074	0.968	3.111(***)	1.632
LM lag	0.556	0.174	0.796	9.632(***)	1.222
Robust LM Lag	0.089	0.893	1.244	2.490	0.228
LM error	0.409	0.028	0.473	7.508(***)	1.756
Robust LM error	0.001	0.746	0.921	0.366	0.762
				b	
Intercept				1694.471(***)	
Income				−9.416(*)	
Education				−11.688(**)	
Uninsured				−5.070	
Rurality				−500.336	
Hospital				195.028	
ρ				0.390(***)	
R^2				0.292	

The number of observations is $n = 92$

[a]The asterisks identify significance at the 0.01, 0.05 and 0.10 level using ***, ** and *, respectively

population increases – an indicator of financial barriers to health care – the percentage of women seeking prenatal care declines and so does the share of kids with a healthy start.

The effect of rurality is quite interesting. In a recent policy brief of the National Rural Health Association (NRHA 2006) it was suggested that health status varies by rurality in a nonlinear fashion: mortality being highest in the most rural places, decreasing with increasing urbanization but shifting upward for the most urbanized central cities. In our study we could not find non-linear effects of rurality.[10] In contradiction to common perceptions of rural areas, we instead find that in Indiana, increasing rurality is associated with decreasing infant mortality, declining teenage pregnancy, and an increased percentage of pregnant women seeking early prenatal care.

Turning now to the important linkage between medical care access and health status, our estimations by and large confirm the results of previous studies. When focusing on hospital care accessibility (Tables 6 and 7), the results suggest that

[10] We experimented with different specifications for the rurality variable, including nonlinearities. However, none of the non-linear specifications yielded significant results.

Table 7 Behavior as a function of accessibility of hospital care (HOSPITAL)[a]

	Smoking	Prenatal	Teenpreg	Healthy
	b	b	b	b
Intercept	58.155(***)	108.027(***)	72.786(***)	20.821(***)
Income	−0.542(***)	−0.261(**)	−0.134	0.136(**)
Education	−0.220(**)	0.302(**)	−1.253(***)	0.118(*)
Uninsured	−0.994(***)	−2.434(***)	0.828	−0.637(***)
Rurality	2.157	16.899(*)	−43.68(**)	−0.018
Hospital	0.504	3.802	6.778	0.210
Diagnostics				
R^2	0.544(***)	0.530(***)	0.400(***)	0.463(***)
Moran's I	2.360(**)	1.628	1.606	0.104
LM Lag	6.068(**)	3.509	0.132	0.312
Robust LM lag	2.000	1.779	2.365	0.614
LM error	4.085(**)	1.744	1.689	0.019
Robust LM error	0.016	0.014	3.923(**)	0.321
	b			
Intercept	48.969(***)			
Income	−0.486(***)			
Education	−0.224(**)			
Uninsured	−0.932(***)			
Rurality	0.727			
Hospital	0.314			
ρ	0.308(***)			
R^2	0.582			

The number of observations is $n = 92$

[a]The asterisks identify significance at the 0.01, 0.05 and 0.10 level using ***, ** and *, respectively

accessibility does not have a beneficial effect for any of the mortality variables or any health behavior variables. Thus, similar to Newhouse and Friedlander (1980) and Thornton (2002) we can conclude that – once socio-economic differences are taken into account – the health status is not affected by accessibility to specialized medical resources.

However, focusing on the impact of primary care, (Tables 4 and 5), a slightly more nuanced picture emerges. The estimations suggest that primary care availability does not have a beneficial impact on health outcomes, at least not when measured via mortality. However, primary care availability does make a difference for health behaviors. Increasing the nurse-to-population ratio significantly increases the percentage of pregnant women seeking early prenatal care and it significantly reduces teenage pregnancy rates. These results thus partially support Starfield et al. (2005) who argue "that primary care improves health [... and ...] that health is better in areas with more primary care physicians" (p. 459). The diagnostics reveal that spatial processes are at work for one health outcome variable (cardiovascular disease mortality of the elderly) and two health behavior variables (smoking prevalence among pregnant women and prenatal care utilization). In each case, Moran's I for the error terms are significant. Moreover, the (robust) Lagrange multiplier tests for the lag are significant and the test statistic exceeds that for the Lagrange multiplier test for the error. Thus, the diagnostics point in the direction of spatial lag models as the proper specification, and implicitly to the existence of spillover effects as discussed in Sect. 2.2. The estimated spatial lag, $\hat{\rho}$, is most pronounced for smoking prevalence during pregnancy, thus a behavior that is easily copied by others. This finding is supported by Nakajima (2007) who finds positive peer effects on youth smoking behavior. Similarly, prenatal care utilization very much depends on knowledge and acceptance, and thus is influenced by those living in close proximity. Our results confirm the existence of such spatial spillovers. Another interesting result is that the only health outcome variable that exhibits spatial spillovers, is cardiovascular disease mortality. Cardiovascular disease is not only sensitive to early diagnosis and treatment, but is also affected by lifestyle choices regarding diet, physical activity, smoking and alcohol consumption. This again points to spatial spillovers that come into play when analyzing health behaviors.

5 Summary and Conclusions

This chapter asks whether poor spatial accessibility leads to poor health outcomes. If people are unwilling or unable to travel long distances for basic preventive and curative care, then physical distance between health care providers and consumers is an important barrier to care with potentially detrimental consequences for a population's health status. Better understanding the linkage between health access and health outcomes is particularly important given that the United States is spending billions on improving health care access.

We tackle this issue by estimating spatial health production functions, using data for Indiana counties. Indiana provides an ideal setting because of its mixed composition, featuring both very rural counties with fewer than 6,000 residents and highly urbanized areas such as the core counties of the Indianapolis-Carmel metropolitan area. The input variables include information on income, education, insurance, degree of rurality and medical resources. Two sets of health production functions are estimated that utilize different specifications of the medical resource variable. One uses an availability measure of the mostly locally consumed primary care services, the other uses a gravity-based spatial accessibility measure for hospital care. The models are estimated for six health outcome variables relating to infants and the elderly, and four health behavior variables. All models are tested for spatial dependencies.

We find that health outcomes, measured by mortality, are not influenced by access to medical resource, whether measured as availability of primary care or distance-based accessibility to hospital care. Thus, similar to other studies, our results suggest that the relationship between health care accessibility and health status is difficult to establish when using mortality to measure health status.[11] In contrast, we find that

[11] Mortality statistics are, however, available publicly and thus are very useful for evaluation purposes.

health behaviors – notably the percentage of pregnant women seeking prenatal care during the first trimester – are positively affected by health care accessibility.

For cardiovascular disease mortality and for two of the health behaviors – smoking during pregnancy and the percentage of pregnant women seeking prenatal care during the first trimester – we find evidence of spatial spillovers. This suggests that cultures and norms guiding health behaviors are not spatially fixed but diffuse across space. From a policy perspective, it is thus important to recognize that efforts to improve health behaviors in one locality will impact health behaviors in neighboring areas as well, and thus eventually trickle through the entire system.

This research is a pilot study of the relationship between spatial accessibility to health care and health outcomes. As such, the regressions reflect the exploratory nature of this study but also highlight a number of problems that arise when dealing with spatially aggregated cross-sectional data. First, mortality and morbidity are influenced by life-time exposure to external conditions rather than contemporaneous exposure. Thus, dealing with the impacts of migration and regional change is important, especially in fast-growing states. Indiana is comparatively stable but the issue will gain importance when extending the focus to the nation. For example, the American Community Survey suggests that only 2.2% of Indiana's 2006 population lived in a different county in the previous year, making Indiana one of the least mobile states in the country. In comparison, the equivalent percentages are 6.2% for Nevada, 5.8% for Arizona, and 4.4% for Florida. Moreover, migration will gain in importance when the focus switches to population groups that are more mobile than infants, and when including the main destination states of post-retirement migration. For example, in states like Florida and Arizona, a large portion of the elderly are newcomers who moved into the states following retirement. Thus, their health status may be strongly influenced by past access to health care in their county/state of origin. In those states it will also be more difficult to assess the demand for health care as it is likely to fluctuate due to retirees' temporary movements (e.g., residing in Arizona during the winter only).

Second, future work should also account for ecological bias resulting from aggregate data by using modern econometric methods proposed in the epidemiological and statistical sciences. Particularly promising is a strategy suggested by Haneuse and Wakefield (2004). They decompose the error term to take into account random effects that may vary by rurality and/or other control variables. In addition, the random effects can then be unstructured or can be assumed to depend on the spatial structure of the data. Unfortunately, methods that use other data sources to provide bounds on the estimated effects cannot be utilized given the paucity of data below the county level for many of the health statistics used in our analysis.

Third, when extending the pilot study to a nationwide analysis, even starker disparities in health status and health care accessibility are expected and more attention should be paid to spatial scale issues. On the one hand, counties are too small to provide reliable data on some health outcomes. For example, in low mortality countries, such as the United States, infant mortality occurs rarely: only 6.5 out of 1,000 babies die before their first birthday. Thus, for small counties the expected number of infant deaths may well be less than one. On the other hand, counties may be too big to identify small-scale spillovers, may have a high degree of internal heterogeneity, and be particularly prone to ecological fallacies. Moreover, it may be necessary to define separate spatial regimes responsive to inherent differences in spatial organization between, for example, the western and eastern portions of the United States.

Finally, future research should also pay attention to the effects of alternative spatial structures. The pilot study reported here is based on contiguity, yet does not take into account asymmetries in interaction. Such asymmetries are very likely in Indiana where the capital, Indianapolis, assumes urban primacy and the major transportation lines radiate to/from the capital. Commuter flows or migration flows may provide guidance for the specification of such alternative spatial structures.

Acknowledgements The research was conducted with partial support from the Purdue Center for Regional Development. The authors would like to thank Eda Unal and Sema Sobu for their research assistance, and three anonymous reviewers for their valuable comments.

References

Allen D, Kamradt J (1991) Relationship of infant mortality to the availability of obstetrical care in Indiana. J Fam Pract 33:609–613

Anderson O, Morrison E (1989) The worth of medical care: a critical review. Med Care Res Rev 46:121–155

Auster R, Leveson I, Sarachek D (1969) The production of health: an exploratory study. J Hum Resour 4:411–436

Chan L, Hart LG, Goodman DC (2006) Geographic access to health care for rural medicare beneficiaries. J Rural Health 22:140–146

Cifuentes J, Bronstein J, Phibbs CS, Phibbs RH, Schmitt SK, Carlo WA (2002) Mortality in low birth weight infants according to level of neonatal care at hospital of birth. Pediatrics 109: 745–751

Corman H, Joyce T, Grossman M (1987) Birth outcome production function in the United States. J Hum Resour 22:339–361

Folland SA, Goodman AC, Stano M (1997) The economics of health and health care. Prentice Hall, Upper Saddle River, NJ

Frankenberger, E (1995) The effects of access to health care on infant mortality in Indonesia. Health Transit Rev 5:143–163

Fuchs, V (1974) Who shall live? Basic Books, New York

GAO (2006) Health professional shortage areas: problems remain with primary care shortage area designations system. GAO–07–84. Washington, DC, October, 2006. http://www.gao.gov/new.items/d0784.pdf

Goyder EC, Botha JL, McNally PG (2000) Inequalities in access to diabetes care: evidence from a historical cohort study. Qual Health Care 9:85–89

Guagliardo MF (2004) Spatial accessibility of primary care: concepts, methods and challenges. Int J Health Geogr 3:3

Gulliford MC, Jack RH, Adams G, Ukoumunne O (2004) Availability and structure of primary medical care services and population health and health care indicators in England. BMC Health Serv Res 4:12

Hadley J (1982) More medical care, better health? Urban Institute, Washington, DC

Haneuse S, Wakefield J (2004) Ecological inference incorporating spatial dependence. In: King G, Rosen O, Tanner MA (eds) Ecological Inference: new methodological strategies. Cambridge University Press, New York, pp 266–301
Joyce T, Racine A, Mocan N (1992) The consequences and costs of maternal substance abuse in New York City: a pooled time series cross section analysis. J Health Econ 11:297–314
Krakauer HI, Jacoby I, Millman M, Lukomnik JE (1996) Physician impact on hospital admission and on mortality rates in the medicare population. Health Serv Res 31:191–211
Lavy V, Strauss J, Thomas D, de Vreyer P (1996) Quality of health care, survival and health outcomes in Ghana. J Health Econ 15:333–357
Luo W, Wang F (2003) Measures of spatial accessibility to health care in a GIS environment: synthesis and a case study in the Chicago region. Environ Plann B 30:865–884
Mansfield CJ, Wilson JL, Kobrinski EJ, Mitchell J (1999) Premature mortality in the United States: the roles of geographic area, socioeconomic status, household types, and availability of medical care. Am J Public Health 89:893–898
McDonald TP, Coburn AF (1988) Predictors of prenatal care utilization. Soc Sci Med 27:167–172
Mokdad AH, Marks JS, Stroup DF, Gberding JL (2004) Actual causes of deaths in the United States, 2000. J Am Med Assoc 291:1238–1245
Nakajima R (2007) Measuring peer effects on youth smoking behavior. Rev Econ Stud 74:897–935
Newhouse JP, Friedlander LJ (1980) The relationship between medical resources and measures of health: some additional evidence. J Hum Resour 15: 200–218
NRHA Policy Brief (2006) Health disparities in rural populations: an introduction. National Rural Health Association May 2006. http://www.nrharural.org/advocacy/sub/policybriefs/HlthDisparity.pdf
PCRD (2006) Rural Hoosiers respond – how are we doing? How can we do better? Purdue Center for Regional Development, Research Paper PCRD–R–3
Penchansky R, Thomas JW (1981) The concept of access. Med Care 19:127–140
Perry B, Gesler W (2000) Physical access to primary health care in Andean Bolivia. Soc Sci Med 50:1177–1188
Rice N, Smith PC (2001) Ethics and geographical equity in health care. J Med Ethics 27:256–261
Shi L, Starfield B (2001) The effect of primary care physician supply and income inequality on mortality among Blacks and Whites in US metropolitan areas. Am J Public Health 91: 1246–1250
Shi L, Macinko J, Starfield B, Wulu J, Regan J, Politzer R (2003) The relationship between primary care, income inequality, and mortality in US states, 1980–1995. J Am Board Fam Pract 16: 412–422
Skinner J (2006) Geography and the use of effective care in the United States. In: Wise DA, Yashiro N (eds) Health care issues in the United States and Japan. University of Chicago Press, Chicago, pp 195–208
Starfield B, Shi L, Macinko J (2005) Contribution of primary care to health systems and health. Milbank Q 83:457–502
Thornton J (2002) Estimating a health production function for the US: some new evidence. Appl Econ 34:59–62
Unal E, Chen S, Waldorf B (2007) Spatial accessibility of health care in Indiana. Purdue University, Department of Agricultural Economics. Working Paper # 07–07. http://agecon.lib.umn.edu/cgi–bin/view.pl
Veugelers PJ, Yip AM, Burge F (2004) Inequalities in health and health services delivery: a multilevel study of primary care and hypertension control. Chronic Dis Can 35:101–107
Waldorf B (2007) What is rural and what is urban in Indiana?" Research Report, Purdue Center for Regional Development. PCRD–R–4. http://www.purdue.edu/dp/pcrd/pdf/PCRD–R–4.pdf, 1 Oct. 2009
Wennberg JE, Fisher E, Skinner J (2002) Geography and the debate over Medicare reform. Health Aff. 2002 Jul–Dec; Suppl Web Exclusives:W96–W114
Wolfe B (1986) Health status and medical expenditures: is there a link? Soc Sci Med 22:993–999
Wyszewianski L (2002) Access to care: remembering old lessons. Health Serv Res 37:1441–1443

Immigrant Women, Preventive Health and Place in Canadian CMAs

Kelly Woltman and K. Bruce Newbold

1 Introduction

Cervical cancer is a disease that affects women of all ages (Health Canada 2002). It is one of the most common malignant diseases in women (Duarte-Franco and Franco 2003), with an estimated 9,900 potential years of life lost due to this disease in Canada in 2003 (National Cancer Institute of Canada 2007). Given its slow progression, identifiable cytological precursors and effective treatments, cervical cancer is also one of the most preventable human cancers (Leyden et al. 2005). With routine cervical cancer screening, the disease is preventable and curable when detected at an early stage (Fehringer et al. 2005; Johnston et al. 2004; Yi 1994). The recent introduction of the HPV vaccine is expected to also have significant health benefits.

Cervical cancer screening commonly uses a Papanicolaou test or smear (referred to subsequently within this paper as Pap test) for early detection. Early detection provides the opportunity to observe any signs of pre-cancerous changes and eliminate abnormal cells before these they become cancerous. According to the Canadian Task Force on Preventive Health Care (formerly the Periodic Health Examination) and guidelines from the National Workshop on Screening for Cancer of the Cervix (Miller et al. 1991; Morrison 1994), screening is recommended following the initiation of sexual activity or at age 18. After two normal smears, routine Pap testing is advised every 3 years until the age of 69.

While more frequent testing may be considered for women at high risk (first intercourse at less than 18 years of age, multiple sexual partners, partner with multiple sexual partners, low socio-economic status) (Health Canada 2002), participation in Pap testing is understood to be the most effective means of decreasing mortality rates from this invasive cancer and is effective in preventing invasive cervical cancer (Eddy 1990; Fehringer et al. 2005; Health Canada 2002; Johnston et al. 2004; Miller et al. 1991).

K. Woltman (✉)
School of Geography and Earth Sciences, McMaster University, 1280 Main Street West, Hamilton, ON L8S 3Z9, Canada,
e-mail: woltmak@mcmaster.ca

A. Páez et al. (eds.), *Progress in Spatial Analysis*, Advances in Spatial Science,
DOI 10.1007/978-3-642-03326-1_17, © Springer-Verlag Berlin Heidelberg 2010

While overall mortality rates from this disease are decreasing (National Cancer Institute of Canada 2007; Eddy 1990; Miller et al. 1991), approximately half of the women who develop invasive cervical cancer have never had a Pap test (Parboosingh et al. 1997). Most notably, failure to participate in Pap testing is the single greatest risk factor for poor outcomes in women who develop cervical cancer (Morrison 1994).

While notably a public health issue, the question of Pap test uptake crosses into the domains of health geography, population geography, and spatial analysis. From a demographic perspective, recent immigrants are typically less likely to be screened for chronic conditions and cancers compared to their longer-term immigrant and native-born counterparts (DesMeules et al. 2004; Goel 1994; Hyman and Guruge 2002; Hyman et al. 2002; Leduc and Proulx 2004; McDonald and Kennedy 2005; Newbold 2005; Woltman and Newbold 2007). Recent immigrant women may be at a higher risk for cervical cancer, primarily because this group has lower rates of Pap testing (McDonald and Kennedy 2007), particularly amongst more recent arrivals (Duarte-Franco and Franco 2003). For example, 27% of immigrant women in Hamilton-Wentworth, Canada, have never had a Pap test, compared to 9% of non-immigrant women (Black and Zsoldos 2003).

While McDonald and Kennedy (2007) noted that the use of Pap smear testing by immigrant women increased with duration of residence in Canada, rates still varied widely. Lack of knowledge, unease, and the cultural incongruity that immigrants experience upon arrival may deter the use of health services (Hyman 2001), especially those services that are not necessarily considered essential by the individual. Physician use and recommendation is also strongly linked to uptake of Pap tests (Grossman 1972; Kenkel 1994). Additional factors associated with a lack of screening include being single, older, low income, certain ethnic backgrounds, (Hyman et al. 2003), low level of education, and speaking neither English nor French (Goel 1994; Bryant et al. 2002; Woltman and Newbold 2007). In particular, Asian origins have been noted to have lower uptake, even after multiple years in Canada (Juon et al. 2003; McDonald and Kennedy 2007). Given Canada's changing demographics, including increased immigrant numbers and a diversity of immigrant sources, a growing number of women may be at risk. Moreover, recent immigrants are increasingly racialized, increasing the likelihood that they face barriers to health care associated with structural (e.g. socio-economic) inequalities and barriers associated with race, culture and language.

The uptake of Pap screening tests is also an inherently geographical problem, linking community, health and population. Indeed, health geography has a long tradition of considering disease diffusion and variations in health status across space. More recent work has focused on context (place) and composition (individual characteristics) and the collective in shaping health and its determinants (Diez Roux 2001, 2002; Ellaway and Macintyre 2001; Macintyre et al. 1993, 2002). As Frohlich et al. (2002) note, such studies are important for determining if health status variations between different communities are a result of individual or aggregate attributes. Research has increasingly attempted to tease out the relative contributions of contextual and compositional effects in examining the complex link between health status, lifestyle behaviours, and context. However, the recent literature (i.e., Macintyre et al. 2002) also argues that to be meaningful, place may function as a residual category unless it is properly and explicitly defined as to how it may influence health.

In the context of immigrant settlement, immigrants tend to be highly concentrated within the nation's largest and most diverse cities. Toronto, Vancouver and Montreal represent the largest immigrant receiving centres in Canada, with 45.7% of the total immigrant population in Canada choosing to live in Toronto, 39.6% in Vancouver and 19% in Montreal in 2001 (Chui et al. 2007). Often settling in affordable and low-income areas (Glazier et al. 2004), recent immigrants may be particularly vulnerable to poor health outcomes subject to limited (or non) use of preventative health services.

Uptake of screening may also vary by place (see, e.g. Ng et al. 2004), and interactions between areas and people may also influence screening participation (Woltman and Newbold 2007). Independent of individual characteristics, it is recognized that an individual's immediate environment may possess both material and social characteristics that are potentially linked to health-seeking behaviours (Diez Roux 2001; Glazier et al. 2004; Ross et al. 2004). For example, neighbourhoods could be sources of important information and support with regard to screening (McDonald and Kennedy 2005). Given that immigrants as a group are less likely to participate in these services, knowledge of these health services may be less likely when immigrants are living closely together (Woltman and Newbold 2007). That is, for example, could women in places with high immigrant concentrations face even greater risk? What are the spatial correlates of uptake, and how does it vary across space? Area level or neighbourhood characteristics might help explain the uptake of preventive health care behaviours.

Notably, the independent importance of individual and neighbourhood factors on the utilization of preventive care has not been investigated. Moreover, immigrant women represent an understudied population. This segment of the Canadian population is becoming an increasingly diverse group, with growing numbers from Asia, Africa, the Caribbean, Latin America and Eastern Europe (Citizenship and Immigration Canada 2001). This is also important in that socio-cultural and racialized barriers may be affecting health care utilization more than ever before. Canada's immigrant population is becoming a more heterogeneous group, which may ultimately lead to further disparities among cultural groups in rates of cervical cancer incidence and mortality (Duarte-Franco and Franco 2003).

The purpose of this paper is therefore to investigate the multilevel characteristics associated with the utilization (lifetime and regular use) of preventive cervical cancer screening in immigrant and native-born women residing in Canada's three largest Census Metropolitan Areas (CMAs). In order to address these objectives, the following questions are investigated:

- Is there evidence of between neighbourhood variation in the utilization of cervical cancer screening?
- Does the neighbourhood concentration of immigrants account for between area differences?

- Does utilization differ between immigrant and native-born women?
- To what extent does CMA moderate the association between immigrant status and utilization?
- Is there evidence of cultural differences?

2 Methods

Data on the use of Pap tests is drawn from Cycle 2.1 (2003) of Statistics Canada's Canadian Community Health Survey (CCHS) master file. The objective of the CCHS is to provide timely, reliable, cross-sectional estimates of health determinants, health status and health system use at sub-provincial levels. A multi-stage stratified cluster design was used to sample household dwellings, which covered approximately 98% of the Canadian population aged 12 and older living in private households. Additional data comes from the 2001 Canadian census public use microdata file, which offers demographic, social and economic information on the population of Canada at various geographical scales.

The current analysis is set at the census tract scale. Using Statistics Canada's postal code conversion file to link with the postal codes of CCHS respondents, the 2001 census was used to provide demographic and socio-economic measures for the census tracts (neighbourhoods) in which respondents were residing. Unsuccessful geocodes were examined on a case by case basis to determine why they did not geocode, with less than 5% of records ultimately discarded due to unsuccessful geocoding.

Defining census tracts as neighbourhoods offer a number of advantages. Importantly, the use of census tracts provides direct linkage to statistical measures provided by Statistics Canada. Although there is disagreement in the literature concerning the best way to capture the concept of "neighbourhood," recent research suggests census tracts are good proxies (Diez Roux 2001; Ross et al. 2004) as compared to socially constructed areas, which are often loosely defined and often lack the ability to link to other statistical data. Indeed, the comparison of several "neighbourhood" units of analysis suggests that census tracts are good proxies for natural neighbourhood boundaries in studies of neighbourhood effects on health (Ross et al. 2004). However, given debate over the definition and division of space (see, e.g. Moon and Brown 1998), we also regard the partitioning of space as worthy of scrutiny and carry out this task through seeing if a particular set of sub-areas – census tracts – are associated with the uptake of Pap tests or not.

Women between the ages of 18 and 69, residing in the Montreal, Toronto and Vancouver CMAs were selected. Two dependent variables are considered: lifetime and regular Pap tests (Hyman et al. 2002). The first dependent variable asks whether the respondent has ever had a Pap test. This variable captures individual *lifetime* uptake of cervical cancer screening, which may include having had a Pap test in countries other than Canada. However, Hyman et al. (2002) found that once initial barriers to screening were overcome, there was less variation between immigrant groups in the proportion of women who engaged in regular screening. Therefore, the second dependent variable asks those who reported ever having had a Pap test whether the respondent has had a Pap test within the past 3 years. The construction of this variable is based upon Canadian screening guidelines, which recommends that women have regular Pap testing at least every 3 years. This variable examines *regular* use of cervical screening services. It is important to note that this variable may not capture "regular" routine screening; for example, this could be the case if a woman's one and only test has been in this 3 year window. Approximately 2.8% of this sample was lost due to non-response (don't know, refused, not stated).

Based on a review of the literature and hypothesized relationships, demographic, health, acculturation and socio-economic variables associated with the uptake of cervical cancer screening were identified (see Table 1). Individual-level variables included age, marital status and cultural origin (based on self-reported cultural/racial origin), self-reported general health, contact with a general practitioner in the past year, and immigrant status. Immigrant status distinguished between recent (resident for less than 10 years) and long-term (resident greater than 10 years) immigrants versus native-born (Canadian-born). As a measure of acculturation, a woman's ability to speak at least one of Canada's official languages was included. Socio-economic characteristics included educational attainment and household income adequacy.

Derived from the census tract profile data from the Canadian census (Statistics Canada 2007), the neighbourhood proportion of immigrants was also included. The percentage of immigrants at the neighbourhood (census tract) level was expressed in increments of ten (i.e., 25% took on a value of 2.5). In addition, a neighbourhood disadvantage index score (NDIS) was derived from five variables including proportion of the total neighbourhood income coming from government transfer payments, proportion of the neighbourhood 15 years and older without a secondary school diploma, mean household income, proportion of families in the neighbourhood with household incomes below the poverty line, and proportion of individuals in the neighbourhood 15 years and older who were unemployed (Boyle and Lipman 2002). These five variables were entered into a principal component analysis. One factor emerged that accounted for approximately 68.0% of the total explained variance. To represent NDIS, a factor regression score was calculated by weighting each of the five variables by its factor loading.

The analyses entailed use of multilevel logistic regression models, with estimation conducted using the MLwiN software. Unlike traditional multivariate methods that require aggregation or disaggregation so that variables can reflect the individual or group level, a multilevel approach can identify relationships among variables measured at both the individual and group levels. This approach is needed to account for the correlation of responses within naturally formed groupings, such as neighbourhoods (Boyle and Lipman 2002). Multilevel models were developed to simultaneously consider i individual females (Level 1) within j neighbourhoods in Montreal, Toronto and Vancouver (Level 2). This model is defined as:

$$\begin{aligned} \text{Logit}\,(\pi ij) &= \beta 0j + \beta i \chi ij \\ \beta 0j &= \beta 0 + \mu 0j \end{aligned} \tag{1}$$

Table 1 Definition and coding of covariates

Variable description	Coding
Age in years	Mean centred
Marital status	
Married, common-law	Reference category
Separated, divorced, widowed	Dummy indicator (1 = yes, 0 = no)
Single	Dummy indicator (1 = yes, 0 = no)
Educational attainment	
Less than high school	Dummy indicator (1 = yes, 0 = no)
High school graduate	Reference category
Post secondary graduate	Dummy indicator (1 = yes, 0 = no)
Household income adequacy	
Low	Dummy indicator (1 = yes, 0 = no)
Middle (lower middle, upper middle quartile)	Reference category
High	Dummy indicator (1 = yes, 0 = no)
Self-reported general health	
Negative (fair, poor)	Reference category
Positive (excellent, very good, good)	Dummy indicator (1 = yes, 0 = no)
CMA	
Montreal	Reference category
Toronto	Dummy indicator (1 = yes, 0 = no)
Vancouver	Dummy indicator (1 = yes, 0 = no)
Neighbourhood proportion of immigrants	
Variable calculated from 2001 census (see text)	(10% increments)
Immigrant status	
Native-born (non-immigrant)	Reference category
Recent immigrant (resident for ≤ 10 years)	Dummy indicator (1 = yes, 0 = no)
Long-term immigrant (resident > 10 years)	Dummy indicator (1 = yes, 0 = no)
Can converse in English and/or French	
Yes	Reference category
No	Dummy indicator (1 = yes, 0 = no)
Consultation with GP/family doctor within the past 12 months	
No	Reference category
Yes	Dummy indicator (1 = yes, 0 = no)
Cultural/racial origin	
White	Reference category
Black	Dummy indicator (1 = yes, 0 = no)
Other Asian (Japanese, Korean)	Dummy indicator (1 = yes, 0 = no)
Filipino	Dummy indicator (1 = yes, 0 = no)
Chinese	Dummy indicator (1 = yes, 0 = no)
South Asian (East Indian, Pakistani, Sri Lankan)	Dummy indicator (1 = yes, 0 = no)
South East Asian (Laotian, Cambodian, Indonesian, Vietnamese)	Dummy indicator (1 = yes, 0 = no)
Other (native, Arab, Afghan, Iranian, self-reported other, multiple races)	Dummy indicator (1 = yes, 0 = no)
Neighbourhood disadvantage index score	
Proportion of the total neighbourhood income coming from government transfer payments	
Proportion of the neighbourhood 15 years and older without a secondary school diploma	
Mean household income (reverse coded)	

In models with two levels of analysis, each level is associated with its own, unexplained residual error. At the individual level, the residual error is constrained to 1 in logistic regression; each successive level is associated with its own error term, which estimates the residual between-neighbourhood variation (Snijders and Bosker 1999). This effectively means residual error is partitioned across levels in multilevel modelling. As highlighted by Boyle and Lipman (2002), the partitioning of responses across neighbourhoods is particularly important because it estimates the potential for measured and unmeasured neighbourhood variables to explain place-to-place variation in cervical cancer screening utilization. As articulated by Snijders and Bosker (1999), the proportion of variance accounted for by neighbourhoods can be calculated using the intra-class correlation coefficient (ICC), which is defined as:

$$\rho = \sigma^2 / \left(\sigma + \pi^2/3\right). \qquad (2)$$

This coefficient is the ratio between the neighbourhood level variation and the total variation (sum of the individual and neighbourhood level variation), where a decline in the ICC indicates that the differences between neighbourhoods have been reduced by the inclusion of explanatory variables (Ross et al. 2004).

Models for lifetime and regular Pap tests were similarly developed to evaluate neighbourhood association with cervical cancer screening service utilization. In each case, a series of five models were developed. The first model created was the null model with no explanatory variables; this serves to estimate the relative importance of individual and neighbourhood effects in accounting for variation in the outcome (Ross et al. 2005). From the null model, additional models were built incrementally, first controlling for age (mean centred), marital status, socio-economic variables, NDIS, health-related covariates, and CMA of residence. Then the neighbourhood proportion of immigrant and immigrant-related variables were added to create the third model. In the fourth model, CMA variables and interactions between CMA and immigrant status were included, along with English/French language ability. With the addition of cultural origin, the full model was created. Odds ratios and associated 95% confidence intervals were estimated.

3 Results

The total (weighted) sample for analysis represented 3,474,352 females aged 18 to 69 residing in the Montreal, Toronto and Vancouver CMAs. The distribution of the sample over the three CMAs is 32% in Montreal, 47% in Toronto and 21% in Vancouver. While the majority of the sample was born in Canada, close to 39% were immigrants. The sample contained a high percentage of immigrants, which was expected given the research focus on the three largest CMAs in Canada which are recognized as the country's largest immigrant receiving centres (Statistics Canada 2007).

3.1 Lifetime Uptake

Overall, approximately 89% of native-born, 61% of recent immigrants (less than 10 years), and 85% of long-term (greater than 10 years) immigrant women reported ever having had a Pap test. Tables 2 and 3 display the multilevel results for lifetime Pap testing. These tables consist of a series of increasingly complex models. The dependent variable is lifetime Pap uptake, or whether the respondent had ever had a Pap smear. Building upon the null model, Model 2 reveals that a higher level of edu-

Table 2 Multilevel logistic regression models: lifetime Pap uptake

Fixed Effects	Null Model β (se)	Model 2 β (se)		Model 3 β (se)	
Intercept	1.87‡ (0.04)	1.50‡ (0.15)		2.02‡ (0.16)	
		OR	(95% CI)	OR	(95% CI)
Age centred		1.04‡	(1.03–1.04)	1.04‡	(1.03–1.05)
Education					
Less than high school		0.53‡	(0.43–0.66)	0.51‡	(0.41–0.64)
Post secondary graduate		1.56‡	(1.33–1.82)	1.80‡	(1.53–2.12)
Income adequacy					
Low		0.74†	(0.60–0.91)	0.89	(0.72–1.11)
High		1.57‡	(1.32–1.87)	1.18	(0.98–1.42)
Marital status					
Separated, widowed, divorced		0.89	(0.71–1.10)	0.70†	(0.56–0.88)
Single		0.48‡	(0.41–0.57)	0.31‡	(0.25–0.37)
Neighbourhood disadvantage index score		0.86‡	(0.80–0.92)	1.04	(0.95–1.13)
Self-reported health					
Positive		1.20	(0.96–1.49)	1.18	(0.94–1.48)
Consultation with GP/family doctor					
Yes		1.95‡	(1.65–2.29)	1.91‡	(1.62–2.26)
CMA					
Toronto		0.81*	(0.67–0.97)	1.79‡	(1.40–2.30)
Vancouver		0.89	(0.73–1.08)	1.73‡	(1.36–2.19)
Neighbourhood proportion of immigrant				0.34‡	(0.20–0.61)
Immigrant status					
Recent immigrant				0.16‡	(0.13–0.20)
Long-term immigrant				0.50‡	(0.41–0.61)

(continued)

Table 2 (continued)

Fixed Effects	Model 4 β (se)		Model 5 β (se)	
Intercept	1.96‡ (0.17)		1.417‡ (0.13)	
	OR	(95% CI)	OR	(95% CI)
Age centred	1.03‡	(1.03–1.04)	1.03‡	(1.02–1.04)
Education				
Less than high school	0.54‡	(0.43–0.67)	0.52‡	(0.42–0.65)
Post secondary graduate	1.75‡	(1.49–2.06)	1.79‡	(1.52–2.12)
Income adequacy				
Low	0.91	(0.73–1.13)	0.94	(0.75–1.17)
High	1.18	(0.98–1.42)	1.16	(0.96–1.39)
Marital status				
Separated, widowed, divorced	0.69†	(0.55–0.86)	0.62‡	(0.49–0.79)
Single	0.30‡	(0.25–0.36)	0.29‡	(0.23–0.33)
Neighbourhood disadvantage index score	1.05	(0.96–1.14)	1.02	(0.93–1.11)
Self-reported health				
Positive	1.17‡	(0.94–1.47)	1.13	(0.89–1.42)
Consultation with GP/family doctor				
Yes	1.93‡	(1.63–2.28)	1.90‡	(1.60–2.25)
CMA				
Toronto	1.93‡	(1.45–2.56)	1.75	(1.32–2.34)
Vancouver	2.55‡	(1.90–3.44)	2.67‡	(1.97–3.62)
Neighbourhood proportion of immigrant	0.33‡	(0.19–0.58)	0.65	(0.36–1.15)
Immigrant status				
Recent immigrant	0.34‡	(0.22–0.52)	0.37‡	(0.24–0.59)
Long-term immigrant	0.60*	(0.40–0.90)	0.67‡	(0.43–1.02)
Cross-level interactions				
Montreal* recent immigrant status	0.57*	(0.34–0.93)	0.80	(0.48–1.34)
Montreal* long-term immigrant status	0.92	(0.57–1.49)	1.14	(0.69–1.87)
Vancouver* recent immigrant status	0.32‡	(0.19–0.54)	0.51*	(0.30–0.89)
Vancouver* long-term immigrant status	0.58	(0.34–0.98)	0.88	(0.50–1.52)
Can converse in English and/or French				
No, neither English nor French	0.53‡	(0.42–0.67)	0.72	(0.51–1.00)
Cultural/racial origin				
Black			1.36	(0.89–2.08)
Other Asian			0.46*	(0.25–0.83)
Filipino			0.48†	(0.30–0.77)
Chinese			0.25‡	(0.19–0.33)
South Asian			0.27‡	(0.20–0.38)
South East Asian			0.24‡	(0.14–0.42)
Latin American			1.63	(0.86–3.11)
Other			1.55‡	(1.14–2.12)

$^{*}p < 0.05$, $^{\dagger}p < 0.01$, $^{\ddagger}p < 0.001$, OR Odds Ratio, 95% CI Confidence Interval

Table 3 Summary of variance (standard error) components, multilevel logistic regression, lifetime Pap uptake

Random Effects	Null Model	Model 2	Model 3	Model 4	Model 5
Level 2, neighbourhood	0.261	0.231	0.125	0.096	0.067
	(0.06)	(0.07)	(0.06)	(0.06)	(0.06)
Level 1, individual	1.00	1.00	1.00	1.00	1.00
Intra-class correlation coefficient (%)	7.35	6.56	3.66	2.84	2.00

cation, higher household income adequacy, and having had contact with a general practitioner within the past year are associated with uptake. On the other hand, being single, achieving less than a high school education, reporting low household income adequacy, residing in Toronto (relative to Montreal) and living in a disadvantaged neighbourhood are negatively associated with uptake. Self-reported health-status is not significantly associated with uptake. As immigrant-specific covariates are considered in Model 3, income adequacy covariates and neighbourhood disadvantage are also reduced to non-significance.

As shown in Table 2 (Model 3), the odds of ever having a Pap test significantly decrease by 0.34 with every 10% increase in the concentration of immigrants. Also, the odds of having a Pap test are 0.16 and 0.50 for recent and long-term immigrant women, respectively, relative to Canadian born women. Similar to being single; being separated, widowed or divorced was negatively associated with uptake, relative to being married or living common-law. Furthermore, the direction of association between Toronto and uptake has reversed: relative to Montreal, women in Toronto and Vancouver are now more likely to have ever had a Pap test.

To examine the extent to which CMA residence moderates the association between immigrant status and uptake, four cross-level interactions are added in Model 4. Relative to non-immigrants residing in Montreal, the results indicate that recent immigrants in Toronto, along with recent and long-term immigrants in Vancouver, are less likely to have ever had a Pap test. However, a number of these effects become non-significant once cultural origin is taken into consideration (Model 5). The association between use and recent immigrants in Vancouver remains significant ($p < 0.05$). In relation to the white reference group, being Chinese, South Asian and other Asian origins decreases the likelihood of Pap testing. Cultural origin also appears to partially explain the effect of neighbourhood concentration of immigrants, wherein this effect became insignificant.

Table 3 highlights evidence of between neighbourhood variations in the lifetime use of Pap testing. According to the null model, the amount of variation attributable to neighbourhoods was approximately 7.4%. Controlling for demographic, socioeconomic, health-related factors and CMA residency, Model 2 explains only a small proportion of between neighbourhood variability. On the other hand, immigrant status, immigrant interactions and cultural origin appear to account for a larger proportion of this variability. For example, Model 3 reveals that the concentration of immigrants at the neighbourhood level and immigrant status exhibit strong associations with uptake. This appears to account for approximately half of the between neighbourhood differences. In the full model (Model 5), the amount of variation attributable to the neighbourhood is decreased by 5.2 percentage points.

3.2 *Regular Pap Testing*

Building upon lifetime uptake, the following section considers whether or not a woman had a Pap test within the past 3 years (regular Pap test). Among women who have participated in Pap testing, approximately 89% of native-born, 93% of recent immigrants, and 86% of long-term immigrant women reported having a similar test within the past 3 years. Regular use among recent immigrants appears to be consistent with that of their native-born and long-term immigrant counterparts, if not higher.

Tables 4 and 5 display the multilevel logistic regression results for regular Pap testing. The dependent variable is having had a Pap test within the past 3 years (regular Pap). After controlling for the null model (Table 4), Model 2 reveals that age exhibits a strong and negative association with regular use, whereas positive health status, contact with a general practitioner and residing in Toronto are strong and positively association with use. However, the effect of Toronto is reduced to non-significance in Model 3. Nevertheless, neighbourhood disadvantage appears to be negatively associated with use. In addition to age and health covariates, neighbourhood concentration of immigrants is positively associated with regular use. The odds of having had a Pap test within the last 3 years increases by 2.21 with every 10% increase in the concentration of immigrants.

Findings also reveal that individual immigrant status and language ability are not significantly associated with regular use. CMA covariates are also insignificant in Model 3 and onwards. In Model 3 of this particular analysis, the potential interactions between CMA and immigrant status were examined in preliminary analyses, but were removed from the model because of insignificance. The effects of culture are examined finally in Model 4. Compared to the white reference group, Chinese cultural origin is significantly associated with use. Model 4 also reveals that the effects of age and neighbourhood disadvantage remain negatively associated with use, whereas positive health, contact with a general practitioner and neighbourhood concentration of immigrant remains positively associated with use.

Table 5 summarizes the variance components of the models discussed above. According to the null model, approximately 3% of the variation is attributable to neighbourhoods. The variance at the neighbourhood-level decreased with the addition of individual- and neighbourhood-level covariates. After controlling for these covariates, the models are able to account for half of the neighbourhood variability; the final model was able to explain 1.5% of the variation between neighbourhoods. While individual characteristics explain much of the variation in neighbourhoods, neighbourhood characteristics were also determinants of utilization.

Table 4 Multilevel logistic regression models: regular Pap testing

Fixed Effects	Null Model	Model 2		Model 3	
	β (se)	β (se)		β (se)	
Intercept	1.99‡ (0.04)	0.79‡ (0.15)		0.70‡ (0.16)	
		OR	95%CI	OR	95%CI
Age centred		0.94‡	(0.94–0.95)	0.94‡	(0.94–0.95)
Education					
Less than high school		0.94	(0.74–1.19)	0.95	(0.75–1.20)
Post secondary graduate		1.09	(0.92–1.30)	1.07	(0.90–1.28)
Income adequacy					
Low		0.92	(0.71–1.19)	0.92	(0.71–1.19)
High		1.17	(0.97–1.40)	1.18	(0.98–1.42)
Marital status					
Separated, widowed, divorced		0.90	(0.75–1.08)	0.91	(0.75–1.09)
Single		0.88	(0.71–1.10)	0.87	(0.70–1.09)
Neighbourhood disadvantage index score		0.92	(0.85–1.00)	0.87†	(0.79–0.95)
Self-reported health					
Positive		1.72‡	(1.40–2.11)	1.73‡	(1.41–2.13)
Consultation with GP/family doctor					
Yes		2.7‡	(2.25–3.23)	1.10‡	(0.92–1.32)
CMA					
Toronto		1.11*	(0.93–1.34)	0.87	(0.68–1.11)
Vancouver		1.15	(0.94–1.40)	0.94	(0.74–1.19)
Neighbourhood proportion of immigrant				2.21†	(1.23–3.97)
Immigrant status					
Recent immigrant				1.12	(0.77–1.62)
Long-term immigrant				1.00	(0.83–1.21)
Can converse in English and/or French					
No, neither English nor French				1.09	(0.67–1.79)
Intercept				0.69‡ (0.16)	
			OR		95%CI
Age centred				0.94‡	(0.94–0.95)
Education					

(continued)

Table 4 (continued)

Fixed Effects	Model 4 β (se)	
Less than high school	0.95	(0.75–1.21)
Post secondary graduate	1.08	(0.91–1.29)
Income adequacy		
Low	0.90	(0.70–1.17)
High	1.18	(0.98–1.42)
Marital status		
Separated, widowed, divorced	0.90	(0.75–1.09)
Single	0.86	(0.69–1.08)
Neighbourhood disadvantage index score	0.87†	(0.80–0.96)
Self-reported health		
Positive	1.75‡	(1.42–2.15)
Consultation with GP/family doctor		
Yes	2.67‡	(2.23–3.20)
CMA		
Toronto	0.88	(0.69–1.13)
Vancouver	0.94	(0.74–1.19)
Neighbourhood proportion of immigrant	2.11*	(1.17–3.82)
Immigrant status		
Recent immigrant	1.09	(0.74–1.63)
Long-term immigrant	0.98	(0.79–1.20)
Can converse in English and/or French		
No, neither English nor French	0.97	(0.58–1.62)
Cultural/racial origin		
Black	1.33	(0.78–2.27)
Other Asian	0.70	(0.35–1.41)
Filipino	0.75	(0.41–1.37)
Chinese	1.57*	(1.01–2.44)
South Asian	0.72	(0.45–1.15)
South East Asian	0.61	(0.24–1.56)
Latin American	1.42	(0.59–3.40)
Other	1.15	(0.77–1.74)

* $p < 0.05$, † $p < 0.01$, ‡ $p < 0.001$, OR Odds Ratio, 95% CI Confidence Interval

Table 5 Summary of variance (standard error) components, multilevel logistic regression, regular Pap use

Random Effects	Null Model	Model 2	Model 3	Model 4
Level 2, neighbourhood	0.103(0.06)	0.06(0.07)	0.053(0.07)	0.050(0.06)
Level 1, individual	1.00	1.00	1.00	1.00
Intra-class correlation coefficient (%)	3.04	1.79	1.59	1.50

4 Discussion

This cross-sectional women's health study has focused on multi-level influences on the lifetime uptake and regular use of cervical cancer screening (Pap testing) services among women in the Montreal, Toronto and Vancouver CMAs. Given the limitation of this study's cross-sectional design, longitudinal information could provide insight into the temporal directions of the associations, and is necessary to better understand whether or not women are having regular Pap tests. A lack of data regarding the role of women's attitudes, beliefs and knowledge regarding preventive health practices also limits this research. In addition, this study relied upon self-reported information about Pap testing, which may be subject to recall bias. Data constraints also meant that this study was not able to test for lack of knowledge of the importance of Pap screening, or a lack of time to undertake screening. However, the CCHS was particularly valuable given the focus on immigrants, and interviews were conducted in over 22 different languages.

Findings reveal that dissimilarities in lifetime uptake exist between the native-born and the foreign-born populations after controlling for age, marital status, socio-economic status, and health covariates. Building upon earlier studies (Goel 1994; Maxwell et al. 2001; Woltman and Newbold 2007), this research has found that recent and long-term immigrant status is strongly and inversely associated with ever having had a Pap test. In other words, recent immigrants are less likely to have had a Pap test, with the likelihood of uptake increasing with duration of residence, in line with findings by McDonald and Kennedy (2007). Possible explanations for the lower uptake amongst recent arrivals include lack of knowledge, lack of time, language barriers and cultural factors (Black and Zsoldos 2003; Hyman and Guruge 2002; Newbold 2005; Woltman and Newbold 2007).

Additional individual-level characteristics such as age, Asian origins, marital status and contact with a general practitioner were found to be associated with uptake, and are consistent with earlier Canadian studies (Bryant et al. 2002; Gupta et al. 2002; Hyman et al. 2002, 2003; Maxwell et al. 2001; Snider et al. 1996). Results also indicate that language ability became insignificant once culture was considered in uptake.

In terms of the factors associated with having had a Pap test within the past 3 years (regular Pap test), neighbourhood disadvantage and the neighbourhood concentration of immigrants appear to be significant predictors. In such cases, the size and relative completeness of an immigrant community may alter uptake rates, with lower rates likely to be associated with less institutionally complete neighbourhoods. Although modest, there was significant between-neighbourhood variation (7.4%), suggesting that policies targeting Pap screening uptake could focus on both people and places. There also appears to be significant differences between neighbourhoods and CMAs in the uptake of cervical cancer screening among recent immigrant arrivals. While results indicate that the association between CMA and cervical cancer screening differs by immigrant status, these interactions lose their statistical significance after controlling for cultural origin. This may be due to differences in the cultural background of immigrants living in these urban centres. For example, Vancouver is home to many immigrants arriving from China, which suggests that uptake may reflect cultural differences. This may also be true at the neighbourhood level where controlling for cultural origin reduces the effect attributable to the neighbourhood concentration of immigrants to non-significance.

Findings also suggest that place is important in the use of regular Pap testing. Approximately 3% of variation in regular use appears to be attributable to between-neighbourhood differences. This suggests that there are discernible differences between neighbourhoods and between people within neighbourhoods (Merlo et al. 2005). In other words, there is moderate evidence to suggest that a possible neighbourhood contextual phenomenon is shaping individual screening behaviour. Factors such as neighbourhood disadvantage and neighbourhood immigrant concentration assist in explaining this variance.

These results provide additional insights into the preventive health behaviours of immigrant and native-born women. After controlling for age, socio-economic and demographic and health-related characteristics, neighbourhood disadvantage and the neighbourhood concentration of immigrants plays a significant role. Among women who have had a least one Pap test, there appears to be a strong and positive association between neighbourhood concentration and regular Pap test use. That is, a higher neighbourhood concentration of immigrants is associated with positive routine screening behaviours among women who already participate in screening. Building upon the determinants of health literature, this research has also found that neighbourhood disadvantage is negatively correlated with cervical cancer screening service use, above and beyond individual socio-economic status.

Chinese origin was associated with having had a Pap test within the past 3 years. Although echoing Hyman et al. (2002) who noted that there was less variation between immigrant groups in the proportion of women who engaged in regular screening once initial barriers to screening were overcome, the direction of this relationship is distinctly different from the first set of analyses (lifetime uptake), which suggested that participating in cervical cancer screening is negatively associated with women of Asian background. This raises two possible explanations. First, the structure of the question that explores the use of Pap tests (and other preventative health care issues) is likely important. That is, the current research focused on both lifetime use (have you ever had a Pap test) and regular use (in the past 3 years). The differential use of these questions in the existing literature may account for disparities in findings. Second, issues that closely reflect the ethnic or cultural makeup of the immigrant population, including diverse issues such as gender roles, trust of western medicine, attitudes and beliefs about reproductive health practices, may create differentials in the use of preventive health care, and ultimately health. There may be a cultural avoidance of invasive medical tests, such as Pap testing (Harlan et al. 1991), which can serve to further isolate a community (Gupta et al. 2002). Additional research is required to better understand the impact of utilization and health-seeking behaviours associated with immigrant status versus ethnic and cultural background, along with contextual factors and individual risk reduction (Harlan et al. 1991; Pickett and Pearl 2001).

Finally, this work raises other issues. First, with the increasing prevalence of HPV vaccines that are targeted toward young women, analysis of the relationship between vaccine awareness, uptake, and continued use of Pap screening will be important. Second, given the potential clustering of immigrants within cities, the geographical availability of doctors offices or clinics for screening, and the role of neighbourhoods and neighbourhood disadvantage in determining Pap screen uptake, we are left within an interesting spatial modelling problem. Although left for future work, research could focus on testing for local spatial autocorrelation of the response variables, expressed as rates of participation at the census tract level. Alternatively, the spatial clustering of the geocoded incidence of participation in Pap testing could be considered.

References

Black M, Zsoldos J (2003) Lay health educators to enhance cancer screening. Summary Report of Focus Groups: Planning with Women from Four Communities. Hamilton Public Health and Community Services, Hamilton, ON, Canada

Boyle MH, Lipman EL (2002) Do places matter? Socioeconomic disadvantage and behavioural problems of children in Canada. J Consult Clin Psychol 70:378–389

Bryant J, Browne AJ, Barton S, Zumbo B (2002) Access to health care: social determinants of preventive cancer screening use in north British Columbia. Soc Indic Res 60:243–262

Chui T, Tran K, Maheux H (2007) Immigration in Canada: a portrait of the foreign-born population, 2006 census. Statistics Canada. http://www12.statcan.ca/english/census06/analysis/immcit/pdf/97–557-XIE2006001.pdf. Accessed 13 Jan 2008

Citizenship and Immigration Canada (2001) Facts and figures, 2000 immigration overview. Ministry of Public Works and Government Services, Strategic Policy, Planning and Research

DesMeules M, Gold J, Kazanjian A, Manuel D et al. (2004) New approaches to immigrant health assessment. Can J Public Health 95:I22–I26

Diez Roux AV (2001) Investigating neighborhood and area effects on health. Am J Public Health 91:1783–1789

Diez Roux AV (2002) Invited commentary: places, people, and health. Am J Epidemiol 155: 516–519

Duarte-Franco E, Franco EL (2003) Cancer of the uterine cervix. Women's Health Surveillance Report. Chapter 12. Available at: http://www.phac-aspc.gc.ca/publicat/whsr-rssf/pdf/WHSR_Chap_12_e.pdf. Accessed 03 May 2006

Eddy DM (1990) Screening for cervical cancer. Ann Intern Med 113:216

Ellaway A, Macintyre S (2001) Women in their place. Gender and perceptions of neighbourhoods and health in the west of Scotland. In: Dyck I, Davis Lewis N, McLafferty S. (eds) Geographies of womens health. Routledge, London, pp 265–281

Fehringer G, Howlett R, Cotterchio M, Klar N, Majpruz-Moat V, Mai V (2005) Comparison of papnicolaou (Pap) test rates across Ontario and factors associated with cervical screening. Can J Public Health 96:140–144
Frohlich K, Potvin L, Gauvin L, Chabot P (2002) Youth smoking initiation: disentangling context from composition. Health Place 8:155–166
Glazier RH, Creatore MI, Gozdyra P, Matheson FI, Steele LS, Boyle E et al. (2004) Geographic methods for understanding and responding to disparities in mammography use in Toronto, Canada. J Gen Intern Med 19:952–961
Goel V (1994) Factors associated with cervical cancer screening: results from the Ontario health survey. Can J Public Health 85:125–127
Gupta A, Kumar A, Stewart DE (2002) Cervical cancer screening among South Asian women in Canada: the role of education and acculturation. Health Care Women Int 23:123–134
Grossman M (1972) On the concept of health capital and the demand for health. J Polit Econ 80:223–225
Harlan LC, Bernsein AB, Kessler LG (1991) Cervical cancer screening: who is not screened and why? Am J Public Health 81:885–890
Health Canada (2002) Cervical cancer screening in Canada: 1998 surveillance report. Cat. No. H39–616/1998E. Ministry of Public Works and Government Services, Canada
Hyman I (2001) Immigration and health. Health Policy Working Paper Series. Working paper 01–05. Ottawa, ON: Health Canada. Available at: http://www.hc-sc.gc.ca/sr-sr/alt_formats/iacb-dgiac/pdf/pubs/hpr-rps/wp-dt/2001–0105-immigration/2001–0105-immigration_e.pdf. Accessed: 22 Mar 2006
Hyman I, Guruge S (2002) A review of theory and health promotion strategies for new immigrant women. Can J Public Health 93:183–187
Hyman I, Singh M, Meana M, George U, Wells L, Stewart DE (2002) Physician-Related Determinants of Cancer Screening among Caribbean Women in Toronto. Ethn Dis 12:268–275
Hyman I, Cameron JI, Singh M, Stewart DE (2003) Physician-related determinants of cervical cancer screening among Chinese and Vietnamese women in Toronto. J Health Care Poor Underserved 14:489–502
Johnston GM, Boyd CJ, MacIsaac MA (2004) Community-based cultural predictors of Pap smear screening in Nova Scotia. Can J Public Health 95:95–98
Juon H-S, Seung-Lee C, Klassen A (2003) Predictors of regular Pap smears among Korean–American women. Prev Med 37:585–592
Kenkel D (1994) The demand for preventative medical care. Appl Econ 26:313–325
Leduc N, Proulx M (2004) Patterns of health services utilization by recent immigrants. J Immigr Health 6:15–27
Leyden W, Manos MM, Geiger AM, Weinmann S et al. (2005) Cervical cancer in women with comprehensive health care access: attributable factors in the screening process. J Natl Cancer Inst 97:675–683
Macintyre S, Maciver S, Sooman A (1993) Area, class and health: should we be focusing on places or people? J Soc Policy 22:213–234
Macintyre S, Ellaway A, Cummins S (2002) Place effects on health: how can we conceptualise, operationalise and measure them? Soc Sci Med 55:125–139
Maxwell CJ, Bancej CM, Snider J, Vik SA (2001) Factors important in promoting cervical cancer screening among Canadian women: findings from the 1996–97 National Population Health Survey (NPHS). Can J Public Health 92:127–133
McDonald JT, Kennedy S (2005) Ethnicity, immigration and cancer screening: evidence from Canadian women. Social and Economic Dimensions of an Aging Population, Working Paper Series, McMaster University 145
McDonald JT, Kennedy S (2007) Cervical cancer screening by immigrant and minority women in Canada. J Immigr Minor Health 9:323–334
Merlo J, Chaix B, Yang M, Lynch J, Rastam L (2005) A brief conceptual tutorial of multilevel analysis in social epidemiology: linking the statistical concept of clustering to the idea of contextual phenomenon. J Epidemiol Community Health 59:443–449

Miller AB, Andersen G, Brisson J, Laidlaw K, Le Pitre N, Malcolmson P et al. (1991) Report of a national workshop on screening for cancer of the cervix. CMAJ 145:1301–1325
Moon G, Brown T (1998) Place, space and health service reform. In: Kearns RA, Gesler WM (eds) Putting health in its Place. Syracuse University Press, Syracuse, NY, pp 270–288
Morrison BJ (1994) Screening for cervical cancer. Canadian Task Force on the Periodic Health Care Examination, Canadian Guide to Clinical Preventive Health Care Ottawa: Health Canada, 870–881
National Cancer Institute of Canada (2007) Canadian cancer statistics 2007. Toronto, ON
Newbold KB (2005) Health status and health care of immigrants in Canada: a longitudinal analysis. J Health Serv Res Policy 10:77–83
Ng E, Wilkins R, Fung M, Berthelot JM (2004) Cervical cancer mortality by neighborhood income in urban Canada from 1971 to 1996. CMAJ 170:1545–1549
Parboosingh EJ, Anderson G, Clarke EA, Inhaber S, Kaegi E, Mills C et al. (1997) Cervical cancer screening: are the 1989 recommendations still valid? CMAJ 1847–1853
Pickett KE, Pearl M (2001) Multilevel analyses of neighbourhood socioeconomic context and health outcomes: a critical review. J Epidemiol Commun Health 55:111–122
Ross NA, Dorling D, Dunn JR, Henriksson G, Glover J, Lynch J et al. (2005) Metropolitan income inequality and working-age mortality: a cross-sectional analysis using comparable data from five countries. J Urban Health 82:101–110
Ross NA, Tremblay SS, Graham K. (2004) Neighbourhood influences on health in Montreal, Can Soc Sci Med 59:1485–1494
Statistics Canada (2007) Population by immigrant status and period of immigration, 2006 counts, for Canada, provinces and territories – 20% sample data (table). Immigration and Citizenship Highlight Tables. 2006 Census. Statistics Canada Catalogue no. 97–557-XWE2006002. Ottawa http://www12.statcan.ca/english/census06/data/highlights/Immigration/Table403.cfm?Lang=E& T=403& GH=4& SC=1& S=99& O=A. Accessed 13 Jan 2008
Snider J, Beauvais J, Levy I, Villeneuve P, Pennock J (1996) Trends in mammography and pap smear utilization in Canada. Chronic Dis Cancer 17:108–117
Snijders T, Bosker R (1999) Multilevel analysis: an introduction to basic and advanced multilevel modeling. Sage, London
Yi JK (1994) Factors associated with cervical cancer screening behaviour among Vietnamese women. J Community Health 19:95–98
Woltman KJ, Newbold KB (2007) Immigrant women and cervical cancer screening uptake. Can J Public Health 98:370–475

Is Growth in the Health Sector Correlated with Later-Life Migration?

Dayton M. Lambert, Michael D. Wilcox, Christopher D. Clark, Brian Murphy, and William M. Park

1 Introduction

The aging population of the United States has long been a subject of debate and inquiry for development planners, policy makers, and researchers. The doubling of the population of Americans older than 65 since 1960 (while the population younger than 65 has grown by only one half) (Fuguitt et al. 2002), has prompted interest in their effect on the economies in which they live (Serow 2003) and their potential as a resource for rural economic development (Fagan 1988; Fagan and Longino 1993; Reeder 1998). Interest in these issues intensified as the baby boomer generation approached retirement age. The retirement of this age cohort is likely to have profound effects on the nation and its economy as this cohort is not only much larger than previous age cohorts, but also healthier and wealthier due to economic growth and advances in the quality of healthcare.

Older Americans increasingly have the means and the motivation to migrate to a different area upon retirement. For example, it is estimated that over the next 18 years, approximately 400,000 retirees each year – with an average of $320,000 to spend on a new home – will choose to relocate beyond their state borders (Vestal 2006). The South and West have been and continue to be popular destinations for these migrants (Serow 2001; He and Schachter 2003), although more are choosing to locate outside of the traditional retirement areas of Florida and Arizona (Vestal 2006). One driving force of this shift is the "halfback" phenomenon in the Southeast where retirees who had previously migrated to the coast are returning halfway back to their ancestral homes by relocating to areas such as the Southern Appalachian mountain regions of eastern Tennessee, western North Carolina, and northern Georgia (Park et al. 2007). Further, later-life migrants are frequently settling in rural places or small towns (Fuguitt et al. 2002). For example, in 2000 a half million more persons above 60 moved into non-metro counties than out of them (Beale 2005). These trends beg the question of how the recent in-migration of

D.M. Lambert (✉)
Department of Agricultural Economics, University of Tennessee, 321 Morgan Hall, Knoxville, TN 37996-4511, USA,
e-mail: dmlambert@utk.edu

A. Páez et al. (eds.), *Progress in Spatial Analysis*, Advances in Spatial Science, 381
DOI 10.1007/978-3-642-03326-1_18, © Springer-Verlag Berlin Heidelberg 2010

older Americans is affecting local economies, particularly in rural areas where the marginal effect of in-migration may be proportionally greater than in more populous urban areas.

Recognizing this opportunity, many state and local governments have turned to attracting later-life migrants as a component of rural economic development policy (Fagan 1988; Fagan and Longino 1993; Reeder 1998). The lure of such a policy to rural communities can largely be attributed to declines in traditional sources of rural economic activity and to the relatively greater impact later-life migrants are likely to have on local government revenues than on expenditures. The relatively high levels of wealth among these migrants, coupled with the absence of school-age children from their households, suggests that they will augment local tax bases by increasing property values and retail spending without increasing education expenditures, which are typically a large component of local government spending (e.g., Mullins and Rosentraub 1992; Serow 2003; Park et al. 2007). Furthermore, these later-life migrants foster a "mailbox economy" where expenditures on local goods and services are financed by income from outside a community as opposed to local, re-circulated dollars (Haas et al. 2006).[1]

Numerous studies have examined the effect of later-life in-migration on local economies (see Serow (2003) for a comprehensive review) and many have considered the effects from a rural economic and community development perspective (e.g., Aday and Miles 1982; Hodge 1991; Mullins and Rosentraub 1992; Rowles and Watkins 1993; Stallmann et al. 1999). Although later-life migrants make significant expenditures in local economies (Haas 1990, p. 388), several studies suggest that these expenditures primarily create low skill, low wage, service sector jobs (e.g., Beale and Fuguitt 1990; Glasgow and Reeder 1990; Reeder and Glasgow 1990; Day and Bartlett 2000).

However, as Day and Bartlett (2000) note, the high skill, high wage, health and medical care service sub-sector is of particular importance to later-life migrants. A number of studies find that the availability of healthcare services is an important factor in destination choice (e.g., Toseland and Rasch 1978; Regnier and Gelwicks 1981; Dwight 1985; Park et al. 2007). In fact, retirement communities often use the availability of local health care services as a marketing tool (Dwight 1985; Dine 1988; Loomis et al. 1989). Thus, the spatial distribution of health care services may play an important role in determining the spatial relocation patterns of later-life migrants. But what is less clear is how the relocation patterns of later life migrants affect the spatial distribution of health care services.

There is an extensive literature addressing the geographic distribution of physicians and other health care resources (e.g., Newhouse et al. 1982; Jiang and Begun 2002; Freed et al. 2004; Mistretta 2007).[2] A wide variety of factors determine

[1] The mailbox economy is a term which refers to the source of income for many retirees. Examples include Social Security, private pensions, etc. all of which are derived from outside the local economy.

[2] A primary concern in this literature is measuring access to physicians or to other health care services (e.g., Joseph and Bantock 1982; Wing and Reynolds 1988; Rosenthal et al. 2005; Pathman et al. 2006).

the distribution of physicians, including; geographical preferences (regions, states, counties, and typologies such as rural versus urban), population (amount, growth and density), physician attributes (specialty, number, and density), patient attributes (income, education levels, age distribution and prevalence of insurance types), local or regional health sector characteristics (wages, number of hospital beds, number, type or concentration of hospitals, and Health Maintenance Organization [HMO] penetration), public expenditures (education and health), and employment measures (unemployment, importance of specific sectors, firm size) (Brasure et al. 1999; Escarce et al. 2000; Jiang and Begun 2002; Rosenthal et al. 2005; Mistretta 2007; Wall and Brown 2007).

This chapter contributes to this literature by analyzing whether an inflow of later-life migrants is correlated with growth in the health care sector as measured by changes in the concentration of health care providers. The attraction of substantial numbers of later-life migrants has the potential to disproportionately increase the demand for health care services. On average, these migrants are often older and have higher expectations of care, along with a greater ability to pay for specific health care services than residents in rural areas (e.g., Park et al. 2007). Moreover, significant later life migration to a particular rural community could allow the community's health care providers to exploit agglomeration economies in the health care sector (Connor et al. 1995; Bates and Santerre 2005) and grow into regional medical centers servicing surrounding rural areas.

However, there are a number of factors that may mitigate or confound the effect of later life migrants on growth in health care services. In general, individuals who do migrate upon retirement are often healthier than other members of their age cohort (Barsby and Cox 1975; Patrick 1980; Sickles and Taubman 1986). Furthermore, as these migrants grow older and their health begins to deteriorate, they often return to their "ancestral homes" to be near family and friends who can serve as caregivers or who can provide comfort after the loss of a spouse (Serow and Charity 1988; Colsher and Wallace 1990). There may also be constraints on the growth of rural health care sectors. A consistent finding reported in the literature examining the spatial distribution of health care services is that the ratio of the number of physicians to total population is considerably higher in suburbs and wealthy urban areas than it is in rural areas and inner cities (Rosenthal et al. 2005; Mistretta 2007). The unevenness of this spatial distribution is largely attributable to the concentration of medical specialists in more densely populated areas where the hospitals, laboratories and other services on which these specialists depend are located (Jiang and Begun 2002). For example, Brasure et al. (1999) found that increments in population density played a significant role in a physician's decision to enter a market, indicating that low populations may serve as a barrier to entry for a wide variety of physician types. Other characteristics of the health care industry, such as mutual ownership or alliances between urban and rural healthcare service providers, may also limit growth in the rural health care sector (Reardon 1996; Ricketts 2000).

Haas and Crandall (1988) explored how the influx of later-life migrants affected rural health care systems using a case study of two counties (one in Florida and one in North Carolina). Their results indicated that a major factor contributing to physicians moving their practice from one area to another was the presence of a population that would require care in their particular specialty. In fact, 61% of physicians surveyed reported that over half of their clientele were over age of 65, and 47% believed that later-life migrants had a direct effect on the growing number of physicians, especially internists, internal medicine sub-specialists and surgical specialists who typically treat older patients (Haas and Crandall 1988). Other studies measured the effect of age distribution (e.g., proportion of population over the age of 65) rather than actual later-life migration. A general conclusion of these studies is that areas where population age is skewed upward generally have the same or more family/general practitioners and fewer specialists (Brasure et al. 1999; Escarce et al. 2000; Jiang and Begun 2002). Jiang and Begun (2002) posit that an urban area with a higher proportion of elderly may be "regarded as economically declining, and thus, less attractive to physicians." However, this result may be sensitive to the fact that only urban areas were included in their analysis.

The relationship between later life migrants and growth in the health care sector is the quintessential "jobs-to-people" or "people-to-jobs" question. There is substantial evidence suggesting that these migrants consider the availability of health care services in selecting their relocation destination. Less clear is the extent to which later life migrants promote growth in the local health care sector. Understanding the effects of later-life migrants on the rural health care sector is important, in part, because of the important role the sector often plays in the community. The health care sector is often a major component of the local economy as it is typically one of the largest employment sectors, often second only to public education, and is important in attracting and retaining retirees and businesses (Doeksen et al. 1996). Also, since the extent of health care services provided in rural areas is often quite limited, in part due to the closure of rural hospitals in the last few decades (Reardon 1996; Capalbo and Heggem 1999), an expansion of these services could have significant welfare impacts on rural residents.

Thus, this chapter examines the "jobs-to-people" side of the question for the Southeastern United States. Specifically, did a change in the number of in-migrating seniors between 1995 and 2000 correlate with a change in the number of persons working in the health care sector from 2000 to 2004?[3] The focus is on the Southeast (AL, AR, GA, LA, MS, NC, SC, and TN), where relatively rapid growth in later-age migrants is occurring and on rural areas, given the economic development aspirations many rural areas hold for later-life migration. We explain changes in the concentration of health professionals from 2000 to 2004 using aggregate, county-level data. A series of spatial lag process models are used to explain growth (or decline) in the concentration of registered nurses (RNs), medical doctors (MDs), and sub-sets of MDs (i.e., office-based surgical specialists, office-based medical specialists, and other office-based specialists) as a function of local demographic, economic, and infrastructural attributes, as well as the influx of migrating cohorts

[3] Due to data limitations, it was not possible to examine the other direction; i.e., to what extent was growth in the health care sector correlated with in-migrating seniors?

from 1995 to 2000. We hypothesize that the relationships between local factors and growth in the concentration of health care professionals may be different between urban core and rural counties. To accommodate this form of spatial heterogeneity, we interact a rural-urban index with main effect factors, therefore allowing marginal effects to vary geographically. Shedding light on how the anticipated increase in later-life migrants during the next decade will affect demand for health care services in particular and local economic development in general will assist policymakers and development professionals seeking ways to diversify the economic portfolio of rural communities.

The remainder of this chapter is organized as follows. The next section describes the data and model used in the empirical analysis. Spatial econometric issues are discussed, including model selection and robust covariance estimation procedures. Results are presented and discussed in the third section. Finally, the chapter concludes with some thoughts on the analysis and on possible future research directions.

2 Data and Empirical Model

The hypothesis that migrating seniors influenced the concentration of health care employment was tested by regressing aggregate county-level control measures and the percent of 1995–2000 in-migrants comprised of individuals in the 35–54, 55–69, and 70 years and above age cohorts (2000 Census) on the change in employment concentration of MDs (2000–2004) and RNs (2000–2003) (Area Resource File 2005) (Table 1). MDs were disaggregated into three sub-professions: office-based surgical specialists, office-based medical specialists, and other office-based specialists (all 2000–2004). Office-based practices include physicians engaged in patient care. This group includes physicians in solo practice, group practice, or other patient care employment. The office-based group also includes physicians in patient services such as those provided by pathologists and radiologists. Surgical specialists are physicians providing colon/rectal surgery, general surgery, neurological surgery, obstetrics-gynecological surgery, ophthalmology, orthopedic surgery, otolaryngology, plastic surgery, thoracic surgery, and urology. Medical specialists include allergy and immunologists, cardiovascular physicians, dermatologists, epidemiologists, gastroenterologists, internal medicine specialists, pediatricians, and pulmonary disease specialists. Other specialists include anesthesiologists, child psychiatrists, radiologists, emergency medicine technicians, forensic pathologists, geneticists, neurologists, public health professionals and general preventative medicine, oncologists, and other unspecified medical specialists. RNs include full-time registered nurses, and nurses working in nursing homes or state general hospitals.

Change in the employment concentration of these professionals was measured using the natural log of the ratio of their location quotients (*LQ*) between 2000 and 2004. The location quotient is constructed relative to employment in each state;

Table 1 Summary statistics[a]

Variable	Abbreviation	Mean	Standard error
ln LQ, MDs 2000–2004	MD	0.2944	0.2222
ln LQ, Office-based medical specialists 2000–2004	OMS	1.0960	0.4304
ln LQ, Office-based surgical specialists 2000–2004	OSS	0.0242	0.4005
ln LQ, Other office-based specialists 2000–2004	OOS	−0.0404	0.4562
ln LQ, RNs 2000–2003	RN	21.8393	0.7601
ln LQ, MDs 2000	MD00	−2.1295	0.2250
ln LQ, Office-based medical specialists 2000	OMS00	−7.0900	0.5080
ln LQ, Office-based surgical specialists 2000	OSS00	−10.5051	0.5989
ln LQ, Other office-based specialists 2000	OOS00	−11.1637	0.6031
ln LQ, RNs 2000	RN00	−30.1596	0.6307
ln Δpopulation density, 1990–2000	dPOPDENS	0.1329	0.0046
ln Median household income, 2000	MEHHY	1.1624	0.0079
ln Hospital beds per capita, 2000	HBPC	−11.4950	0.4410
ln Health expenditure/population, 1997	HEXP	−5.4239	0.2536
% 25+ with high school diploma, 2000	HS	70.5116	0.2690
% employment in agriculture, 2000	PERAG	4.4275	0.1385
% employment in construction, 2000	PERCON	8.3875	0.0927
% employment in manufacturing, 2000	PERMAN	21.6071	0.3214
Unemployment rate, 2000	UNEMP	5.6422	0.1003
% commuting, 2000	COMM	34.0660	0.6799
% white, 2000	WHT	71.7672	0.7421
% population 65+, 1999	POPO65	13.3459	0.1156
% in-migrants, 30–54, 1995–2000	IN3054	32.9744	0.1751
% in-migrants, 55–69, 1995–2000	IN5569	9.6109	0.1657
% in-migrants, 70+, 1995–2000	IN70UP	5.1015	0.0870
Rurality index (RI), 2000	RI	0.4835	0.0051

[a]N = 688. States included in the sample are AL, AR, GA, LA, MS, NC, SC, and TN

$$
\begin{aligned}
{}^{k}\Delta LQ_{2000}^{2004} = {} & \ln\left[\left(e_{i,2004}^{s,k}\Big/e_{i,2004}^{s}\right)\Big/\left(E_{i,2004}^{k,s}\Big/E_{i,2004}^{s}\right)\right] \\
& - \ln\left[\left(e_{i,2000}^{s,k}\Big/e_{i,2000}^{s}\right)\Big/\left(E_{i,2000}^{k,s}\Big/E_{i,2000}^{s}\right)\right]
\end{aligned} \tag{1}
$$

where k = MDs, RNs, office-based surgical specialists, office-based medical specialists, and other office-based specialists; e^k is the number of persons occupying the kth profession in the ith county; e is total employment in the ith county; $E^{k,s}$ is the number of individuals in the kth profession in the sth state; and E^s is total employment in the sth state. This perspective appreciates potential heterogeneity between states due to policy or other unobservable factors specific to a particular state.

The effect of local control factors and migrating cohorts on change in the employment concentration of these professions was estimated using a linear model:

$$
\begin{aligned}
{}^{k}\Delta LQ_{2000}^{2004} = {} & f\left(W^{k}\Delta LQ_{2000}^{2004}, {}^{k}LQ_{2000}, IS_{2000}, LM_{2000},\right. \\
& \left.\times HS_{2000}, DC_{2000}, SC_{2000}, \Delta MIG_{1995-2000}^{g}, RI_{2000}\right)
\end{aligned} \tag{2}
$$

where $W^k \Delta LQ_{2000}^{2004}$ is the spatial lag of the 2000–2004 change in employment concentration; *IS* are industry structure variables; *LM* are labor market characteristics; *HS* are local health care resources; *DC* are demographic characteristics; *SC* are settlement attributes; $\Delta MIG_{1995-2000}^{g}$ is the proportion of the gth in-migrating cohort relative to all in-migrants locating in county i between 1995–2000; and *RI* is Waldorf's (2006) rurality index (*RI*) constructed using 2000 census data. We *a priori* hypothesize that county-level change in each measure is a function of change in neighboring counties by specifying (2) as a spatial lag process model (Anselin and Florax 1995, discussed below). Model selection procedures explicitly test these assumptions (discussed below).

The *RI* measure is composed of population, population density, the percent of the population designated as rural or urban according to the US Census, and the distance of a county to a metropolitan county categorized using the Office of Management and Budget (OMB) urban core-non-core county classification system. All other explanatory variables in (2) were interacted with the rurality index to gauge the influence of local control factors and cohort in-migration on changes in employment concentration in the health profession across an urban-rural continuum.

2.1 Migration Cohorts

The proportion (%) of total in-migrants to a county between 1995 and 2000 comprised of individuals in one of three different age cohorts (35–54, 55–69, and 70+) was used to measure the impact of in-migration by age cohort on changes in employment concentration in the medical profession between 2000 and 2004. Changes in net migration between 1995 and 2000 for the three migrating age cohorts were calculated using data from the 2000 decennial US Census (Fig. 1). A significantly higher percentage of metropolitan counties had positive net migration ("net inflow") for the 35–54 and 70+ age cohorts, while the reverse was true for the 55–69 age cohort, with a higher percentage of non-metropolitan counties experiencing net inflow than was the case for metropolitan counties.

The primary focus is on later-life migration and its correlation with growth in the health sector. This is not to be confused with retiree migration, though the two terms have been used interchangeably since the 1980s (Haas et al. 2006). Here, life course is represented by age and not by employment status. The three cohorts were selected to reflect heterogeneous health status (House et al. 2005), and therefore demand for health services by progressively more aged cohorts (Wolinsky et al. 1986). The 55–69 and 70+ age cohorts represent the migrating senior populations. The 35–54 cohort represents the "Baby Boomer" generation all of whom (including those born during 1961–1964) aged into the cohort by 2000.

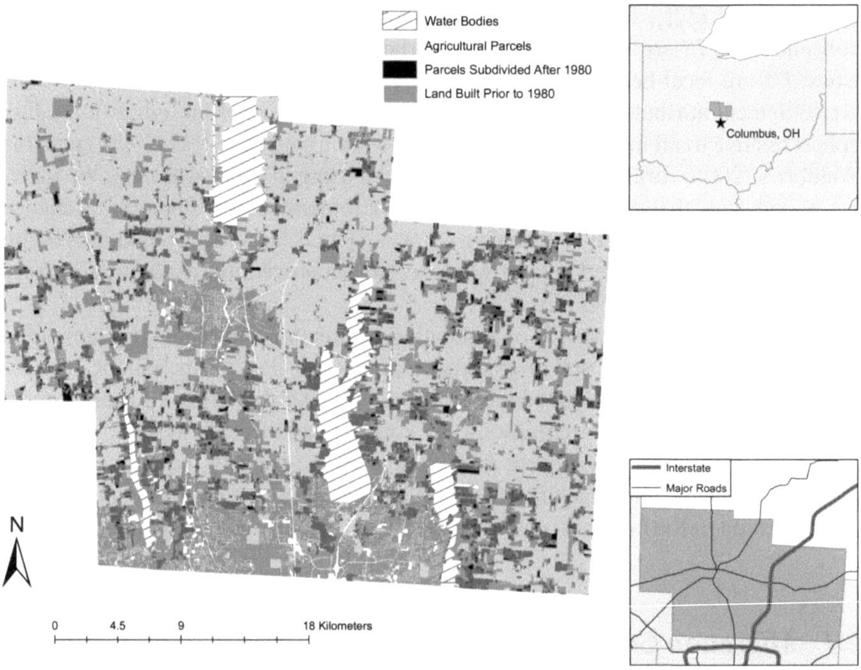

Fig. 1 Distribution of quantile proportions of total in-migrants composed of individuals in the 55–69 (top panel) and 70+ age cohorts (bottom panel)

2.2 Control Variables

Five sets of control measures are included to reflect conditions at the beginning of the period (Table 1). First, *industry structure*, as measured by the proportions of total county employment employed in agriculture, construction, and manufacturing, comes from the Regional Economic Information System files from the Bureau of Economic Analysis.

The second set of control measures is comprised of *labor market characteristics*, specifically, the unemployment rate, the natural log of median household income, and the proportion of the population older than 25 with a high school diploma – all from the 2005 Area Resource Files data compilation. Counties with relatively higher median household incomes may be attractive to healthcare providers, although the effect of higher incomes on changes in the relative concentration of healthcare providers is somewhat ambiguous. County unemployment rate was used to control for local job market characteristics. The proportion of the population above 25 with a high school degree reflects human capital potential and in urban analysis has been linked to economic growth.

The third set of control measures, *settlement characteristics*, includes the percent of the work-force commuting out of the county and the natural log of the change in population density from 1990 to 2000. Population density enters the model as a change variable to control for regional economic and demographic inertia that may influence the location choice of medical professionals.[4] Higher out-commuting rates indicate greater relative economic activity in neighboring counties, likely implying low growth in healthcare employment. Alternatively, greater increases in population density should lead to more rapid growth in the healthcare sector, *ceteris paribus*.

The fourth set of control measures are *demographic characteristics*. These factors are represented by the percentage of the population that is white and the percentage of the population that was 65 or older in 1999. In the past, the percentage of population aged 65 or older might reflect a history of population decline. However, since many rural counties have become magnets for later-life migrants, a high proportion of citizens at or beyond retirement age are now perhaps just as likely to reflect county attractiveness as a retirement destination to this age group. While population loss is unlikely to promote an increase in healthcare employment, the presence of an aging population may serve to attract healthcare providers.

The last set of control measures is comprised of *health care resources*. The natural log of hospital beds per capita at the beginning of the period was used as a measure of access to health care and of the relative size of the county hospital network. All else equal, it is hypothesized that growth in the MD and RN professions will be positively correlated with this measure, but the magnitude of the relationship may vary with each occupation. For RN's and office-based surgeons, the relationship is expected to be strongest. Registered nurses care for short- and long-term patients. Surgical specialists depend on specialized equipment and other infrastructure provided by hospitals. Medical specialists tend to concentrate in more densely populated counties providing human and physical capital, which supports technologies used by these specialists as well as offering a larger market for the high fixed cost services they provide.

The aggregated MD measure represents all MD's including general practitioners, pediatricians, and the specialized MD's. While specialists may be more limited in rural areas without access to scale economies, the lower infrastructure requirements for other MD's (e.g., general practitioners) may mean that proximity to a medical hub is not as important. For difficult medical problems, patients will usually be referred to a regional medical center by their local personal care physician. The total number of hospital beds and the 2000 Census population were obtained from the ARF (2005) data base. The second measure – total county health care expenditures per capita (1997) – was extracted from the Census of Government Files (1997). Holding other factors constant, it is hypothesized that counties with relatively higher health care expenditures per capita will experience growth in the MD and RN professions as these expenditures may reflect the establishment or maintenance of a more extensive health care infrastructure, specialized medical technologies, and demand for affordable health care.

[4] The rurality index controls for 2000 "base" level effects since it is determined in part by population density, *inter alia* additional demographic measures of settlement concentration. Because of this, 2000 population concentration measures were not included in the regression models.

2.3 *Spatial Econometric Issues*

Recent years have witnessed an increasing number of applied studies in geography, economics, and regional science in which the spatial dimension of population and economic growth are incorporated in regression models (e.g., Bao et al. 2004; Moreno et al. 2004; Boarnet et al. 2005; Cohen and Paul 2005; Lambert et al. 2006; McGranahan et al. 2006; Cho et al. 2007; Monchuk et al. 2007; Wojan et al. 2008). This surge was fueled by recent theoretical developments in spatial econometrics along with better access to spatial data and the increased availability of easy-to-use computational tools.

Most regional growth studies use a spatial process model going back to Whittle (1954), in which an endogenous variable is specified to depend on spatial interactions between cross-sectional units plus a disturbance term. The interactions are modeled as a weighted average of nearby cross-sectional units, and the endogenous variable comprising the interactions is usually referred to as a spatially lagged variable. The weights are grouped in a matrix identifying neighborhood connections, which forms the distinctive core of spatial process models. The model is termed a spatial autoregressive lag model in the terminology of Anselin and Florax (1995). Whittle's spatial autoregressive lag model (SAR) was popularized and extended by Cliff and Ord (1973, 1981), who distinguished models in which the disturbances follow a spatial autoregressive process. The general model, which contains a spatially lagged endogenous variable, as well as spatially autoregressive disturbances in addition to exogenous variables, is called a spatial autoregressive model with autoregressive (AR) disturbance of order (1,1) (SARAR) (Anselin 1988; Anselin and Florax 1995); $\mathbf{y} = \rho\mathbf{W}_1\mathbf{y} + \mathbf{X}\boldsymbol{\beta} + \boldsymbol{\varepsilon}$, $\boldsymbol{\varepsilon} = \lambda\mathbf{W}_2\boldsymbol{\varepsilon} + \mathbf{u}, \mathbf{u} \sim \text{iid}(\mathbf{0}, \boldsymbol{\Omega})$, where $\mathbf{W}_1$ and $\mathbf{W}_2$ are (possibly identical) matrices defining interrelationships between spatial units, and $\text{E}[\mathbf{uu}'] = \boldsymbol{\Omega}$.

Spatial process models are typically estimated using maximum likelihood or generalized moment (GM) procedures. A GM approach is used here because we have no reason to believe that the errors generated by our models are normally distributed. A county with a given change in employment or business establishment growth (y_i) may be surrounded by other counties

$$\sum\nolimits_{j,i\neq j}^{n} w_{ij}\, y_j \tag{3}$$

with similar growth rates. Feedback between spatial units may be significant; meaning that growth in one county is dependent on or explained by growth in surrounding counties. Significant interaction suggests information spillovers, thick labor markets, or forward-backward economic linkages across space (Anselin 2002; Moreno et al. 2004). Agglomerative effects, as we use the term here, imply some form of regional clustering or "spillover" due to centripetal effects (Fujita et al. 1999), suggesting the presence of "thick" labor markets or access to relatively larger demand markets. Significant positive spillover is consistent with this outcome. On the other hand, significant negative spillover may suggest that high occupational growth in one region is negatively correlated with growth in surrounding regions (i.e., centrifugal effects), or immobile factors are at work (e.g., barriers to labor are high, or resources are fixed), which may be indicative of "deglomeration" effects. These hypotheses are tested by the significance of the autoregressive parameter ρ.

In this application, the lag process is modeled using a row-standardized first-order queen contiguity matrix, which identifies local neighborhoods of counties. However, the parametric and structural assumptions about the error process are relaxed. Spatial error occurs when omitted variables follow a spatial structure such that $\mathbf{\Omega} \neq \sigma_u{}^2\mathbf{I}$ (Anselin 1988). Non-spherical errors may be simultaneously caused by heteroskedasticity or autocorrelated error processes, and are usually linked to heterogeneity associated with cross-sectional spatial units (Kelejian and Prucha 2007). Inclusion of fixed effects is one approach to tackle this problem. But when the data is a cross-section, and in cases where the causes of spatial heterogeneity cannot be identified as discrete units (such as census blocks or states), the researcher must specify spatial structure vis-à-vis $\mathbf{W}_2$, often-times with little in the way of theoretical guidance.[5,6] Instead, we take a non-parametric approach motivated by Conley (1999) and Kelejian and Prucha (2007) using spatial heteroskedastic-spatial autocorrelation robust (spatial HAC) covariance matrices to model potential spatial error dependence in the regressions.

2.4 Heteroskedastic-Spatial Autocorrelation Robust Standard Error Estimation

The approach taken by Conley (1999) and Kelejian and Prucha (2007) extends the Newey-West class of heteroskedastic-autocorrelation consistent (HAC) covariance matrices developed for time series analysis to dependence between cross-sectional units. Recall the asymptotic covariance matrix of the general method of moments (GMM) estimator:

$$AsyVar(\boldsymbol{\beta}_{\text{GMM}}) = (\mathbf{M}'\mathbf{P}\mathbf{M})^{-1}\mathbf{M}'\mathbf{Q}(\mathbf{Q}'\mathbf{Q})^{-1}\Psi(\mathbf{Q}'\mathbf{Q})^{-1}\mathbf{Q}'\mathbf{M}(\mathbf{M}'\mathbf{P}\mathbf{M})^{-1}\text{[7]}. \quad (4)$$

In the case of the spatial lag process model estimated with instrumental variables (SAR-IV, i.e., the GMM estimator), $\mathbf{M} = [\mathbf{Wy}, \mathbf{X}]$ (spatially lagged dependent

[5] In the case of the lag model, the relationship between $\mathbf{W}_1\mathbf{y}$ and $\mathbf{y}$ is usually much clearer. Hypotheses about how agents or spatial units react to and interact with one another can be guided by the choice of elements in $\mathbf{W}_1$ (e.g., Bao et al. 2004).

[6] In many empirical studies, the spatial autoregressive parameter (λ) is considered a nuisance parameter, suggesting that the main advantage gained from its estimation is one of efficiency rather than theoretically informed information. In other studies, some researchers assume that the error parameter explains "knowledge spillovers" due to unobserved heterogeneity across spatial units (e.g., Cohen and Paul 2005). In our approach, we assume the former interpretation of the parameter describing the spatial error process.

[7] We multiply the covariance by $n/(n-k)$ to correct for bias.

variable, and exogenous variables, respectively), and $\mathbf{Q} = [\mathbf{X}, \mathbf{WX}, \mathbf{W}^2\mathbf{X}]$ (instrumental variables, including the spatially lagged exogenous variables with higher-order lags, respectively), and $\mathbf{P} = \mathbf{Q}(\mathbf{Q}'\mathbf{Q})^{-1}\mathbf{Q}'$. $\boldsymbol{\Psi}$ is a relational matrix[8] that generates weighted averages of the cross-products of residuals based on a non-parametric kernel density estimator $K(d_{ij}/d_{\max})$ that determines cross-product pairs (i, j) over a certain distance ($d_{\max}$) at a decaying rate. The individual elements of this matrix are (Kelejian and Prucha 2007); $\psi_{kl} = \sum_i \sum_j q_{ik}\, q_{jl}\, K\left(d_{ij}/d_{\max}\right) \varepsilon_i \varepsilon_j$.

The properties of $K(d_{ij}/d_{\max})$ are such that the function is bounded and symmetric, real and continuous, and must integrate to one (Kelejian and Prucha 2007). Typical candidate functions include Parzen, Bartlett, Epanechnikov, or bi-square kernels (Kelejian and Prucha 2007; Anselin and Lozano-Gracia 2007). In our application we use the Barlett density function; $[K(d_{ij}/d_{\max}) = (1 - d_{ij}/d_{\max})]$.[9] Note that when $K(d_{ij}/d_{\max}) = 0$ for all $d_{ij} > d_{\max}$, and $K(d_{ij}/d_{\max}) = 1$ for $d_{ij} = 0$.

We apply an adaptive kernel function where $d_{\max}$ changes with respect to each cross-sectional unit. For every observation i, the vector of distances between i and all other observations are sorted in ascending order. The number of neighbors surrounding i is identified by a contiguity matrix. This value is used as a cutoff point to identify $d_{\max}$, the last distance entry in the truncated vector corresponding to spatial unit i. This mechanism permits $K(d_{ij}/d_{\max})$ to expand or contract across cross-sectional units, conditional on the number of neighbors surrounding a given observation, and thereby re-weighting residual cross-products according to a localized neighborhood structure. The weight attributed to counties not adjacent to county i is zero. In this study, the road distance (in miles) between county seats was used as the distance measure between counties.

2.5 Model Specification

The goal of the model specification search was to (i) determine the appropriateness of the spatial lag model, and (ii) to determine whether use of the spatial HAC estimator was necessary as indicated by spatial structure in the residuals of (2). Given these criteria, there were two possible estimation methods (Ordinary Least Squares or SAR-IV), and two possible covariance structures (a bias-corrected Huber-White "sandwich" estimator or the spatial HAC covariance matrix), yielding four potential specifications.

A stepwise procedure was used to determine which combination of estimation procedures and covariance matrix estimation was appropriate, given evaluation of the location quotient associated with a medical profession. In the first step, a Wald test was used to test the hypothesis that change in the location quotient for a given group in a county was a function of change in surrounding counties (e.g., a test for the significance of $W^k \Delta LQ_{2000}^{2004}$ as measured by ρ). Given the results of this

[8] This matrix is also called the "spectral density" matrix in the usual GMM terminology.

[9] Experimentation with alternative kernel structures yielded no substantial differences between the standard errors of the lag model.

test, (2) was estimated with OLS ($\rho = 0$) or SAR-IV ($|\rho| \neq 0$). In the second step, the semivariogram of the residuals from (2) (based on OLS or SAR-IV) was estimated to detect spatial structure in the disturbance terms. Application of the spatial HAC covariance matrix was determined by visual inspection of each semivariogram. A variety of functional forms could be applied to describe semivariogram structures, including exponential, spherical, or Gaussian. However, there may be situations where spatial structure cannot aptly be described using the typical continuous functions. For example, over a certain range a semivariogram might follow an exponential pattern increasing to a certain range, only to drop and continue in sinusoidal fashion. Although nonlinear piecewise algorithms could be applied in such a situation, they are beyond the scope of this study. Alternatively, visual inspection is oftentimes enough to determine the presence (or absence) of spatial structure in the semivariogram. This approach is subjective, but errs on the conservative side since any visual evidence of structure would suggest use of the spatial HAC estimator.

3 Results and Discussion

The first section presents the model selection diagnostics. Discussion of the relevant control variables in the models follow. The focus then turns to discussion of the correlation between in-migrating cohorts from 1995 to 2000 and growth in health care providers from 2000 to 2004. A discussion of some interesting relationships pertaining to the spatial heterogeneity of employment growth in certain health care sectors follows. A brief section on the correlated growth of office-based surgical and medical specialists concludes the results and discussion section.

3.1 *Model Specification Results*

Change in the local concentration of both office-based medical specialists and surgeons were correlated with similar changes in surrounding counties over the 2000–2004 periods (Table 2). The semivariogram of the respective SAR-IV residuals suggests spatial structure in the disturbance terms (Fig. 2). For office-based medical specialists and surgeons the standard errors of the SAR-IV model were estimated using the SHAC covariance matrix. Change in the other office-based MDs over the period was not correlated with change in neighboring counties (Table 2). For this sub-group, (2) was estimated with OLS. Residual analysis suggested some degree of spatial error autocorrelation (Fig. 2). In these cases, standard errors were estimated with the SHAC covariance matrix. Considering MD's as a group, the hypothesis that the change in the concentration of these professionals in a given county was correlated with change in surrounding counties was not tenable. Inspection of the OLS residual semivariograms for the MD regression suggests little (if any) spatial structure in the residuals. Therefore, standard errors were estimated

Table 2 Model specification

Model	Error structure[a] OLS	IV-SAR	Spatial lag significant?	Estimation method	Covariance estimation[c]
1. MD	No	No	No (0.07)[b]	OLS	Huber-White
2. RN	No	No	No (0.32)	OLS	Huber-White
3. Office-based Med. Specialists	Yes	Yes	Yes (2.94)	IV-SAR	SHAC[d]
4. Office-based surgical MDs	Yes	Yes	Yes (3.97)	IV-SAR	SHAC
5. Other office-based specialists	Yes	Yes	No (1.04)	OLS	SHAC

[a]Conclusions are drawn from a semivariogram was estimated using the OLS or IV-SAR residuals.
[b]Wald test (in parentheses) is based on the spatial lag coefficient, and is a $\chi^2(1)$ variate. Critical value for 10% (5%) level is 2.71 (3.84).
[c]All heteroskedastic-robust covariance matrices were multiplied by $n/(n-k)$ to correct for bias.
[d]Spatial HAC

with the usual Huber-White heteroskedastic-robust estimator with the bias correction factor of $n/(n-k)$. The same conclusions were drawn for the RN growth model. The spatial lag autoregressive coefficient was not significantly different from zero, and spatial structure in the OLS residuals was not evident.

3.2 *Important Control Variables*

Holding other factors constant, the concentration of office-based medical specialists increased more rapidly in counties with relatively high median household incomes (Table 3). Local increases in the concentration of professional RNs were positively correlated with the per capita number of hospital beds in a county, as expected. But in counties where the concentration of RNs were high at the beginning of the period, the 2000–2004 change in the location quotient for this group was lower, suggesting a crowding-out effect with respect to employment opportunities.

3.3 *The Relationship Between In-Migrating Seniors and Concentration of Medical Professionals Appears Limited*

In general, we find little evidence to support the hypothesis that in-migrating seniors were correlated with local concentration of medical professionals in the Southeastern US from 2000 to 2004 (Table 3). The main effect of in-migrating 70-year plus seniors was inversely related to change in the concentration of MD's. However, inspection of the interaction term suggests that the association is spatially

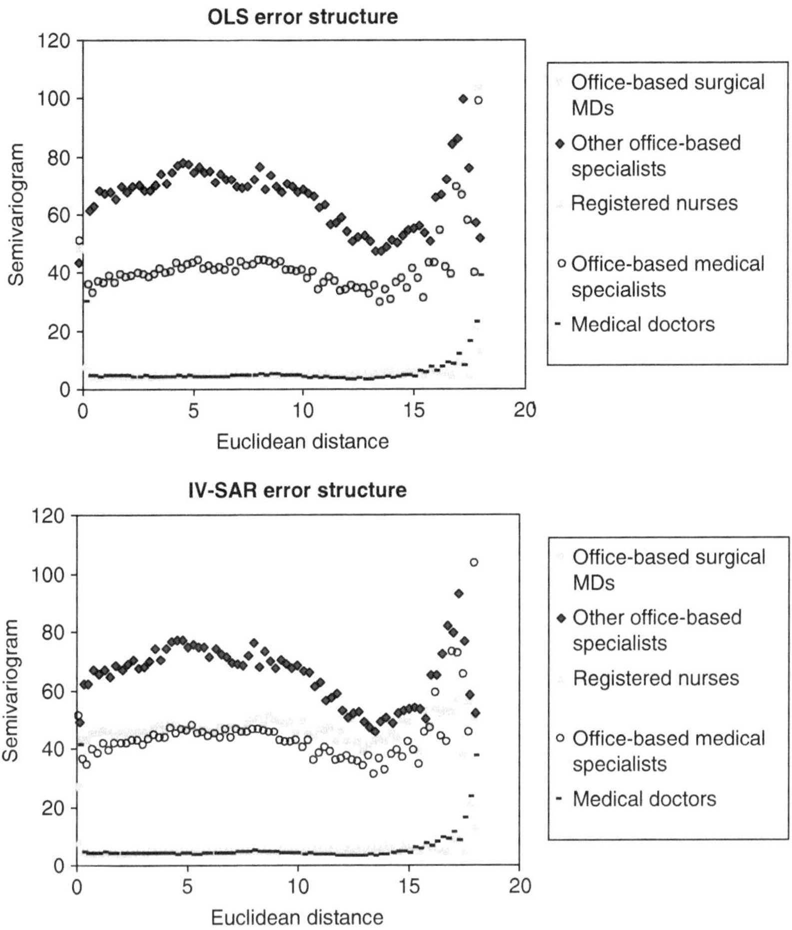

Fig. 2 Semivariograms of residual error structure

heterogeneous. Inflow of migrating 70+ seniors becomes positive moving away from urban centers to more rural locations (Fig. 3). There is also some indication that counties with relatively older populations (as measured by the percent of the population above 65 in 1999) experienced increased concentration of medical specialists, but that the effect is spatially heterogeneous and diminishes moving away from urban centers to less populated (and more rural) counties (Fig. 3).

The point on the rural-urban continuum at which the effect of in-migrating 70+ seniors was associated with increases in the concentration of MDs was determined to be 0.46, suggesting that in-migration of this cohort into urban-rural transition counties and more rural areas correlates with growth of MD professionals. An urban-rural distinction is evident. Moving closer to urban centers, the effect of this cohort becomes increasingly negative (Fig. 4). On the other hand, the opposite

Table 3 Regression results

Variable	MDs	MD specialists	Other specialists	Surgical specialists	RNs
Wy		−0.132*		0.119**	
Constant	17.139	−44.413	−40.806	−43.848	−3.597
ln LQ, 2000	−0.302	−0.447	0.177	−0.326	−0.935***
dPOPDENS	16.012	−0.964	−11.844	19.034	−11.276
MEDHHY	11.144	32.352*	−21.419	−7.415	−3.389
HBPC	−0.144	−0.084	0.017	−0.090	1.201***
HEXP	0.095	−0.086	0.142	0.181	−0.022
HS	−0.338	−0.142	0.601	0.463	0.097
PERAG	0.887**	0.370	0.747	0.480	0.345
PERCON	−0.454	−0.931	0.214	0.685	0.994
PERMAN	−0.097	−0.091	0.532**	0.552**	0.103
UNEMP	−0.273	2.021*	0.683	1.242	−0.865
COMM	0.049	0.078	−0.010	0.157**	−0.005
WHT	−0.087	0.106	0.039	−0.130	0.039
POPO65	0.919	2.034***	−0.799	0.945	−0.676
IN3054	−0.205	−0.503	0.588	−0.186	0.162
IN5569	0.186	−0.399	−0.437	−0.440	0.671
IN70UP	−1.254**	−0.776	1.113	0.775	−0.162
RI	−38.379	113.969	93.162	79.551	22.084
RI × ln LQ, 2000	−0.767	−0.187	−1.107***	−0.265	−0.147
RI × dPOPDENS	−34.795	0.519	33.240	−39.162	15.644
RI × MEDHHY	−28.804*	−49.169	57.919	10.731	4.507
RI × HBPC	0.359*	0.311	0.123	0.559	−0.033
RI × HEXP	−0.243*	0.195	−0.333	−0.472	0.149
RI × HS	0.778	−0.113	−1.531	−0.816	−0.195
RI × PERAG	−2.143**	−1.257	−2.529**	−1.412	−0.778
RI × PERCON	1.040	2.181	−0.059	−1.158	−1.879*
RI × PERMAN	0.125	0.051	−1.636***	−1.125**	−0.217
RI × UNEMP	0.312	−4.644**	−1.781	−2.688	1.696
RI × COMM	−0.121	−0.275	−0.016	−0.359	−0.071
RI × WHT	0.175	−0.306	−0.129	0.266	−0.007
RI × POPO65	−1.548	−3.919**	2.258	−2.197	0.884
RI × IN3054	0.608	0.926	−1.133	0.597	−0.357
RI × IN5569	−0.707	0.897	0.690	0.888	−0.943
RI × IN70UP	2.733**	1.282	−2.462	−1.366	0.385
Adj. R^2	0.48	0.30	0.24	0.27	0.30

*,**,*** significant at the 1%, 5%, and 10% levels

spatial pattern is observed looking at the association between counties with relatively older populations and changes in the concentration of office-based medical doctors. The urban–rural "switching point" occurred in slightly more remote counties (0.52), and the effect was opposite of that observed with MD professionals and in-migration of the 70+ cohort (Fig. 4).

Location quotient for MDs and the % of the population 65+

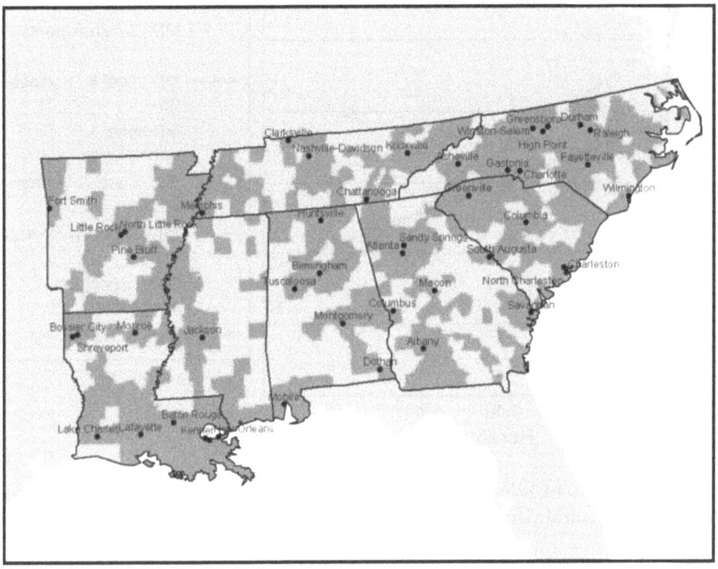

Location quotient for office-based medical specialists and 70+ in-migrants

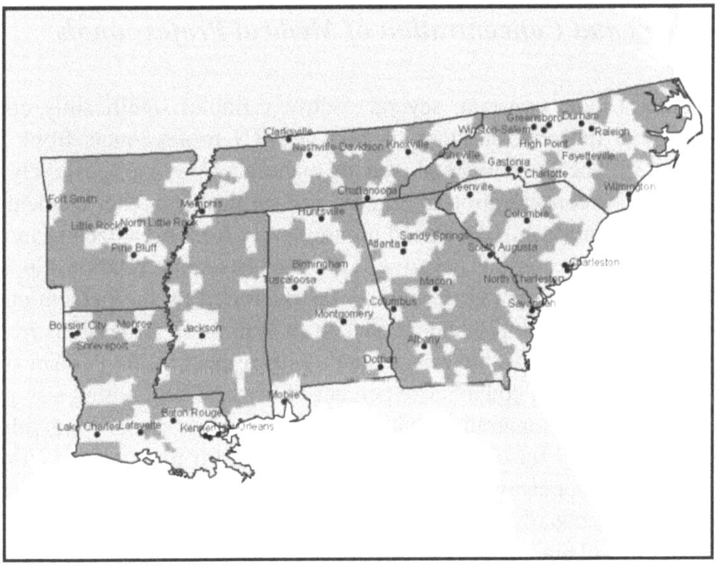

Fig. 3 Top panel, shaded counties are those with rurality indices ≤ 0.52; bottom panel, counties with rurality indices ≥ 0.49. Both are associated with positive change in the professional concentration of MD's and the office-based MD sub-group

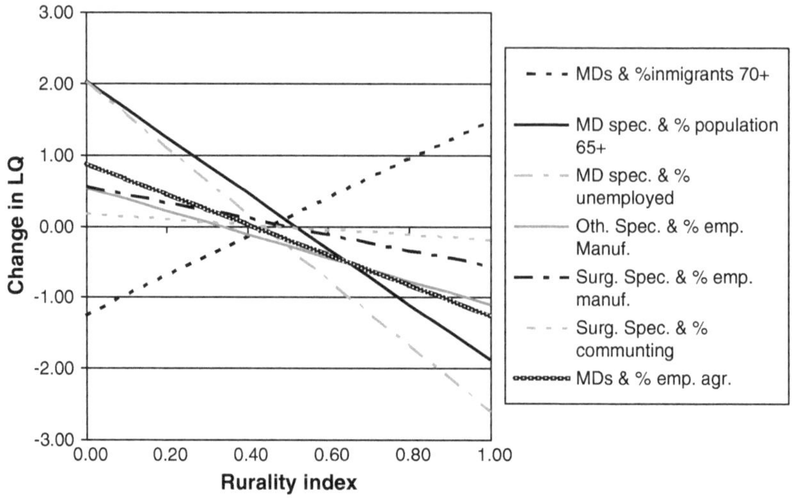

Fig. 4 Marginal effects of selected demographic and socio-economic variables on changes in location quotients measuring different medical professions across a rural–urban continuum

3.4 Spatial Heterogeneity of Other Demographic and Economic Factors and Concentration of Medical Professionals

Holding other factors constant, several control variables significantly correlated with changes in the concentration of MD and RN professionals from 2000 to 2004 (Table 3). The percent employed in manufacturing was positively related with growth in the concentration of office-based surgical and other medical professionals, but the relationship decreased moving away from urbanized to more rural regions, eventually becoming negative. This suggests some relationship between agglomeration due to location economies (as measured by the percent employed in manufacturing), and employment concentration in these professions from 2000 to 2004. A similar relationship was evident with respect to the percent of workers commuting outside a county. The percent of workers commuting also provides some indication of agglomeration due to urbanization economies – individuals living in one community (or county) and traveling round-trip to work in an adjacent county, or even farther away. The relationship was positive in more urban areas, but decreased moving towards more remote rural counties.

The county-level unemployment rate was correlated with growth in the concentration of office-based medical specialists, but the relationship was relatively weak ($P = 0.07$). Employment concentration of MDs increased in counties with relatively more persons employed in the agricultural sector, but the relationship decreased moving away from the urban core to more rural counties.

3.5 Agglomeration and Deglomeration of Office-Based Surgical and Medical Specialists

There was significant spatial lag autocorrelation (AR) explaining change in the employment concentration of surgical specialists and medical specialists, but the relationships were not strong (Table 3). The positive AR coefficient explaining changes in the concentration of office-based surgeons seems to be consistent with the notion that professionals providing surgical care are attracted to locations that might be characterized as regional medical centers. Hospitals are an excellent example of an industry that benefits from pecuniary externalities gained by locating in close proximity to other hospitals, *inter alia* the business services, logistical support, coordinating infrastructure, and human capital (in terms of physician networks and nursing pools). In general, provision of state-of-the art surgical care requires highly specialized technology and skilled personnel to maintain and operate equipment. For example, locating in metropolitan statistical areas (the "urban core") lowers costs of maintaining expensive apparatus, holding other factors constant, by virtue of urbanization economies associated with these regions. Such agglomeration forces are characterized by the relatively small (but significant) lag coefficient associated with the office-based surgeon concentration equation, and suggests that office-based surgeons are attracted to areas where the regional density of surgical providers is relatively high.

A "deglomeration" effect was evident in the equation explaining change in the concentration of medical specialists from 2000 to 2004, although the relationship was weak with a significance level of only 7%. The negative pattern of lag autocorrelation possibly suggests a crowding-out effect of employment opportunities in this medical profession. That is, competition for patients seeking special medical care may be stiff. Therefore, individuals in this medical profession can increase their demand threshold by locating in non-adjacent counties.

4 Conclusions

The objective of this empirical analysis was to determine whether aggregate changes in the concentration of health care professionals was associated with previous migration of seniors in the Southeastern United States. Empirical research, anecdotal information, case studies, and common sense suggest that as the population of a given town, county, or region ages, demand for health services will increase. In particular, one would expect that places attractive to migrating seniors would experience growth in the health care sector, which would in turn increase the number of individuals working in the profession. This line of reasoning has implications for local development planners and policymakers with respect to meeting the demands of not only local citizens, but also newly arriving baby boomers who usually bring with them different preferences, expectations regarding health care in particular, and lifestyle choice in general. Without the necessary infrastructure in place, including physical and human capital, provision of such amenities will remain a challenge. We found limited evidence supporting the hypothesis that later-life migrants have influenced employment growth in this sector from 2000 to 2004. In-migration of 70+ seniors from 1995 to 2000 is positively correlated with changes in the concentration of medical doctors. The relationship is heterogeneous, and strengthens moving away from urban-core areas. Other control variables appear to be more strongly related to observed changes.

The results were somewhat surprising, and are not without limitations. We expected that the in-flow of senior migrants would be positively correlated with increases in the concentration of medical professionals. And while there may be other factors that might explain changes in the concentration of medical professionals over the time period, inclusion of such variables would probably do little to change the statistical relationship observed between changes in the concentration of health care professionals and senior in-migration.

One explanation may be that an aggregate, regional perspective may paint a picture with too broad a stroke. Case-study comparisons focusing on established retirement communities versus emerging ones may generate different conclusions about the impact of later-life migrants on growth in the health sector (e.g., Park et al. 2007). Such an approach lends itself to looking at sector changes supplemented by input-output analysis.

Second, the location quotients used to measure changes in the concentration of health professionals may be inadequate as a measure of overall growth in demand for health care services. Other measures might include relative measures of HMO's, or raw "physical" measures such as hospital beds or surgical units. The time period covered in the analysis is also relatively short. A longer series may provide a better grasp of the dynamics between migrating cohorts and change in the medical sector, but acquiring such data is expensive.

Third, it is extremely difficult to untangle the effects of general population change and migration to that of the impact on later-life migrations, given the data at hand. While we controlled for age cohorts in the analysis, this may not be adequate. Although generally using fewer health services (except for perhaps women in their child bearing years), younger migrants may out-number older migrants, thereby potentially nullifying age-related effects. Similarly, out-migration may have an opposite effect, removing the supply of clients needed by health care workers to remain in business. Therefore, in some counties the net effect of in-migration may be countered by out-migration, dampening the net effect of newcomers on the demand for health care providers. The model used here could be improved by constructing a ratio of in- to out-migrating persons to control for this effect.

Fourth, future research could focus on rural areas only. While the model did control for rural–urban differences, it may be reasonable to assume that metropolitan areas are already well-equipped with medical resources and would therefore be better positioned to recruit new staff. On the other hand, it may take rural areas longer to build up medical staff, especially in places lacking certain natural or physical amenities.

Lastly, the regional coverage may have limited the results. Inclusion of more states in the analysis would introduce more variability into the models, which would increase the likelihood of identifying statistically meaningful relationships. But inclusion of more counties into the analysis would do little to alter the results obtained in our subset of states, which was the main focus of the study.

References

Aday RH, Miles LA (1982) Long-term impacts of rural migration of the elderly: implications for research. Gerontol 22:331–336

Anselin L (1988) Spatial econometrics: methods and models. Kluwer, London

Anselin L (2002) Under the hood: issues in the specification and interpretation of spatial regression models. Agric Econ 27:247–267

Anselin L, Florax RJGM (1995) New directions in spatial econometrics. Springer, Berlin

Anselin L, Lozano-Gracia N (2007) Error in variables and spatial effects in hedonic house price models of ambient air quality. Empir Econ 34:5–34

Area Resource File (ARF) (2005) US Department of Health and Human Services, Health Resources and Services Administration. Bureau of Health Professions, Rockville, MD

Bao S, Henry M, Barkley D (2004) Identifying urban-rural linkages: tests for spatial effects in the Carlino-Mills model. In: Anselin L, Florax RJGM, Rey SJ (eds) Advances in spatial econometrics: methodology, tools and applications. Springer, Berlin, 321–333

Barsby S, Cox DR (1975) Interstate migration of the elderly: an economic analysis. Lexington Books, Lexington, MA

Bates LJ, Santerre RE (2005) Do agglomeration economies exist in the hospital services industry. East Econ J 31:617–628

Beale C (2005) Rural America as a retirement destination. Amber Waves (June)

Beale CL, Fuguitt GV (1990) Decade of pessimistic nonmetro population trends ends on optimistic note. Rural Dev Perspect 20:4–18

Boarnet MG, Chalermpong S, Geho E (2005) Specification issues in models of population and employment growth. Pap Reg Sci 84:21–46

Brasure M, Stearns SC, Norton EC, Ricketts T (1999) Competitive behavior in local physician markets. Med Care Res Rev 56:395–414

Capalbo SM, Heggem CN (1999) Innovations in the delivery of health care services to rural communities: telemedicine and limited-service hospitals. Rural Dev Perspec 14:8–15

Cho S-H, Kim SG, Clark CD, Park WM (2007) Spatial analysis of rural economic development using a locally weighted regression model. Agric Res Econ Rev 36:24–38

Cliff AD, Ord JK (1981) Spatial processes. Pion, London

Cliff AD, Ord JK (1973) Spatial autocorrelation. Pion, London

Cohen JP, Paul CM (2005) Agglomeration economies and industry location decisions: the impacts of spatial and industrial spillovers. Reg Sci Urban Econ 35:215–237

Colsher PL, Wallace RB (1990) Health and social antecedents of relocation in rural elderly persons. J Gerontol Soc Sci 45:S32–S38

Conley TG (1999) GMM Estimation with cross-sectional dependence. J Econ 92:1–45

Connor RA, Hillson SD, Krawelski JE (1995) Competition, professional synergism, and the geographic distribution of rural physicians. Med Care 33:1067–1078

Day FA, Barlett JM (2000) Economic impact of retirement migration on the Texas Hill Country. J Appl Gerontol 19:78–94

Dine DD (1988) Demand for retirement housing accommodates industry growth. Mod Healthc 18: 56–60.

Doeksen GA, Johnson T, Willoughby C (1996) Measuring the economic importance of the health sector on a local economy: a brief review and procedures to measure local impacts. Southern Rural Development Center, Starkville, MS

Dwight MB (1985) Affluent elderly want to live where quality care's readily available. Mod Healthc April:74–76

Escarce JJ, Polsky D, Wozniak GD, Kletke PR (2000) HMO growth and the geographical redistribution of generalist and specialist physicians, 1987–1997. Health Serv Res 35:825–848

Fagan M (1988) Attracting retirees for economic development. Center for Economic Development and Business Research, Jacksonville State University

Fagan M, Longino, Jr CF (1993) Migrating retirees: a source for economic development. Econ Dev Q 7:98–106

Freed GL, Nahra TA, Wheeler JR (2004) Relation of per capita income and gross domestic product to the supply and distribution of pediatricians in the United States. J Pediatr 144:723–728

Fuguitt GV, Beale CL, Tordella SJ (2002) Recent trends in older population change and migration for nonmetro areas, 1970–2000. Rural Am 17:11–19

Fujita M, Krugman P, Venables AJ (1999) The spatial economy: cities, regions, and international trade. MIT, Cambridge

Glasgow N, Reeder RJ (1990) Economic and fiscal implications of nonmetropolitan retirement migration. J Appl Gerontol 9:433–451

Haas WH III (1990) Retirement migration: boon or burden? J Appl Gerontol 9:387–392

Haas WH, Crandall LA (1988) Physicians' view of retirement migrants' impact on rural medical practice. Gerontol 28:663–666

Haas WH III, Bradley DE, Longino CF Jr, Stoller EP, Serow WJ (2006) In retirement migration, who counts? A methodological question with economic policy implications. Gerontol 46:815–820

He W, Schachter JP (2003) Internal migration of the older population: 1995 to 2000. Census 2000 Special Report, US Census Bureau

Hodge G (1991) The economic impact of retirees on smaller communities. Res Aging 13:39–54

House JS, Lantz, PM, Herd P (2005) Continuity and change in the social stratification of aging and health over the life course. J Gerontol B Psychol Sci Soc Sci 60:15–26

Jiang HJ, Begun JW (2002) Dynamics of change in local physician supply: an ecological perspective. Soc Sci Med 54:1525–1541

Joseph AE, Bantock PR (1982) Measuring potential physical accessibility to general practitioners in rural areas: a method and case study. Soc Sci Med 16:85–90

Kelejian HH, Prucha IR (2007) HAC estimation in a spatial framework. J Econom 140:131–154

Lambert DM, McNamara K, Garret M (2006) Food industry investment flows: implications for rural development. Rev Reg Stud 36:140–162

Loomis LM, Sorce P, Tyler PR (1989) A lifestyle analysis of healthy retirees and their interest in moving to a retirement community. J Hous Elder 5:19–35

McGranahan DA, Wojan TR, Lambert DM (2006) Rural growth as creative enterprise. Paper presented at the 53rd Annual North American Meetings of the Regional Science Association International, Toronto, Canada, Nov 16–18

Mistretta MJ (2007) Differential effects of economic factors on specialist and family physician distribution in Illinois: a county-level analysis. J Rural Health 23:215–221

Monchuk DC, Miranowski JA, Hayes DJ, Babcock BA (2007) An Analysis of Regional Economic Growth in the U. S. Midwest. Rev Agric Econ 29:17–39

Moreno R, López-Bazo E, Vayá E, Artis M (2004) External effects and cost of production. In: Anselin L, Florax RJGM, Rey SJ (eds) Advances in spatial econometrics: methodology, tools and applications. Springer, Berlin, pp 297–317

Mullins D, Rosentraub M (1992) Fiscal pressure? The impact of elder recruitment on local expenditures. Urban Aff Q 28:337–354

Newhouse JP, Williams AP, Bennett BW, Schwartz WB (1982) Does the geographical distribution of physicians reflect market failure. Bell J Econ 13:493–505

Park WM, Clark CD, Lambert DM, Wilcox MD (2007) The long-term impacts of retiree in-migration on rural areas: a case study of Cumberland County, Tennessee. The University of Tennessee Institute for Public Service, Knoxville
Pathman DE, Ricketts TC III, Konrad TR (2006) How adults' access to outpatient physician services relates to the local supply of primary care physicians in the rural southeast. Health Serv Res 41:79–102
Patrick CH (1980) Health and migration of the elderly. Res Aging 2:233–241
Reardon J (1996) The presence of hospital systems in rural areas. J Econ Issues 30:859–876
Reeder RJ (1998) Retiree-attraction policies for rural development. Agriculture Information Bulletin No. 741. U.S. Dept. of Agriculture, Economic Research Service, Washington, DC
Reeder RJ, Glasgow N (1990) Nonmetro retirement counties: strengths and weaknesses. Rural Dev Perspect 20:15–30
Regnier V, Gelwicks LE (1981) Preferred supportive services for middle to higher income retirement housing. Gerontol 21:54–58
Ricketts TC (2000) The changing nature of rural health care. Annu Rev Public Health 21:639–657
Rosenthal MB, Zaslavsky A, Newhouse JP (2005) The geographic distribution of physicians revisited. Health Serv Res 40:1931–1952
Rowles GD, Watkins JF (1993) Elderly migration and development in small communities. Growth Change 24:509–538
Serow WJ (2001) Retirement migration counties in the southeastern United States: geographic, demographic, and economic correlates. Gerontol 41:220–227
Serow WJ (2003) Economic consequences of retiree concentrations: a review of North American studies. Gerontol 43:897–903
Serow WJ, Charity DA (1988) Return migration of the elderly in the United States: recent trends. Res Aging 10:155–168
Sickles RC, Taubman P (1986) An analysis of the health and retirement status of the elderly. Econ 54:1339–1356
Stallmann JL, Deller SC, Shields M (1999) The economic and fiscal impact of aging retirees on a small rural region. Gerontol 39:599–610
Toseland R, Rasch J (1978) Factors contributing to older persons satisfaction with their communities. Gerontol 18:395–402
Vestal C (2006) Retirees boosting states' rural economies. Available at: www.stateline.org, 15 Sept. 2007
Waldorf B (2006) A continuous multi-dimensional measure of rurality: moving beyond threshold measures. Paper presented at the Annual Meetings of the Association of Agricultural Economics, Long Beach, CA, July
Wall TP, Brown LJ (2007) The urban and rural distribution of dentists, 2000. J Am Dent Assoc 138:1003–1011
Wing P, Reynolds C (1988) The availability of physician services: a geographic analysis. Health Serv Res 23:649–67
Whittle P (1954) On the stationary process in the plane. Biometrika 41:434–439
Wojan TR, Lambert D, McGranahan DA (2008) Emoting with their feet: Bohemian attraction to creative milieu. J Econ Geogr 7:711–736
Wolinsky FD, Mosely II RR, Coe RM (1986) A cohort analysis of the use of health services by elderly Americans. J Health Soc Behav 27:209–219

Part V
Regional Applications

Evolution of the Influence of Geography on the Location of Production in Spain (1930–2005)

Coro Chasco Yrigoyen and Ana M. López García

1 Introduction

In recent years, there has been a growing interest in the geographic aspects of development or the question of where economic activities take place. There is an extensive literature in urban economics, location theory and economic agglomeration. In fact, many economic activities are concentrated geographically and most people in advanced countries or regions live in densely populated metropolitan areas. The main issue is how to explain this concentration. Most of the references assume two approaches, *first nature* (Sachs 2000) and *second nature* (Krugman 1993, 1999; Venables 2003), which are also identified as Sachs' (first nature) and Krugman's approach (second nature). Krugman's New Economic Geography abstracts from natural conditions. It states that agglomerations can be explained by second nature alone (i.e. by man-made agglomeration economies due to increasing returns to scale and transportation costs), which arises endogenously in the economic process.

However, real world agglomeration is possibly caused by both first and second nature. In this case, it would be interesting to compute the exact influence of both types of agglomeration advantages on economic distribution across space. In fact,

> if first nature is important, existing agglomerations are likely to be very stable since they are tied to specific places. At the same time, attempts to form new agglomerations at places without geographic advantages might fail. If, however, first nature does not matter much, agglomerations are footloose and can emerge and break down at any location lending much more power to regional policy. (Roos 2005)

In this chapter, our aim is to examine the influence of geographic features on the location of production in Spain. In other words, we focus on quantifying how much of the geographic pattern of GDP can be attributed to only exogenous first nature elements (physical and political geography), how much can be derived from

---

C. Chasco Yrigoyen (✉)
Dpto. Economía Aplicada, Facultad de Ciencias Económicas y Empresariales, Universidad Autónoma de Madrid, Carretera de Colmenar Viejo Km. 15.500, Madrid 28049, Spain,
e-mail: coro.chasco@uam.es

A. Páez et al. (eds.), *Progress in Spatial Analysis*, Advances in Spatial Science,
DOI 10.1007/978-3-642-03326-1_19,

endogenous second nature factors (man-made agglomeration economies) and how much is due to the interaction of both effects. Specifically we disentangle the two net effects empirically, as well as their mixed effect, for the Spanish case analyzing their evolution during the twentieth century.

For this purpose, we follow a methodology based on Roos (2005) for Germany. He proposes to employ an analysis of variance (ANOVA) to infer the unobservable importance of first nature indirectly in a stepwise procedure. In order to disentangle first and second nature effects empirically, we control for second nature because every locational endowment will be reinforced and overlaid by second nature advantages. In a dynamic context, we also estimate how much agglomeration can by explained by both gross and net second nature with the aim of isolating the importance of first nature alone.

Whereas this method seems quite clear and direct, we demonstrate that results could be biased if some potential econometric questions are not properly taken into account; e.g. multicollinearity, relevant missing variables, endogeneity, spatial autocorrelation and spatial heterogeneity.

In fact, in many countries GDP density is strongly polarized on two subspaces, core and periphery, displaying spatial heterogeneity. In the particular case of Spain, the core is located in the coastal plus Madrid provinces and the periphery is constituted by the hinterland. If we consider the Spanish territory as a whole, we find that at most, 88% of GDP's spatial variation can be explained by direct and indirect effects of geography during the twentieth century. This result contrasts with Roos' findings for Germany (72%) pointing out the main role played by geography in Spain. After controlling for agglomeration economies and the interaction effect of first–second nature, the net influence of natural geography is only about 6–7% nowadays. Nevertheless, some of these results could be significantly biased for the group of inland provinces, in which only a 70% of agglomeration is explained by geography, being the mixed effects the most determining almost along the whole period. For this reason, we propose to take into account spatial effects explicitly in the models.

2 Theoretical Principles and Background

Since it would be impossible to summarize in any simple way the rich range of conclusions from the studies related to this matter, next we highlight some of the most significant for our econometric analysis.

2.1 First Nature

First nature factors are also called "pure geography" (Henderson 1999). They are natural features such as climate or resource endowments, which are *exogenous* to the economy. Since nature endows all places with specific features, one obvious explanation to the concentration of population and firms in some regions is that they must have some natural advantage. On the contrary, sparseness and depopulation is very often related to absolute endowment disadvantages – lack of natural resources, bad climate, poor land quality, cold temperatures and propensity to disease – and/or long distances from the core economic centers, which penalizes either the relative prices of different goods or the relative profitability of different activities. Although Venables (1999) states that the degree of geographic determinism should not be exaggerated, it is clear that the impact of physical geography on development appears to derive from key relationships between climate and disease, climate and agricultural productivity, and also between location and technology transfer.

The main question is how much geography still matters for economic development. Gallup et al. (1999) find that location and climate have sizable effects on population density, as well as on income levels and growth rates – or even economic policy choice – through their effects on transport cost, disease burdens and agricultural productivity, among other channels. In particular, these authors regress the population density on geography variables such as distances to the coast and waterways, several measures of elevation, soil quality, availability of water and climate. In the international sample used, those factors explain 73% of the observed variability of the population density.[1] Nevertheless as stated in Roos (2005), this estimation might grossly exaggerate the importance of first nature due to the large number of independent variables used, which could lead to *multicollinearity*. Besides, he explains that there are other potential *missing variables* that are crucial in explaining the uneven distribution of population in the world. This is the case of institutional, historical, cultural and economic conditions, which are so diverse on the global level that they threaten the consistency of the geography estimates.

On their side, Ellison and Glaeser (1997) and Kim (1999) think that a substantial portion of the observed geographic concentration of industries is affected by a wide range of natural advantages. In another paper, Ellison and Glaeser (1999) found that – apart from interfirm spillovers – geography is an important determinant of agglomeration, accounting for 50–86% of the observed variability. However, it can also been criticized that these figures are likely to overstate the importance of geography because of the broad definition of first nature. In fact they measure first nature with labor and capital endowments, such as labor costs, labor qualification and the size of the consumer market. Nevertheless, neither the regional endowments with mobile factors nor the prices of these factors are really exogenous. On the contrary, there might be a reverse causation – *simultaneity* – running from the presence of a particular industry in a region to the region's endowment with labor or capital. Actually if it is true that human and economic agglomerations can be explained by an accidental accumulation of favorable natural features, it is also true that households and firms interact on product and labor markets. If these markets are spatially segmented we expect economic activity taking place where people live, but at the same time we also expect people living where economic activity takes place.

[1] See other similar applications for Peru (Escobal and Torero 2005) and China (Ravallion 2007).

Consequently, it seems difficult to isolate the net influence of first nature on agglomeration since it is tightly joint to other factors belonging to what is called "second nature."

2.2 Second Nature

Second nature factors are man-made "agglomeration economies," i.e. interaction between economic agents among themselves (rather than the interaction between agents and nature), as well as knowledge and information spillovers, economies of intra-industry specialization, labor market economies or economies of scale in industry-specific public services, product differentiation and market size effects. Second nature, which is *endogeneous* to the economy, emphasizes the efficiency gains from proximity since interactions between economic agents (firms and consumers) are more efficient in densely packed areas than when people are widely dispersed (Kanbur and Venables 2007). These agglomeration forces can therefore create virtuous circles of self-reinforcing development in some regions while others lag behind. In this same direction, Fujita et al. (1999) demonstrate that the increasing returns to scale of some productive activities could be one of the keys to explain spatial economic inequality.

Venables (1999) shows that second nature represents investment in transport and communication infrastructure, as well as its maintenance linking coastal to hinterland regions. In effect, although there is an association – in some places – among coastal locations, urbanization and growth, it is also true that investment in transportation and communication infrastructure linking coastal and interior areas facilitates hinterland development. It is known that access to hinterland resources is a geographic challenge to be overcome by infrastructure investment. Therefore, again we find a close connection between first and second nature. On the one hand, first nature geography constitutes an initial advantage that becomes usually amplified by second nature agglomeration forces. On the other hand, it is also known that the adverse effects of geography on economic growth can be overcome by different factors (Henderson 1999). As Krugman (1993) argues, first nature advantages generally tend to create second nature advantages through cumulative processes. These are decisive to explain the concentration of population that has taken place both during and after the industrialization process.

Even more, the new economic geography follows the new trade theory by showing how second nature effects can lead to a highly differentiated spatial organization of economic activity, even when the underlying physical geography is undifferentiated (Gallup et al. 1999). Krugman's theory shows that agglomerations can be explained by *second nature alone* (net second nature).

It seems clear that first and second nature have an obvious incidence on agglomeration. Nevertheless it is necessary to compute the contribution of each net component as well as the first–second nature mixed effect. As stated before, this is the main aim of this chapter.

2.3 The Spanish Case

Referring to the particular case of Spain, Dobado (2006) coincides with Venables and Roos when considering first and second nature as non-contradictory but complementary, since real-world agglomerations are caused by both forces. In his opinion, the authentic peculiarity of Spanish regions – when compared to others in Southern Europe – consists in the existence of a large group of provinces with very low levels of population and GDP concentration close to another minor group with high densities. This is the so called duality core-periphery that, in the Spanish case is clearly conditioned by significant geographical – first nature – differences. The "core" is constituted by Madrid and the coastal provinces, which in general terms, exhibit low altitude, humid climate and few extension, and concentrate the highest levels of GDP per area. The "periphery" is located in a depopulated hinterland, with more extreme temperatures than in the core and an abrupt topography.

Tirado et al. (2003) and Rosés (2003) analyze the role played by scale economies – second nature – on industrial agglomeration in Spain. They think that the major industrial concentration around Barcelona at the end of the nineteenth century was the result of both some initial natural advantages and a cumulative causation process linked to the increasing role of scale economies in production. They coincide with Krugman and Livas (1996) in considering that the protectionist policy – in the first decades of the twentieth century – weakened Barcelona's role in favor of capital cities located in geographical centers (Madrid and Saragossa). Transport costs from these core cities to domestic consumers could be minimized reinforcing the agglomeration tendencies and avoiding dispersion.

Viladecans (2004) also explains the uneven location of manufacturing activities in Spain as a result of two types of agglomeration economies, i.e. urbanization and localization economies. She states that the effect of specialization in one sector on a geographical area – localization economies – is a determining factor in the location of firms belonging to that sector. More precisely, the geographical distribution of most of the industrial sectors is influenced, to some extent, by the productive environment.

Ayuda et al. (2005) analyze the combined influence of first and second nature forces in population concentration as a two-step process. In effect, while geography can be expected to play a very important role in the Spanish pre-industrial economy, increasing returns seem to be the driving force of population concentration in the industrializing period. These authors explain that only those regions with particularly favorable resources for the location of certain types of industry could generate their own growth dynamics based on such comparative advantages. They compute the importance of natural or situational advantages on population density in the Spanish provinces at five different moments since 1787 to 2000. It covers the pre-industrial situation, the Spanish industrialization, the development process and the moments referred to a mature modern economy. The main results underscore the importance of geographical factors in explaining the distribution of the Spanish population in the last two centuries. Historically, the highest population densities have been found in the maritime or non-mountainous provinces, as well as in those areas with the highest annual rainfall.

Considering all this, it is clear that geographic considerations should be taken into account in empirical – and theoretical – studies of cross-country (or region) economic concentration. It is also evident that the term "geography" should be split into first and second nature, since it includes not only natural advantages but also the scale economies or efficiency gains derived from proximity. Moreover, there is a combined or mixed effect of first–second nature on agglomeration that should be isolated to quantify to what extent natural endowments and man-made agglomeration economies mutually interacts. We can also conclude that from the concrete econometric modeling point of view, we must explicitly consider some potential problems, such as multicollinearity, relevant missing variables, endogeneity and spatial effects, if we want to reach reliable conclusions.

3 Data and Model

3.1 Data

It is our aim to explain agglomeration from first and second nature elements. Hence, we must define first what we understand for agglomeration and geography to find the appropriate indicators. Differently to Rosenthal and Strange (2001), we do not want to determine the degree of agglomeration but how geography – in general terms – influences the spatial distribution of production activities. Regarding the endogenous variable, several measures have been used in the literature. This is the case of population, which has been applied to evaluate consumption, mainly when relying on the hypothesis that "firms follow people" (e.g. Graves 1979; Cragg and Kahn 1997; Knapp et al. 2001, for the US). Others, such as employment or GDP, are production indicators that would depend on the hypothesis that "people follow jobs" (e.g. Freeman 2001, in the US; Roos 2005, in Germany). Ciccone and Hall (1996) and Rappaport and Sachs (2003) decide on using population and employment densities as measures of agglomeration because they think that economic activity takes place where people live, and vice versa. Dobado (2004) proposes several indicators in absolute terms (area, GDP, population) or relative to the area (GDP or population density). In these last cases, agglomeration is conceived as the spatial concentration of not only production activities but also both workers/citizens.

In order to make better comparisons with Roos' computations for German regions, we opt to use the relative GDP density – GDP per km^2 – as the endogenous variable. He argues that this variable is more appropriate than population or employment densities to determine how geography influences the distribution of economic activity across a territory.[2] In this way, Delgado and Sánchez (1998) use

[2] However, it must be said that this indicator has important drawbacks. On the one hand, regions that are specialized in high-value added sectors will automatically display greater GDP values, while it could not necessarily reflect in the true level of spatial agglomeration of firms and workers.

the same variable to compute the evolution of income density in Spain. Since area is constant in each region every time, the evolution of this variable only depends on the quantity of the generated GDP.

Formally, the endogenous variable is defined as follows:

$$\log(gd_i) = \log \frac{Y_i/A_i}{\sum_i Y_i/A_i} = \frac{\log\left[Y_i/\sum_i Y_i\right]}{\log\left[A_i/\sum_i A_i\right]} \tag{1}$$

where Y is GDP and A_i is the area of region i. The relative GDP density of a region is its GDP density relative to the average density of all regions or, equivalently, the ratio of its share of GDP relative to the share of the country's total area. If log (gd_i) is equal to zero, region i's GDP share is equal to its area share. If it is larger (smaller) than zero, the region has a concentration of economic activity above (below) the average.

Next, we define some good indicators to measure first and second nature effects. About first nature, we are interested in those geographical characteristics that are related to the distribution of economic activity. In general, this is the case of natural endowment, physical geography, relative location and political geography. Examples of natural endowment positively related to GDP density are agriculture, minerals, natural resources, good soil and water supply (Gallup et al. 1999; Rappaport and Sachs 2003). Some of these authors, as well as Rappaport (2000), Limão and Venables (2001) and Roos (2005), also include certain kind of physical geography indicators, such as altitude, latitude, distance to the coast and waterways, lying to the seashore (or being landlocked), navigable rivers and climate. Location is another geographical feature affecting agglomeration, which has been represented as relative distance to core – or other – regions or simply by the latitude–longitude Earth coordinates.

Following Ayuda et al. (2005) and Dobado (2004), we have chosen the annual rainfall (*rainfall*) as a good proxy for agricultural potential, due to such dry conditions that are predominant in the Mediterranean regions (see in Table 1 a full description of the variables). We have also considered some climate variables, such as temperature (*temmin*, *temaver*, *temmax*, *tembel0*, *overcast*) and altitude (*altit*), as well as maritime length (*maritlim*, *coast*). We expect negative values for extreme temperatures and high altitudes, but a positive relationship between seashore extension and GDP density. Besides, we have included longitude and latitude, which are the X–Y Earth coordinates (*xcoo*, *ycoo*). As we will prove further, in Spain at present, being an Eastern Mediterranean region constitutes a relative advantage than lying to the Cantabric or the Atlantic seashores. However the North–South direction seems to be no longer significant in terms of agglomeration.

On the other hand, the level of GDP per km^2 in a region like Madrid is possibly overstated because many workers commute everyday from neighboring Castilian provinces; as a result, the level of agglomeration in these Castilian provinces would be understated. In addition, it is known that first and second nature factors have different effects in different industries, as stated in Alonso-Villar et al. (2004). Using aggregate GDP does not allow analyzing this issue properly.

Table 1 Variable list for the Spanish provinces

Variable	Description	Units	Font	Period
Gd	GDP per area	Euros/sq. m.	FBBVA, FUNCAS	1930–2005
Capit	Capital city	0–1	Self elaboration	–
Altit	Altitude or elevation	meters	INE	–
Temmin	Minimum temperature	Celsius	INE	1997–2005[a]
Temaver	Average temperature	Celsius	INE	1997–2005[a]
Temmax	Maximum temperature	Celsius	INE	1997–2005[a]
Tembel0	Equal or below zero Celsius temperature	# Days	INE	1997–2005[a]
Rainfall	Total annual precipitation	Millimeter	INE	1997–2005[a]
Overcast	Overcast	# Days	INE	1997–2005[a]
Maritlim	Maritime limit	0–1	Self elaboration	–
Coast	Seashore length	Kilometers	INE	–
Xcoo	Longitude (X-coordinate)	Grades	Self elaboration	–
Ycoo	Latitude (Y-coordinate)	Grades	Self elaboration	–
Pop	Population	People	FBBVA, FUNCAS	1930–2005
Prod	GDP per employee	Euros	FBBVA, FUNCAS	1930–2005

[a]Average of the period, *INE* Spanish National Institute for Statistics, *FBBVA* Foundation of the *Bilbao Vizcaya Argentaria* Bank

Political geography has also been highlighted by Mathias (1980), McCallum (1995) and Roos (2005) who consider that agglomeration is positive or negatively affected by containing a capital city or being a border region, respectively. In this case, we have considered a dummy variable to indicate the presence of a capital city in a region (*capit*). Similarly to the German regions (Roos 2005), the Spanish autonomies concentrate a lot of legislative and executive power in their capital cities. This is why provinces with a capital city should have better access to information about regional government investment and decision plans (Ades and Glaeser 1995; Funck 1995; Ayuda et al. 2005).

In order to measure man-made agglomeration economies (second nature) we have also followed Roos (2005) what allows us to make better comparisons with this case. He chose total population (*pop*) and labor productivity (*prod*) since on aggregate levels both variables can capture many agglomeration economies, i.e. informational spillovers and labor market economies. Population could be considered as an indirect measure of agglomeration economies. In effect, as stated in Henderson (1988) if agglomeration economies exist in an area, labor productivity should rise in the level of population (employment). Other indicators, such as population density (proposed in Gallup et al. 1999), provide not so clear relationship with GDP density (e.g. some densely/sparsely populated areas are rich whereas others are poor, which are the cases of Western Europe/New Zealand and Indonesia/African Sahel, respectively).

3.2 Model

Three forces operate in forming agglomerations: an unobservable direct effect of first nature, a first nature effect working through induced agglomeration economies and a direct effect of second nature, which would exist even without any first nature forces. In order to get a better knowledge of these effects, Roos (2005) states a methodology based on analysis of variance (ANOVA). The total variance V of the dependent variable can be decomposed into four parts:

$$V = V_u + V_f + V_s + V_{fs} \tag{2}$$

where V is the total variance of the dependent variable, V_u is the unexplained variance, V_f is the variance explained by first nature alone, V_s is the variance explained by second nature alone and V_{fs} is the variance explained by a combination of both forces.

ANOVA is employed to infer the unobservable importance of first nature alone indirectly, as well as to assess about the relative importance of first and second nature forces. It is a four-step process that proceeds as follows:

1. Since man-made agglomeration effects (second nature) are usually triggered by natural advantages (first nature), we must first identify the net from the gross second nature effect. For this purpose, we regress two gross second nature variables on first nature. These regressions explain how much of the gross second nature effects are caused by purely first nature. By mean of the residuals of the regressions, we filter the net from the gross second nature variables.
2. We estimate how much of GDP per area variance can be explained by gross $(V_s + V_{fs})$ and net (V_s) second nature advantages. These calculations can be derived from the results of two regressions of GDP density on both gross and net second nature variables.
3. We estimate how much of GDP per area variance can be explained jointly by first and second nature $(V_f + V_s + V_{fs})$. The total effect of first and second nature can be obtained from a regression, using first and net second nature variables as explanatory variables.
4. We calculate the difference between the result in step 3 (total effect of first and second nature) and step 2 (total effect of second nature), which is the importance of first nature alone (V_f) on GDP per area.

Next, we will explain the whole process in depth.

Since first and second nature are interrelated, in a *first step* it is necessary to disentangle the second nature variables (population and GDP per worker) empirically. For that purpose, we can regress them on geography and take the residuals $\hat{\pi}$

and $\hat{\delta}$ as variables of net second nature forces:

$$\begin{aligned}\log\left(pop_i\right) &= \gamma_0 + \sum_{k=1}^{K} \gamma_k f_{ki} + \pi_i \\ \log\left(prod_i\right) &= \rho_0 + \sum_{1}^{K} \rho_k f_{ki} + \delta_i\end{aligned} \tag{3}$$

where pop_i and $prod_i$ are total population and GDP per worker in region i, f_{ki} is the group of k geography variables, γ, ρ are coefficients and π, δ are the error terms of the regressions.

While variables $s_{mi} = \{\log(pop_i), \log(prod_i)\}$ are "gross" second nature variables, the residuals of these regressions $\hat{s}_{mi} = \left\{\hat{\pi}_i, \hat{\delta}_i\right\}$ could be taken as geography-filtered or net second nature forces. The introduction of these sets of variables, s_{mi}, $\hat{s}_{mi}$, as explanatory variables will allow to evaluate the total influence of gross and net second nature variables on GDP density.

In a *second step* we can compute the effects of total – both gross and net – second nature variables on GDP per area. In this fashion, the gross second nature variables influence is obtained with the estimation of the following equation:

$$\log\left(gd_i\right) = \alpha_0 + \sum_{m=1}^{M} \phi_m s_{mi} + \varepsilon_i \tag{4}$$

The resulting determination coefficient indicates this gross effect of second nature:

$$R^2_{gs} = \frac{\left(V_s + V_{fs}\right)}{V} \tag{5}$$

Regarding the net effect of second nature on GDP per area, it is derived from the estimation of the following equation:

$$\log\left(gd_i\right) = \alpha_0 + \sum_{m=1}^{M} \phi_m \hat{s}_{mi} + \varepsilon_i \tag{6}$$

The net effect of second nature on agglomeration can be expressed as:

$$R^2_{ns} = \frac{V_s}{V} \tag{7}$$

Therefore, the mixed effect of the interaction between first and second nature on GDP density can be extracted as follows:

$$\frac{V_{fs}}{V} = R^2_{gs} - R^2_{ns} \tag{8}$$

In the *third step*, we measure the total effect of first and second nature on GDP per area. We could simply include, in another equation, the gross second nature variables as regressors together with a set of first nature indicators. However, this could bias the estimates of the first nature coefficients since first nature also has an effect on the second nature variables. In order to adjust the later for the former, we specify a regression of GDP per area on first and *net* second nature variables, which avoids the stochastic regressors problem:

$$\log(gd_i) = \alpha_0 + \sum_{k=1}^{K} \phi_k f_{ki} + \sum_{m=1}^{M} \phi_m \hat{s}_{mi} + \varepsilon_i \tag{9}$$

The joint importance of first and second nature is measured by the corresponding determination coefficient:

$$R^2_{f+s} = \frac{V_f + V_{fs} + V_s}{V} \tag{10}$$

In the *fourth step*, we derive the net importance of first nature on GDP density from the results of the previous estimations:

$$\frac{V_f}{V} = R^2_{f+s} - R^2_{gs} \tag{11}$$

The estimation of (4), (6) and (9) by Ordinary Least Squares (OLS) could lead to biased results due to the presence of endogeneity on some of the explanatory variables and/or spatial effects on the residuals. Roos (2005) and Gallup et al. (1999) only consider the first problem but omit the second.

In effect, on the one hand endogeneity in a regressor can lead to a well-known simultaneity bias in the OLS estimates. Even in the pure-geography variables there could be different degrees of exogeneity. Physical geography variables (temperature, coast, etc.) can be considered as exogeneous since they do no depend on underlying economic forces. However political geography could have more endogeneous elements; e.g. the location of state capitals, though do not change very often, are possibly the result of the economic importance of the corresponding city. Moreover, the second nature variables (population and productivity) are much more endogenous and simultaneously determined with GDP density.

On the other hand, spatial autocorrelation and/or spatial heterogeneity in the OLS residuals are also causes of misspecification problems in the regression (see Anselin 1988 for a complete view of this topic). They must be tested and corrected, as will be shown hereafter.

4 Evolution of the Spatial Distribution of GDP per Area

In this section, we explore the geographic dimension of GDP per area for the continental Spanish provinces (47 provinces in total). We have excluded those provinces without geographical connection: the Balearic and Canary Islands and the African cities, Ceuta and Melilla.[3] In the case of the African cities, they are administrative regions not comparable in size with the others (population and GDP densities are extremely high). In order to explore these issues, we need a data set consistently defined over the century. For that purpose, we have used the GDP, employment and population series proposed by Alcaide (2003), for 1930 to 2000, and Alcalde and Alcalde (2007), for 2000 to 2005. The data on area are extracted from the Spanish Office for Statistics (INE) databank.[4]

Actually, we have selected five periods: 1930, 1950, 1970, 1990 and 2005, since they constitute good references for our analysis, corresponding to relevant facts related to Spanish economic history (Table 2). In effect, in 1930 Spain put an end to General Primo de Rivera's dictatorship. The economy enjoyed a prosperous moment thanks to a large public expenditure. Road and rail networks improved driving force to the development of industry and employment. At that moment, there were some industrialized enclaves, especially in the Axis Madrid-North-Barcelona, as well as other provinces in the Cantabric and Mediterranean Coast (Fig. 1). However, during the mid-1930s and 1940s the economic crisis and the Civil War stopped this process leading to an autarkical regime and recession. In 1950, approximately in the middle of General Franco's dictatorship, Spain had experienced a ruralization process with an increasing participation of agricultural sector. Rationing of food, commodities and energetic resources expelled the Spanish population from cities to rural places.

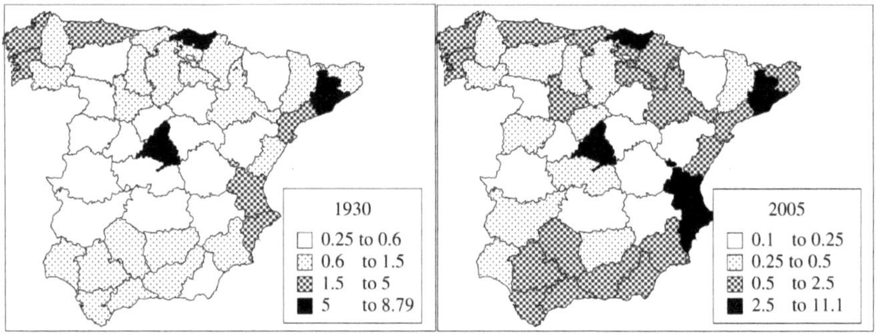

Fig. 1 Choropleth maps of relative GDP per area (1 = national GDP/km^2).
The variables have been classified with a method called "natural breaks," which allow identifying breakpoints between classes using Jenks optimization (Jenks and Caspall 1971). This method is rather complex, but basically it minimizes the sum of the variance within each of the classes, finding groupings and patterns inherent in the data.

[3] In spatial econometric applications, some authors prefer to exclude those Spanish regions without neighbours (e.g. Márquez and Hewings 2003), since it is politically debatable how to connect them to the rest of the system.

[4] This data are available in the INE webpage: http://www.ine.es

Table 2 Descriptive Statistics of Relative GDP per area.

Variable	Mean	Pearson CV	Minimum	Q1	Median	Q3	Maximum
GDP 1930	1.38	1.30	0.26	0.48	0.74	1.37	8.78
GDP 1950	1.42	1.42	0.20	0.48	0.72	1.39	9.03
GDP 1970	1.50	1.66	0.14	0.32	0.54	1.41	10.62
GDP 1990	1.45	1.64	0.10	0.27	0.51	1.40	10.89
GDP 2005	1.44	1.61	0.10	0.26	0.57	1.41	11.01

GDP relative GDP per area (1 = national GDP per km^2), *CV* coefficient of variation, *Q1, Q3* first and third quartiles, 1 = national GDP per km^2

During the 1950s and 1960s, the incipient political and economic openness set the basis for a decisive industrialization and tertiarization process. The Development Plans produced economic prosperity and liberalization, leading to new economic poles in Galicia, Castile, Andalusia, Aragón and Extremadura. This processes joint to a new great exodus from rural zones to industrial and urbanized areas – inland and abroad – helped to equilibrate the traditional inequality in the distribution of wealth across the Spanish territory. In 1970, close to the ending of Franco's regime, Spain was no more rural but urban.

By the beginning of the 1990s, Spain is one of the democracies belonging to the Economic European Community. In the late 1980s, a strict plan of economic stabilization, based on a traumatic industrial restructuring and liberalization customs, reformed the Spanish economy. The transfer of funds proceeding from the EEC made possible an ambitious policy of public investments in infrastructures. Nevertheless, income disparities across the Spanish regions still remained and even deepened. In 2005, economic development depicted a peculiar structure similar to a star, with its center in Madrid and the axis in the peripheral areas: the vast Mediterranean metropolitan areas, coastal Andalusia and Seville, coastal Galicia and the Cantabric regions. In addition, inside this big star, there was a vast rural desert, only broken by a few urban oases, like Valladolid, Saragossa, Badajoz, Burgos, Álava and Navarre.

Figure 2 plots the density functions for Spain-log relative GPD per km^2. These density plots may be interpreted as the continuous equivalent of a histogram in which the number of intervals has been set to infinity and then to the continuum. From the definition of the data, 0 on the horizontal axis indicates Spanish average GDP, 2 indicates twice this average, and so on.

This figure shows the evolution of the dependent variable over time from 1930 until 2005. It is an interesting graph in which the distributions are more or less bimodal with a second mode around two standard deviational units above the mean. The distributions in 1930 and 1950 are quite similar and non-normally distributed (the Jarque-Bera normality test rejects log-normality with 95% of confidence, as shown in Table 3). Both exhibit a main skewed mode just on the mean and a slight minor mode two standard deviational units above the mean. Nevertheless, the central mass of the distribution significantly decreased in 1970 to reach the

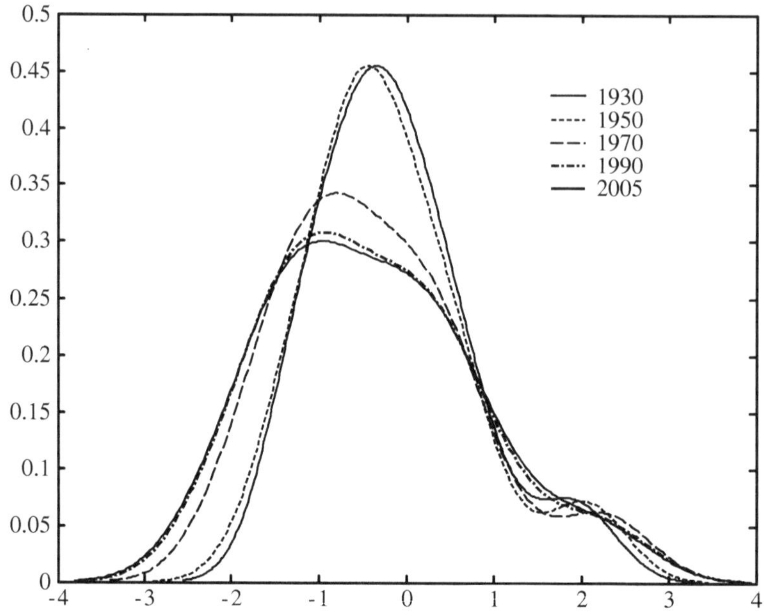

Fig. 2 Kernel density estimates of log relative GDP per area

Table 3 Normality and spatial autocorrelation tests of log relative GDP per area

Variable	1930	1950	1970	1990	2005
Jarque Bera normality test	5.95**	7.42**	5.08*	3.00	2.56
Moran's I spatial autocorrelation test	0.20**	0.17**	0.19**	0.18**	0.18**

**Significant at 5%, *significant at 10%. Inference for Moran's I test is based on the permutation approach (999 permutations)

lowest point in the 2005. Log-normality could be accepted, though only at 0.28 level. In the last decades, the main mode moves around one standard deviational unit below the mean whereas the second mode allocates throughout the second half of the distribution, particularly around two standard deviational units above the mean.

That is to say, compared with 1930 and 1950, more regions reported in 1970, 1990 and 2005, GDP either 50% of the Spanish average or almost twice the Spanish average. Moreover these modes situated below and above the Spanish average may reflect the existence of two groups of provinces with GDP density converging toward a lower and higher GDP density levels than the rest of provinces, respectively. The progressive deconcentration of probability mass from 100% can be interpreted as evidence for slight divergence. As stated before, in 1930 and 1950 Spain was mainly an underdeveloped rural country, only depicted by few economic poles located in the traditional thriving regions. GDP was more or less uniformly distributed across the country with these exceptions, which constitute a second mode around two standard deviations above the mean.

During the following decades, the strong economic development and profound social changes deepened this picture leading to a spillover process that principally benefited other contiguous regions. Economic prosperity caught up the whole country but not with the same intensity. As shown in Fig. 2, different modes in 2005 suggest dissimilar growth velocities inside a country which is more or less divided into two subspaces. On the one hand, coastal (and Madrid) thriving regions constitute a more homogeneous area in terms of economic development, though traditional enclaves (the Bask Country, Catalonia, Navarre and Madrid) still remain the leaders (second mode). On the other hand, the hinterland lagging regions are becoming a vast rural wasteland with the exception of some provinces (mainly the region capitals), which absorbs most of the GDP generated in this subspace (first mode).

This result is similar to others in the literature of Spanish regions and urban areas (see, e.g. Goerlich et al. 2002; Garrido 2002; Márquez and Hewings 2003; Pulido and López 2003; Dobado 2006; Mella and Chasco 2006). Nevertheless, it contrasts somehow with the results shown in Roos for the German regions in 2000, which show a skewed non-normal distribution with a prominent second mode about 1.5 deviational units above the mean.

As well, during the whole period we can also find some kind of general spatial trend – spatial autocorrelation – in GDP per area, as shown in Fig. 1: from the inland (low GDP density) to the coastal provinces (high GDP density), with the exception of Madrid. In the given period, the GDP per area distributions display a significant degree of spatial autocorrelation (Table 3): the magnitude of the Moran's I tests[5] are high and significant at $p < 0.05$, which is above its expected value under the null hypothesis of no spatial autocorrelation, $E[I] = -0.02$ (approximately in all the cases). Inference is based on the permutation approach (999 permutations), since not all the series distributes normally (Anselin 1995). Though we should be cautious because it is a large sample test, this result suggests that the evolution of production distribution appears to be somewhat clustered in nature. That is, provinces with very relatively high/low production density levels tend to be located near other provinces with high/low production density levels more often than would be expected as a result of purely random factors. If this is the case, then each province should not be viewed as an independent observation.

Figure 3 provides a more disaggregated view of the nature of spatial autocorrelation for production density by means of a Moran scatterplot (Anselin 1996), which plots the standardized log-relative production density of a province (LG) against its spatial lag (also standardized), W_LG. A province's spatial lag is a weighted average of the productions of its neighboring provinces, with the weights being obtained from a row-standardized spatial weight matrix (**W**). The four different quadrants of

[5] We have specified the spatial weights matrix, **W**, such that each element is set equal to 1 if province j has a common border with i, and 0 otherwise. Similar results have been observed with other specifications. These include an inverse distance matrix (such that each element w_{ij} is set equal to the inverse of the squared distance between provinces i and j), and a matrix obtained from a 200 km distance threshold to define a province's neighborhood set (as stated in Rey and Montouri 1999).

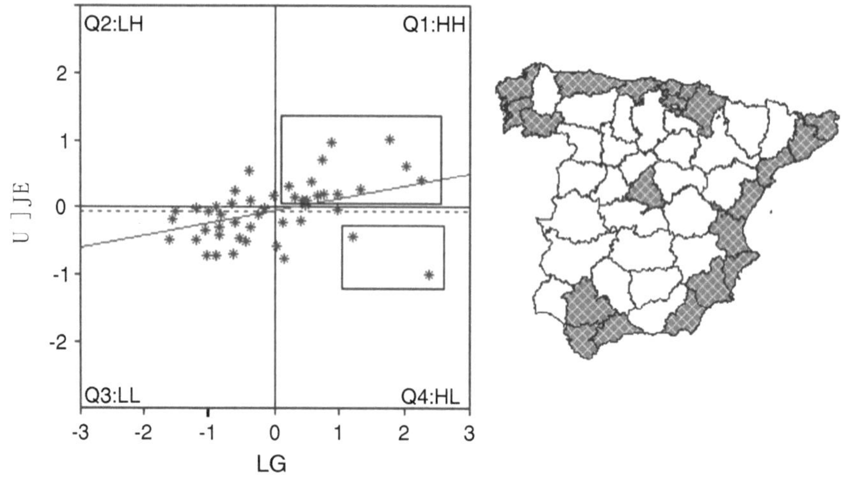

Fig. 3 Moran scatterplot of log relative GDP per area in 2005 (left). Map with the selection of provinces ever located in Quadrant 1, plus Madrid and Valencia

the scatterplot identify four types of local spatial association between a province and its neighbors: HH ("High-High"), LL ("Low-Low"), LH ("Low-High") and HL ("High-Low").

In Quadrant 1, the Moran scatterplot represents those high-GDP density provinces that are surrounded by high-GDP density neighbors, which have been highlighted in the map. It can be appreciated that they are all mainly located in the Coastal limits of the country. We have also selected Madrid and Valencia, located in Quadrant 4, in which we can find the group of high production density provinces surrounded by low production density neighbors. Quadrants 2 and 4 represent negative spatial dependence, while Quadrants 1 and 3 belong to positive forms of spatial dependence.

In the map we have selected all the provinces ever located in Quadrant 1 (high-high association) during the considered periods (1930, 1950, 1970, 1990 and 2005). We have also included Madrid and Valencia due to the major level of agglomeration effects detected around these regions (OECD 2000; Peeters and Chasco 2006). Therefore, the Moran scatterplot reveal the presence of spatial heterogeneity in the form of two clusters of production density in Spain: the coastal provinces, with the spatial discontinuity of Madrid (higher production density) and the hinterland (lower production density).

These results agree with the bimodal distributions shown in Fig. 2, which reflect a situation of two groups of provinces with GDP density levels converging toward a lower and higher GDP density levels than the rest of provinces, respectively. That is to say, spatial autocorrelation and spatial heterogeneity are two effects that must be tested when modeling GDP density since they could lead to biased coefficients if they are not adequately taken into account.

5 Influence of Geography on the Location of Production

In this section, we apply the ANOVA methodology proposed in Roos (2005) for German regions in 2000. In our case, we present a dynamic analysis for the last century testing for not only endogeneity but also spatial effects in the residuals. As stated before, it is a four-step analysis that proceeds as follows: (1) we filter gross second nature indicators from first nature interrelations; (2) we estimate how much of GDP per area variance can be explained by gross ($V_s + V_{fs}$) and net (V_s) second nature advantages; (3) we estimate how much of GDP per area variance can be explained jointly by gross first and second nature ($V_f + V_s + V_{fs}$); and (4) we calculate the difference between the result in step three and two, which is the importance of first nature alone (V_f).

5.1 Filtering Gross Second Nature from First Nature Elements

In order to disentangle empirically the second nature variables (population and GDP per worker) from first nature interactions, we proceed to regress them on geography and take the residuals as variables of net second nature forces (see (3)). Table 4 presents the results of the final regressions of the second nature variables on first nature, after elimination of insignificant variables.[6]

Measured by R^2, we can say that, in average during the twentieth century, gross first nature – itself and interactions with second nature – explains about 54% of the second nature's spatial variation, reaching to 60% in the last decade (though this measure could be overstated due to a certain degree of multicollinarity present in the productivity equations). The fit of both population and labor productivity equations are good (even when the capital dummy is excluded) and higher than those found in Roos' application for Germany (43% for population and 9% for productivity). This supports the idea that – contrary to Germany – Spain is a country with different climatic zones, which there are places more or less favorable to live in. For an international sample, Gallup et al. (1999) computed in 73% the contribution of geography to population density. However, as shown before, this estimation might be exaggerated due to the high degree of multicollinearity present in their model.

In general, the capital dummy has the largest influence on both second nature variables. Particularly in the population equations, it has an increasing impact[7] that

[6] We follow a general-to-specific modeling strategy. In a first regression, we include the complete set of first nature variables. In a step-by-step sequenced process, we exclude the variable with the lowest t-statistic and estimate the remaining equation again. This procedure is repeated until all coefficients are significantly different from zero at the 10% level.

[7] In semi-logarithmic equations, the dependent variable changes by $[\exp(b) - 1] \cdot 100\%$ if the explanatory variable changes from zero to one unit, where b is the explanatory variable coefficient.

Table 4 Second nature on first nature OLS regression results

Depend. Variable	Log(pop)					Log(prod)				
	1930	1950	1970	1990	2005	1930	1950	1970	1990	2005
Constant	13.3***	13.4***	13.4***	13.4***	13.6***	2.77***	3.98***	6.20***	8.47***	9.55***
Capit	0.41**	0.47**	0.63***	0.77***	0.77**	0.15**	0.12*	0.14**	0.09**	
Altit								−0.0003*		−2e–4**
Temmin	0.04*	0.04*	0.08***	0.07**	0.08***			0.04**	0.02**	
Temaver						−0.10***	−0.07***	−0.06***	−0.03*	−0.04***
Temmax										
Rainfall	0.001**	0.001**	0.001**	0.001*						
Tembel0						−0.008***	−0.007***			−0.003***
Overcast	−0.01***	−0.01***	−0.01**	−0.10**	−0.01**					
Maritlim						−3e–4*		−4e–4*		
Coast				0.33*	0.50**					−0.08**
Xcoo						5.2e–7***	4.9e–7***	6.1e–7***	4.9e–7***	3.5e–7***
Ycoo										
R2	0.48	0.49	0.56	0.63	0.65	0.51	0.39	0.57	0.47	0.60
R2_wc	0.37	0.37	0.42	0.47	0.49	0.25	0.35	0.50	0.41	0.60
Multic. #	10.3	10.3	10.3	10.8	6.7	34.2	31.9	44.5	34.9	34.8
Net 2nd	pi_{30}	pi_{50}	pi_{70}	pi_{90}	pi_{05}	del_{30}	del_{50}	del_{70}	del_{90}	del_{05}

***Significant at 0.01, **significant at 0.05, *significant at 0.1, *Log(pop)* log population, *Log(prod)* log labor productivity, *Multic.* # multicollinarity number, R2_wc R^2 of the regressions without *capit*, *pi, del* residuals of (3)

ranges from 51% in 1930 to 116% in 2005. Nevertheless, during the last two decades this variable looses its power on labor productivity. It is as if capital cities are – in general – more capable of attracting people at cost to productivity.

We should also highlight the recent influence of the coast dummy. In 2005, changes from zero to one (non-coastal to coastal) cause a population increase of 65% but a labor productivity decrease of 8%. This apparently contradictory result – population increase joint to productivity reduction in coastal regions – could be explained by the existence in most Mediterranean provinces of a predominant less-productive "sun and beach" tourism activity and certain hand-worker intensive industries.

Finally, since we find significant relations between second nature and geography, we can conclude that both forces interact. Therefore, we have filtered the residuals of these ten regressions, *pi*, *del*, which will be considered as net second nature forces.

5.2 Second Nature Effects on GDP per Area

In this step, we compute second nature effects on GDP per area with the estimation of two equations. Firstly, we regress the log-relative GDP per area on population and labor productivity. The resulting determination coefficient will indicate the second nature gross effect $R^2_{gs} = (V_s + V_{fs})/V$. Secondly, the second nature net effect on GDP per area is obtained from the estimation of this variable on the residuals, *pi*, *del*, derived from the last estimations, with the help of the corresponding determination coefficient $R^2_{ns} = V_s/V$.

As stated in Roos (2005), one problem is that the second nature variables are endogenous and simultaneously determined with GDP. This might lead to the well-known simultaneity bias in the regressions violating the necessary conditions to obtain estimates with good properties. The instrumental variables estimation is the standard approach to overcome the consequences of simultaneity, i.e. bias, inefficiency and inconsistency on OLS-estimators.

The principle of the IV estimation is based on the existence of a set of instruments that are strongly correlated to the original endogenous variables but asymptotically uncorrelated to the error term.

Once these instruments are identified, they are used to construct a proxy for the explanatory endogenous variables, which consists of their predicted values in a regression on both the instruments and the exogenous variables. However, it is very difficult to find such instruments because most socioeconomic variables will be endogenous as well. In the standard simultaneous equations framework, the instruments are the "excluded" exogenous variables.

In our case, in order to decide whether we need IV estimation, we have first analyzed the potential system feedbacks between the dependent variable, log-relative GDP per area, and the four second nature explanatory variables, i.e. population, labor productivity and the OLS residuals (*pi, del*) found in Table 4 estimations. For this purpose, we have used the *Durbin-Wu-Haussman* (DWH) test, which is an "exogeneity test" (Anselin 1999) that compares the IV and OLS estimates assuming the former are consistent. Although consistent, in small samples the IV estimates may be inferior to OLS in terms of mean squared error. This test reports the confidence level at which consistency of OLS estimates can be rejected. In fact, it is an F test with $(k^*,\ n-k-k^*)$ degrees of freedom on the null hypothesis of exogeneity of a k^* subset of the total k explanatory variables, with n as the number of observations (for technical issues, see Davidson and Mckinnon 1993).[8] Since we need to estimate IV equations to perform this test, we must first decide the set of adequate instruments for each potential stochastic regressor. As stated above, they should be correlated to the original endogenous variables but asymptotically uncorrelated to the error term.

Roos proposes to use mainly time-lagged variables as instruments, since they are highly correlated with the actual variables but also non-contemporary correlated with the errors.[9] Besides, we have also considered other space and/or time lagged second nature variables as well as "excluded" first nature explanatory variables. In all cases, we have selected only those instruments more correlated with the corresponding endogenous regressor and less correlated with OLS error terms.[10] In Table 5, we have shown the instruments definitely used in each equation, as well as the results of the *Durbin-Wu-Haussman* (DWH) test.

Results show a high degree of simultaneity in some of the second nature regressors with respect to log-relative GDP per area. This is the case of log-population, for 1970 and 1990 equations, and log-labor productivity, for 1950 and 1990 equations. Regarding net second nature variables, population series (*pi*) are mainly exogenous, though productivity variables (*del*) exhibit clear endogeneity except for 2005. As a consequence, both (4) and (6) must be estimated by IV for all the periods, with the exception of 2005, which is the only case of total absence of endogeneity in the regressors.

In Table 6, we show the estimation results of (4) and (6), in which log-relative GDP per area is regressed on gross and net second nature variables, respectively. Being aware of the potential drawback coming from the asymptotic considerations of all statistical inference for IV estimates (which may not be very reliable for small data sets), we have computed the so-called asymptotic t-tests as a ratio of the estimate to its asymptotic standard error.

---

[8] As shown in Anselin (1999), DWH test is consistent with spatially autocorrelated OLS residuals.

[9] Non-contemporary dependence between regressors and the error terms lead to asymptotically unbiased estimators only in absence of temporal autocorrelation. However, in our case it is difficult to suppose time independence between the error terms what could somewhat affect our results.

[10] The goodness of the instruments can be proved with the help of the Sargan test, which contrasts the null hypothesis that a group of s instruments of q regressors is valid. This is a Chi-2 test with $(s–q)$ degrees of freedom that rejects the null when at least one of the instruments is correlated with the error term (Sargan 1964). In our case, we can clearly accept the null with a confidence level of 0.99. All the computations can be obtained upon request from the authors.

Table 5 Instruments and endogeneity tests in second nature effect regressions

Gross second		Instruments	DWH	Net second		Instruments	DWH
Log (pop)	1930	pi30, tembel0	3.5*	pi	1930	pi50	0.0
	1950	lpo30	0.2		1950	pi30	0.1
	1970	lpo50	61***		1970	pi50	13***
	1990	lpo70	5.0**		1990	pi70	0.0
	2005	lpo90	0.7		2005	pi90	0.8
Log (prod)	1930	del30, lpr70	0.8		1930	del50, lpr30	12***
	1950	lpr30, del50	3.2*		1950	del30, lpr50	16***
	1970	lpr50, del70	2.1	del	1970	del50, lpr70, lpr50	3.3**
	1990	lpr70	9.1***		1990	del70, lpr90	5.8**
	2005	lpr90, del05, xcoo	1.2		2005	del90, lpr05, lpr90	0.6

Log(pop) log population, *Log(prod)* log labor productivity, *pi* residual of the regression of log population on first nature variables, *del*: residual of the regression of log labor productivity on first nature variables, *tembel0* # days with temperatures below zero Celsius, *xcoo* X-coordinate, *DWH* Durbin-Wu-Haussman exogeneity test, ***significant at 0.01, **significant at 0.05, *significant at 0.1

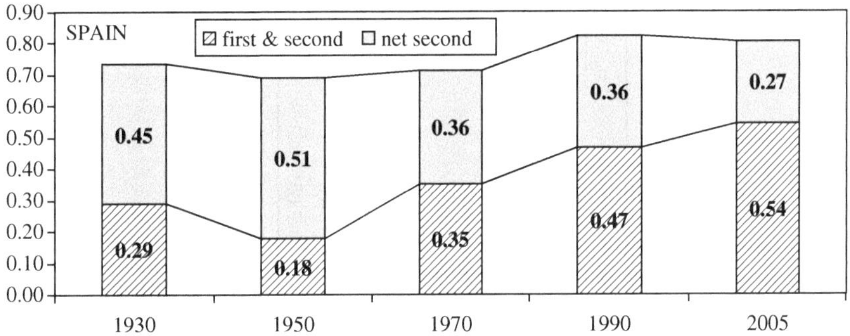

Fig. 4 Evolution of the impact of second nature forces on GDP density

As stated in Anselin (1988, p. 244), in the IV estimation approach the residuals have a zero mean, so than the standard variance decomposition can be obtained and a determination coefficient can be computed in the usual manner (the ratio of the variance of the predicted values over the variance of the observed values for the dependent variable). Consequently, the five regressions on population and productivity provide a determination coefficient R^2_{gs} between 0.69 (year 1950) and 0.82 (year 1990), which is the share of GDP density variance that is explained by gross second nature effects. The estimation of the other five equations yield $0.27 \leq R^2_{ns} \leq 0.51$, which is the importance of net second nature on GDP density. Regarding the mixed effect of the interaction between first and second nature on GDP density (R^2_{fs}), it can be extracted as the difference between R^2_{gs} and R^2_{ns} (Equation 8). Figure 4 summarizes the results for the estimations of Table 6.

To some extent, second nature has increased its importance on GDP density in Spain during the last century, accounting for 0.74 in 1930 to 0.81 in 2005. Roos

Table 6 OLS regression results of GDP per area on second nature variables

Period	Gross second nature					Net second nature				
	1930	1950	1970	1990	2005	1930	1950	1970	1990	2005
Estimation	IV	IV	IV	IV	OLS	IV	IV	IV	IV	OLS
Constant	−14.6**	−16.1**	−23.1**	−36.3**	−45.2**	−0.14	−0.16	−0.33*	−0.39*	−0.39**
Log(pop)	0.92**	0.72**	0.77**	0.91**	0.99**					
Log(prod)	1.97**	2.16**	2.41**	2.91**	3.56**					
Pi						0.58	0.51	0.83**	0.89**	0.98**
Del						2.85**	2.96**	3.11*	3.54*	2.88
R-squared	0.74	0.69	0.71	0.82	0.81	0.45	0.51	0.36	0.36	0.27
Sp. Chow	34.4**	40.2**	35.3**	27.4**	10.4**	75.9**	80.5**	106.1**	107.2**	105.9**
LM (sp.er.)	7.49**	14.4**	22.7**	20.7**	7.03**	5.19*	8.25*	7.67	7.58**	5.99*

Log(pop) log population, *Log(prod)* log labor productivity, *pi* residual of the regression of log population on first nature variables, *del* residual of the regression of log labor productivity on first nature variables, *Sp. Chow* spatial Chow test, *LM (sp.er.)* Lagrange Multiplier test for spatial error autocorrelation, ** significant at 0.01, * significant at 0.05

found that only 65% of German GDP density in 2000 was caused by gross second nature. He decomposed it into a mixed-indirect effect (29%) and a net-direct effect (36%). In Spain, net second nature forces reach the maximum effect in 1950 (0.51) and progressively decline to 0.27 in 2005. Pertaining to the interaction effect of physical geography and agglomeration economies, it registers a growing trend from 0.29 (1930) to 0.54 (2005), almost doubling – at this moment – Roos' results for Germany. This result shows the more and more importance of the interaction between economic agents and nature as determinants of GDP density. This is clear in certain economic activities related with tourism, which has been the main engine of Spanish economy since the 1960s.

The final line of diagnostics in Table 6 reports an asymptotic LM test for spatial error autocorrelation[11] (Anselin 1999). In addition, we have also tested for spatial heterogeneity in the errors, in the form of two subspaces, as detected before for GDP density distributions (Fig. 3), i.e. higher/lower GDP density provinces (coast/hinterland, respectively). For this purpose, we use the spatial Chow test proposed by Anselin (1990), in which the null hypothesis states that the coefficients are the same in all regimes. It is based on an asymptotic Wald statistic, distributed as a χ^2 distribution with $[(m-1)\cdot k]$ degrees of freedom (m being the number of regimes). In Table 6, the null hypothesis on the joint equality of coefficients is clearly rejected by the Chow-Wald test in all the regressions, i.e. their values are sufficiently extreme for a distribution with three degrees of freedom. Therefore, both spatial effects are present in the regressions on second nature variables demonstrating the existence of non-randomness in the error terms. It is known that sometimes, spatial autocorrelation in the residuals may be induced by a strong spatial heterogeneity that is not correctly modeled by spatial dependence specifications (Brunsdon et al. 1999).

Consequently, in order to capture the polarization pattern previously observed in the distribution of GDP density among the Spanish provinces, we allow cross-region parameter variation in a *spatial regimes model* with two subspaces corresponding to coastal provinces (plus Madrid) and the rest of inland provinces.[12] There are 21 provinces included in the higher GDP density group (coast) and 26 provinces included in the lower GDP density group (hinterland).

As shown in Table 7, spatial instability has important effects on the determination coefficients. In general terms, they are higher in the coastal subspace than in the hinterland, mainly for net second nature. In Fig. 5, we have graphed the dynamics experienced by both groups. Differences in GDP density inside the leading group are much due to net agglomeration economies, whereas differences in lower GDP density group depend more on mixed effects (interaction between geography and man-made agglomerations).

---

[11] This test has been constructed in the same fashion as in Burridge (1980). The spatial weight matrix is specified as in foot note 7.

[12] We have also estimated a groupwise heteroskedastic error model. In general, both GLS and LM estimations produce signifficant variance coefficients in each subspace, but cannot absorb all the heteroskedasticity and spatial dependence still present in the residuals.

Table 7 OLS regression results of GDP/area on second nature in two spatial regimes

Period		Gross second nature					Net second nature				
		1930	1950	1970	1990	2005	1930	1950	1970	1990	2005
Estimation		IV	IV	IV	IV	OLS	IV	IV	IV	IV	OLS
Const	Inland	−10**	−13**	−14**	−22**	−20*	−0.7**	−0.7**	−1.1**	−1.2**	−1.2**
	Coast	−11**	−11**	−19**	−31**	−40**	0.5**	0.5**	0.5**	0.5**	−0.5**
Log (pop)	Inland	0.60**	0.60**	0.50**	0.62**	0.69**					
	Coast	0.68**	0.45**	0.57**	0.63**	0.71**					
Log (prod)	Inland	1.40**	1.72**	1.17*	1.59*	1.16					
	Coast	1.66**	1.93**	2.21**	2.80**	3.47**					
	Inland						0.13	0.28	0.27	0.37	0.33
pi	Coast						0.86**	0.77**	0.84**	0.89**	0.95**
Del	Inland						0.86	1.63*	0.95	1.42**	0.23
	Coast						1.81**	1.94**	2.71*	2.52*	2.07
R2	Inland	0.69	0.70	0.44	0.59	0.56	0.30	0.46	0.09	0.15	0.06
	Coast	0.77	0.71	0.73	0.86	0.76	0.71	0.71	0.72	0.74	0.64
LM (sp. er.)		0.69	1.43	6.5*	7.6**	0.14	0.95	0.48	0.28	0.21	0.41

log(pop) log population, *log(prod)* log labor productivity, *pi* residual of the regression of log population on first nature variables, *del* residual of the regression of log labor productivity on first nature variables, *LM (sp er.)* Lagrange Multiplier test for spatial error autocorrelation (for 2005, it is the LM-EL test), ** significant at 0.01, * significant at 0.05.

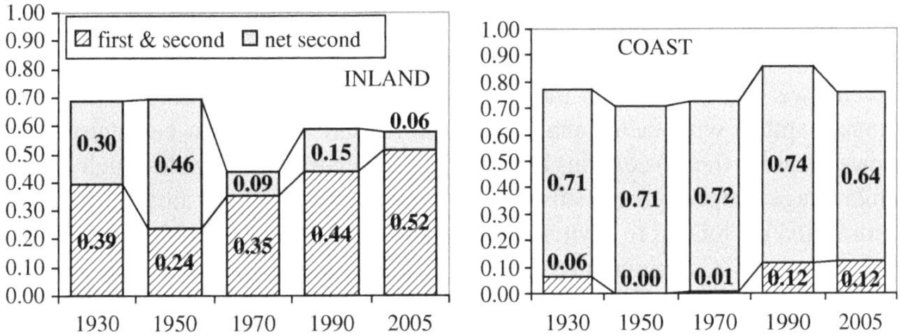

Fig. 5 Evolution of the impact of second nature on GDP density in two regimes

On its side, spatial autocorrelation in the residuals disappear in all the equations (the LM tests are not significant) with the exception of gross second nature in 1970 and 1990. As a result, in most cases, the spatial regimes model controls for the presence of both spatial effects in second nature equations. This result confirms our initial hypothesis about the importance of taking into account spatial instability in GDP density distributions.

The influence of space on GDP density is certainly conspicuous. It leads to the so-called "two Spains," which are no longer split along the usual North *versus* South partition. In this case, we find a relevant geographical division: on the one hand, the coastal provinces plus Madrid, in which population and production focuses and on the other hand, an even more depopulated and sparse hinterland.

5.3 *First and Second Nature Joint Effect on GDP per Area*

We estimate how much of GDP per area variance can be explained jointly by gross first and second nature ($V_f + V_s + V_{fs}$). As in (9), we include a set of first nature indicators together with the net second nature variables (*pi, del*) as regressors. The joint importance of first and second nature is then measured by $R^2{}_{f+s} = (V_f + V_{fs} + V_s)/V$.

Thus from the set of the country's – physical and political – geography variables (Table 1) we must choose only those that are both related to the distribution of GDP density and not correlated with net second nature forces. As in Table 3, we pursue a general-to-specific modeling strategy in a first regression of GDP density on the complete set of 13 geography variables and the 2 net second nature variables. This procedure is repeated until all coefficients are all significantly different from zero at the 10% level. We find that only eight geographic variables fulfill the cited requirements in all periods: regional capital, altitude, minimum temperature, average temperature, # days with below 0 °C temperature, # days above 25°C temperature, total rainfall and X-coordinate.

The regressions of GDP density on the complete set of ten variables lead to high multicollinearity which inflates the determination coefficients. To avoid this problem, we opted for group the seven physical geography variables (excluding regional capital) with factor analysis.[13] The rotated factors can be interpreted as follows: factor 1 (*temp*) contains high scores of temperature variables, such as minimum/average temperature (positive), # days with below 0 °C and altitude (negative). Factor 2 (*dry*) is related to dryness, with high scores in total rainfall (negative) and # days above 25°C temperature (positive). Regarding factor 3 (*east*), it is mainly based on East-West orientation (X-coordinate). The regressions of GDP density on the two net second nature variables, three geography factors and the regional capital show much lower multicollinearity number, between 1.94 (1950) and 2.04 (1930), well below the acceptable limit of 20/30 (Anselin 1995).

Again, we should test for the presence of endogeneity in the second nature variables since they could be simultaneously determined by GDP density. In this case, using the instruments shown in Table 5, we find that all second nature variables obtain significant DWH values except in the period 2005. Thus, we apply IV method with the exception of in 2005, in which OLS is used (Table 8). As we can see, the joint contribution of first and second nature to GDP density remains constant (88–89%) across the twentieth century. That is to say, almost a 90% of the agglomeration pattern has been constantly explained by natural geography and agglomeration economies together, remaining the other 10% unexplained by these factors.

Once more, though there is no remaining spatial autocorrelation in the error terms, the spatial Chow test points out the problem of spatial instability in the coefficients. The estimation of the spatial regimes models illustrates the differences between the two subspaces. Hence, the joint contribution of total geography is significantly lower in the inland provinces, much similar to Roos' figures for Germany (72%).

All coefficients have the expected signs. Results show the great importance of net second nature variables (population and productivity) on GDP density, which are significant for all the periods and spatial regimes. Among physical geography, temperature has the largest influence; e.g. in 1930, it increased the relative GDP density 68% reaching to 112% in 2005.

Regional capital is also a very influential variable and it obtains its main scores after 1990, from which Spanish regions ("autonomies") where officially recognized (34% in 1930, 101% in 2005). Similar to the German case, Spain is now a decentralized state with 17 regions that have a lot of legislative and executive power concentrated in the regional capital.

The results explain the growing influence of this variable on economic activity. Geographical orientation has also registered a rising tendency during the last century; i.e. Eastern locations are prone to record more GDP density than Western ones.

Regarding the spatial regimes, we find some interesting variations. In the group of inland provinces, regional capital is – by far – the most important determinant

[13] Factors have been extracted using principal components and rotated with Varimax method.

Table 8 First and second nature joint effect on GDP density

Period		1930		1950		1970		1990		2005	
Estimation		IV		IV		IV		IV		OLS	
Model		Basic	Spat. reg.	Basic	Spat. reg.	Basic	Spat. reg.	Basic	Spat. reg.	Basic	Spat. reg.
Const.	Sp.	−0.23**		−0.35**		−0.50**		−0.62**		−0.62**	
	In.		−0.61**		−0.56**		−0.88**		−0.84**		−0.92**
	Co.		0.25		0.28		−0.01		0.11		−0.01
Capital	Sp.	0.29*		0.54**		0.53**		0.71**		0.70**	
	In.		0.47*		0.59**		0.68**		0.94**		0.90**
	Co.		−0.05		0.10		0.24		0.21		0.26
Factor 1	Sp.	0.52**		0.49**		0.61**		0.73**		0.75**	
Temp	In.		0.24**		0.37***		0.33**		0.63**		0.55**
	Co.		0.25		0.08		0.38*		0.30*		0.41**
Factor 2	Sp.	−0.31**		−0.35**		−0.37**		−0.23**		−0.17*	
Dry	In.		−0.07		−0.23*		−0.35*		−0.34*		−0.21
	Co.		−0.34**		−0.31**		−0.26**		−0.18**		−0.12
Factor 3	Sp.	0.16*		0.19**		0.23**		0.21**		0.23**	
East	In.		0.02		0.12		0.03		0.14		0.11
	Co.		0.15*		0.16**		0.22**		0.18**		0.23**
Pi	Sp.	0.70**		0.57**		0.75**		0.84**		0.96**	
	In.		0.44		0.41*		0.54*		0.61**		0.59**
	Co.		0.36**		0.81**		0.77**		1.00**		1.08**

(continued)

Table 8 (continued)

Period		1930		1950		1970		1990		2005	
Estimation		IV		IV		IV		IV		OLS	
Model		Basic	Spat. reg.	Basic	Spat. reg.	Basic	Spat. reg.	Basic	Spat. reg.	Basic	Spat. reg.
Del	Sp.	1.87**		1.99**		2.57**		2.52**		2.71**	
	In.		1.43**		1.45**		1.57*		2.07**		2.09
	Co.		2.03**		1.82**		2.74**		1.93*		1.72
R-squared	Sp.	0.88		0.89		0.88		0.89		0.88	
	In.		0.74		0.75		0.63		0.72		0.66
	Co.		0.93		0.91		0.92		0.89		0.86
Multicol. #		2.04	4.84	1.94	4.49	1.99	4.68	1.94	4.56	1.94	4.53
Sp.Chowt			23.9**		22.3**		22.8**		28.1**		22.5**
LM (sp. er.)		1.75	0.54	1.97	0.00	1.82	0.07	1.00	0.25	0.46	0.30

sp. Spain, *in*. inland, *co*. coast, *temp* temperature, *dry* dryness, *east* West-East orientation, *pi del* residuals of the regressions of log population and log labor productivity on first nature variables, respectively, *multicol*. # multicollinearity number, *LM (sp.er.)* Lagrange Multiplier test for spatial error autocorrelation (for 2005, it is the LM-EL test), **significant at 0.05, *significant at 0.01

particularly from 1990, increasing GDP density by about 150%. It is followed by temperatures, since natural conditions differ considerably across the inland provinces, while Eastern orientation is not significant at all.

This outcome makes clear the situation of the progressively depopulated interior of the country. That is to say, location of production in the hinterland depends mainly on natural and political conditions. In these provinces agglomeration takes place mainly close to capitals and big cities, where the executive power and services concentrate producing employment and welfare. Concerning the coastal (plus Madrid) subspace, temperatures and dryness are the variables that exert the maximum influence on GDP density. In this area, longitude has gained more weight on GDP density illustrating the present advantage of the long Mediterranean urban areas with respect to the declining Cantabric-Atlantic axis (Le Gallo and Chasco 2008).

5.4 *First Nature Net Effect on GDP per Area*

If we calculate the difference between the determination coefficient in Tables 8 and 6 (Table 7 for spatial regimes) we obtain the importance of first nature alone (V_f) for the whole Spain: $V_f/V = R^2_{f+s} - R^2_{gs}$. In Fig. 6 we show the complete ANOVA

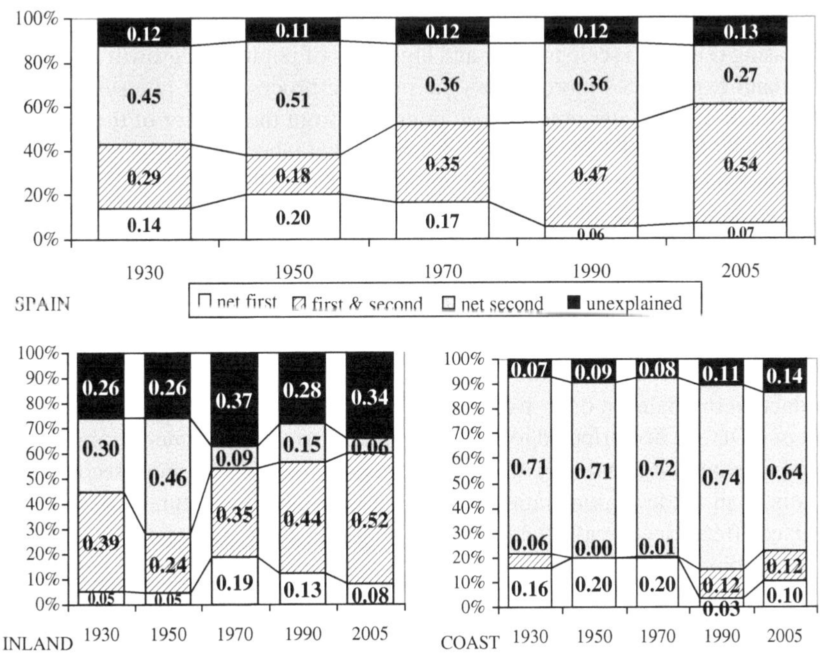

Fig. 6 Evolution of the variance decomposition of regressions in Table 8

decomposition for both the whole country and each of the spatial regimes. The total variation that can be assigned to the net effect of first nature ranks from 14% (in 1930) to 6–7% in 1990 and 2005, respectively. This result is almost coincident with Roos' who found a 7.1% for Germany. Nevertheless, it changes a bit when considering the two spatial regimes. Net first nature has had – in general – more influence on GDP density in the coastal provinces than in the inland, though they are leveling in both regimes at present (about 8–10%).

Independently of natural conditions, man-made agglomeration economies play an important role in the distribution of economic activity across Spanish territory. Nevertheless, this role is much significant in the coast than in the hinterland. In effect, since the coastal provinces share similar natural conditions, differences in GDP density are much due to interactions between economic agents among themselves than between agents and nature. In this subspace, first and second nature exerts basically a net influence. In contrast, the hinterland shows wider disparities in terms of physical geography – abrupt topography and continental weather – what confer more weight to mixed first–second nature; i.e. second nature forces are likely to overlay and to strengthen the forces of first nature. As a general rule, gross (net and mixed) first nature has increased its influence in Spain with time and it constitutes a 60% of GDP density distribution at present. However it is truer in the hinterland than in the group of higher GDP density provinces, in which first nature global effect has maintained practically stable during the last century in only a 22% of GDP per area.

Therefore, similar as in Gallup et al. (1999) Spanish economy is likely to bifurcate on two pathways. The coast plus Madrid metropolitan area experiences decreasing returns to scale in labor and high rates of population growth whereas the hinterland experiences more or less the opposite process. The two systems interact through ever-greater pressures on migration from the interior of the country to Madrid and the coast. This result demonstrates that when analyzing agglomeration in Spain, this dichotomous reality should not be avoided.

6 Conclusions

In this chapter, we examine the influence of geographic features on the location of production in Spain. In other words, we quantify how much of the geographic pattern of GDP can be attributed to only exogenous first nature elements (physical and political geography) and how much can be derived from endogenous second nature factors (man-made agglomeration economies), in which first nature also operates as a mixed effect. Specifically we disentangle the contribution of each net component of geography with the aim of isolating the importance of first nature alone. If first nature is relevant, existing agglomerations will be very stable and attempts to create new agglomerations at places without geographic advantages, might possibly fail. On the contrary, if first nature does not matter much, regional policies could have more success in creating agglomerations anywhere.

For this purpose, we follow we follow a methodological approach based on Roos (2005) for Germany. He proposes to employ an analysis of variance (ANOVA) to infer the unobservable importance of first nature indirectly in a stepwise procedure. We also estimate how much of agglomeration can be explained by geography elements in a dynamic context, analyzing their evolution during the twentieth century. We demonstrate that results could be biased if some potential econometric questions are not properly taken into account; e.g. multicollinearity, relevant missing variables, endogeneity, spatial autocorrelation and spatial heterogeneity. Even so, this methodology has some important limitations. On the one hand, the definition of the endogenous variable as aggregated "GDP per area" implies more a concept of agglomeration of value added rather than the spatial concentration of workers/citizens. In addition, aggregate GDP does not allow analyzing the dissimilar effects of geography on different industries. On the other hand, the cited potential econometric problems are not always easy to solve, mainly multicollinearity and endogeneity, what could somewhat bias the results.

The main outcome of our study reveals that production is not randomly distributed across Spanish regions. In an exploratory spatial data analysis we find that GDP density has been historically bifurcated on two pathways, core and periphery, i.e. the coast plus Madrid metropolitan area and the hinterland, respectively. Even more, during the twentieth century this polarization has deepened leading to a new configuration of the so-called "two Spains." Therefore, we have estimated our models testing for and considering explicitly these spatial regimes.

Thus considering the Spanish territory as a whole we find that at most, 88% of GDP's spatial variation can be explained by direct and indirect effects of geography during the entire period (1930–2005). These figures remain significantly far from those found in Roos for Germany (72%), pointing out the major role played by geography in Spain. After controlling for agglomeration economies and the interaction effect of first-second nature, the net influence of natural geography ranks from 20% (in 1950) to 6–7% nowadays. On the other side, whereas in 1930, we find a prevalence of net second nature (e.g. transport and communication, according to Venables 1999), in the end of the period, second nature agglomeration forces were amplified by first nature geography. Therefore, there is a close connection between second and first nature. Krugman (1993) argues that first nature advantages generally tend to create second nature advantages through cumulative processes. Nevertheless, at least in the period of analysis and from a global point of view, net second nature seems to have been the initial advantages that were amplified by first nature forces. In effect, since the 1960s, the implantation of the model of "sun and beach" tourism as the main engine for the Spanish economy has benefited those existing agglomerations with better natural conditions.

However, the influence of geography varies significantly from one spatial regime to the other. For example, in the group of inland provinces only a 70% of agglomeration (in average, during the whole period) is explained jointly by first and second nature forces, being the mixed effect quite strong, though with some variations in time (from 24% in 1950 to 52% in 2005). On its side, among the group of core provinces (coast plus Madrid), which share quite common physical geography characteristics, net second nature give the highest contribution to agglomeration, around 70% along the whole period and first plus second effects record a 90% of total agglomeration. In addition, the start point (year 1930) is very different from one group to another and the long-term implications are conditioned by this initial situation. While the coast group maintains the structure of the variance decomposition along time, the inland group has lost an important part of net second effects. That is to say, "the evolution of the influence of geography on the location of production in Spain (1930–2005)" is quite different for these two groups. On one side, the core-coast group has not relevant changes during the period. On the other side, the progressive irrelevance of net second forces in the inland group is the cause of the increasing effects of first-second forces. From a political point of view and according to Venables (1999), we can conclude that inland provinces would need more investment in transport and communication infrastructures. In particular, they are necessary to connect the periphery-inland territories to the core-coastal ones.

In conclusion, independently of the interest of these findings for the Spanish regional analysis, we recommend taking into account spatial autocorrelation and heterogeneity explicitly in Roos' methodology, since the core-periphery pattern is strongly present in most regions of the world. If they are not properly taken into account, results could be biased and rich information would be ignored.

Acknowledgements Coro Chasco acknowledges financial support from the Spanish Ministry of Education and Science SEJ2006–02328/ECON and SEJ2006–14277-C04–01. The comments received by three anonymous referees are also gratefully acknowledged.

References

Ades AK, Glaeser, EL (1995) Trade and circuses: explaining urban giants. Q J Econ 110:195–227

Alcaide J (2003) Evolución económica de las regiones y provincias españolas en el siglo XX. Fundación BBVA, Madrid

Alcalde J, Alcalde P (2007) Balance económico regional (autonomías y provincias) años 2000 a 2005. FUNCAS, Madrid

Alonso-Villar O, Chamorro-Rivas JM, González-Cerdeira X (2004) Agglomeration economies in manufacturing industries: the case of Spain. Appl Econ 36:2103–2116

Anselin L (1988) Spatial econometrics: methods and models. Kluwer, Dordrecht

Anselin L (1990) Spatial dependence and spatial structural instability in applied regression analysis. J Reg Sci 30:185–207

Anselin L (1995) Space Stat Version 1.80: users' guide. Regional Research Institute, West Virginia University, Morgantown, WV

Anselin L (1996) The Moran scatterplot as an ESDA tool to assess local instability in spatial association. In: Fischer M, Scholten H, Unwin D (eds) Spatial analytical perspectives on GIS. Taylor and Francis, London, UK, pp 111–126

Anselin L (1999) Spatial data analysis with SpaceStat™ and ArcView®. Workbook, 3rd edn. Department of Agricultural and Consumer Economics. University of Illinois at Urbana-Champaign., Urbana, IL

Ayuda MI, Collantes F, Pinilla V (2005) From locational fundamentals to increasing returns: the spatial concentration of population in Spain, 1787–2000. Documento de Trabajo 2005–05, Facultad de Ciencias Económicas y Empresariales, Universidad de Zaragoza

Brunsdon C, Fotheringham AS, Charlton M (1999) Some notes on parametric significance tests for geographically weighted regression. J Reg Sci 39:497–524
Burridge P (1980) On the Cliff-Ord test for spatial autocorrelation. J R Stat Soc B 42:107–108
Ciccone A, Hall RE (1996) Productivity and the density of economic activity. Am Econ Rev 86: 54–70
Cragg M, Kahn M (1997) New estimates of climate demand: evidence from location choice. J Urban Econ 42:261–284
Davidson R, Mckinnon JG (1993) Estimation and inference in econometrics. Oxford University Press, New York
Delgado M, Sánchez J (1998) Las desigualdades territoriales en el Estado Español: 1955–1995. Revista de Estudios Regionales 51:61–89
Dobado R (2004) Un legado peculiar: la geografía. In: Llopis E (ed) El legado económico del Antiguo Régimen en España. Editorial Crítica, Barcelona, pp 43–94
Dobado R (2006) Geografía y desigualdad económica y demográfica de las provincias españolas (siglos XIX y XX). Investigaciones de Historia Económica 5:133–170
Ellison G, Glaeser EL (1997) Geographic concentration in U.S. manufacturing industries: a dartboard approach. J Polit Econ 105:889–927
Ellison G, Glaeser EL (1999) The geographic concentration of industry: does natural advantage explain agglomeration? Am Econ Rev Pap Proc 89:311–316
Escobal J, Torero M (2005) Adverse geography and differences in welfare in Perú. In: Kanbur R, Venables AJ (eds) Spatial inequality and development. Oxford University Press, Oxford, pp 77–122
Freeman DG (2001) Sources of fluctuations in regional growth. Ann Reg Sci 35:249–266
Fujita M, Krugman P, Venables AJ (1999) The spatial economy. cities, regions, and international trade. MIT, Cambridge
Funck RH (1995) Competition among locations: objectives, instruments, strategies, perspectives. In: Giersch H (ed) Urban agglomeration and economic growth. Springer, Heidelberg, pp 227–255
Gallup JL, Sachs JD, Mellinger AD (1999) Geography and economic development. Int Reg Sci Rev 22:179–232
Garrido R (2002) Cambio estructural y desarrollo regional en España. Pirámide, Madrid
Goerlich F, Mas M, Pérez F (2002) Concentración, convergencia y desigualdad regional en España. Papeles de Economía Española 93:17–36
Graves PE (1979) A life-cycle empirical analysis of migration and climate, by race. J Urban Econ 6:135–147
Henderson JV (1988) Urban development – theory; fact, and illusion. Oxford University Press, Oxford
Henderson JV (1999) Overcoming the adverse effects of geography: infrastructure, health and agricultural policies. Int Reg Sci Rev 22:233–237
Jenks GF, Caspall FC (1971) Error on choroplethic maps: definition, measurement, reduction. Ann Assoc Am Geogr 61:217–244
Kanbur R, Venables AJ (2007) Spatial disparities and economic development. In: Held D, Kaya A (eds) Global inequality. Polity, London (UK), pp 204–215
Kim S (1999) Regions, resources, and economic geography: sources of US regional comparative advantage, 1880–1987. Reg Sci Urban Econ 29:1–32
Knapp TA, White NE, Clark DE (2001) A nested logit approach to household mobility. J Reg Sci 41:1–22
Krugman P (1993) First nature, second nature, and metropolitan location. J Reg Sci 33:129–144
Krugman P (1999) The role of geography in development. Int Reg Sci Rev 22:142–161
Krugman P, Livas R (1996) Trade policy and third world metropolis. J Dev Eco 49:137–150
Le Gallo J, Chasco C (2008) Spatial analysis of urban growth In Spain, 1900–2001. Empir Econ 34:59–80
Limão N, Venables AJ (2001) Infrastructure, geographical disadvantage, transport costs, and trade. World Bank Econ Rev 15:451–479

Márquez MA, Hewings GJD (2003) Geographical competition between regional economies: the case of Spain. Ann Reg Sci 37:559–580
Mathias K (1980) Wirtschaftsgeographie des Saarlandes. Buchverlag Saarbrücker Zeitung
McCallum J (1995) National borders matter: Canada–US regional trade patterns. Am Econ Rev 85:615–623
Mella JM, Chasco C (2006) Urban growth and territorial dynamics: a spatial-econometric analysis of Spain. In: Reggiani A, Nijkamp P (eds) Spatial dynamics, networks and modeling. Edward Elgar, New York, pp 219–260
OECD (2000) Small and medium-sized enterprises: local strength, global reach. OECD Observer, June 2000, Organisation for Economic Cooperation and Development, Paris, France
Peeters L, Chasco C (2006) Ecological inference and spatial heterogeneity: an entropy-based distributionally weighted regression approach. Pap Reg Sci 85:257–276
Pulido A, López AM (2003) Madrid: Economía dinámica. Economistas 95:21–26
Rappaport J (2000) Is the speed of convergence constant? Federal Reserve Bank of Kansas City Working Paper 99–13 (August)
Rappaport J, Sachs J (2003) The United States as a coastal nation. J Econ Growth 8:5–46
Ravallion M (2007) Geographic inequity in a decentralized anti-poverty program: a case study of China. Policy Research Working Paper Series 4303, The World Bank
Rey S, Montouri B (1999) US regional income convergence: a spatial econometric perspective. Reg Stud 33:143–156
Roos MWM (2005) How important is geography for agglomeration? J Econ Geogr 5:605–620
Rosenthal S, Strange WC (2001) The determinants of agglomeration. J Urban Econ 50:191–229
Rosés JR (2003) Why isn't the whole of Spain industrialized? New Economic Geography and early industrialization, 1797–1910. J Econ Hist 64:995–1022
Sachs J (2000) Tropical underdevelopment. CID Working Paper 57
Sargan JD (1964) Wages and prices in the United Kingdom: a study of econometric methodology. In: Hart PE, Mill G, Whitaker JK (eds) Econometric analysis for national economic planning. Butterworths, London, pp 25–63
Tirado DA, Paluzie E, Pons J (2003) Industrial agglomerations and wage gradients: the Spanish economy in the interwar period. Document de Treball de la Facultat de Ciències Econòmiques i Empresarials de la Universitat de Barcelona, E03/103
Venables AJ (1999) But why does geography matter, and which geography matters? Int Reg Sci Rev 22:238–241
Venables AJ (2003) Spatial disparities in developing countries: cities, regions and international trade. Available for download at: http://www.econ.ox.ac.uk/members/tony.venables/unpub_papers.html#spatrktv
Viladecans E (2004) Agglomeration economies and industrial location: city level evidence. J Econ Geogr 4:565–582

Comparative Spatial Dynamics of Regional Systems

Sergio J. Rey and Xinyue Ye

1 Introduction

Research on the question of regional income convergence has gone through two phases over the past two decades. The first generation of regional convergence studies began to appear as growth theorists turned their attention away from international analyses of country growth patterns having discovered the region as a new unit of analysis (Barro and Sala-i-Martin 1991). This change in scale had a key advantage of increasing (in some cases substantially) the number of cross-sectional observations available for model estimation and hypothesis testing. While the scale of the analysis shifted, these first generation studies relied on the same underlying theoretical and empirical frameworks used in the international literature.

In the second phase, the underlying geographical dimensions of the data in convergence studies began to attract attention (Rey and Montouri 1999). This was reflected in several developments. The first saw the increasing application of the methods of spatial econometrics and spatial data analysis to regional case studies. These applications have generated abundant evidence that the spatial effects of dependence and heterogeneity tend to be the rule rather than the exception in practice, and as such their consideration should form a crucial component of empirical analysis. Thus the second generation of regional convergence studies is those characterized by concerns with spatial effects.

Both phases of regional convergence research have yielded an enormous literature of empirical studies.[1] At the same time, the spatially explicit methods applied in the second phase of this literature have been designed for cross-sectional data sets while the convergence question itself has both spatial and temporal dimensions. It is not at all clear if these spatial methods require adjustment when applied

[1]For recent overviews of this literature see Rey and Le Gallo (2008), Rey and Janikas (2006), Abreu et al. (2005) and Fingleton (2003).

Sergio J. Rey (✉)
Department of Geography, San Diego State University and School of Geographical Sciences, Arizona State University, P.O. Box 875302, Tempe, AZ 85287, USA,
e-mail: Sergio.Rey@asu.edu

A. Páez et al. (eds.), *Progress in Spatial Analysis*, Advances in Spatial Science,
DOI 10.1007/978-3-642-03326-1_20,

in a dynamic context. Moreover, despite the richness of this literature, relatively few studies have compared the rates of convergence and inequality across different national systems. What few comparative studies that have appeared have focused on the more advanced economies of Europe and the United States (Boltho 1989). Also, these comparative studies are clearly first generation in their approach to regional data as space is largely ignored. Thus, although we are gaining an understanding of the role of spatial effects in the analysis of inequality and convergence, we currently do not know if these effects are present in the same way at different stages of economic development.

This chapter seeks to contribute to the literature by addressing these gaps. We do so by drawing on some recently development methods of exploratory space–time data analysis (ESTDA) (Rey 2001; Rey and Janikas 2006; Janikas 2007) to develop a framework for the comparative analysis of spatial income inequality dynamics between different economic systems. We apply this framework to a case study involving the United States and China, two large economies at different stages of development.

Our exploratory approach is designed to identify interesting patterns in the spatial and temporal dimensions of the regional growth series. This is in response to the criticisms made of formal growth theories, which rest on restrictive assumptions about representative economies and randomness in space that are largely at odds with the characteristics of regional data (Fingleton 2004). In order to develop a more spatially explicit growth theory it is first necessary to develop operational measures that capture the spatial dynamics inherent in regional datasets. We see the exploratory methods we suggest in what follows as an initial step towards these ends.

This chapter addresses these issues through the integration of recent advances in distribution dynamics and spatial pattern analysis. Some novel approaches for inference are suggested to complement the descriptive approaches in the existing literature, as well as to provide new bases for comparative analysis. While the substantive focus of the research is on regional inequality dynamics, the methodological issues examined are relevant to the study of a wide class of phenomena that have spatial and temporal dimensions. These new statistical measurements also create opportunities for novel scientific visualization and new research hypothesis. As such, this project is among the efforts for more powerful analytical methods for spatiotemporal data, which has been viewed as a critical need in research in geography and regional science (Rigby and Willmott 1998).

In the remainder of the chapter, we first discuss the motivations for this research from theoretical, methodological and empirical perspectives, which give rise to comparative spatial dynamics analysis of regional systems. Next, we present a comparative exploratory space–time analysis of regional income dynamics over the 1978–1998 period in China and the United States. The chapter closes with a summary and concluding comments.

2 Theoretical and Methodological Motivations

With the dramatic improvement in computer technology and the increase in volumes of geographically referenced socioeconomic data, the importance of space to many socioeconomic processes has been gaining a growing recognition (Egenhofer and Golledge 1997; Peuquet 2002; Goodchild and Janelle 2004). At the same time, the study of economic inequality and convergence continues to attract enormous attention and it has generated a dynamic academic landscape where geography and other social sciences interact (Sassen 1994; Krugman 1999; Gruber and Gaines 2001). This interest has been reflected in spatial and temporal thinking of this research domain, that is, analyzing spatial patterns of economic convergence and the dynamics of geographical inequality (Rey 2004a). However, the literatures of spatial pattern analysis (form analysis) and time series analysis (process analysis) are mainly separated.

> While geographers have always been custodians of knowledge about form, arguably the custodians of process have been the substantive sciences of geology, ecology, hydrology, epidemiology, demography, economics, etc. A concern for process is therefore likely to change the landscape of GIScience dramatically, requiring much closer interaction with these sciences. (Goodchild 2006, p. 4)

2.1 *Space, Time and Regional Inequality*

Longstanding concerns with spatial inequality, its temporal persistence and causative processes, have generated lasting discussions and fascinating debates among adherents of the various schools of economic development, such as neoclassical growth theory, endogenous growth theory and new economic geography (Barro and Sala-i-Martin 1991; Aghion and Howitt 1998; Fujita and Krugman 2004).

There is increasing awareness of the importance space in the empirical analysis of growth and convergence (Rey and Montouri 1999; Fingleton 2004; Yamamoto 2006) together with a recognition that the existing growth theories do not fully treat the rich spatial patterns encountered in empirical work. Recent work in economic geography has also been criticized for failing to deal with the major problems of development and inequality, as well as for fuzzy concepts, shaky evidence, and policy irrelevance (Hamnett 2003). Hence, Bode and Rey (2006) call for "further research on integrating space into formal theoretical models of growth and convergence as well as on developing the next generation of analytical methods needed to implement those models" as "the preconditions for reliable policy recommendations, one of the primary goals of economic research."

Exploratory data analysis (EDA) has evolved from a small sub-field to an important part of the methodological domain. Haining and Wise (1997) define EDA as "to identify data properties for purposes of: pattern detection in data, hypothesis formulation from data, some aspects of model assessment." After the incorporation of spatial properties of data, exploratory spatial data analysis (ESDA) aims are "detecting spatial patterns in data, formulating hypotheses based on the geography of the data, assessing spatial models" (Haining and Wise 1997). ESDA is a powerful body of techniques to visualize spatial distributions and detect patterns of spatial association (Anselin 1993), often revealing complex spatial phenomenon not identified otherwise (Le Gallo et al. 2003). Hence, the development of new methods of ESDA has stimulated a number of research efforts (Anselin and Getis 1992; Longley et al. 2001; Getis et al. 2004; Rey and Anselin 2006).

Both spatial and temporal attributes of data are important, but existing approaches focus primarily on one of these attributes. For example, researchers have relied on either spatial analysis or time series methods though regional inequality dynamics has both temporal and spatial dimensions underlying empirical analysis (Rey 2004a). It is clear that new methods are needed to truly integrate space and time. Goodchild (2004, 2006) suggests this is a major research priority for the processes that define the Earth's dynamics. To consider both dimensions jointly, requires extending EDA (and exploratory temporal data analysis) for space, and at the same time incorporating time into ESDA (Rey et al. 2005).

2.2 Distribution Dynamics and Spatial Pattern Analysis

Barro and Sala-i-Martin (1991, 1992) and Sala-i-Martin (1996) discuss two types of convergence in growth empirics: σ and β convergence. The former reflects the decline of the dispersion of income across the economic units over time; the latter indicates the negative partial correlation between the growth rate in income over time and its initial level. Quah (1993) argues that these two empirical strategies might be misleading because of the arbitrary assumptions about the dynamics as a whole. Distribution dynamics refer to the difference among the overall shape characteristics of the regional income distribution and the evolution of these characteristics over time, as well as the amount of internal mixing or rank mobility taking place within these same distributions. Quah (1996) comments that the distribution dynamics empirics will lead to new theories on economic growth and convergence.

In response, a number of EDA techniques have been applied to regional income distributions. Using Markov chain techniques, Quah documents the degree to which this instability characterizes the data. Markov chains have been applied to study steady-state trends (Magrini 1999), modality (Quah 1996) and rank mobility (Hammond and Thompson 2002). Stochastic kernels are considered as extensions of the Markov chain to a continuous field. Bianchi (1997) employs Markov chain approach in the analysis of modality and the application in the internal mixing is carried out by Tsionas (2000).

Some recent work points out that the dominant focus in the empirical literature on shape regularities may be masking some interesting patterns that are internal to those distributions (Overman and Ioannides 2001; Ioannides and Overman 2004). Based on a critical review of empirical approaches and methodological advances in spatial econometrics and spatial statistics, Rey and Janikas (2005) highlight the important roles of spatial dependence, spatial heterogeneity, and spatial scale in the analysis of regional income distribution dynamics. Rey (2001, 2004b) suggests a series of spatial empirics for distributional dynamics, such as spatial Markov, regional cohesion of rank mobility and spatial decomposition of rank dynamics. To characterize complex map patterns has always been a challenge for spatial analysis (Getis and Boots 1978; Boots and Getis 1988; Okabe et al. 2000). Geometric indicators and graphical depiction have been used to summarize spatial patterns, such as Weber's Triangle, the Gravity Model, and Central Place Theory, among others (Mu 2004). Geometric criteria are also applied to identify spatial structure through a spatial weight matrix (Anselin 1988; Getis and Aldstadt 2004; Aldstadt and Getis 2006). For instance, Aldstadt and Getis (2006) demonstrate that spatial association varies in distance/direction and clusters are irregular in shape. We suggest that these findings can be revisited with perspectives from computational geometry where methods have been developed to meet fast algorithmic requirements for geometric computing (Mulmuley 1994; O'Rourke, 1994; Chazelle 1995; Eppstein 2005). Recent progress in statistical shape analysis (Goodall and Mardia 1999) reveals great potential for studying shape variations at microscale such as human brains (Mardia and Dryden 1999) to the Voronoi polygons examination of the central place theory (Dryden and Mardia 1998). As commented by Goodchild (2006), "... GIScience is applicable to varying degrees in any space, ... such as the three-dimensional space of the human brain, ... At the same time, advances made in the study of other spaces may be suitable sources of cross-fertilization in GIScience. Perhaps the next decade will see a much greater degree of interaction between GIScience and the sciences of other spaces, and much more productive collaboration." While analytical cartography and computational geometry can generate in-depth visualization and summary of location and spatial pattern, they largely ignore dynamic effects.

This chapter hopes to contribute to the cross-fertilization of distribution dynamics and spatial pattern analysis, through summarizing and comparing the geometry of various spaces of regional economic growth.

In this regard, several interesting research questions are examined:

1. To what extent is economic growth associated with spatial context, dependence, or heterogeneity?
2. Are regions with similar economic growth trends clustered?
3. How stable are certain spatial patterns (structures) over time? Are they clustered over time?

3 Empirical Motivation: Regional Inequality in China and the United States

Because of their growing importance in the world economic system, China and the United States have been the center of numerous debates about economic growth and regional convergence. Despite this rich empirical literature, comparative analysis of regional inequality dynamics between the two economies remains largely unexplored, let alone the underlying geographical dimensions of regional growth processes (Rey and Janikas 2005; Janikas 2007). Moreover, applications of comparative analysis between different economic systems are currently lacking an inferential basis.

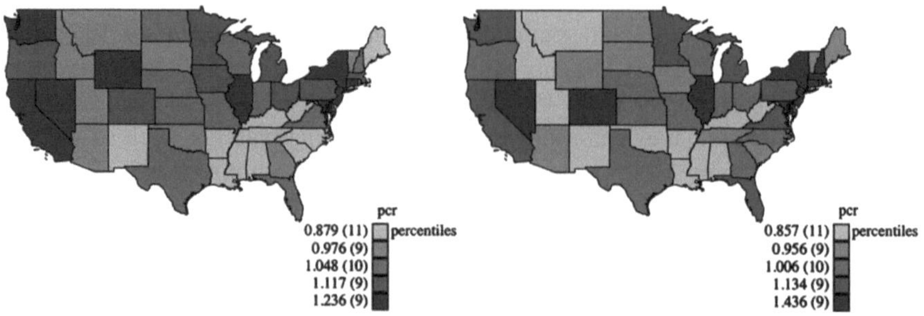

Fig. 1 Per capita incomes in the United States, 1978 and 1998

Regional inequality has generated lasting debates among the convergence, divergence, inverted-U, and Neo-Marxist uneven development schools (Pritchett 1997; Fujita et al. 1999; Puga 1999; Tsionas 2000; Rey and Janikas 2005). The debate on the trajectories and mechanisms of regional development has been focused over the scope and consequences of regional policies and the extent and sources of regional inequality (Sidaway and Simon 1990; Fan and Casetti 1994; Wei and Ye 2004; Ye and Wei 2005), which is reflected in numerous empirical studies of specific nations and continents (Rey and Janikas 2005). However, the findings are mixed and sometimes conflicting (Ye and Wei 2005).

Many studies have been conducted on the US experience, and most of them conclude that regional convergence has been very strong, with two persistent regional clusters: the Northeast-Mid Atlantic cluster of high income states and Southeast cluster of low income states (Barro and Sala-i-Martin 1991, 1992; Fan and Casetti 1994; Bernard and Jones 1996; Vohra 1996; Rey and Montouri 1999; Tomljanovich and Vogelsang 2002; Sommeiller 2007), as shown in Fig. 1.

Since the late 1970s, China has been undergoing economic reforms introducing market mechanisms and opening its economy to the outside world. The reform process, however, was spatially uneven and has traditionally emphasized coastal development (Lyons 1991; Lin 1997; Wei 2000, 2009; Benjamin et al. 2005; Wei and Ye 2009), as shown in Fig. 2.[2] Starting in the mid-1990s, the Chinese government began to make more efforts on development of poorer regions and reduction of spatial inequalities through launching western development strategies and, recently, providing incentives for developing rural areas. While some maintain that globalization and liberalization have brought wealth to transitional countries like China, others argue that the transition in former socialist countries is characterized by partial reform, path dependency, and geographical unevenness, and have recorded

[2] While China's official GDP statistics are sometimes regarded as of questionable quality (Rawski 2001), the NSB (National Statistical Bureau of China) published adjusted GDP data to deal with both overestimates and underestimates of provincial GDP data for the years before 2004 (Fan and Sun 2008). For a recent discussion justifying using per capita GDP as a valid and reliable indicator of provincial economic development and well-being in China, see Fan and Sun (2008).

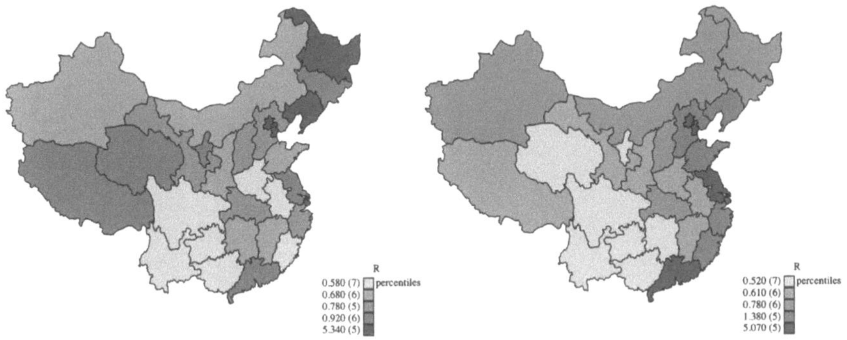

Fig. 2 Per capita incomes in China, 1978 and 1998

persistent or rising income gaps and spatial inequalities (Wei and Ye 2004; Ye and Wei 2005).

Not only the presence of spatial dependence presents a challenge to the use of statistical inference, but the partitioning of the economic units into either a 29-region system (29 land provinces in China) or 48-region system (48 states in the United States) raises another concern very similar to the modifiable areal unit problem (MAUP) (Openshaw and Alvanides 1999). Rey (2004a) finds that the regional inequality decomposition fundamentally changes both quantitatively and qualitatively when its spatial partition scheme (regionalization scheme) varies. It is important to check whether the difference among regional systems is sensitive to both how the observations are partitioned into each system and how they are spatially distributed within each system. However, this issue has been largely neglected in previous comparative studies.

In the following sections, comparative space–time analysis of regional systems will be conducted using the case study of China and the United States. The two datasets are relative per capita income over the 1978–1998 period at the province (China) and state (the United States) levels. The two data sets are comparable regarding regional inequality because the states (United States) and provinces (China) are self-contained and well-functioning units which form the theoretical structure for spatial interaction models in spatial economy (Fan and Casetti 1994).

4 Comparative Spatial Dynamics

4.1 Inequality and Spatial Dependence

Many inequality measures have been introduced and discussed in the literature. In regional inequality analysis, a popular measure is Theil's inequality measure (Theil 1967). Attention is first directed towards the relationship between regional

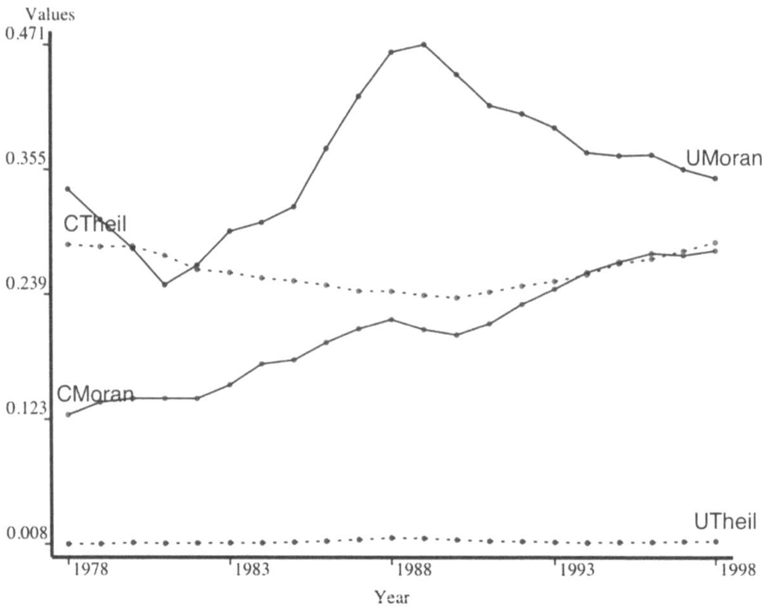

Fig. 3 Convergence and spatial independence in the United States and China

income inequality and spatial dependence over time using the global Theil and Moran's I (Fig. 3), which shed light on the debates between competing economic growth theories and policies in these two distinct economic systems (Rey 2004a). There is a U shape for regional inequality over time, with spatial clustering trends in China while there is an obvious inverted U shape for spatial dependence with relatively stable (or slightly inverted-U shape) regional inequality in the United States.

In studies of regional income inequality, the decompositional property has been exploited to investigate the extent to which global Theil is attributable to inequality between or within different partitions of the observational units. This approach can provide a deeper understanding of global inequality (Rey 2004a). Two common regionalization schemes in China are Three Belts or Six Macro Regions (Fig. 4). Three Belts are the eastern, central, and western economic belts while Six Macro Regions refer to six main geographic regions (North-West, North, North-East, South-West, Central-South, East). There are four census regions in the United States: Northeast, Midwest, South, and West. Eight BEA (Bureau of Economic Analysis) Regions are New England, Mideast, Great Lakes, Plains, Southeast, Southwest, Rocky Mountain, Far West (Fig. 5).

As revealed by Fig. 6, intra-regional inequality dominates the overall disparity in China for most of the time regardless of the regionalization system while the dominance status in the United States will generally either be granted to inter-regional (Eight BEA Regions) or intra-regional inequality (Four Census Regions). The inter-regional inequality share always grows in China while in the United States case this

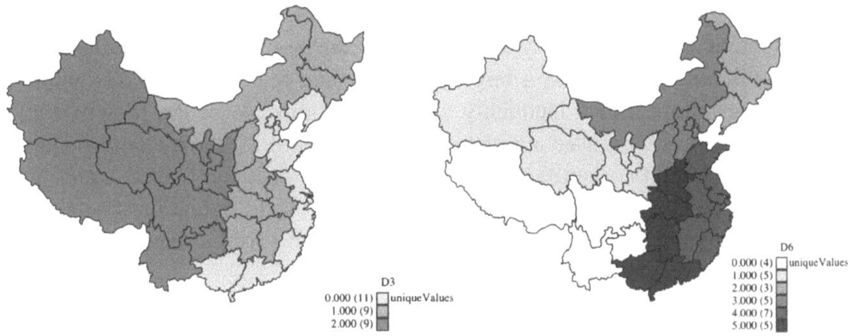

Fig. 4 Regionalization system in China

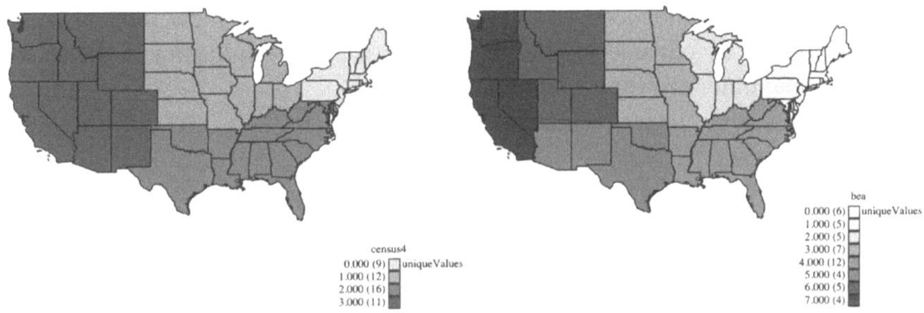

Fig. 5 Regionalization system in the United States

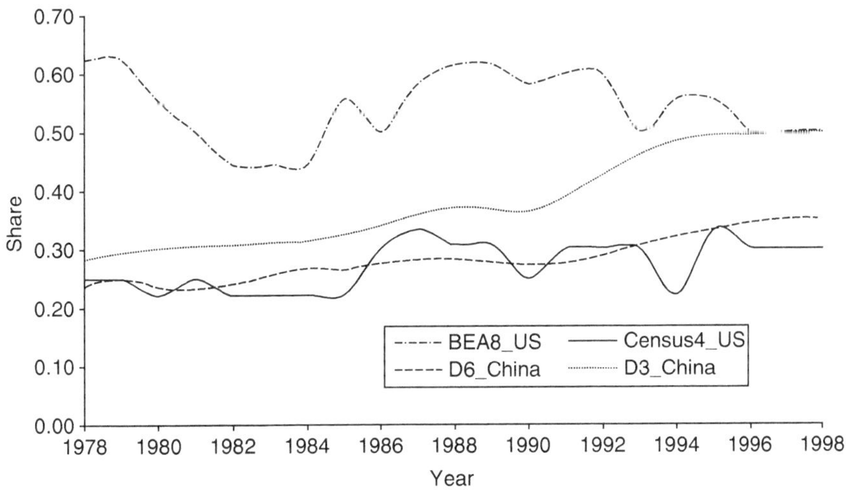

Fig. 6 Inter-regional inequality share in China and the United States

component fluctuates substantially in the same time period. The choice of regionalization system matters in both systems. In China, the more aggregate regionalization scheme (three belts) leads to a larger share of inter-regional inequality while in the US case intra-regional inequality grows with the scale of the regionalization scheme. China has witnessed a widening difference of the inter-regional inequality shares between the two partition schemes over time while the United States has a narrowing gap. The above studies have illuminated to some extent the spatial structure underlying the dynamics of regional inequality at various stages of economic development.

4.2 Distance-Based Local Markov Transition

Local indicators of spatial autocorrelation (LISA) show a disaggregated view at the nature of spatial dependence (Anselin 1995). We can embed these indicators in a dynamic context by considering the movement of a given indicator in the scatter plot over some time interval.

At a given time, t, the coordinate of each unit i's LISA is $(y_{i,t}, yl_{i,t})$ with:

$$yl_{i,t} = \sum_{j=1}^{n} w_{i,j} y_{j,t}$$

Given this, $D_{i,t,t+1}$ is economic unit i's LISA transition from time t to $t + 1$, measured by the segment length of $[(y_{i,t-1}, yl_{i,t-1}), (y_{i,t}, yl_{i,t})]$.

In a similar vein to what is done in Markov models of income distributions, we can discretize the values of the indicators to consider transitions across the classes of a scatter plot over time. The four classes are High-Low (first quadrat), Low-High (second quadrat), Low-Low (third quadrat) and High-Low (fourth quadrat). Besides four types of intraclass transitions, 12 types of inter-class transitions can be identified based on the four classes.

Because this discretization considers only class transitions it may treat transitions of different magnitudes as equal in constructing the LISA transition probability matrix. We suggest using a threshold distance to address this issue. We can set the threshold to be some value such as the average of all the transition distances on the Moran scatter plot, which is on the conservative side. With this threshold, there are two inter-class transitions in the left view of Fig. 7 and both of them move from

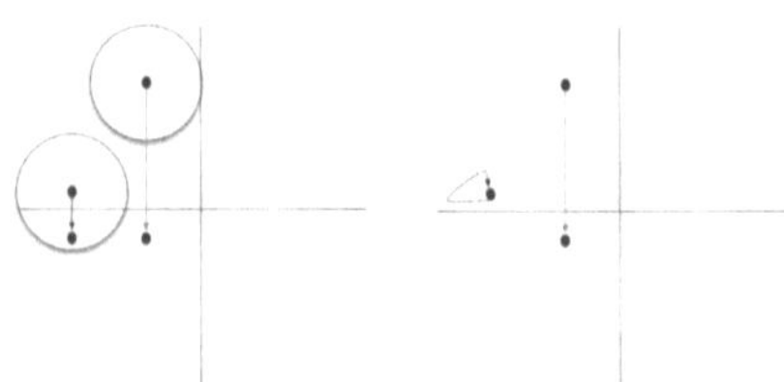

Fig. 7 Local Moran Markov transition

Table 1 Local Moran transition matrix in China (ND/D)

	HH	LH	LL	HL
HH	82/82	0/0	0/0	0/0
LH	3/2	47/48	1/1	0/0
LL	0/0	1/1	397/397	2/2
HL	1/1	0/0	2/1	44/45

Table 2 Local Moran transition matrix in the United States (ND/D)

	HH	LH	LL	HL
HH	223/228	9/6	0/0	6/4
LH	6/3	141/146	9/7	0/0
LL	0/0	5/2	356/362	7/4
HL	3/2	0/0	8/6	187/190

Low-High section to Low-Low section on the Moran scatter plot. An inter-class transition is significant only if its distance is larger than the threshold, otherwise the transition is treated as an intra-class transition and will be considered to stay in the original class, as shown on the right view of Fig. 7.

We use these thresholds to construct Tables 1 and 2 which reveal that China has more significant transitions in local Markov matrix. For instance, the 356/362 located in the LL–LL position of Table 2 indicates that 356 transitions are considered intra-class movements before the distance (ND) threshold is applied while six more transitions will be treated as intra-class movements because their lengths are shorter than the average movement (D). This is contrasted with the case of LH to HH transitions where six original transitions occur, but three of these involve movements that are shorter than the threshold distance and are therefore treated as intra-class movements (LH–LH). We return to the use of the threshold based transitions in a comparative analysis later in this chapter.

4.3 LISA Time Path

The LISA Time Path Plot takes a continuous view of these transition to illustrate the pair-wise movement of an economic unit (observation)'s value and its spatial lag over time (Rey et al. 2005). The path of observation i over time can be written as $[(y_{i,1}, yl_{i,1}),(y_{i,2}, yl_{i,2}),\ldots,(y_{i,T}, yl_{i,T})]$. $y_{i,t}$ is per capita income of province/state i at time t and $\mathrm{yl}_{i,t}$ is its spatial lag at time t. This graph is helpful in identifying the stability levels of local growth across a given structural process on the Moran scatter plot. Since individual aspects of the contemporaneous process can be dissected by interval gaps, the length and tortuosity of the time path are summarized for each economic unit, as follows:

$$\Gamma_i = \frac{N * \sum_{t=1}^{T-1} d\left(L_{i,t}, L_{i,t+1}\right)}{\sum_{i=1}^{N} \sum_{t=1}^{T-1} d\left(L_{i,t}, L_{i,t+1}\right)} \tag{1}$$

where: $L_{i,t}$ is the location of economic unit i on the Moran scatter plot at time t, which is $(y_{i,t}, yl_{i,t})$. $d(L_{i,t}, L_{i,t+1})$ is the distance (movement) between the locations of economic unit i at time t and $t+1$. N is the number of spatial units. If an economic unit's movement over time is longer than the average, Γ_i will be larger than 1, and vice versa.

$$\Delta_i = \frac{\sum_{t=1}^{T-1} d(L_{i,t}, L_{i,t+1})}{d\left(L_{i,1}, L_{i,T}\right)} \tag{2}$$

where Δ_i is the economic unit i's tortuosity on the Moran scatter plot over time. A larger Δ_i indicates a more tortuous movement on the graph.

A scalar instability measure of dynamic LISA is:

$$\Lambda_i = \frac{N * \sigma_i}{\sum_{i=1}^{N} \sigma_i} \tag{3}$$

where σ_i is the standard deviation of economic unit i's interval segment lengths of LISA time path.

Figure 8 contrasts the LISA time paths of all the provinces/states in China and the United States at the same scale. It reveals that China has much more dispersed spatial dynamics. These patterns can be furthered analyzed in several ways. Tables 3 and 4 report the three suggested indicators to capture the continuous nature of the LISA time paths for each province and state. They are also mapped on Figs. 11 and 12. The lop left view is the geographical distribution of Γ_i values (length); top right view is for Δ_i (tortuosity); bottom left view displays Λ_i (instability); and the bottom right view is for space–time integration ratio, which will be discussed in the following section. China's rich provinces (coastal) tend to be more dynamic (top left view), more tortuous (top right view) and more stable (bottom left view) while the Northeast-Mid Atlantic cluster of high income states are more dynamic, less tortuous and more instable reflected by these three types of values compared to the rest

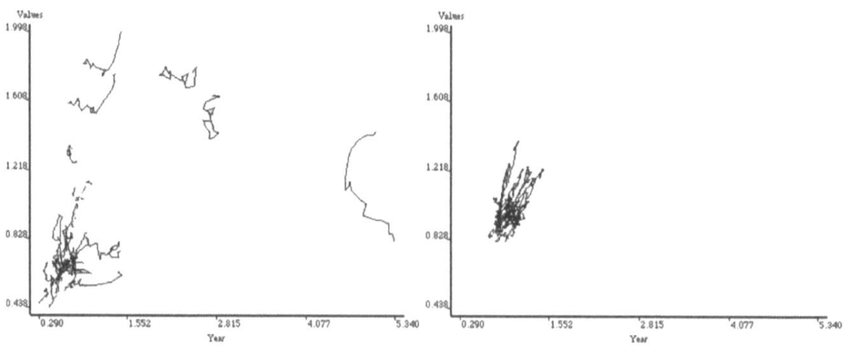

Fig. 8 LISA time path (*left*: China; *right*: the United States)

Table 3 Spatial dynamics in China

Province	Length	Tourtuosity	Instability	Spatial joins	Similar dynamics	ST integration
AH	0.65	0.83	0.68	6	1	0.17
BJ	2.17	1.91	3.06	2	0	0.0
FJ	1.05	1.23	0.44	3	2	0.67
GS	1.08	0.83	0.79	6	4	0.67
GD	1.18	1.32	0.48	4	1	0.25
GX	0.45	0.48	0.78	4	1	0.25
GZ	0.41	0.43	0.92	4	3	0.75
HEB	0.58	0.65	1.88	7	1	0.14
HL	0.86	1.10	0.58	2	0	0.00
HEN	0.76	0.50	1.89	6	0	0.00
HUB	0.63	0.58	1.84	6	1	0.17
HUN	0.35	0.41	0.67	6	3	0.50
NM	0.91	0.76	0.97	8	5	0.63
JS	1.32	1.41	0.55	4	2	0.50
JX	0.55	0.67	0.51	6	0	0.00
JL	1.12	1.09	1.01	3	2	0.67
NX	1.12	0.97	0.77	3	3	1.00
QH	1.58	1.10	0.57	4	3	0.75
SN	0.81	0.67	1.33	7	4	0.57
SD	0.77	0.75	0.52	4	2	0.50
SH	1.92	2.45	0.84	2	0	0.00
SX	0.86	0.73	1.05	4	2	0.50
SC	0.36	0.46	0.73	8	5	0.63
TJ	1.68	1.79	1.28	2	0	0.00
XJ	0.82	1.04	0.83	3	0	0.00
XZ	2.13	1.67	1.40	4	2	0.20
YN	0.51	0.64	1.22	4	1	0.25
ZJ	1.20	1.48	0.49	5	2	0.40

of their systems. As mentioned above, a longer movement on the Moran scatter plot suggests a more mobile local spatial dependence over time. A more tortuous path indicates a more fluctuating local spatial dependence evolution in direction while a large variance among the segments of LISA time path demonstrates a more fluctuating local spatial dependence evolution. The maximum and minimum of Γ_i are 2.17 (Beijing) and 0.35 (Hunan) in China, 2.78 (North Dakota) and 0.48 (Alabama) in the United States. The maximum and minimum of Δ_i are 2.45 (Shanghai) and 0.41 (Hunan) in China, 5.69 (Arkansas) and 0.31 (North Carolina) in the Untied States. The maximum and minimum of Λ_i are 3.06 (Beijing) and 0.44 (Fujian) in China, 3.45 (North Dakota) and 0.34 (Alabama) in the United States.

Table 4 Spatial dynamics in the United States

State	Length	Tourtuosity	Instability	Spatial joins joins	Similar dynamics	ST integration dynamics
AL	0.48	0.49	0.34	4	2	0.50
AZ	0.79	0.92	0.67	5	3	0.60
AR	0.73	5.69	0.70	6	2	0.33
CA	0.84	0.39	1.06	3	1	0.33
CO	1.02	0.73	0.73	7	3	0.43
CT	1.48	0.42	1.51	3	3	1.00
DE	1.08	1.12	0.82	3	3	1.00
FL	0.83	0.74	0.74	2	0	0.00
GA	0.63	0.33	0.67	9	4	0.44
ID	1.10	0.58	0.91	12	6	0.50
IL	0.59	0.84	0.57	5	4	0.80
IN	0.63	0.71	0.71	4	4	1.00
IO	1.20	0.73	1.23	6	3	0.50
KA	0.83	1.45	0.80	4	1	0.25
KN	0.57	4.43	0.54	7	3	0.43
LO	1.02	3.17	1.19	3	1	0.33
ME	1.53	0.66	1.28	1	1	1.00
MD	0.84	1.53	0.74	4	3	0.75
MA	1.37	0.40	1.34	5	5	1.00
MI	1.08	0.75	1.03	3	3	1.00
MN	1.43	0.99	1.37	4	0	0.00
MS	0.58	1.10	0.49	4	2	0.50
MO	0.62	1.42	0.53	8	0	0.00
MT	1.71	0.60	1.01	4	3	0.75
NE	1.12	1.18	1.10	6	2	0.33
NV	0.77	0.46	1.16	5	3	0.60
NH	1.54	0.57	1.37	3	3	1.00
NJ	1.03	0.53	1.07	3	3	1.00
NM	0.86	1.05	0.69	5	5	1.00
NY	1.03	0.46	1.03	5	5	1.00
NC	0.64	0.31	0.75	4	4	1.00
ND	2.78	1.21	3.45	3	1	0.33
OH	0.51	0.54	0.63	5	3	0.60
OK	1.16	0.97	1.55	6	4	0.67
OR	0.81	0.50	1.13	4	3	0.75
PA	0.62	1.39	0.45	6	4	0.67
RI	1.42	0.43	1.46	2	2	1.00
SC	0.67	0.35	0.75	2	2	1.00
SD	1.65	1.08	1.89	6	1	0.17
TN	0.60	0.50	0.51	8	6	0.75
TX	1.02	1.64	1.36	4	3	0.75
UT	0.80	0.70	0.61	6	6	1.00
VT	1.15	0.48	1.04	3	3	1.00
VA	0.69	0.63	0.66	5	3	0.60
WA	1.03	0.85	0.90	2	2	1.00
WV	0.64	0.62	0.60	5	2	0.40
WI	0.70	0.82	0.72	4	3	0.75
WY	1.77	0.53	2.10	6	3	0.50

4.4 Space–Time Covariance Matrix

The spatial dynamics can also be summarized using graph theoretic concepts. More specifically, the pairwise temporal covariance between economies can be used to define a network which is then visualized geographically. The covariance structure of incomes is portrayed as the links between the centroids of each polygon. Various levels of correlations can be visualized differently to more distinctly identify cross-sectional relationships. Covariance links can be mapped based on specified criterion (Rey and Janikas 2006). A network graph identifies the spatial joins displaying similar income growth trends with a focal economy (Fig. 9). A spider graph reflects all the other economic units, which share the similar temporal dynamics with the focal region (Fig. 10). Two regions are defined to be similar if the correlation of their time series is above national average. These two graphs show the spatial distribution of temporal dynamics, which is usually masked by the national growth trend.[3]

At the macro level there is some evidence that the spatial dynamics are more integrated in the United States than China (Fig. 9), which is also revealed by Fig. 8. However, this macro structure can many times mask a great deal of turbulence at the micro level reflecting movements of individual economies up and down the statistical distribution. Moving from the macro perspective we can check whether there are particular economies in each system that display interesting dynamics. This is illustrated in the spider graph in Fig. 10, which focuses on two regions: Zhejiang Province in China; California in the United States. The spider graph considers the integration of each focal region (Zhejiang and California) with their respective national systems and identifies the specific regions with which they share common dynamics, as reflected in high standardized pairwise temporal correlations. These are indicated by edges connecting each focal region to the dynamically similar region. Those similar regions that are also spatially contiguous to the focal region are indicated by thicker edges.

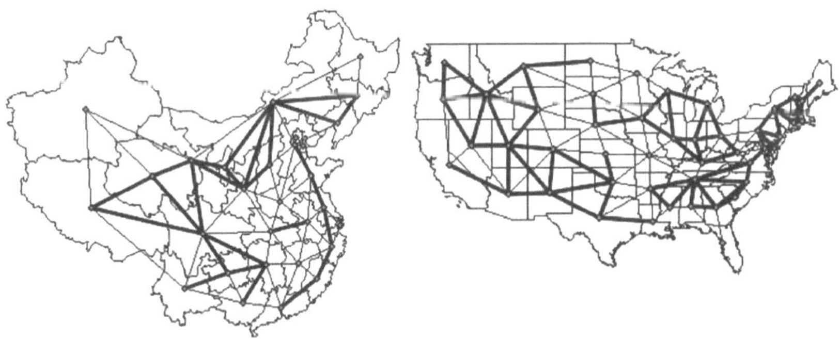

Fig. 9 Covariance networks in China and the United States (*thick segments* indicate similar temporal linkages)

[3] The national trends in both system have been removed from the regional data sets here as the income values are expressed as percentages of the national means for the given year.

Fig. 10 Spider graphs of Zhejiang province (China) and California (the United States) (the *links* indicate similar temporal linkages and the *thicker segments* highlight spatial joins)

Based on the spider graph, we consider the spatial joins each unit has and ask whether these joins are linking economies that display similar temporal dynamics with the focal unit. We take the ratio of the number of spatial joins with strong temporal linkages over the number of total spatial joins of the focal unit, with a value of 1 indicating very strong space–time integration for the focal economy, while a 0 suggests very weak linkages between the focal unit and its spatial lag (Tables 3 and 4). Table 3 reports that only one province in China (Ningxia) has similar temporal dynamics with all of its neighbors while five provinces are dissimilar with any of their neighbors. It can be observed that 16 states in the United States have similar temporal dynamics with all of their neighbors while only two states are dissimilar with any of their neighbors (Table 4). All the focal values are mapped on the bottom right views of Figs. 11 and 12. China has a z-value of 1.533 for the global Moran of these integration statistics, while for the United States the clustering of these integration measures is much stronger (z-value of 4.782), which further indicates the United States has a much more spatially integrated economic system.

4.5 Inferential Issues

Once the new space–time indicators are developed, an extensive set of inferential approaches are needed to evaluate their sampling distributions for comparative analysis between two regional systems. The inferential approaches suggested here rely on random labeling and spatial permutations of the relative values for two maps (two regional systems) simultaneously. The relative mobility in a Classic (and Local Moran) Markov transition matrix can be expressed as:

$$\tau = 1 - \frac{\sum_i P_{i,i}}{k} \quad (4)$$

where $P_{i,i}$ is the diagonal element of a Markov possibility transition matrix P and k is the number of total classes. If there are no crossclass transitions, all of the diagonal elements are 1 and τ is 0. The more dramatic the inter-class mobility, the

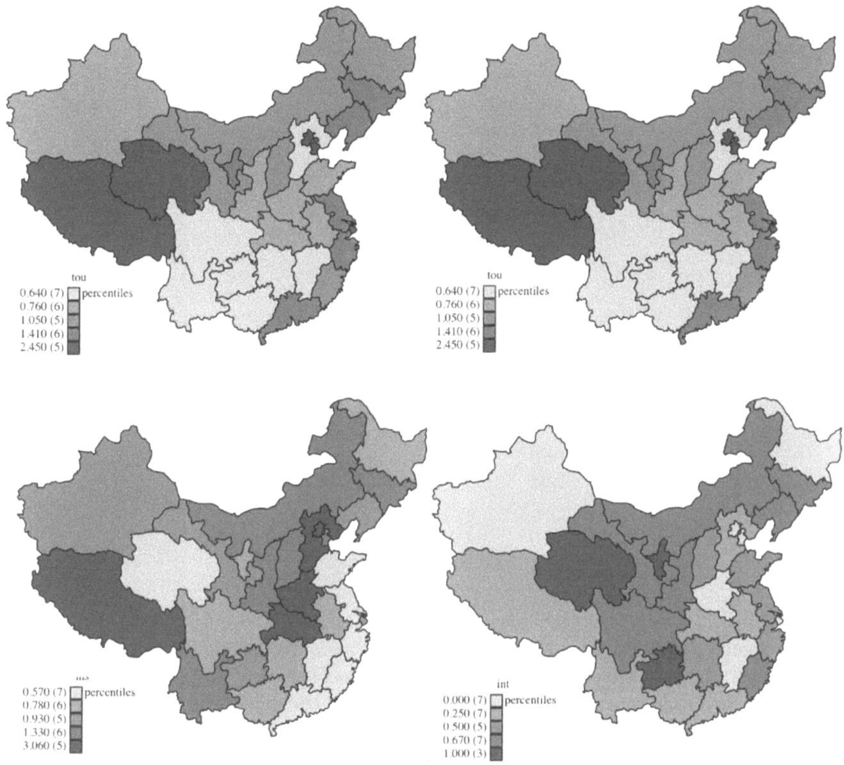

Fig. 11 Spatial dynamics in China (*top left view*: the length of LISA time paths (1); *top right view*: the tortuosity of LISA time paths (2); *bottom left view*: the instability of LISA time paths (3); *bottom right view*: space–time integration ratio of temporal dynamics)

larger the value of τ. The maximum value of τ is 1, which means none of the states (provinces) stays in the same income class over time for Classic Markov and all of the local economic structures (Low-Low, Low-High, High-High and High-Low) have changed during the transition period for Local Moran Markov.

The steps of the inferential approach follow:

1. Calculate the difference of relative mobility of Classic (Or Local Moran) Markov for China (τ_c) and the United States (τ_u)
2. Randomly reassign all the relative per capita incomes to new locations regardless of their original systems (For instance, New York State's relative per capita income might be reassigned to a province in China while Shanghai's income might be assigned to California)[4]

[4] Under the random labeling approach, the null hypothesis is that mobility rates are the same in the two countries.

Fig. 12 Spatial dynamics in the United States (*top left view*: the length of LISA time paths (1); *top right view*: the tortuosity of LISA time paths (2); *bottom left view*: the instability of LISA time paths (3); *bottom right view*: space–time integration ratio of temporal dynamics)

3. Calculate τ for the two maps and calculate the difference of relative mobilities between them
4. Repeat steps 2 and 3, M times
5. The actual indicator obtained at step 1 can then be compared against the expected value to check whether the difference is significant between the two regional systems

By extending the relative mobility analysis to include an inferential component we find that both of the relative mobilities of the Classic and Local Moran Markov transition matrices are relatively small in China and the United States (Table 5). However, while the Classic Markov reports no statistically significant difference between China and the United States (the pseudo p value is 0.499), the difference regarding the local Moran Markov statistic is significant with the pseudo p value equaling 0.067 after 1,000 permutations. This means that local economic structure in China is generally more stable than that in the United States. The diagonal values in Tables 6 and 7 show that China is more stable than the United States in most of the transition probabilities.

From a spatial perspective, China has relatively pronounced clusters for the provinces with similar economic growth trends (Fig. 13). Two spatial clusters in China warrant attention: the poverty trap composed of six inland provinces (lightest

Table 5 Relative mobility of classic and local Moran Markov in China and the United States

	Classic	Local Moran
China	0.134	0.017
The United States	0.136	0.055

Table 6 Local Moran transition probability matrix in China

	HH	LH	LL	HL
HH	1.000	0.000	0.000	0.000
LH	0.059	0.992	0.020	0.000
LL	0.000	0.003	0.993	0.005
HL	0.021	0.000	0.043	0.936

Table 7 Local Moran transition probability matrix in the United States

	HH	LH	LL	HL
HH	0.937	0.038	0.000	0.025
LH	0.038	0.904	0.058	0.000
LL	0.000	0.014	0.967	0.019
HL	0.015	0.000	0.040	0.944

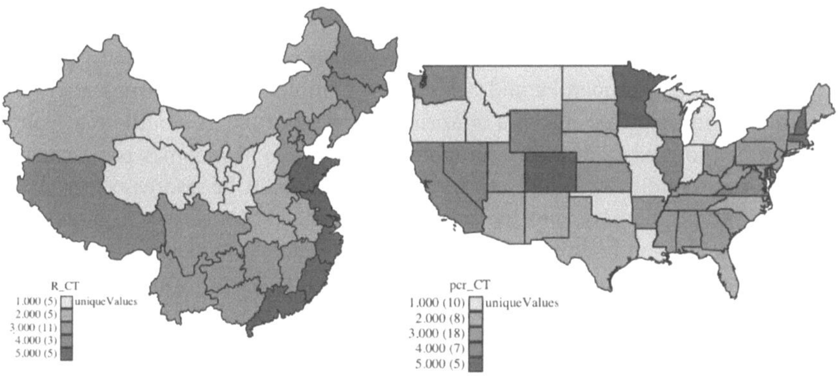

Fig. 13 Convergence classification in China and the United States

units) and the rich club along the coast (darkest units). The two figures report average convergence/divergence direction for each province/state. The lightest (gray-one) unit's initial income is below the national average and its average shift over time is to move further down. The gray-two unit's initial income is below the average and its average shift is upwards. The gray-three unit's initial income is around the average and stays in the same class over time. The gray-four unit's initial income is above the average while its average shift is downwards. The darkest (gray-five) unit's initial income is above the average and on average it moves up.

5 Summary

This chapter suggests new exploratory measures to integrate the distributional, temporal and spatial dimensions of regional economic growth, unfolding the complex spatial patterns of economic inequality evolution. Through linking distribution/kernel dynamics to geometric summaries of the map/diagram, the exploratory space–time analysis of geographical dynamics is conducted in a macro-meso-micro framework. Coupled with this is work on developing inferential methods for comparative analysis.

From a theoretical perspective, the new space–time measures hold the potential of generating some novel hypotheses about the nature of space in an economic system's evolution. Comparative analyses will help to narrow the gap between growth theories and their empirical testing to some extent. This will better our understanding of the role of space among different regional economic systems. From a policy perspective, the development of spatially explicit indicators will provide policy makers and urban planners with new tools to design and target poverty eradication programs by identifying irregular spatial economic performance, which have long been on the government agendas in various economic systems. The developed methods are also expected to have implications in areas such as comparative space–time dynamics of land use evolution, disease diffusion, crime hot spots, socioeconomic inequalities, among others.

While these new statistics show promise, much more work needs to be done on their theoretical properties and a number of implementation issues also need to be further investigated. From a confirmatory perspective, it would be a fruitful avenue to explore whether the geographical distribution of incomes is a structural driver in economic growth and convergence. This could be addressed by incorporating a graph measure into the economic growth or convergence model. The regression results can be used to validate the exploratory analysis, and at the same time exploratory analysis may provide useful indicators for confirmatory analysis.

References

Abreu M, de Groot HLF, Florax RJGM (2005) Space and growth: a survey of empirical evidence and methods. Région et Développement 21:13–44

Aghion P, Howitt P (1998) Endogenous growth theory. MIT, Cambridge

Aldstadt J, Getis A (2006) Using AMOEBA to create a spatial weights matrix and identify spatial clusters. Geogr Anal 38:327–343

Anselin L (1988) Spatial econometrics: methods and models. Kluwer, Dordrecht

Anselin L (1993) Exploratory spatial data analysis and geographic information systems. Technical Report 1, Regional Research Institute, West Virginia University

Anselin L (1995) Local indicators of spatial association - LISA. Geogr Anal 27:93–115

Anselin L, Getis A (1992) Spatial statistical analysis and geographic information systems. Ann Reg Sci 26:19–33

Barro R, Sala-i-Martin X (1991) Convergence across states and regions. Brookings Pap Econ Act 1:107–182

Barro R, Sala-i-Martin X (1992) Convergence. J Polit Econ 100:223–251
Benjamin D, Brandt L, Giles J (2005) The evolution of income inequality in rural China. Econ Dev Cult Change 53:769–824
Bernard A, Jones C (1996) Productivity and convergence across U.S. states and industries. Empir Econ 21:113–135
Bianchi M (1997) Testing for convergence: Evidence from nonparametric multimodality tests. J Appl Econom 12:393–409
Bode E, Rey SJ (2006) The spatial dimension of economic growth and convergence. Pap Reg Sci 85:171–176
Boltho A (1989) European and United States regional differentials: a note. Oxf Rev Econ Pol 5:105–115
Boots B, Getis A (1988) Point pattern analysis, vol 8, Scientific Geography Series. Sage, Newbury Park, CA
Chazelle B (1995) Computational geometry: a retrospective. In: Du DZ, Hwang F (eds) Computing in Euclidean geometry, 2nd edn. World Scientific, Washington, DC, pp 22–46
Dryden IL, Mardia K (1998) Statistical shape analysis. Wiley, Indianapolis, IN
Egenhofer MJ, Golledge RG (1997) Spatial and temporal reasoning in Geographic Information Systems. Oxford University Press, New York
Eppstein D (2005) Geometry in action: a collection of applications of computational geometry. http://www.ics.uci.edu/~eppstein/geom.html, 5 Oct. 2009
Fan C, Casetti E (1994) The spatial and temporal dynamics of US regional income inequality, 1950–1989. Ann Reg Sci 28:177–196
Fan CC, Sun M (2008) Regional inequality in China, 1978–2006. Eurasian Geogr Econ 49:1–20
Fingleton B (ed) (2003) European regional growth. Springer, Berlin
Fingleton B (2004) Theoretical economic geography and spatial econometrics: bridging the gap between theory and reality. In: Getis A, Mur J, Zoeller H (eds) Spatial econometrics and spatial statistics. Palgrave, Hampshire, pp 8–27
Fujita M, Krugman P (2004) The new economic geography: past, present and the future. Pap Reg Sci 83:139–164
Fujita M, Krugman P, Venables AJ (1999) The spatial economy. MIT, Cambridge, MA
Getis A, Aldstadt J (2004) Constructing the spatial weights matrix using a local statistic. Geogr Anal 36:90–104
Getis A, Boots B (1978) Models of spatial processes: an approach to the study of point, line, and area patterns. Cambridge University Press, Cambridge, UK
Getis A, Mur J, Zoeller H (eds) (2004) Spatial econometrics and spatial statistics. Palgrave, Hampshire
Goodall C, Mardia K (1999) Projective shape analysis. J Comput Graph Stat 8:143–168
Goodchild M (2006) Geographical information science: fifteen years later. In: Fisher P (ed) Classics from IJGIS: twenty years of the International Journal of Geographical Information Science and Systems, vol 2. CRC, Boca Raton, FL, pp 107–133
Goodchild M, Janelle D (2004) Thinking spatially in the social sciences. In: Goodchild M, Janelle D (eds) Spatially integrated social science: examples in best practice. Oxford University Press, New York, pp 3–22
Goodchild M (2004) GIScience: geography, form, and process. Ann Assoc Am Geogr 94:709–714
Gruber L, Gaines BJ (2001) Globalization and political conflict: the long-term prognosis. Manuscript, The Harris School, University of Chicago
Haining R, Wise S (1997) Exploratory spatial data analysis. NCGIA Core Curriculum in GIScience
Hammond G, Thompson E (2002) Mobility and modality trends in U.S. state personal income. Reg Stud 36:375–387
Hamnett C (2003) Contemporary human geography: fiddling while Rome burns? Geoforum 34:1–3
Ioannides YM, Overman, HG (2004) Spatial evolution of the US urban system. J Econ Geogr 4:131–156
Janikas M (2007) Comparative regional income dynamics: clustering, scale, and geocomputation. Ph.D. thesis, University of California, Santa Barbara and San Diego State University

Krugman P (1999) The role of geography in development. Int Reg Sci Rev 22:142–161
Le Gallo J, Ertur C (2003) Exploratory spatial data analysis of the distribution of regional Per Capita GDP in Europe, 1980–1995. Pap Reg Sci 82:175–201
Lin G (1997) Red capitalism in South China. UBC, Vancouver
Longley P, Goodchild M, Maguire D, Rhind D (eds) (2001) Geographical information systems and science. Wiley, New York
Lyons T (1991) Interprovincial disparities in China. Econ Dev Cult Change 39:471–506
Magrini S (1999) The evolution of income disparities among the regions of the European Union. Reg Sci Urban Econ 29:257–281
Mardia K, Dryden IL (1999) The complex Watson distribution and shape analysis. J R Stat Soc Series B Stat Methodol 61:913–926
Mu L (2004) Polygon characterization with the multiplicatively weighted Voronoi diagram. Prof Geogr 56:223–239
Mulmuley K (1994) Computational geometry: an introduction through randomized algorithms. Prentice-Hall, Englewood Cliffs
Okabe A, Boots B, Sugihara K, Chiu SN (2000) Spatial tessellations: concepts and applications of Voronoi diagrams. Wiley, New York
Openshaw S, Alvanides S (1999) Applying geocomputation to the analysis of spatial distributions. In: Longley P, Goodchild M, Maguire D, Rhind D (eds) Geographic information systems: principles and technical issues, vol I. Wiley, New York, pp 267–282
O'Rourke J (1994) Computational geometry in C. Cambridge University Press, Cambridge
Overman HG, Ioannides YM (2001) The cross-sectional evolution of the US city size distribution. J Urban Econ 49:543–566
Peuquet DJ (2002) Representations of space and time. Guilford, New York
Pritchett L (1997) Divergence, big time. J Econ Perspect 11:3–17
Puga D (1999) The rise and fall of regional inequalities. Eur Econ Rev 43:303–334
Quah DT (1993) Empirical cross-section dynamics in economic growth. Eur Econ Rev 37:426–434
Quah DT (1996) Twin peaks: growth and convergence in models of distribution dynamics. Econ J 106:1045–1055
Rawski T (2001) What's happening to China's GDP statistics? China Econ Rev 12:347–354
Rey S, Anselin L (2006) Recent advances in software for spatial analysis in the social sciences. Geogr Anal 38:1–4
Rey S, Janikas M, Smirnov O (2005) Exploratory geovisualization of spatial dynamics. Geocomputation 2005 Proceedings (CDROM)
Rey S (2001) Spatial empirics for regional economic growth and convergence. Geogr Anal 33: 195–214
Rey S (2004a) Spatial analysis of regional income inequality. In: Goodchild M, Janelle D (eds) Spatially integrated social science: examples in best practice. Oxford University Press, Oxford, pp 280–299
Rey S (2004b) Spatial dependence in the evolution of regional income distributions. In: Getis A, Mur J, Zoeller H (eds) Spatial econometrics and spatial statistics. Palgrave, Hampshire, pp 194–213
Rey S, Janikas M (2005) Regional convergence, inequality, and space. J Econ Geogr 5:155–176
Rey S, Janikas M (2006) STARS: space-time analysis of regional systems. Geogr Anal 38:67–86
Rey S, Le Gallo J (2008) Spatial analysis of economic growth convergence. In: Mills TC, Patterson K (eds) Handbook of econometrics. Palgrave, Hampshire
Rey S, Montouri BD (1999) U.S. regional income convergence: a spatial econometric perspective. Reg Stud 33:143–156
Rigby DL, Willmott C (1998) Infrastructure needs in geography and regional science. A Report to the National Science Foundation. http://www.nsf.gov/sbe/bcs/geograph/geoinfra.htm
Sala-i-Martin X (1996) The classical approach to convergence analysis. Econ J 106:1019–1036
Sassen S (1994) Cities in a world economy. Pine Forge, Thousan Oaks

Sidaway JD, Simon D (1990) Spatial policies and uneven development in the "Marxist-Leninist" states of the third world. In Simon D (ed) Third World regional development. Paul Chapman, London, pp 24–38
Sommeiller E (2007) Regional income inequality in the United States, 1913–2003. Ph.D. thesis, University of Delaware
Theil H (1967) Economics and information theory. North Holland, Amsterdam
Tomljanovich M, Vogelsang TJ (2002) Are U.S. regions converging? Using new econometric methods to examine old issues. Empir Econ 27:49–62
Tsionas EG (2000) Regional growth and convergence: evidence from the United States. Reg Stud 34:231–238
Vohra R (1996) How fast do we grow? Growth Change 27:47–54
Wei YHD (2000) Regional development in China: states, globalization and inequality. Routledge, London
Wei YHD (2007) Regional development in China: transitional institutions, embedded globalization, and hybrid economies. Eurasian Geogr Econ 48:16–36
Wei YHD, Ye X (2004) Regional inequality in China: A case study of Zhejiang province. Tijdschr Econ Soc Geogr (J Econ Soc Geogr) 95:44–60
Wei YHD, Ye X (2009) Beyond convergence: Space, scale, and regional inequality in China. Tijdschrift voor Economische en Sociale Geografie. 100:59–80
Yamamoto D (2006) Beyond convergence: regional income disparities in the United States and Japan, 1955–2001. Ph.D. thesis, University of Minnesota
Ye X, Wei YHD (2005) Geospatial analysis of regional development in China: The case of Zhejiang province and the Wenzhou model. Eurasian Geogr Econ 46:342–361

Growth and Spatial Dependence in Europe

Wilfried Koch

1 Introduction

The convergence of European regions has been largely discussed in the empirical literature during the last decade. Two observations are often emphasized. First, the convergence rate among European regions appears to be very slow (Barro and Sala-i-Martin 1991, 1995; Armstrong 1995; Sala-i-Martin 1996a,b). Second, the tools used in the regional science literature show that the geographical distribution of European per capita GDP is highly clustered and characterized by global and local autocorrelation (Armstrong 1995; Ertur et al. 2007; López-Bazo et al. 1999; Le Gallo and Ertur 2003 with a UE15 regional database and Ertur and Koch 2006, with a UE27 enlarged regional database). Many other studies also provide evidence of global and local spatial autocorrelation as Rey and Montouri (1999) for US State data on per capita income throughout the period 1929–1994, Ying (2000) for growth rates of production in the Chinese provinces since the late 1970s, and Conley and Ligon (2002). These authors also develop an empirical approach that explicitly allows for interdependence among countries, and they underline the importance of cross-country spillovers in explaining growth using an international database.

Other empirical studies also show the importance of geography in the diffusion of knowledge and R&D: Keller (2002) suggests that the international diffusion of technology is geographically localized, in the sense that the productivity effects of R&D decline with the geographic distance between countries. Audretsch and Feldman (1996), Jaffe (1989), Acs et al. (1992, 1994), Feldman (1994a,b) and Anselin et al. (1997) have identified the existence of spatially-mediated knowledge spillovers of R&D or academic research effects.

In this context, this paper presents the spatially augmented Solow model developed by Ertur and Koch (2007) that includes technological interdependence among regions in the structural model in order to take into account this global and local

W. Koch
Laboratoire d'Economie et de Gestion, LEG-UMR 5118, Université de Bourgogne,
Pôle d'Economie et de Gestion, BP 21611, 21066 Dijon Cedex, France,
e-mail: wilfried.koch@u-bourgogne.fr

A. Páez et al. (eds.), *Progress in Spatial Analysis*, Advances in Spatial Science,
DOI 10.1007/978-3-642-03326-1_21,

spatial autocorrelation and these neighborhood effects on growth and convergence. Thus, we consider the Solow model (Solow 1956; Swan 1956) with physical capital externalities as suggested by Romer (1986), Krugman (1991a,b) and Grossman and Helpman (1991), among others, who have focused on the role that spillovers of economic knowledge across agents and firms play in generating increasing returns and ultimately economic growth. We also add spatial externalities in the model in order to take into account spatial knowledge spillovers and technological interdependence between regions.

More specifically, in the next section, we assume that technical progress depends on the stock of per worker physical capital, which represents the stock of knowledge as in Romer (1986), in the home region and depends on the stock of knowledge in the neighboring regions, which spills on the technical progress of the home region so as the regions are geographically close. This model leads to an equation for the steady state income level as well as a spatial conditional convergence equation. Then, we present the database and the spatial weights matrix, which is used to model spatial connections between all regions in the sample. Next, we estimate the effects of investment rate, population growth and location on the per worker real income at steady state using a spatial econometric specification. We also estimate the magnitude of physical capital externalities at steady state, which is usually not identified in the literature. Finally, we assess the role played by technological interdependence in growth and convergence processes. To this end, we estimate a spatial version of the conditional convergence equation which leads to a convergence speed close to 2% as generally found in the literature. The last section concludes.

2 A Spatially Augmented Neoclassical Growth Model

2.1 *Production Function and Spatial Externalities*

In this section, we present the Ertur and Koch (2007) model of growth with physical capital externalities and spatial externalities, which implies technological interdependence in Europe between N regions denoted by $i = 1,\ldots,N$. Let us consider an aggregate Cobb-Douglas production function exhibiting constant returns to scale in labor and reproducible physical capital of the form, in region i at time t:

$$Y_i(t) = A_i(t)\, K_i(t)^{\alpha}\, L_i(t)^{1-\alpha} \tag{1}$$

with the standards notations: $Y_i(t)$ the output, $K_i(t)$ the level of reproducible physical capital, $L_i(t)$ the level of labor and $A_i(t)$ the aggregate level of technology defined as:

$$A_i(t) = \Omega(t)\, k_i(t)^{\varphi} \prod_{j \neq i}^{N} A_j(t)^{\gamma w_{ij}} \tag{2}$$

The function describing the aggregate level of technology $A_i(t)$ of any region i depends on three terms. First, as in the neoclassical growth model, we assume that a part of technological progress is exogenous and identical to all regions: $\Omega(t) = \Omega(0)\,e^{\mu t}$ where μ is its constant growth rate. Second, we assume that each region's aggregate level of technology increases with the aggregate level of per worker physical capital $k_i(t) = K_i(t)/L_i(t)$ available in that region.[1] The parameter φ, with $0 \leq \varphi < 1$, describes the strength of home externalities generated by the physical capital accumulation process. Therefore, we follow the well-known Arrow (1962) and Romer's (1986) treatment of knowledge spillover from capital investment. In addition, in the third term, we assume that there are regional externalities emanating from knowledge accumulation in the other regions, which spills over from these neighboring regions j to the considered region i and improves its production efficiency. The regional technological interdependence implied by these regional externalities is measured by the parameter $\gamma \geq 0$. This parameter is assumed identical for each region but the net effect of these spatial externalities on the level of productivity of the firms in a region i depends on the relative spatial connectivity between this region and its neighbors. We represent the technological interdependence between a region i and all the regions belonging to its neighborhood by the spatial friction parameters w_{ij}, for $j = 1, \ldots, N$ and $j \neq i$. These parameters are non negative, non stochastic and finite; we have $0 \leq w_{ij} \leq 1$ and $w_{ij} = 0$ if $i = j$. Moreover, it is assumed that:

$$\sum_{j \neq i}^{N} w_{ij} = 1$$

for $i = 1, \ldots, N$.[2] The more a given region i is connected to its neighbors, the higher w_{ij} is and the more region i benefits from spatial externalities.

Resolving (2) for $A_i(t)$ and replacing the result in the production function (1) written per worker, we obtain:

$$y_i(t) = \Omega(t)^{\frac{1}{1-\gamma}}\, k_i(t)^{u_{ii}} \prod_{j \neq i}^{N} k_j(t)^{u_{ij}} \qquad (3)$$

with:

$$u_{ii} = \alpha + \varphi\left(1 + \sum_{r=1}^{\infty} \gamma^r w_{ii}^{(r)}\right) \quad \text{and} \quad u_{ij} = \varphi \sum_{r=1}^{\infty} \gamma^r w_{ij}^{(r)}$$

with $w_{ij}^{(r)}$ the element of the line i and the column j of the matrix W to the power of r, and $y_i(t) = Y_i(t)/L_i(t)$ the level of per worker output. We note that if there

[1] We assume that all knowledge is embodied in per worker physical capital and not in the level of capital in order to avoid the scale effects (Jones 1995).

[2] This hypothesis allows us to assume a relative spatial connectivity between all regions in order to underline the importance of the geographical neighborhood for economic growth. Moreover, it allows avoiding spatial scale effects and then explosive growth.

are no physical capital externalities, that is $\varphi = 0$, we have $u_{ii} = \alpha$ and $u_{ij} = 0$, and then the production function is written as usual. Finally, in order to warrant local convergence and then avoid explosive or endogenous growth, we suppose that social returns are decreasing: $\alpha + \varphi / (1 - \gamma) < 1$.[3]

As in the textbook neoclassical growth model, we assume that a constant fraction of output s_i is saved and that the labor exogenously grows at the rate n_i for a region i. We also suppose that there is a constant and identical annual rate of depreciation of physical capital for all regions, denoted by δ. We can derive the expression of the per worker output at steady state for an economy i:[4]

$$\begin{aligned}\ln y_i(t)^* = {} & \frac{1}{1-\alpha-\varphi}\ln\Omega(t) + \frac{\alpha+\varphi}{1-\alpha-\varphi}\ln s_i - \frac{\alpha+\varphi}{1-\alpha-\varphi}\ln(n_i+g+\delta) \\ & - \frac{\alpha\gamma}{1-\alpha-\varphi}\sum_{j\neq i}^{N} w_{ij}\ln s_j + \frac{\alpha\gamma}{1-\alpha-\varphi}\sum_{j\neq i}^{N} w_{ij}\ln\left(n_j+g+\delta\right) \\ & + \frac{\gamma(1-\alpha)}{1-\alpha-\varphi}\sum_{j\neq i}^{N} w_{ij}\ln y_j(t)^* \end{aligned} \tag{4}$$

This spatially augmented neoclassical growth model has the same qualitative predictions as the textbook Solow growth model[5] pertaining to the influence of the own saving rate and the own population growth rate on the per worker real income at steady state of a region i. First, the per worker real income at steady state for a region i depends positively on its own saving rate and negatively on its own population growth rate. Second, it can also be shown that the per worker real income for a region i depends positively on saving rates of neighboring regions and negatively on their population growth rates.[6]

2.2 *Transitional Dynamics and Local Convergence*

This model predicts that per worker income in a given region converges to that region's steady state value. Writing the fundamental dynamic equation of Solow including the production function (3), we obtain:

$$\frac{k_i(t)}{k_i(t)} = s_i\Omega(t)^{\frac{1}{1-\gamma}} k_i(t)^{-(1-u_{ii})}\prod_{j\neq i}^{N} k_j(t)^{u_{ij}} - (n_i+\delta) \tag{5}$$

---

[3] See Ertur and Koch (2007) for the proof.

[4] The balanced rate of growth is $g = \mu/[(1-\alpha)(1-\gamma)-\varphi]$.

[5] Note that when $\gamma = 0$, we have the model elaborated by Romer (1986) with $\alpha + \varphi < 1$ and when $\gamma = 0$ and $\varphi = 0$, we have the Solow growth model.

[6] In fact, this equation is written in implicit form. When we write this equation in explicit form, it is possible to evaluate elasticities of the variables. See Ertur and Koch (2007) for more details about the predictions of this model.

The main element behind the convergence result in this model is also diminishing returns to reproducible capital. Physical capital externalities and technological interdependence only slow down the decrease of marginal productivity of physical capital, therefore the convergence result is still valid under the hypothesis $\alpha + \varphi/(1-\gamma) < 1$, in contrast with endogenous growth models where marginal productivity of physical capital is constant.

In addition, our model provides quantitative predictions about the speed of convergence to steady state. As in the literature, the transitional dynamics can be quantified by using a log linearisation of (5) around steady state, for $i = 1, \ldots, N$:

$$\frac{d \ln k_i(t)}{dt} = -(1-u_{ii})(n_i + g + \delta)[\ln k_i(t) - \ln k_i^*] + \sum_{j \neq i}^{N} u_{ij}(n_i + g + \delta)[\ln k_j(t) - \ln k_j^*] \tag{6}$$

We obtain a system of differential linear equations. Let us note $\chi_i = [\ln k_i(t) - \ln k_i^*]$ and $\chi_i(t) = d \ln k_i(t)/dt$, for $i = 1, \ldots, N$, we obtain in matrix form:

$$\dot{\chi}(t) = \mathbf{J}\chi(t) \tag{7}$$

where:

$$\mathbf{J} = -(1-\alpha)\,\mathrm{diag}(n+g+\delta) + \varphi \mathrm{diag}(n+g+\delta)(\mathbf{I} - \gamma\mathbf{W})^{-1} \tag{8}$$

is the matrix of the system, with diag $(n+g+\delta)$ the diagonal matrix with the terms $(n_i + g + \delta)$.[7] The general solution of the system can be written in the following matrix form: $\chi(t) = \mathbf{VDb}$, where $\mathbf{D}$ is the diagonal matrix with the terms $e^{\lambda_J t}$ where λ_J are the eigenvalues of the matrix $\mathbf{J}$, $\mathbf{V}$ the matrix of characteristic vectors associated with the eigenvalues of $\mathbf{J}$ and $\mathbf{b}$ a vector of constant which we can evaluate with the initial condition. Indeed, since the matrix $\mathbf{J}$ is d-stable, its eigenvalues are negatives and so: $\chi(0) = \mathbf{Vb}$, then: $\mathbf{b} = \mathbf{V}^{-1}\chi(0)$. Finally the general solution can be written in the following form: $\chi(t) = \mathbf{VDV}^{-1}\chi(0)$, or:

$$\ln \mathbf{k}(t) - \ln \mathbf{k}^* = \mathbf{VDV}^{-1}[\ln \mathbf{k}(0) - \ln \mathbf{k}^*] \tag{9}$$

and subtracting both sides by ln $\mathbf{k}(0)$ and rearranging terms:

$$\ln \mathbf{k}(t) - \ln \mathbf{k}(0) = -\left(\mathbf{I} - \mathbf{VDV}^{-1}\right)\ln \mathbf{k}(0) + \left(\mathbf{I} - \mathbf{VDV}^{-1}\right)\ln \mathbf{k}^* \tag{10}$$

Replacing ln $\mathbf{k}^*$ by its expression in matrix form:

$$\ln \mathbf{k}^* = [(1-\alpha)\mathbf{I} - \varphi(\mathbf{I} - \gamma\mathbf{W})^{-1}][(\mathbf{I} - \gamma\mathbf{W})^{-1}\Omega + \mathbf{S}] \tag{11}$$

[7] See Ertur and Koch (2007) for a proof of local convergence.

where $\mathbf{S}$ is the $(N \times 1)$ vector of logarithms of saving rate divided by the effective rate of depreciation, we obtain after rearranging terms:

$$\begin{aligned}
&\ln \mathbf{k}(t) - \ln \mathbf{k}(0) = -\left(\mathbf{I} - \mathbf{VDV}^{-1}\right) \ln \mathbf{k}(0) \\
&+ \frac{\varphi}{1-\alpha}\left(\mathbf{I} - \mathbf{VDV}^{-1}\right)(\mathbf{I} - \gamma \mathbf{W})^{-1} \ln \mathbf{k}(0) \\
&+ \frac{1}{1-\alpha}\left(\mathbf{I} - \mathbf{VDV}^{-1}\right)(\mathbf{I} - \gamma \mathbf{W})^{-1} \Omega + \frac{1}{1-\alpha}\left(\mathbf{I} - \mathbf{VDV}^{-1}\right) \mathbf{S} \\
&+ \frac{\varphi}{1-\alpha}\left(\mathbf{I} - \mathbf{VDV}^{-1}\right)(\mathbf{I} - \gamma \mathbf{W})^{-1}\left(\mathbf{I} - \mathbf{VDV}^{-1}\right)^{-1}\left[\ln \mathbf{k}(t) - \ln \mathbf{k}(0)\right] \qquad (12)
\end{aligned}$$

This equation shows that the convergence process of a region i is more complicated than the usual equation in the literature since it depends not only on usual variables as initial level of per worker output, the saving rate and the population growth rate, but also on the same variables in the neighboring regions. It also depends on the rate of growth of these neighboring regions reflecting global technological interdependence. However, we can note that if there are no physical capital externalities, that is $\varphi = 0$, this equation reduces to the traditional conditional convergence equation except for the constant term. Another case is of interest: when we consider the case of unconditional convergence process, we have $n_i = n$ for all $i = 1, \ldots, N$, and then the eigenvalues of the matrix $\mathbf{J}$ can be rewritten in function of the eigenvalues of the $\mathbf{W}$ matrix denoted by λ_W. Indeed, we have:

$$\lambda_J = -\left(1 - \alpha - \frac{\varphi}{1 - \gamma \lambda_W}\right)(n + g + \delta) \qquad (13)$$

3 Data and Spatial Weight Matrix

All data are extracted from the Cambridge database. More precisely, we consider 204 European regions belonging to 17 countries over the 1977–2000 period at NUTS2 level for Belgium (11), Denmark (1), Germany (31), Greece (13), Spain (16), France (22), Ireland (2), Italy (20), Luxembourg (1), the Netherlands (12), Austria (9), Portugal (1), Finland (6), Sweden (8), United Kingdom (37), Norway (7), Switzerland (7). We measure n as the average growth rate of the working-age population (ages 15–64), per worker real income is measured by the GVA (Gross Value Added) divided by the number of workers, and finally the saving rate s is measured as the average share of gross investment in GVA.

The Markov-matrix $\mathbf{W}$, containing the terms w_{ij}, corresponds to the so called spatial weights matrix commonly used in spatial econometrics to model spatial interdependence between regions or countries (Anselin 1988). More precisely, each region is connected to a set of neighboring regions by means of a purely spatial pattern introduced exogenously in $\mathbf{W}$. The elements w_{ii} on the diagonal are set to zero whereas the elements w_{ij} indicate the way the region i is spatially connected to the region j. In order to normalize the outside influence upon each region, the weights matrix is standardized such that the elements of a row sum up to one. For the variable $\mathbf{x}$, this transformation means that the expression $\mathbf{Wx}$, called the spatial lag variable, is simply the weighted average of the neighboring observations.

Various matrices are considered in the literature: a simple binary contiguity matrix, a binary spatial weights matrix with a distance-based critical cut-off, above which spatial interactions are assumed negligible, more sophisticated generalized distance-based spatial weights matrices with or without a critical cut-off. The notion of distance can be quite general and different functional forms based on distance decay can be used (for example inverse distance, inverse squared distance, negative exponential etc.). The critical cut-off can be the same for all regions or can be defined to be specific to each region leading in the latter case, for example, to k-nearest neighbors weights matrices when the critical cut-off for each region is determined so that each region has the same number of neighbors.

It is important to stress that the connectivity terms w_{ij} should be exogenous to the model to avoid the identification problems raised by Manski (1993) in social sciences. This is the reason why we consider pure geographical distance, more precisely great circle distance between centroid, which is indeed strictly exogenous; the functional form we consider is simply the k-nearest neighbors weights matrix $\mathbf{W}(k)$ with the general term defined as follows in standardized form $[w(k)_{ij}]$:

$$w(k)_{ij} = w(k)^*_{ij} / \sum w(k)^*_{ij} \quad \text{with} \quad w(k)^*_{ij} = \begin{cases} 0 \text{ if } i = j \\ 1 \text{ if } d_{ij} d_i(k) \\ 0 \text{ if } d_{ij} > d_i(k) \end{cases} \tag{14}$$

where d_{ij} is the great circle distance between regional centroid and $d_i(k)$ is a critical cut-off distance defined for each region i. More precisely, $d_i(k)$ is the k-th order smallest distance between regions i and j so that each region i has exactly k neighbors. In this analysis, we consider $k = 10$, 15 and 20.

4 Analysis and Results

4.1 Empirical Model and Spatial Econometric Framework

In this section, we follow Mankiw et al. (1992) in order to evaluate the impact of saving, population growth and location on real income. Taking (4), we find that the per worker real income along the balanced growth path, at a given time ($t = 0$ for simplicity) is:

$$\ln\left[\frac{Y_i}{L_i}\right] = \beta_0 + \beta_1 \ln s_i + \beta_2 \ln(n_i + g + \delta) + \theta_1 \sum_{j \neq i}^{N} w_{ij} \ln s_j$$
$$+ \theta_2 \sum_{j \neq i}^{N} w_{ij} \ln\left(n_j + g + \delta\right) + \rho \sum_{j \neq i}^{N} w_{ij} \ln\left[\frac{Y_j}{L_j}\right] + \varepsilon_i \quad (15)$$

where $\frac{1}{1-\alpha-\varphi}\ln\Omega(0) = \beta_0 + \varepsilon_i$ for $i = 1, \ldots, N$, with β_0 a constant and ε_i a region-specific shock since the term $\Omega(0)$ reflects not just technology but also resource endowments, climate, and so on..., and then it may differ across regions. We also suppose that $g + \delta = 0.05$ as used in the literature since Mankiw et al. (1992) and Romer (1989). We have finally the following theoretical constraints between coefficients: $\beta_1 = -\beta_2 = (\alpha + \varphi) / (1 - \alpha - \varphi)$ and $\theta_1 = -\theta_2 = (\alpha + \gamma) / (1 - \alpha - \varphi)$. Equation (15) is our basic econometric specification in this section.

In the spatial econometrics literature, this kind of specification, including the spatial lags of both endogenous and exogenous variables, is referred to as the spatial Durbin model (see Anselin 1988, 2001), we have in matrix form:

$$\mathbf{y} = \mathbf{X}\beta + \mathbf{W}\mathbf{X}\theta + \rho\mathbf{W}\mathbf{y} + \varepsilon \quad (16)$$

where $\mathbf{y}$ is the $(N \times 1)$ vector of logarithms of real income per worker, $\mathbf{X}$ the $(N \times 3)$ matrix with the constant term, the vectors of logarithms of investment rate and the logarithms of physical capital effective rates of depreciation, $\mathbf{W}$ the $(N \times N)$ spatial weights matrix, $\beta' = [\beta_0, \beta_1, \beta_2]$, $\theta' = [\theta_1, \theta_2]$ and the spatial autocorrelation coefficient is $\rho = \gamma(1-\alpha) / (1 - \alpha - \varphi)$.[8] ε is the $(N \times 1)$ vector of errors supposed identically and normally distributed so that $\varepsilon \sim N\left(0, \sigma^2\mathbf{I}\right)$.

In the first column of Table 1, we estimate the textbook Solow growth model using the White heteroskedasticity consistent covariance matrix estimator. The coefficients of saving and population growth have the predicted signs. However, the coefficients are weakly significant and the effect of saving rate is lower than as expected. The overidentifying restriction is not rejected and the estimated capital share is close to 0.2 the lower bound generally admitted for this parameter.

The Solow growth model is however misspecified since it omits variables due to technological interdependence and physical capital externalities.

Indeed, we can write the spatially augmented Solow growth model in the following matrix form:

$$\mathbf{y} = \frac{\alpha}{1-\alpha}\mathbf{S} + \frac{\varphi}{1-\alpha}(\mathbf{I} - \gamma\mathbf{W})^{-1}\ln\mathbf{k}^* + (\mathbf{I} - \gamma\mathbf{W})^{-1}\varepsilon \quad (17)$$

with $\mathbf{S}$ the $(N \times 1)$ vector of logarithms of investment rate divided by the effective rate of depreciation. Therefore the error term in the Solow growth model contains

[8] In practice, the spatially lagged constant is not included in $\mathbf{WX}$, since there is an identification problem for row-standardized $\mathbf{W}$ (the spatial lag of a constant is the same as the original variable).

Table 1 OLS and spatial error model (level model)

Model	OLS	SEM-MLE		
		W10	W15	W20
Unrestricted regression				
Constant	10.256	10.256	10.071	9.678
	(0.000)	(0.000)	(0.000)	(0.000)
$\ln s_i$	0.292	0.262	0.262	0.269
	(0.074)	(0.068)	(0.057)	(0.053)
$\ln(n_i + 0.05)$	−0.135	−0.077	−0.115	−0.136
	(0.566)	(0.666)	(0.522)	(0.448)
γ	–	0.860	0.902	0.943
		(0.000)	(0.000)	(0.000)
Restricted regression				
Constant	9.862	9.795	9.715	9.378
	(0.000)	(0.000)	(0.000)	(0.000)
$\ln s_i - \ln(n_i + 0.05)$	0.245	0.199	0.215	0.225
	(0.101)	(0.121)	(0.083)	(0.069)
γ	–	0.863	0.898	0.941
		(0.000)	(0.000)	(0.000)
Test of restriction	0.237	0.939	0.591	0.473
(Wald/LR/PMP)	(0.627)	(0.333)	(0.442)	(0.491)
Implied α	0.197	0.166	0.177	0.184
	(0.000)	(0.000)	(0.000)	(0.000)

p-values are in parentheses; p-values for the implied parameters are computed using the delta method. The White heteroskedasticity consistent covariance matrix estimator is used for statistical inference in the OLS estimation. LR is the likelihood ratio test. PMP stands for posterior model probability

omitted information since we can rewrite it:

$$\varepsilon_{Solow} = \frac{\varphi}{1-\alpha}(\mathbf{I} - \gamma\mathbf{W})^{-1} \ln \mathbf{k}^* + (\mathbf{I} - \gamma\mathbf{W})^{-1} \varepsilon \tag{18}$$

We also note the presence of spatial autocorrelation in the error term even if there is no physical capital externalities, and then the presence of technological interactions between all countries through the inverse spatial transformation $(\mathbf{I} - \gamma\mathbf{W})^{-1}$.

In Table 3, we estimate the spatially augmented Solow growth model with the maximum likelihood estimation method and the bayesian heteroskedastic MCMC estimation method.[9] Many aspects of the results support the model. First, all the coefficients have the predicted signs and the spatial autocorrelation coefficient, ρ, is highly positively significant. Second, the coefficients of saving rates of the region i and its neighboring regions j are significant. Third, the joint theoretical restriction $\beta_1 = -\beta_2$ and $\theta_2 = -\theta_1$ is not rejected. Finally the α implied by the coefficients

[9] James LeSage provides a function to estimate this model in his Econometric Toolbox for Matlab (http://www.spatial-econometrics.com). The regularity conditions of the maximum likelihood estimators are described in Lee (2004) and the bayesian heteroskedastic MCMC estimation method is developed by LeSage (1997). Endogeneity problem of explanatory variables is an important issue in this literature and could be taken into account in future research using recent development of spatial econometrics like Fingleton and Le Gallo (2008).

Table 2 OLS and spatial error model (level model)

Model	SEM-Bayesian Heter.		
	W10	W15	W20
Unrestricted regression			
Constant	10.399	10.313	10.295
	(0.000)	(0.000)	(0.000)
$\ln s_i$	0.142	0.146	0.147
	(0.181)	(0.167)	(0.167)
$\ln(n_i + 0.05)$	0.016	−0.013	−0.017
	(0.482)	(0.459)	(0.452)
γ	0.780	0.806	0.834
	(0.000)	(0.000)	(0.000)
Restricted regression			
Constant	10.010	9.986	9.969
	(0.000)	(0.000)	(0.000)
$\ln s_i - \ln(n_i + 0.05)$	0.090	0.102	0.109
	(0.263)	(0.225)	(0.205)
γ	0.780	0.808	0.836
	(0.000)	(0.000)	(0.000)
Test of restriction	*rest./unrest.*		
(Wald/LR/PMP)	0.70/0.30	0.61/0.39	0.64/0.36
Implied α	0.082	0.093	0.098

See Table 1 for notes

in the constrained regression is significantly close to one-third as expected. The coefficient γ, representing the strength of spatial externalities, is very strong since it is higher than 1. This result shows the importance of spatial externalities in the distribution of income in Europe. In contrast, the φ estimated is negative but non-significant which indicates that there are not physical capital externalities in the European regions. This result is consistent with the evidence against the importance of permanent within-industry knowledge spillovers for growth at the regional and urban level (see Glaeser et al. 1992, for instance). More specifically, we can test the absence of physical capital externalities represented by φ since $\varphi = 0$ implies in the specification (15) the following expression:

$$\ln\left[\frac{Y_i}{L_i}\right] = \beta_0' + \beta_1' \ln s_i + \beta_2' \ln(n_i + g + \delta) + \theta_1' \sum_{j \neq i}^{N} w_{ij} \ln s_j + \theta_2' \sum_{j \neq i}^{N} w_{ij} \ln\left(n_j + g + \delta\right) + \gamma \sum_{j \neq i}^{N} w_{ij} \ln\left[\frac{Y_j}{L_j}\right] + \varepsilon_i \quad (19)$$

Table 3 Spatial Durbin model (level model)

Model	SDM-MLE			SDM-Bayesian Heter.		
	W10	W15	W20	W10	W15	W20
Unrestricted regression						
Constant	1.628	1.407	0.722	1.469	1.579	1.701
	(0.198)	(0.347)	(0.689)	(0.101)	(0.128)	(0.160)
$\ln s_i$	0.303	0.295	0.307	0.174	0.187	0.202
	(0.037)	(0.032)	(0.027)	(0.083)	(0.061)	(0.041)
$\ln(n_i + 0.05)$	−0.102	−0.145	−0.178	−0.021	−0.074	−0.124
	(0.569)	(0.417)	(0.321)	(0.444)	(0.333)	(0.249)
$W \ln s_j$	−0.504	−0.645	−0.762	−0.355	−0.534	−0.601
	(0.059)	(0.021)	(0.016)	(0.067)	(0.013)	(0.014)
$W \ln(n_j + 0.05)$	0.330	0.502	0.486	0.157	0.415	0.656
	(0.409)	(0.298)	(0.404)	(0.334)	(0.175)	(0.111)
ρ	0.872	0.907	0.943	0.862	0.884	0.916
	(0.000)	(0.000)	(0.000)	(0.000)	(0.000)	(0.000)
Common factor test	1.897	3.767	3.941	rest./unrest.		
(LR/PMP)	(0.387)	(0.152)	(0.139)	1.00/0.0	1.00/0.0	1.00/0.0
Restricted regression						
Constant	1.597	1.430	1.112	1.574	1.595	1.339
	(0.001)	(0.002)	(0.006)	(0.000)	(0.000)	(0.000)
$\ln s_i - \ln(n_i + 0.05)$	0.233	0.248	0.264	0.142	0.166	0.192
	(0.074)	(0.046)	(0.034)	(0.111)	(0.062)	(0.041)
$W (\ln s_i - \ln(n_i + 0.05))$	−0.431	−0.598	−0.684	−0.299	−0.512	−0.627
	(0.057)	(0.012)	(0.010)	(0.070)	(0.007)	(0.003)
ρ	0.867	0.903	0.942	0.862	0.883	0.919
	(0.000)	(0.000)	(0.000)	(0.000)	(0.000)	(0.000)
Common factor test	1.716	3.735	3.853	rest./unrest.		
(LR/PMP)	(0.190)	(0.053)	(0.050)	1.00/0.0	1.00/0.0	1.00/0.0
Test of restriction	1.120	0.623	0.562	rest./unrest.		
(LR/PMP)	(0.571)	(0.732)	(0.755)	0.57/0.43	0.55/0.45	0.55/0.45
Implied α	0.332	0.398	0.421	0.257	0.367	0.406
	(0.005)	(0.000)	(0.000)			
Implied φ	−0.143	−0.200	−0.212	−0.133	−0.224	−0.245
	(0.115)	(0.016)	(0.010)			
Implied γ	1.052	1.203	1.286	1.016	1.196	1.298
	(0.000)	(0.000)	(0.000)			

p-values are in parentheses; p-values for the implied parameters are computed using the delta method. LR is the likelihood ratio test. PMP stands for posterior model probability

with $\beta_1' = -\beta_2' = \alpha/(1-\alpha)$, $\theta_2' = -\theta_1' = \alpha\gamma/(1-\alpha)$ hence $\theta_1' + \beta_1'\gamma = 0$ and $\theta_2' + \beta_2'\gamma = 0$. Specification (19) is the so-called constrained spatial Durbin model which is formally equivalent to a spatial error model written in matrix form:

$$\mathbf{y} = \mathbf{X}\beta' + \varepsilon_{Solow} \quad \text{and} \quad \varepsilon_{Solow} = \gamma \mathbf{W} \varepsilon_{Solow} + \varepsilon \tag{20}$$

where $\beta' = [\beta'_0, \beta'_1, \beta'_2]$ and ε_{Solow} is the same as above with $\varphi = 0$. Hence, we have the textbook Solow growth model with spatial autocorrelation in the errors terms.

We estimate the spatial error model in the subsequent columns of Tables 1 and 2 using the maximum likelihood estimation method and the bayesian heteroskedastic MCMC estimation method. We note that the coefficients have the predicted signs and the spatial autocorrelation coefficient in error term, γ, is also highly positively significant. We can test the non-linear restrictions with the common factor test (Burridge 1981) using the likelihood ratio test and the posterior model probability (PMP). The LR tests cannot reject the non-linear restrictions and the PMP tests conclude in favor of the restricted model against the unrestricted model. This direct test also supports the absence of physical capital externalities.

Finally, we should note that these regressions based on the methodology proposed by Mankiw et al. (1992) are valid only if the regions are at their steady states or if deviations from steady state are random. So, as already shown by Jones (1997) with international data, most of the regions in Europe have probably not reached their steady-state level. Therefore, in order to study more precisely the distribution of real income per worker in Europe, we must take into account out-of-steady-state dynamics with a spatial conditional convergence.

4.2 A Spatial Conditional Convergence Model

The spatial convergence model cannot be estimated directly with (12). In this section, we assume, with the results of the previous section, that there are no physical capital externalities ($\varphi = 0$). This implies that the matrix **J** reduces to a diagonal matrix with the terms $-(1-\alpha)(n+g+\delta)$ on its diagonal.[10]

As a result, the resolution is now identical to the traditional problem in the growth literature. Indeed, for each region $i = 1, \ldots, N$, the (6) can be rewritten for the per worker income:[11]

$$\frac{d \ln y_i(t)}{dt} = \frac{\mu}{1-\gamma} - (1-\alpha)(n+g+\delta)\left[\ln y_i(t) - \ln y_i^*\right] \tag{21}$$

The solution for $\ln y_i(t)$, subtracting $\ln y_i(0)$, the per worker real income at some initial date, from both sides, is:

[10] If the physical capital externalities are different from 0, Ertur and Koch (2007) propose to simplify the system assuming that the gaps of economics with respect to their own steady states are proportionate. They give a local version of the spatial β-convergence model displayed in this paper.

[11] We also suppose that the speed of convergence is identical for all regions as in the traditional literature about conditional convergence (Barro and Sala-i-Martin 1991, 1992, 1995; Mankiw et al. 1992). In fact, in the Solow growth model, each speed of convergence depends on each country because of the population rates of growth n_i in its expression. See Durlauf et al. (2001), Ertur et al. (2006) or Ertur and Koch (2007) for local version of the Solow growth model.

$$\ln y_i(t) - \ln y_i(0) = \left(1 - e^{-\lambda t}\right) \frac{\mu}{1-\gamma} \frac{1}{\lambda} - \left(1 - e^{-\lambda t}\right) \ln y_i(0) + \left(1 - e^{-\lambda t}\right) \ln y_i^* \tag{22}$$

The model predicts conditional convergence since the growth of per worker real income is a negative function of the initial level of income per worker, but only after controlling for the determinants of the steady-state. Rewrite (22) in matrix form:

$$\ln \mathbf{y}(t) - \ln \mathbf{y}(0) = \left(1 - e^{-\lambda t}\right) [\mathbf{C} - \ln \mathbf{y}(0) + \ln \mathbf{y}^*]$$

where $\ln\mathbf{y}(0)$ is the $(N \times 1)$ vector of the logarithms of initial level of real income per worker, $\ln \mathbf{y}^*$ is the $(N \times 1)$ vector of the logarithms of real income per worker at steady-state, $\mathbf{C}$ is the $(N \times 1)$ vector of constant. Introducing (4) in matrix form:

$$\ln \mathbf{y}^* = (\mathbf{I} - \gamma \mathbf{W})^{-1} \left[\frac{1}{1-\alpha} \Omega + \frac{\alpha}{1-\alpha} \mathbf{S} - \frac{\alpha\gamma}{1-\alpha} \mathbf{WS} \right]$$

where $\mathbf{S}$ is the $(N \times 1)$ vector of logarithms of saving rate divided by the effective rate of depreciation, premultiplying both sides by the inverse of $(\mathbf{I} - \rho \mathbf{W})^{-1}$ and rearranging terms we obtain:

$$\begin{aligned} \ln \mathbf{y}(t) - \ln \mathbf{y}(0) = {} & \left(1 - e^{-\lambda t}\right) \left(\mathbf{C} + \frac{1}{1-\alpha} \Omega\right) - \left(1 - e^{-\lambda t}\right) \ln \mathbf{y}(0) \\ & + \gamma \left(1 - e^{-\lambda t}\right) \mathbf{W} \ln \mathbf{y}(0) + \frac{\alpha}{1-\alpha} \left(1 - e^{-\lambda t}\right) \mathbf{S} \\ & - \frac{\alpha\gamma}{1-\alpha} \left(1 - e^{-\lambda t}\right) \mathbf{WS} + \gamma \mathbf{W} [\ln \mathbf{y}(t) - \ln \mathbf{y}(0)] \end{aligned} \tag{23}$$

Finally, dividing by T on both sides, we can rewrite this equation for a region i:

$$\begin{aligned} \frac{\ln y_i(t) - \ln y_i(0)}{T} = {} & \beta_0 + \beta_1 \ln y_i(0) + \beta_2 \ln s_i + \beta_3 \ln(n_i + g + \delta) \\ & + \theta_2 \sum_{j \neq i}^{N} w_{ij} \ln s_j + \theta_3 \sum_{j \neq i}^{N} w_{ij} \ln\left(n_j + g + \delta\right) \\ & + \theta_1 \sum_{j \neq i}^{N} w_{ij} \ln y_j(0) + \gamma \sum_{j \neq i}^{N} w_{ij} \frac{\ln y_j(t) - \ln y_j(0)}{T} + \varepsilon_i \end{aligned} \tag{24}$$

where:

$$\beta_0 = \left(1 - e^{-\lambda T}\right) \left(\frac{\mu}{1-\gamma} \frac{1}{\lambda} + \frac{1}{1-\alpha} \Omega(T) \right)$$

is a constant, and:

$$\beta_1 = -\frac{\left(1-e^{-\lambda T}\right)}{T}, \quad \beta_2 = -\beta_3 = \frac{\left(1-e^{-\lambda T}\right)}{T}\frac{\alpha}{1-\alpha}, \quad \theta_1 = \frac{\left(1-e^{-\lambda T}\right)}{T}\gamma,$$

$$\theta_3 = -\theta_2 = \frac{\left(1-e^{-\lambda T}\right)}{T}\frac{\alpha\gamma}{1-\alpha}$$

In matrix form, we have the constrained spatial Durbin model which is estimated as the model in the previous section. We note that this empirical specification is very close to empirical studies in the recent growth literature using geographical data and applying the appropriate spatial econometric tools (see for example Ertur et al. 2007; Fingleton 1999; Le Gallo et al. 2003). However, the model in this paper is directly linked to the theoretical model.

In the first column of Table 4, we estimate a model of unconditional convergence. The results show that there is conditional convergence between European regions since the coefficient on the initial level of per worker income is negative and strongly significant. Therefore, there is tendency for poor regions to grow faster on average than rich regions in Europe. Note that this result is different to the traditional result in the literature about the failure of income convergence in international cross-countries (De Long 1988; Romer 1987; Mankiw et al. 1992). We estimate the convergence predictions of the textbook Solow model in the second column of Table 4. We report regressions of growth rate over the period 1977 to 2000 on the logarithm of per worker income in 1977, controlling for investment rate and growth of working-age population. The coefficient on the initial level of per worker income is also significantly negative; in other words, there is strong evidence of conditional convergence. The results also support the predicted signs of investment rate and working-age population growth rate. However, the speed of convergence associated with both estimations is close to 0.7% far below 2% usually found in the convergence literature (Barro and Sala-i-Martin 1995 for instance) suggesting that

Table 4 OLS and spatial error model (convergence model)

Model	OLS-un.	OLS-cond.
Unrestricted regression		
Constant	0.085	0.073
	(0.000)	(0.000)
$\ln y_{1977}$	−0.007	−0.007
	(0.000)	(0.000)
$\ln s_i$	–	0.019
		(0.000)
$\ln(n_i + 0.05)$	–	−0.013
		(0.105)
γ	–	–
Implied λ	0.008	0.007
	(0.000)	(0.000)

p-values are in parentheses; p-values for the implied parameters are computed using the delta method. The White heteroskedasticity consistent covariance matrix estimator is used for statistical inference in the OLS estimation

Table 5 OLS and spatial error model (convergence model)

Model	SEM-MLE			SEM-Bayesian Heter.		
	W10	W15	W20	W10	W15	W20
Unrestricted regression						
Constant	0.114	0.115	0.114	0.109	0.109	0.105
	(0.000)	(0.000)	(0.000)	(0.000)	(0.000)	(0.000)
$\ln y_{1977}$	−0.011	−0.011	−0.011	0.009	−0.009	−0.009
	(0.000)	(0.000)	(0.000)	(0.000)	(0.000)	(0.000)
$\ln s_i$	0.028	0.027	0.025	0.026	0.026	0.023
	(0.000)	(0.000)	(0.000)	(0.000)	(0.000)	(0.000)
$\ln(n_i + 0.05)$	−0.017	−0.017	−0.016	−0.011	−0.012	−0.011
	(0.001)	(0.000)	(0.001)	(0.075)	(0.059)	(0.066)
γ	0.668	0.736	0.762	0.589	0.650	0.661
	(0.000)	(0.000)	(0.000)	(0.000)	(0.000)	(0.000)
Implied λ	0.012	0.013	0.013	0.010	0.010	0.010
	(0.000)	(0.000)	(0.000)			

See Table 5 for notes

the process of convergence is indeed very weak. SEM versions of the conditional convergence model are in Table 5.

The textbook Solow model is misspecified since it omits variables due to regional technological interdependence. Therefore, as in the previous section, the error terms of the Solow model contains omitted information and are spatially autocorrelated. In Table 6, we estimate the spatially augmented Solow model. Many aspects of the results support this model. First, all the coefficients are significant and have the predicted signs. The spatial autocorrelation coefficient ρ is highly positively significant which shows the importance of the role played by regional technological interdependence on the convergence process. Second, the coefficient on the initial level of per worker income is significantly negative, so there is strong evidence of conditional convergence after controlling for those variables determining the steady state according to the spatially augmented Solow model says. Third, the λ implied by the coefficient on the initial level of income is about 1.4% which is closer to the value usually found about the speed of convergence in the literature. However, the common factor test is strongly rejected whatever the test strategy (LR or PMP) or the spatial weights matrix used. The theoretical non-linear constraints are then rejected by the data, so we cannot conclude precisely about the assumption of the absence of physical capital externalities ($\varphi = 0$). The spatial error model implied by this hypothesis fits the data well since all the coefficients are significant, have the predicted signs and the implied λ is about 1.2%, a value less by those implied by the spatial Durbin model.

5 Conclusion

In this chapter, we considered a neoclassical growth model, which explicitly takes into account technological interdependence between regions under the form of spatial externalities. The qualitative predictions of this spatially augmented Solow

Table 6 Spatial Durbin model (convergence model)

Model	SDM-MLE			SDM-Bayersian Heter.		
	W10	W15	W20	W10	W15	W20
Unrestricted regression						
Constant	−0.001	−0.036	−0.048	0.016	−0.014	−0.012
	(0.979)	(0.415)	(0.389)	(0.310)	(0.368)	(0.405)
$\ln y_{1977}$	−0.012	−0.012	−0.013	−0.010	−0.011	−0.010
	(0.000)	(0.000)	(0.000)	(0.000)	(0.000)	(0.000)
$\ln s_i$	0.031	0.027	0.024	0.032	0.023	0.026
	(0.000)	(0.000)	(0.000)	(0.000)	(0.000)	(0.000)
$\ln(n_i + 0.05)$	−0.019	−0.018	−0.016	−0.008	−0.008	−0.007
	(0.000)	(0.000)	(0.001)	(0.098)	(0.107)	(0.130)
$W \ln y_{1977}$	0.010	−0.011	0.012	0.009	0.010	0.010
	(0.000)	(0.000)	(0.000)	(0.000)	(0.000)	(0.000)
$W \ln s_j$	−0.041	−0.041	−0.036	−0.040	−0.041	−0.038
	(0.000)	(0.000)	(0.000)	(0.000)	(0.000)	(0.000)
$W \ln (n_j + 0.05)$	0.015	0.006	0.002	0.012	0.005	0.007
	(0.165)	(0.672)	(0.922)	(0.125)	(0.334)	(0.332)
ρ	0.447	0.459	0.499	0.500	0.483	0.519
	(0.000)	(0.000)	(0.000)	(0.000)	(0.000)	(0.001)
Common factor test	18.665	16.584	10.323	rest./unrest.		
(LR/PMP)	(0.000)	(0.001)	(0.016)	0.00/1.00	0.00/1.00	0.00/1.00
Implied λ	0.014	0.014	0.015	0.012	0.012	0.012
	(0.000)	(0.000)	(0.000)			

p-values are in parentheses; p-values for the implied parameters are computed using the delta method. LR is the likelihood ratio test. PMP stands for posterior model probability

model provided a better understanding of the important role played by geographical location and neighborhood effects in the growth and convergence processes. In addition, the econometric model leads to estimates of structural parameters close to predicted values. The estimated capital share parameter is close to one-third, but the physical capital externalities are not significant, so that we can conclude to absence of Marshallian externalities in European Regions. This result is close to those found in the literature as Glaeser et al. (1992) for instance. The strong value of the technological parameter is consistent with the high spatial autocorrelation usually found in the regional science literature and also shows the important role played by technological interdependence in the economic growth and income distribution processes.

Our results are then important to better understand the phenomena of spatial autocorrelation generally found in the spatial distribution of income and in the regional economic growth and convergence. Moreover, the empirical consequences show that the traditional econometric results are misspecified, since they omit spatially autocorrelated errors and spatially autoregressive variable.

Acknowledgements I would like to thank Kristian Behrens, Alain Desdoigts, Cem Ertur, Julie Le Gallo, Diego Legros as well as participants at the Workshop on Spatial Econometrics, Kiel, Germany, April 2005 and at the 45th European Congress of the Regional Science Association, Amsterdam, Août 2005, for valuable comments and suggestions. The usual disclaimer applies.

References

Acs ZJ, Audretsch DB, Feldman MP (1992) Real effects of academic rechearch: comment. Am Econ Rev 82:363–367

Acs ZJ, Audretsch DB, Feldman MP (1994) R&D spillovers and recipient firm size. Rev Econ Stat 76:336–340

Anselin L (1988) Spatial econometric: methods and model. Kluwer, Dordrecht

Anselin L (2001) Spatial econometrics. In: Baltagi B (ed) Companion to econometrics. Basil Blackwell, Oxford, pp 310–330

Anselin L, Varga A, Acs Z (1997) Local geographic spillovers between university research and high technology institutions. J Urban Econ 42:422–448

Armstrong H (1995) An appraisal of the evidence from cross-sectional analysis of the regional growth process within the European Union. In: Armstrong H, Vickerman R (eds) Convergence and divergence among European Union. Pion, London, pp 40–65

Arrow K (1962) The economic implications of learning by doing. Rev Econ Stud 29:155–173

Audretsch DB, Feldman MP (1996) R&D Spillovers and the geography of innovation and production. Am Econ Rev 86:630–640

Barro RJ, Sala-i-Martin X (1991) Convergence across states and regions. Brookings Pap Econ Act 107–182

Barro RJ, Sala-i-Martin X (1992) Convergence. J Polit Econ 100:223–251

Barro RJ, Sala-i-Martin X (1995) Economic growth theory. McGraw-Hill, Boston

Burridge P (1981) Testing for a common factor in a spatial autoregressive model. Environ Plann Series A 13:795–800

Conley TG, Ligon E (2002) Economic distance and cross-country spillovers. J Econ Growth 7:157–187

De Long JB (1988) Productivity growth, convergence and welfare: comment. Am Econ Rev 78:1138–1154

Durlauf SN, Kourtellos A, Minkin A (2001) The local Solow growth model. Eur Econ Rev 45:928–940

Ertur C, Koch W (2006) Regional disparities in the European Union and the enlargement process: an exploratory spatial data analysis, 1995–2000. Ann Reg Sci 40:723–765

Ertur C, Koch W (2007) Growth, technological interdependence and spatial externalities: theory and evidence. J Appl Econom 22: 1033–1062

Ertur C, Le Gallo J, Baumont C (2006) The European regional convergence process, 1980–1995: do spatial regimes and spatial dependence matter? Int Reg Sci Rev 29:2–34

Ertur C, LeSage J, Le Gallo J (2007) Local versus global convergence in Europe: a Bayesian spatial econometrics approach. Rev Reg Stud 37:82–108

Feldman MP (1994a) Knowledge complementary and innovation. Small Bus Econ 6:363–372

Feldman MP (1994b) The geography of innovation. Kluwer, Boston

Fingleton B (1999) Estimates of time to economic convergence: an analysis of regions of European Union. Int Econ Rev 22:5–34

Fingleton B, Le Gallo J (2008) Estimating spatial models with endogenous variables, a spatial lag and spatially dependent disturbances: finite sample properties. Pap Reg Sci 87:319–339

Glaeser EL, Kallal HD, Scheinkman JA, Shleifer A (1992) Growth in cities. J Polit Econ 100:1126–1152

Grossman G, Helpman E (1991) Innovation and growth in the global economy. MIT, Cambridge, MA
Jaffe AB (1989) Real effects of academic research. Am Econ Rev 79:957–970
Jones CI (1995) R&D-based models of economic growth. J Polit Econ 103:759–784
Jones CI (1997) Convergence revisited. J Econ Growth 2:131–153
Krugman P (1991a) Increasing returns and economic geography. J Polit Econ 99:483–499
Krugman P (1991b) Geography and trade. MIT, Cambridge, MA
Lee LF (2004) Asymptotic distributions of quasi-maximum likelihood estimators for spatial autoregressive models. Econometrica 72:1899–1925
Le Gallo J, Ertur C (2003) Exploratory spatial data analysis of the distribution of regional per capita GDP in Europe, 1980–1995. Pap Reg Sci 82:175–201
Le Gallo J, Ertur C, Baumont C (2003) A spatial econometric analysis of convergence across European regions, 1980–1995. In: Fingleton B (ed) European regional growth. Springer, Berlin pp 99–129
LeSage JP (1997) Bayesian estimation of spatial autoregressive models. Int Reg Sci Rev 20:113–129
López-Bazo E, Vayá E, Mora AJ, Suriñach J (1999) Regional economic dynamics and convergence in the European Union. Ann Reg Sci 33:343–370
Mankiw NG, Romer D, Weil DN (1992) A contribution to the empirics of economic growth. Q J Econ 107:407–437
Manski CF (1993) Identification of endogenous social effects: the reflection problem. Rev Econ Stud 60:531–542
Rey SJ, Montouri BD (1999) U.S. regional income convergence: a spatial econometric perspective. Reg Stud 33:145–156
Romer PM (1986) Increasing returns and long run growth. J Polit Econ 94:1002–1037
Romer PM (1989) Capital accumulation in the theory of long run growth. In: Barro RJ (ed) Modern business cycle theory. Harvard University Press, Cambridge, MA, pp 51–127
Sala-i-Martin X (1996a) Regional cohesion: evidence and theories of regional growth and convergence. Eur Econ Rev 40:1325–1352
Sala-i-Martin X (1996b) The classical approach to convergence analysis. Econ J 106:1019–1036
Solow RM (1956) A contribution to the theory of economic growth. Q J Econ 70:65–94
Swan TW (1956) Economic growth and capital accumulation. Econ Rec 32:334–361
Ying LG (2000) Measuring the spillover effects: some Chinese evidence. Pap Reg Sci 79:75–89
Romer PM (1987) Crazy explanations for the productivity slowdown. In: Fischer S (ed) NBER macroeconomics annual. MIT, Cambridge, pp 163–202

Author Index

Subject Index

MIX
Papier aus verantwortungsvollen Quellen
Paper from responsible sources
FSC® C105338
FSC
www.fsc.org

If you have any concerns about our products,
you can contact us on
ProductSafety@springernature.com

In case Publisher is established outside the EU,
the EU authorized representative is:
Springer Nature Customer Service Center GmbH
Europaplatz 3, 69115 Heidelberg, Germany

Printed by Libri Plureos GmbH
in Hamburg, Germany